Recent Advances in Oxidative Stress Associated Chronic Diseases Volume 2

Kaviyarasu Kasinathan ·
Azhaguchamy Muthukumaran
Editors

Recent Advances in Oxidative Stress Associated Chronic Diseases Volume 2

A Review of The Health Benefits and Risks of The Substance

Springer

Editors
Kaviyarasu Kasinathan
The University of South Africa
Roodepoort, Gauteng, South Africa

Azhaguchamy Muthukumaran
Department of Biotechnology
School of Biochemical and Processing Engineering
Kalasalingam Academy of Research and Education
Krishnankoil, Tamil Nadu, India

ISBN 978-981-95-3757-0 ISBN 978-981-95-3758-7 (eBook)
https://doi.org/10.1007/978-981-95-3758-7

This Springer imprint is published by the registered company Springer Nature Singapore Pte Ltd.
The registered company address is: 152 Beach Road, #21-01/04 Gateway East, Singapore 189721, Singapore

Contents

Contributors

Leggins Abraham Department of Physics, Sri Ramakrishna Engineering College, Coimbatore, Tamil Nadu, India

Rathi Muthaiyan Ahalliya Department of Biochemistry, Karpagam Academy of Higher Education, Coimbatore, Tamil Nadu, India;
Department of Biotechnology, Karpagam Academy of Higher Education, Coimbatore, Tamil Nadu, India

Vaishnavi Ajith Department of Biotechnology, School of Bioengineering, College of Engineering and Technology, SRM Institute of Science and Technology, Kattankulathur, Chengalpattu District, Tamil Nadu, India

N. M. I. Alhaji PG & Research Department of Chemistry, Khadir Mohideen College (Affiliated to Bharathidasan University, Tiruchirappalli), Adirampattinam, Tamil Nadu, India

Devipriya Anbumani Nano and Energy Biosciences Laboratory, Department of Biotechnology, Thiruvalluvar University, Serkkadu, Vellore, Tamilnadu, India

T. Aravinth Karthik Department of Biotechnology, School of Bioengineering, College of Engineering and Technology, SRM Institute of Science and Technology, Kattankulathur, Chengalpattu District, Tamil Nadu, India

A. Ayeshamariam PG & Research Department of Physics, Khadir Mohideen College (Affiliated to Bharathidasan University, Tiruchirappalli), Adirampattinam, Tamil Nadu, India

J. C. Balachandar Dr. Bala's School of Fitness & Preventive Medicine, Faridabad, Delhi-NCR, India

Poornasundari Baluchamy Department of Zoology, Thiruvalluvar University, Serkkadu, Vellore, Tamil Nadu, India

Yash Banabakode Department of Biotechnology, School of Bioengineering, College of Engineering and Technology, SRM Institute of Science and Technology, Kattankulathur, Chengalpattu District, Tamil Nadu, India

S. Beer Mohamed Department of Materials Science, School of Technology, Central University of Tamil Nadu, Thiruvarur, Tamil Nadu, India

Rhea Biju Luke Department of Biotechnology, School of Bioengineering, College of Engineering and Technology, SRM Institute of Science and Technology, Kattankulathur, Chengalpattu District, Tamil Nadu, India

Pandian Chithamparam Department of Biotechnology, Kalasalingam Academy of Research and Education, Krishnankoil, Tamil Nadu, India

R. Deepika Department of Biotechnology, School of Bioengineering, College of Engineering and Technology, SRM Institute of Science and Technology, Kattankulathur, Chengalpattu District, Tamil Nadu, India

Kayal Vizhi Dhandapani Nano and Energy Biosciences Laboratory, Department of Biotechnology, Thiruvalluvar University, Serkkadu, Vellore, Tamilnadu, India

Sana Firdous Department of Biotechnology, School of Bioengineering, College of Engineering and Technology, SRM Institute of Science and Technology, Kattankulathur, Chengalpattu District, Tamil Nadu, India

B. Mythili Gnanamangai Department of Biotechnology, K.S. Rangasamy College of Technology, Tiruchengode, Tamil Nadu, India

Swayam Gupta Department of Biotechnology, School of Bioengineering, College of Engineering and Technology, SRM Institute of Science and Technology, Kattankulathur, Chengalpattu District, Tamil Nadu, India

Shaikh Zaahid Iqbal Department of Biotechnology, School of Bioengineering, College of Engineering and Technology, SRM Institute of Science and Technology, Kattankulathur, Chengalpattu District, Tamil Nadu, India

S. Irshad Ahamed Department of General Surgery, Pondicherry Institute of Medical Sciences, Puducherry, Tamil Nadu, India

M. Ismail Fathima Department of Physics, Velammal College of Engineering and Technology, Virahanoor, Madurai, Tamil Nadu, India

Rangasamy Kavitha Deparment of Biochemistry, Bharathiar University, Coimbatore, Tamil Nadu, India

M. Lavanya Department of Biotechnology, Kalasalingam Academy of Research and Education, Krishnankoil, Tamil Nadu, India

Karpaga Swathi Marieswaran Department of Biotechnology, Kalasalingam Academy of Research and Education, Krishnankoil, Tamil Nadu, India

Sivakumar S. Moni Department of Pharmaceutics, College of Pharmacy and Health Research Centre, Jazan University, Jazan, Kingdom of Saudi Arabia

M. Mouhli Department of Physics, Sacred Heart College (Autonomous), Tirupattur, Tamil Nadu, India

Dharanidharan Murugan Department of Biotechnology, Kalasalingam Academy of Research and Education, Krishnankoil, Tamil Nadu, India

Azhaguchamy Muthukumaran Department of Biotechnology, Kalasalingam Academy of Research and Education, Krishnankoil, Tamil Nadu, India

Madhuri Patel Department of Biotechnology, School of Bioengineering, College of Engineering and Technology, SRM Institute of Science and Technology, Kattankulathur, Chengalpattu District, Tamil Nadu, India

K. Prabha Department of Food Technology, K.S. Rangasamy College of Technology, Tiruchengode, Tamil Nadu, India

Meenakshi Prakash Department of Biotechnology, School of Bioengineering, College of Engineering and Technology, SRM Institute of Science and Technology, Kattankulathur, Chengalpattu District, Tamil Nadu, India

Sharine Priscilla Department of Biotechnology, School of Bioengineering, College of Engineering and Technology, SRM Institute of Science and Technology, Kattankulathur, Chengalpattu District, Tamil Nadu, India

Harsath Ram Govindan Radhakrishnan Department of Biotechnology, Kalasalingam Academy of Research and Education, Krishnankoil, Tamil Nadu, India

D. Rajkumar Department of Physics, Sacred Heart College (Autonomous), Tirupattur, Tamil Nadu, India

R. Ramesh Department of Physics, Sacred Heart College (Autonomous), Tirupattur, Tamil Nadu, India

Babujanarthanam Ranganathan Nano and Energy Biosciences Laboratory, Department of Biotechnology, Thiruvalluvar University, Serkkadu, Vellore, Tamilnadu, India

J. S. Reshma Department of Biotechnology, School of Bioengineering, College of Engineering and Technology, SRM Institute of Science and Technology, Kattankulathur, Chengalpattu District, Tamil Nadu, India

Mariaamalraj Samykannu Department of Biotechnology, Sethu Institute of Technology, Pulloor, Kariapatti, Tamil Nadu, India

Daniel Raja Femil Selta Department of Biochemistry, Sri Ramakrishna College of Arts and Science for Women, Coimbatore, Tamil Nadu, India

Sahabudeen Sheik Mohideen Department of Biotechnology, School of Bioengineering, College of Engineering and Technology, SRM Institute of Science and Technology, Kattankulathur, Chengalpattu District, Tamil Nadu, India

S. Victor Vasanth Singh Department of Biotechnology, Kalasalingam Academy of Research and Education, Krishnankoil, Tamil Nadu, India

Janarthanan Sivasankar Department of Biotechnology, School of Bioengineering, College of Engineering and Technology, SRM Institute of Science and Technology, Kattankulathur, Chengalpattu District, Tamil Nadu, India

H. Sowmya Lakshmi Department of Physics, Sacred Heart College (Autonomous), Tirupattur, Tamil Nadu, India

S. Srivaishnavi Department of Physics, Sacred Heart College (Autonomous), Tirupattur, Tamil Nadu, India

S. Subhashini School of Physiotherapy, Dhanalakshmi Srinivasan University, Tiruchirappalli, Tamil Nadu, India

Arivoli Subramaniyan Department of Zoology, Thiruvalluvar University, Serkkadu, Vellore, Tamil Nadu, India

C. Thirupathy Department of Physics, Sacred Heart College (Autonomous), Tirupattur, Tamil Nadu, India

Rebeccas Thomas Department of Biotechnology, School of Bioengineering, College of Engineering and Technology, SRM Institute of Science and Technology, Kattankulathur, Chengalpattu District, Tamil Nadu, India

Abbreviations

$(CH_3)_2S$	Dimethyl sulfide
·OH	Hydroxyl radical
$_1O^2$	Singlet oxygen
$^1\Delta g$	Singlet oxygen
3-MST	3-mercaptopyruvate sulfur transferase
$3O_2$	Reactive triplet ground state
4HB	4-hydroxybenzoate
4-HNE	4-hydroxy-2-nonenal
4HNE	4-hydroxynonenal
8-OHdG	8-hydroxy-2′-deoxyguanosine
8-oxo-dG	8-Oxo-2′-Deoxyguanosine
AAPH	2,2′-Azobis(2-amidinopropane) dihydrochloride
AAT	Alpha-1 antitrypsin
AAT	Aspartate aminotransferase
AAV	Adeno-associated virus
ABA	Abscisic acid
ABC1	ABC transporter
ABTS	2,2′-azino-bis(3-ethylbenzothiazoline-6-sulfonic acid)
ABTS	3-ethylbenzothiazoline-6-sulfonic acid
ACOG	American college of obstetricians and gynecologists
ACPA	Anti-citrullinated protein antibodies
ACPA	Anti-citrullinated protein antibody
aCSF	Artificial cerebrospinal fluid
AD	Alzheimer's disease
ADMA	Asymmetric dimethyl arginine
ADME	Absorption, distribution, metabolism, and excretion
ADP	Adenosine diphosphate

AF	Atrial fibrillation
AGE	Advanced glycation end-product
AgNPs	Silver nanoparticles
ALA	Alpha-linolenic acid
ALA	Aminolevulinic acid
ALA	α-Linolenic acid
ALD	Alcoholic liver disease
ALEs	Advanced lipoxidation end products
ALS	Amyotrophic lateral sclerosis
AMD	Age-related macular degeneration
AMR	Antimicrobial resistance
ANI	Average nucleotide identity
ANN	Artificial neural networks
AOPPs	Advanced oxidation protein products
AO_X	Alternative oxidase
AP	Apurinic/apyrimidinic
APX	Ascorbate peroxidase
ARE	Antioxidant response element
AsA	Ascorbic acid
AT1	Angiotensin II receptor type 1
ATF6	Activating transcription factor 6
ATM	Ataxia-telangiectasia mutated
ATP	Adenosine triphosphate
AUC	Area under curve
AuNMs	Gold-based nanomaterials
Aβ	Amyloid-beta
BAC	Blood alcohol concentration
BBB	Blood—brain barrier
BER	Base excision repair
BER and NER	Excisions and nucleotides
bFGF	Fibroblast growth factor primary fibroblast growth factor
BH4	Tetrahydrobiopterin
BHA	Butylated 4-hydroxyanisole
BHA	Butylated hydroxyl anisole
BHT	Butylated 4-hydroxytoluene
BMD	Bone mineral density
BOD	Biochemical oxygen demand
BP	Blood pressure
CAT	Catalase
CAT	Enzyme catalase
CBD	Cannabidiol
CBS	Cystathionine β-synthase
CCl_4	Carbon tetrachloride
CDC	Centres for disease control and prevention

CDK2	Cyclin-dependent kinase 2
CDKN1A, GADD45B, MAFF	Mitochondrial oxidative stress-related gene signature
CDO	Cysteine dioxygenase
CF	Cystic fibrosis
cGMP	3′,5′-cyclic guanosine monophosphate
CH_3SH	Methyl mercaptan
CHD	Coronary heart disease
CHIPS	Control of hypertension in pregnancy study
CHMs	Chinese herbal medicines
CIM	Critical illness myopathy
CKD	Chronic kidney disease
CKD/AKI	Chronic or acute
Cl^-	Chloride ions
CMD	Coronary microvascular dysfunction
CMs	Compromised cardiomyocytes
CNC	Cap ‘n’ collar
-COOH	Carboxyl
COPD	Chronic obstructive pulmonary disease
COS	Carbonyl sulfide
COX	Cyclooxygenase
COX-2	Cyclooxygenase-2
CPD	Cyclobutane pyrimidine dimers
CRC	Colorectal cancer
CRP	C-reactive protein
CRP and TNF-α	Inflammatory biomarkers
crtI	Phytoene desaturase
CSE	Cystathionine γ-lyase
CT	Computed tomography
CTGF	Connective tissue growth factor
CUD	Cannabis use disorder
CUPRAC	Cupric reducing antioxidant capacity
CVD	Cardiovascular disease
CYP450s	Cytochrome P450 monooxygenases
CYPs	Cytochrome P450s
CYS 34	Free thiol groups
CysSSH	Cysteine persulfide
cytochrome bc1 complex	Complex III
D_2	Ergocalciferol
D_3	Cholecalciferol
DAG	Diacylglycerol
DAMPs	Danger-related molecular patterns
DAS28	Disease activity score
DbCM	Diabetic cardiomyopathy
$DCFH_2$-DA	2′7′-Dichlorodihydrofluorescein Diacetate

DCW	Dry cell weight
ddsA gene	Decaprenyl diphosphate synthase gene
DHA	Dehydroascorbic acid
DHA	Docosahexaenoic acid
DHAR	Dehydroascorbate reductase
DHE	Dihydroethidium
DISC	Death-inducing signaling complex
DM	Diabetes mellitus
DMAPP	Dimethylallyl diphosphate
DMAPP	Dimethylallyl pyrophosphate
DMARDs	Disease-modifying anti-rheumatic drugs
DMD	Mainly duchenne
DMF	Dimethyl fumarate
DMS	Dimethyl sulfide
DNA	Deoxyribonucleic Acid
DOX	Doxorubicin
DPCs	DNA-protein crosslinks
DPP	Decaprenyl diphosphate
DPPH	2,2-diphenyl-1-picrylhydrazyl
dRP	5′-deoxyribose phosphate residue
DSB	Double-strand break
DSC	Differential scanning calorimetry
DTNB	5,5′-dithiobis-2-nitrobenzoic acid
Dxs	1-deoxy-D-xylulose-5-phosphate synthase
EA	Ellagic acid
EAMs	Extra-articular manifestations
EAs	Ergot alkaloids
ECG	Electrocardiogram
ECSOD	Extracellular fluids
EDHFs	Endothelium-derived hyperpolarizing
E-DNA	Electrochemical DNA
EDRFs	Endothelium-derived relaxing
EDTA	Ethylenediaminetetraacetic acid
EGCG	Epigallocatechin-3-gallate
EGF	Epidermal growth factor
EGFR	Epidermal growth factor receptor
ELEM	Leukoencephalomalacia
ELISA	Enzyme-linked immunosorbent assay
eNOS	Endothelial
eNOS	Endothelial nitric oxide synthase
EPA	Eicosapentaenoic acid
EPR	Extended producer responsibility
EPS	Exopolysaccharides
EPS	Microbial exopolysaccharides
EPSPS	5-enolpyruvylshikimate-3-phosphate synthase

EPSs	Extracellular polymeric substances
ER	Endoplasmic reticulum
ERK1/2	Extracellular signal-regulated protein kinases 1 and 2
ERKs	Transmit signals from the membrane to the nucleus, promoting either survival
ERT	Enzyme replacement therapy
ERTs	Enzyme replacement therapies
ES	Embryonic stem
ESKD	End-stage kidney disease
ESR	Erythrocyte sedimentation rate
ET	Electron transfer method
ETC	Electron transport chain
ETC	Mitochondrial electron transport chain
FA	Ferulic acid
FAD	Flavin adenine dinucleotide
Fdx	Ferredoxin
FDX—FTR—TRX	Ferredoxin-thioredoxin
Fe^{2+}-TPTZ	Ferrous form
Fe^{3+}-TPTZ	Ferric-tripyridyl triazine
FET	ROS form via forward
FEV1	Forced expiratory volume in one second
FFA	Free fatty acids
FLS	Fibroblast-like synoviocytes
FMN	Flavin mononucleotide
FR	Free radicals
FRAP	Ferric reducing antioxidant power
FRASC	Ferric reducing ability of plasma with stabilized ascorbate
FUM	Fumarase
FUM	Fumonisins
FVC	Forced vital capacity
GA	Gallic acid
GFP	Green fluorescent protein
GGPP	Geranylgeranyl diphosphate synthase
GI	Gastrointestinal
GOX	Glycolate oxidase
GPP	Geranyl pyrophosphate
GPx	Glutathione peroxidase
GR	Glutathione reductase
GS	Glutamine synthetase
GS^-	Thiolate anion
GSH	Glutathione
GSH	Glycine, cysteine, and glutamate assay
GSNO	Nitrosoglutathione

GSNOR	S-nitrosoglutathione reductase
GSSG	Glutathione disulfide
GSSG	Reduced and oxidized GSH
GSSH	Glutathione persulfide
GST	Glutathione S-transferase
H/R	Hypoxia and hypoxia-reoxygenation
H_2O_2	Hydrogen peroxide
H_2S	Hydrogen sulfide
HAT	Hydrogen atom transfer
HBV	Hepatitis B
HCC	Hepatocellular carcinoma
HCL	Hydrachloric acid
HCO_3^-	Bicarbonate
HCV	Hepatitis C
HD	Huntington's disease
HDAC	Histone deacetylase
HDACs	Inhibiting histone deacetylases
HDL	High-density lipoprotein
HDP	Hypertensive conditions of pregnancy
HFrEF	Heart failure with reduced ejection fraction
HGF or VEGF	Bioactive molecules such as growth factors
HHP	High hydrostatic pressure
HIF-1	Hypoxia-inducible Factor-1
HIF-1α	Hypoxia inducible Factor 1α
HNE	4-hydroxy-2-nonenal
HNE	Hydro-mechanical neonatal encephalopathy
HNO	Nitroxyl
HO-1	Haem oxygenase-1
HOCl	Hydrochlorous acid
HOCl	Hypochlorous acid
HPLC	High-performance liquid chromatography
HSA	Human serum albumin
HSC	Hepatic stellate cell
HSF1	Heat shock factor 1
HSPs	Heat shock proteins
HTT	Genetic mutation in the huntingtin
HεdC	Heptanone-etheno-2′-deoxycytidine
IBD	Inflammatory bowel disease
IBS	Irritable bowel syndrome
IC50	Half-maximal inhibitory concentration
ICH	Intracerebral haemorrhage
IGF-I	Insulin-like growth factor-1
IHD	Ischemic heart disease
IL	Iliotibialis
iNOS	Inducible

iNOS	Inducible nitric oxide synthase
IONs	Iron oxide nanoparticles
IPP	Isopentenyl pyrophosphate
IRE1	Inositol requiring enzyme-1
Itga-8	Integrin subunit alpha 8
IVC	Integrated virtual care
IVF	In-vitro fertilization
JNK	Jun N-terminal kinase
K_1	Phylloquinone
K_2	Menaquinone
Keap1	Kelch-like ECH-associated protein 1
L-AA	L-ascorbic acid
LCA	Life cycle assessments
LC-MS	Liquid chromatography-mass spectrometry
LCY-β	Lycopene β-cyclase
LDL	Low density lipoprotein
LDPE	Low-density polyethylene
LOH	Harmful alcohols
LOOH	Lipid hydroperoxide
LOOH	Lowering lipid hydroperoxides
LOONO	Forming alkyl peroxynitrate
LOX	Lipoxygenases
LPO	Lipid peroxidation
LPO	Lipoxygenase
LPS	Lipopolysaccharide
MAGE	Multiplex automated genome engineering
MAMPs/PAMPs	Microbes or pathogens molecular patterns
MAO	Monoamine oxidase
MAPK/PI3K/AKT	Mitogen-Activated Protein Kinase/ Phosphoinositide 3-Kinase/Protein kinase B
MAPKs	Mitogen-activated protein kinases
MCP	Monocyte chemoattractant proteins
MCP-1	Monocyte chemotactic protein-1
MCU	Mitochondrial calcium uniporter
MDA	Malondialdehyde
MDA	Malondialdehyde assay
MDA	The malondialdehyde
MD-CE/ME	Microdialysis—capillary electrophoresis/ microchip electrophoresis
MDD	Major depressive disorder
MDHA	Mono-dehydroascorbate
MDHAR	Monodehydroascorbate reductase
MD-LC	Microdialysis—liquid chromatography
MDs	Muscular dystrophies
MeJA	Methyl jasmonate

MeOH	Methanol
MEP	2-C-methyl-D-erythritol 4-phosphate
METLIN	Metabolite and chemical entity database
MFC-Bs	Microbial fuel cell-based biosensors
mHtt	Misfolded mutant huntingtin protein
mHTT	Mutant huntingtin
MI	Maximum impulse
MI	Myocardial infarction
MIP	Macrophage inflammatory proteins
miRNAs	MicroRNAs
MLR	Monocyte-to-lymphocyte ratio
mM	Milli mole
MMIF	Macrophage migration inhibitory factor
MM-LDL	Minimally oxidized LDL
MMPs	Matrix metalloproteinases
MnSOD	Mitochondrial matrix
MnSODs	Manganese superoxide dismutases
MOFs	Metal-organic frameworks
MPO	Myeloperoxidase
MPST	Mercaptopyruvate sulfurtransferase
mPTP	Mitochondrial permeability transition pore
MRI	Magnetic resonance imaging
mRNA	Messenger RNA
MRSA	Methicillin-resistant staphylococcus aureus
MS	Multiple sclerosis
MSH	Mycothiol
Msr	Methionine sulfoxide reductase
MT	Metallothionein
mtDNA	Mitochondrial DNA
mtETC	Mitochondrial electron transport chain
mTOR	Mammalian target of rapamycin
MVA	Mevalonate
N_2O_3	Dinitrogen trioxide
NAC	N-acetylcysteine
N-acetyl-5-methoxytryptamine	Melatonin
NAC-N-	Acetylcysteine
NADH dehydrogenase	Complex I
NADPH	Nicotinamide adenine dinucleotide phosphate
NAFL	Non-alcoholic fatty liver
NAFLD	Non-alcoholic fatty liver disease
$NaHCO_3$	Sodium bicarbonate
NASH	Non-alcoholic steatohepatitis
NBT	Nitro blue tetrazolium
Nc	Neocuproine

N-DCs	Neutrophils transdifferentiate into dendritic cell-like phenotypes
NDG	Neurodegenerative
NDs	NADH dehydrogenases
NEG	Non-enzymatic glycosylation
NER	Nucleotide excision repair
NF-kB	Nuclear factor kappa B
NF-κB	Nuclear factor kappa-light-chain-enhancer of activated B cells
NF-κB	Nuclear factor-κB
NLR	Neutrophil-to-lymphocyte ratio
NNAT	Neuronatin
nNOS	Neuronal
nNOS	Neuronal NOS
NO^-	Nitric oxide
NO^+	Nitrosyl cations
NO_2^-	Nitrite
NO^{2+}	Nitronium ions
$NO_2^\bullet$	Nitrogen dioxide
NO_3^-	Nitrate
NOFNiR	Forming nitrite reductase
NOS	Nitric oxide synthase
NOS2	Nitrosative synthase-2
NOS^{-1}	NO synthase-like
NOX	NADPH oxidases
NOX	Oxidase
NOX2	gp91phox
NOX2, NOX4	NADPH oxidases
NP-SH	Non-protein thiols
NR	Nitrate reductase
Nrf2	Nuclear factor erythroid 2-related factor 2
NRO	Non-radical oxidants
NSCLC	Non-small cell lung cancer
O_2	Oxygen
$O_2^{\bullet-}$	Superoxide
$O_2^{\bullet-}$	Superoxide anions
OBS	Oxidative balance score
OFR	Oxygen-free radical
OGG1	8-oxoguanine DNA glycosylase 1
OGTT	Oral glucose tolerance test
-OH	Hydroxyl
$ONOO^-$	Peroxynitrite
$ONOOCO_2$	Nitroso peroxocarbonate
OPP	Pentose phosphate
ORAC	Oxygen radical absorbance capacity

ORM	Oxygen-reactive molecules
OTA	Ochratoxin A
OVX	Ovariectomy
ox-LDL	Distinctive features of oxidatively modified LDL
ox-LDL	Oxidatively modified LDL
OXPHOS	Oxidative phosphorylation
oxPTM	Oxidative posttranslational modification
OxS	Oxidative stress
PAHS	Polycyclic aromatic hydrocarbons
PAI	Plasminogen activator inhibitor
PAL	Phenylalanine ammonium lyase
PAMPs	Pathogen-related molecular patterns
PBM	Photobiomodulation
PBMCs	Peripheral blood mononuclear cells
PBS	Phosphate-buffered saline
pCA	p-coumaric
PCa	Prostate cancer
PCS	Phytochelatin synthase
PD	Parkinson's disease
PDGF	Platelet-derived growth factor
PDIs	Protein disulfide isomerases
PDMPs	Prescription drug monitoring programs
PDS	Phytoene desaturase
PEF	Pulse electric field
PEOz	Polymeric scaffolds made from poly (2-ethyl-2- oxazoline)
PERK	Protein kinase R-like endoplasmic reticulum kinase
PERP polyphenol-enriched plants	Polyphenols
PG	Propyl gallate
PGPB	Plant growth-promoting bacteria
PGSA-g-EG	Poly (glycerol-co-sebacate) acrylate ethylene glycol
pHBA	P-hydroxybenzoic
Pi	Inorganic phosphate
PI3K	Phosphoinositide 3-kinase
PIH	Pregnancy—induced hypertension
PK	Protein kinases
PKC	Protein kinase C
PLA	Polylactic acid
PLCL	Poly(l-lactide-co-caprolactone)
PLGA	Poly(lactic-co-glycolic acid)
PM	Pectoralis major
POD	Peroxidases

PON 1	Paraoxonase 1
PON-1	Paroxonase-1
PP	Polypropylene
PPO	Protoporphyrinogen oxidase
PPP	Pentose phosphate pathway
Prdx6	Peroxiredoxin-6
pre-COPD	Symptoms of the disease developing the condition
PROs	Patient-reported outcomes
PR_X	Peroxiredoxin
PSH	Protein-SH
PSII and PSI	Photosynthesis through light—dependent photosystems
PTEN	Phosphatase and TENsin homolog
PTEN	Phosphatase and Tensin Homolog on chromosome 10
PTM	Post—translational modification
PTP	Tyrosine phosphatases
PUFA	Polyunsaturated fatty acids
PUFAs	Proteins, and polyunsaturated fats
RA	Rheumatoid arthritis
rAAT	Recombinant human alpha-1 antitrypsin
rADI	Recombinant arginine deiminase
RAGEs	Receptor for advanced glycation end-products
RANETS	Regulated activation normal T cells expressed and secreted
RANTES	Regulated upon Activation, Normal T-cell Expressed and presumably Secreted
RAS	Renin-angiotensin system
RBOHD	Mitochondrial burst oxidase homolog D
RCS	Reactive carbonyl species
RDI	Dietary reference intake
REACH	European union regulation concerning the registration, evaluation, authorization, and restriction of chemicals
RET	Reverse electron transfer
RET	Reverse electron transport
rhASM	Recombinant acid sphingomyelinase
rhDNase	Recombinant human DNase
rhGAA	Recombinant human alpha-glucosidase
r-HuEPO	Recombinant human erythropoietin
Riboflavin	Vitamin B2
RNS	Reactive nitrogen species
$RO^{\bullet}$	Alkoxyl radical
RONs	Recepteur d'Origine nantais

RONSS	Reactive oxygen, nitrogen, and sulfur species
$ROO^{\bullet}$	Peroxyl Radical
ROS	Reactive oxygen species
ROS/RNS	Oxygen and nitrogen free radicals
$R\text{-}S(O)NH_2$	Sulfenic amide
$R\text{-}S(O_2)\ NH_2$	Sulfonamide
R-S•	Sulfur-centered radicals
RS•	Thiol radicals
RSI	Reactive species interactome
RSM	Response surface methodology
RSNO	S-nitroso thiols
R-SOH	Sulfenic acid
RSOH	Tosulfenic acid
RSOO•	Peroxysulfur Radicals
RSS	Peroxysulfur Radicals
RSS	Reactive sulfur species
RSSH	Persulfides
RTM	Resin transfer moulding
RuBP	Produces urea ribulose bisphosphate
S	Sulfur
S_2^-	Sulfide
$S_2O_3^{2-}$	Thiosulfates
SAPK	Stress-activated protein kinase
SASP	Senescence-associated secretory phenotype
SCFAs	Short chain fatty acids
SCN5A	Splicing of the Na^+ voltage-gated channel alpha subunit 5
SDF-1	Stromal cell-derived factor-1
SeNPs	Selenium nanoparticles
SET	Single electron transfer
sGC	Guanylate cyclase
-SH	Cysteine thiol
-SH	Sulfhydryl group
-SH	Thiol
SHS	Sodium hydrosulphide
SLE	Systemic lupus erythematosus
SNO's	S-nitrosothiols
SNP	Single nucleotide polymorphism
SNP	Sodium nitroprusside
SO_2	Sulfur dioxide
SO_3	Sulfur trioxide
SO_4^{2-}	Sulfur or sulfate
SOD	Peroxisomal superoxide dismutase
SOD	Superoxide dismutase
SOD, CAT, GPx	Erythrocyte antioxidant enzymes

SOD1	SOD isoforms are localized in one-of-a-kind cellular cubicles: cytosolic
SOD2	SOD isoforms are localized in one-of-a-kind cellular cubicles: mitochondrial
SOD3	SOD isoforms are localized in one-of-a-kind cellular cubicles: extracellular
-SOH	Sulfenic acids
-SOH	Sulfinylation
SPAs	Synthetic phenolic antioxidants
SPIONs	Superparamagnetic ions
SQR	Sulfide quinone oxidoreductase
SR	Sarcoplasmic reticulum
SSB	Single-strand break
SSBs	Small single base lesions
STC	Sterigmatocystin
SUB	Substance use disorder
SUD	Substance use disorders
S_x^{2-}	Polysulfides
Sypl2	Synaptophysin-like protein 2
T1D	Type 1 diabetes
T2DM	Type 2 diabetes mellitus
T3	Triiodothyronine
T4	Hormone thyroxine
TAC	Total antioxidant capacity
TBA	Thiobarbituric acid
TBARS	Reduced TBA-reactive chemicals
TBARS	Thiobarbituric acid reactive substances
TBHQ	Tert-butylhydroquinone
TBI	Traumatic brain injury
TCA	Tricarboxylic acid
TCA	Trichloroacetic acid
TE	Trolox equivalents
TEAC	Trolox equivalent antioxidant capacity
TGA	Thermogravimetric analysis
TLC	Thin-layer chromatography
TME	The tumor microenvironment
TME	Tumor microenvironment
TMJ	Temporomandibular joint
TNB	2-nitro-5-thiobenzoate
TNF	Tumour necrosis factor
TNF-α, IL-1β	Astrocytes release cytokines
tPA	Tssue plasminogen activator
TPTZ	2,4,6-tripyridyl-s-triazine
TRC	Tumor-repopulating cell
tRNA	Transfer RNA

Trx	Thioredoxin
TrxR	Thioredoxin reductase
UC	Ulcerative colitis
UCP	Uncoupling protein
UPR	Unfolding protein response
UVA	Ultraviolet A radiation
UVB	Ultraviolet B radiation
UVC	Ultraviolet C radiation
VEGF	Vascular endothelial growth factor
VLDL	Very low-density lipoprotein
WCBs	Whole-cell bacterial biosensors
WHO	The World Health Organization
WMS	Whole metagenome sequencing
WUE	Crop water use efficiency
Xo	Xanthine oxidase
YAP	Yes-associated protein
ZEA	Zaralenone
$\Delta\Psi m$	Mitochondrial membrane potential

Oxidative Stress and Pathophysiology of Chronic Diseases

Daniel Raja Femil Selta, Leggins Abraham, Rathi Muthaiyan Ahalliya, and Rangasamy Kavitha

Abstract Oxidative stress, where there is an imbalance between the production of reactive oxygen species (ROS) and the body's antioxidant defenses, is a key contributor to the pathophysiology of many chronic diseases. This process is involved in cellular injury, inflammation, and dysfunction, which are typical of diseases like cardiovascular disease, diabetes, neurodegenerative disorders, and cancer. Under these circumstances, prolonged oxidative stress drives the development of tissue injury, amplifies inflammatory reactions, and modifies cellular signal transduction. In addition to causing DNA damage and protein oxidation, oxidative stress also impacts lipid metabolism, which leads to atherosclerotic plaque formation and insulin resistance. The resultant effects of cumulative oxidative damage induce the modification of normal physiological mechanisms and the establishment of disease processes. This review discusses the pathways through which oxidative stress promotes the pathophysiology of chronic diseases and identifies potential therapeutic interventions, such as antioxidant therapies and lifestyle modifications, to reduce oxidative damage and enhance disease outcomes. The connection between oxidative stress and chronic diseases offers insights into new preventative and treatment strategies for controlling these ubiquitous health disorders.

D. R. F. Selta (✉)
Department of Biochemistry, Sri Ramakrishna College of Arts and Science for Women, Coimbatore, Tamil Nadu, India
e-mail: femidaniel97@gmail.com

L. Abraham
Department of Physics, Sri Ramakrishna Engineering College, Coimbatore, Tamil Nadu, India

R. M. Ahalliya
Department of Biochemistry, Karpagam Academy of Higher Education, Coimbatore, Tamil Nadu, India

Department of Biotechnology, Karpagam Academy of Higher Education, Coimbatore, Tamil Nadu, India

R. Kavitha
Deparment of Biochemistry, Bharathiar University, Coimbatore, Tamil Nadu, India

K. Kasinathan and A. Muthukumaran (eds.), *Recent Advances in Oxidative Stress Associated Chronic Diseases Volume 2*, https://doi.org/10.1007/978-981-95-3758-7_1

Keywords Oxidative stress · Reactive oxygen species (ROS) · Inflammation · Chronic diseases · Antioxidant therapy

1 Introduction

Chronic diseases, such as diabetes, cardiovascular disorders, and respiratory conditions, have become the leading contributors to global deaths and health complications, with their impact escalating rapidly across diverse populations and economic strata [1]. A significant challenge lies in the fact that many individuals with chronic conditions exhibit inadequate self-management practices, often linked to reduced confidence in their ability to enact behavioral changes (low self-efficacy) and limited motivation to engage in health-promoting activities. Such patterns frequently result in worsened clinical outcomes and a decline in overall well-being. Compounding this issue, multimorbidity—the coexistence of multiple chronic illnesses in a single individual—is increasingly common, accelerating physical and cognitive decline, raising mortality risks, and imposing substantial financial strain on healthcare systems. Ultimately, the strategies patients employ to address their health needs can either foster successful self-management or perpetuate a cycle of deteriorating health and weakened self-assurance. Free radicals refer to unstable molecules characterized by one or more unpaired electrons in their outermost orbital, rendering them chemically reactive. In biological systems, the most prevalent types include reactive oxygen species (ROS) and reactive nitrogen species (RNS), originating from oxygen and nitrogen, respectively. These molecules form via electron transfer processes, where atoms or molecules gain or lose electrons during metabolic reactions [29]. ROS contribute to critical cellular activities such as cell growth, specialization, movement, programmed cell death (apoptosis), and tissue necrosis. At controlled levels, ROS and RNS are crucial for normal physiological operations, including maintaining redox equilibrium and modulating transcription factors. However, excessive ROS accumulation overwhelms the body's antioxidant defense, destabilizing redox balance and triggering oxidative stress. This state damages essential biological components like DNA, proteins, and cell membranes. Oxidative stress—a mismatch between free radical production and the body's ability to neutralize them—is closely tied to the pathogenesis of chronic illnesses and age-related decline [23]. Biomarkers of oxidative damage, which reflect oxidative stress levels, are elevated in aging populations and individuals with lifestyle risk factors such as nutritionally deficient diets, tobacco use, excessive alcohol intake, sedentary behavior, and hereditary predispositions. To counteract this stress, cellular systems employ a sophisticated network of antioxidant mechanisms. These include enzymatic defenses like superoxide dismutases (SODs), catalase (CAT), glutathione peroxidases (GPXs), thioredoxin (Trx), and non-enzymatic antioxidants with low molecular weight. Collectively, these systems modulate oxidative stress to maintain homeostasis. An effective antioxidant defense—which integrates both internally produced (endogenous) and

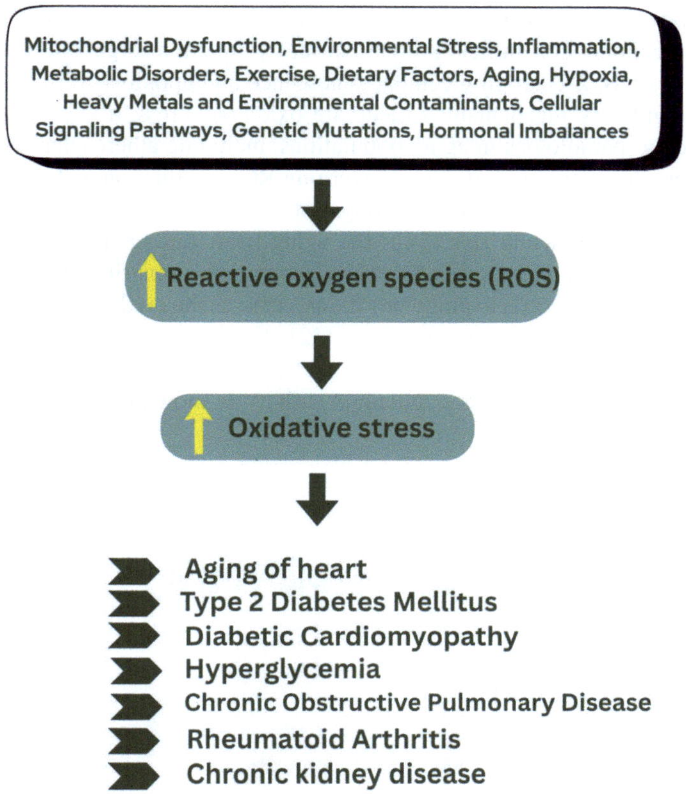

Fig. 1 Increased oxidative stress is a fate of pathophysiology in chronic diseases

externally sourced (exogenous) antioxidants—safeguards cellular integrity against ROS—induced harm, mitigating molecular and structural damage as shown in Fig. 1.

2 Oxidative Stress and Aging of Heart

Historically, reactive oxygen species (ROS) were viewed as toxic metabolic byproducts that drove cellular damage and disease. Contemporary research, however, highlights their dual role: beyond causing molecular harm, ROS act as signaling molecules, modulating pathways critical for cellular communication [52]. Striking an equilibrium in ROS generation is vital for mitigating cellular injury and supporting longevity. For example, controlled ROS levels can bolster cellular adaptability to stressors like metabolic shifts, mechanical strain, or oxidative challenges [40]. Transient oxidative stress, such as during ischemia–reperfusion events, may even confer cytoprotective benefits via pathways like mTOR or Wnt signaling. Conversely, excessive ROS accumulation inflicts mitochondrial dysfunction, organelle impairment, and DNA lesions, accelerating cellular senescence and apoptosis [20]. Disruptions in redox regulatory mechanisms exacerbate oxidative stress, a key contributor to cardiovascular pathologies in aging populations, including atherosclerosis [46, 67]. Aging alters the expression of genes governing ROS production and detoxification in both human and rodent hearts, elevating systemic and cardiac oxidative stress and heightening cardiovascular risk [54]. The aging heart further displays diminished antioxidant defenses and amplified oxidant generation, impairing stress adaptation [37]. Mitochondria, central to aging processes, serve as the predominant ROS source via oxidative phosphorylation byproducts [32]. This section also explores endogenous ROS—generating systems, such as NADPH oxidase (NOX) enzymes, nitric oxide synthase, and endoplasmic reticulum activity as shown in Fig. 2, respectively.

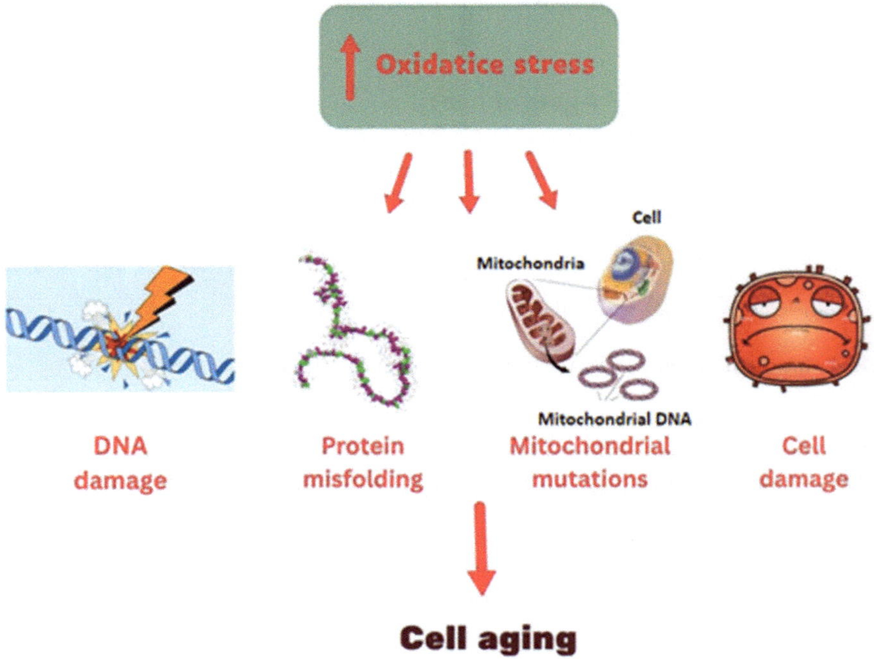

Fig. 2 Increased oxidative stress in aging process

3 Reactive Oxygen Species and Homeostasis in Type 2 Diabetes Mellitus

Disruption of redox balance represents a frequent pathological state, characterized by a transition from beneficial oxidative signaling (*eustress*) to harmful oxidative stress (*distress*) [59, 60]. Such disturbances can compromise glucose metabolism and destabilize regulatory mechanisms. Excessive ROS generation undermines insulin responsiveness in peripheral tissues and induces dysfunction in insulin-producing pancreatic β-cells [68]. Studies in both laboratory and clinical settings demonstrate that higher ROS concentrations correspond with reduced insulin sensitivity [51]. Oxidative stress triggers serine phosphorylation and degradation of insulin receptor substrate-1 (IRS1) through mechanisms not involving the proteasome, though this phenomenon shows limited association with the metabolic dysregulation caused by oxidative stress in insulin pathways [66]. These findings underscore the centrality of redox processes in insulin signaling, leading to the classification of type 2 diabetes as a disorder of redox imbalance [48]. Key components of insulin signaling pathways—including Akt, FOXO proteins, PTEN, PTP1B, and JNKs—are modulated by redox conditions, enabling multiple regulatory opportunities via oxidative modifications. Elevated oxidant levels stimulate JNK activity, a primary driver of impaired insulin signaling. Additionally, oxidative stress induces apoptosis in pancreatic β-cells, further reducing insulin production. The susceptibility of these cells to oxidative damage stems from their comparatively weak antioxidant defenses relative to other tissues [45].

4 Hyperglycemia and Oxidative Stress

The pathogenesis of type 2 diabetes mellitus (T2DM) is complicated by the paradoxical role of glucose, the central metabolic driver of the disease, which exerts toxic effects in affected individuals—a process termed glucose toxicity or glucose—induced impaired responsiveness. Under conditions of prolonged hyperglycemia, insulin-producing pancreatic β-cells undergo oxidative stress [45]. Clinical interventions that lower blood glucose concentrations in diabetic patients enhance both insulin responsiveness and secretion. In contrast, healthy individuals exhibit improved insulin sensitivity and β-cell performance under sustained glucose administration, highlighting the pathological specificity of glucose toxicity. Chronically elevated glucose levels drive excessive ROS generation via mitochondrial electron transport chain activity [38, 47], while simultaneously promoting the synthesis of advanced glycation end-products (AGEs). Experimental studies in bovine aortic endothelial cells demonstrate that normalization of hyperglycemia—elevated mitochondrial ROS mitigates glucose—triggered pathways such as PKC activation, AGE accumulation, sorbitol buildup, and NF-κB induction. In diabetic conditions, AGE-RAGE signaling activates NADPH oxidase-1 (NOX1), exacerbating oxidative stress

[9]. Furthermore, AGE-RAGE interactions upregulate NF-κB expression [6], and sustained NF-κB activity fosters chronic inflammation, a key feature of advanced diabetes.

5 Homeostatic Setpoints and Glucose Tolerance

The emergence of glucose intolerance and diabetes mellitus reflects an alteration in the body's homeostatic equilibrium. Research suggests oxidative stress and persistent inflammation may drive a recalibration of these metabolic thresholds [60]. However, this relationship poses a causality paradox, as obesity or insulin resistance does not universally progress to glucose intolerance. Initial β-cell dysfunction, potentially influenced by genetic predisposition, can heighten hepatic glucose output, diminish cellular glucose absorption, and accelerate fat breakdown. These effects elevate circulating glucose and non-esterified fatty acid (NEFA) concentrations, initiating a self-perpetuating cycle that intensifies insulin resistance and β-cell impairment. Excess glucose and NEFA stimulate interleukin-1β secretion in pancreatic islets, promoting cytokine-mediated β-cell apoptosis [35]. As previously noted, hyperglycemia directly damages cellular systems. Similarly, elevated NEFA levels adversely impact insulin-responsive tissues and β-cells: NEFA exposure inhibits insulin signaling downstream of receptor binding, suppresses glucose utilization and glycogen storage, and compromises β-cell efficacy [55]. Prolonged NEFA elevation induces lipotoxic stress across multiple cell types, including β-cells, ultimately degrading their functionality and survival [58].

6 Oxidative Stress and Diabetic Cardiomyopathy

6.1 The Myocardium as a Predilection Site of Oxidative Stress

In healthy physiological states, a tightly regulated equilibrium exists between reactive oxygen species (ROS) generation and clearance, ensuring stable ROS concentrations necessary for cellular processes. Redox-sensitive mechanisms are vital for preserving cardiac homeostasis and function, where baseline ROS levels modulate excitation—contraction coupling, adaptive hypertrophy, and myocardial stability [56]. However, transient or sustained ROS overproduction induces cardiac oxidative stress, interfering with signaling cascades and oxidizing cellular structures. Such disturbances cause functional impairments and may activate necrotic or apoptotic cell death pathways [39]. Cellular dysfunction arises from oxidative damage to proteins, lipid peroxidation, DNA strand breaks, and dysregulated microRNA activity [42, 57]. ROS also stimulate stress-responsive pathways, including JNKs,

ERK1/2, p38 MAP kinases, NF-κB, and PKC. Overabundant ROS or highly reactive oxidants (e.g., hydroxyl radicals) can trigger pathological cardiac remodeling, marked by dysregulated calcium signaling, cardiomyocyte enlargement, excessive fibrosis, and apoptosis—a process notably amplified in diabetic hearts [21]. Oxidative stress is intricately connected to endoplasmic reticulum (ER) stress and the unfolded protein response (UPR). Disrupted redox balance during oxidative stress compromises protein folding, resulting in misfolded protein accumulation [11]. Lipotoxic and inflammatory insults further impair cardiac ER function, exacerbating ER stress, and UPR activation [30, 31]. While appropriately regulated UPR supports adaptive mechanisms to restore secretory pathway integrity, prolonged or severe ER stress drives maladaptive UPR, culminating in apoptotic cell death [27]. Obesity, metabolic syndrome, and diabetes are closely linked to systemic low-grade inflammation and sustained oxidative stress [44]. In type 2 diabetes, elevated markers of chronic inflammation—such as C-reactive protein—are frequently observed [8, 13]. Similarly, oxidative stress indicators like malondialdehyde, TBARS, hydroxynonenal (HNE), and 8-hydroxyguanosine are heightened in these metabolic disorders [8]. The heart's susceptibility to oxidative damage stems from its intense metabolic demands and abundant mitochondria. Mitochondria account for ~ 25–30% of cellular volume in most mammals, containing ~ 92 protein subunits encoded by nuclear and mitochondrial DNA, organized into five multi-protein complexes [41]. During oxidative phosphorylation, 0.2–2% of electrons escape the respiratory chain, reacting with oxygen to generate superoxide (O_2^-)—a process amplified in diabetic cardiac tissue. Additional ROS sources include NADPH oxidases, xanthine oxidase, cytochrome P450 enzymes, and uncoupled nitric oxide synthase [64].

Hyperglycemia-driven mitochondrial ROS overproduction exacerbates oxidative stress. Four primary pathways mediate hyperglycemia-driven cellular injury: (1) increased polyol pathway flux, (2) enhanced AGE synthesis and RAGE activation, (3) stimulation of protein kinase C isoforms, and (4) upregulated hexosamine pathway activity—all stemming from mitochondrial ROS excess [22]. Both diabetic humans and animal models exhibit reduced activity of glyceraldehyde 3-phosphate dehydrogenase (GAPDH), a key glycolytic enzyme impaired under hyperglycemia [22]. GAPDH suppression causes glycolytic intermediate buildup, diverting substrates into pathogenic pathways. Notably, mitochondrial superoxide overproduction inhibits GAPDH via poly(ADP-ribosyl)ation by PARP, which is triggered by superoxide-induced DNA damage. This inhibition can be counteracted by uncoupling protein-1 or mitochondrial SOD2, underscoring mitochondrial superoxide's central role [15]. Elevated intracellular glucose levels are a defining feature of diabetic microvascular dysfunction [49]. Within the myocardium, endothelial cells represent the predominant cell type, accounting for 54% of cardiac cell nuclei, whereas cardiomyocytes constitute approximately 31% [49]. Unlike microvascular complications, mitochondrial reactive oxygen species (ROS) overproduction in diabetic macrovasculature and cardiac tissue mainly stems from enhanced fatty acid oxidation, driven by insulin resistance in specific metabolic pathways. Glycogen synthase kinase-3β (GSK-3β) is pivotal in regulating cardiac glucose metabolism. As previously discussed,

ROS disrupt insulin signaling, and in both type 1 and type 2 diabetic cardiomyocytes, impaired insulin signaling hinders Akt-mediated phosphorylation of GSK-3β, altering energy metabolism. Diabetic hearts exhibit diminished glucose uptake and heightened fatty acid breakdown, promoting lipid buildup—a phenomenon termed cardiac lipotoxicity [43]. This lipid overload can trigger myocardial inflammation and oxidative injury [43]. Cardiac lipotoxicity, characterized by excessive lipid uptake, oxidation, and storage, further correlates with mitochondrial impairment [63]. Studies using genetically modified mice with cardiomyocyte-specific overexpression of long-chain acyl-CoA synthetase 1 demonstrated elevated fatty acid absorption, mitochondrial structural abnormalities, and amplified ROS generation [63]. Thus, mitochondrial ROS overproduction arises from both hyperglycemia and lipotoxicity-driven fatty acid oxidation in cardiomyocytes [63]. Free fatty acid-induced ROS mirrors glucose-triggered ROS in suppressing GAPDH activity, thereby exacerbating hexosamine pathway dysregulation, protein kinase C activation, and AGE accumulation [16]. Furthermore, the Randle cycle (glucose-fatty acid cycle) emphasizes the interplay between fatty acid and glucose metabolism, positioning fatty acid oxidation as a key modulator of glycolysis and pyruvate dehydrogenase function [43].

7 Oxidative Stress in Chronic Obstructive Pulmonary Disease

7.1 Effects of Oxidative Stress in COPD

Oxidative stress biomarkers in the pulmonary system can be non-invasively assessed through breath analysis. Ethane, a volatile compound generated during lipid peroxidation, is detectable via gas chromatography and shows elevated levels in the breath of COPD patients, with concentrations correlating with clinical progression [47]. Exhaled breath condensate serves to quantify markers such as hydrogen peroxide (H_2O_2), malondialdehyde, 4-hydroxy-2-nonenal (4-HNE), and 8-isoprostane, all of which are significantly higher in COPD patients compared to smokers without COPD and healthy controls [4]. Notably, H_2O_2 and 8-isoprostane levels surge during acute exacerbations [7]. These biomarkers persist at higher levels even in ex-smokers, suggesting they are generated internally by activated immune cells within lung tissue. Heightened oxidative and nitrative stress promotes the formation of superoxide anions and nitric oxide (NO), which react rapidly to produce peroxynitrites—a highly reactive oxidant. Peroxynitrite, measurable in COPD patients' breath condensate through selective oxidation of the fluorescent probe 2′,7′-dichlorofluorescein, correlates with disease severity as shown in Fig. 3. It induces nitration of tyrosine residues in proteins, leading to elevated nitrotyrosine levels observed in sputum and lung biopsies of COPD patients [53]. Furthermore, superoxide anion levels are

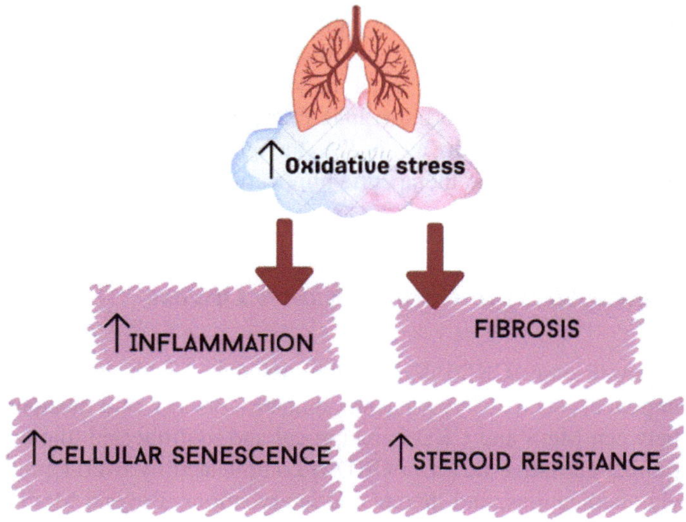

Fig. 3 Increased oxidative stress in chronic obstructive pulmonary disease

increased in the skeletal muscle of these patients, contributing to muscle weakness and wasting, hallmark features of advanced COPD [3].

7.2 Oxidative Stress in Rheumatoid Arthritis

Rheumatoid arthritis (RA) progression is closely tied to heightened oxidative stress, diminished antioxidant capacity, and dysfunctional antioxidant systems [12]. Excessive reactive nitrogen species (RNS) and reactive oxygen species (ROS) contribute to RA pathogenesis by surpassing physiological thresholds, inducing harm to cellular components such as membranes, proteins, and nucleic acids [18]. Notably, mitochondrial ROS generation in whole blood and monocytes is five times higher in RA patients compared to healthy individuals, underscoring oxidative stress as a central driver of disease mechanisms [50]. T-cell activation stimulates cytokine release, promoting oxidative stress, which antioxidants like glutathione help neutralize. Vitamins C and E, essential non-enzymatic antioxidants, support immune responses, tissue regeneration, and membrane integrity. However, research suggests systemic depletion of antioxidants in RA. For instance, studies report decreased serum vitamin C [14] and plasma vitamin E [2] in RA patients, reflecting compromised antioxidant defenses. Furthermore, RA patients exhibit diminished levels of tocopherols, β-carotene, retinols, and sulfhydryl (SH) groups, as well as reduced activity of enzymes like glutathione reductase and superoxide dismutase [33]. Enzymatic activity data, however, remain inconsistent—studies show variable alterations in superoxide dismutase, glutathione peroxidase, catalase, and NADPH oxidase [50].

Nitric oxide (NO·) levels inversely correlate with glutathione concentrations, potentially compensating for elevated nitrogen dioxide (NO_2·) generation [65]. Impaired endothelial NO· synthase activity, a hallmark of vascular injury, is also linked to atherosclerosis, a common comorbidity in RA [26]. These findings collectively highlight disrupted antioxidant pathways in RA, though further investigation is needed to clarify the mechanisms and therapeutic implications.

7.3 Oxidative Stress in Chronic Kidney Disease

Chronic kidney disease (CKD) is intricately associated with metabolic syndrome (MetS) components such as hypertension, diabetes, obesity, and dyslipidemia. The interplay between MetS and CKD is reciprocal and multifaceted, though isolating MetS as a direct causative factor in CKD is complicated by overlapping variables like lifestyle, medications, and therapeutic interventions. Beyond MetS, CKD risk is amplified by nephrotoxin exposure, acute kidney injury, smoking, aging, and systemic redox imbalance [62]. A major contributor to reactive oxygen species (ROS) generation is mitochondrial dysfunction, particularly involving the electron transport chain within the oxidative phosphorylation (OXPHOS) system. In CKD, impaired mitochondrial function drives excessive ROS production, amplifying oxidative damage. Overexpression of OXPHOS-related genes in CKD patients further aggravates this stress [25]. Enzymatic sources of ROS, including NADPH oxidase (Nox), xanthine oxidase (XO), and lipoxygenases, are also upregulated in CKD [10]. Specific Nox isoforms are implicated in renal pathologies such as nephrolithiasis, hypertension, and acute kidney injury [28]. Notably, the Nox4 isoform, abundant in renal tissue, is a dominant ROS generator in chronic conditions like diabetic nephropathy [24]. Enhanced Nox-derived superoxide levels in early CKD stages are linked to renal microvascular impairment [17, 19]. Nitric oxide (NO), essential for renal hemodynamics, regulates blood pressure, sodium excretion, and adaptation to dietary salt fluctuations. It also neutralizes oxidative stress by scavenging superoxide radicals. In CKD, however, NO bioavailability is diminished due to reduced L-arginine (its precursor) and elevated inhibitors like asymmetric dimethylarginine [5]. This NO deficiency, combined with superoxide-mediated inactivation, heightens renal vascular resistance, promoting hypertensive kidney damage and cardiovascular complications in CKD. In CKD, antioxidant defense mechanisms are often compromised. For example, superoxide dismutase (SOD), a key free radical scavenger, is found at lower levels in CKD patients. Additionally, genetic polymorphisms of glutathione-S-transferase, a major antioxidant enzyme, have been linked to increased oxidative stress in end-stage renal disease (ESRD) patients [61]. Furthermore, CKD patients exhibit reduced plasma concentrations of antioxidant enzymes such as catalase and glutathione peroxidase [36], as well as lower levels of intracellular glutathione and thiols [34]. These impairments in antioxidant systems exacerbate the oxidative stress seen in CKD, contributing to disease progression and complications.

8 Conclusion

Oxidative stress is a fundamental factor in the pathophysiology of many chronic diseases, leading to its development and progression by inducing inflammation, cell damage and malfunction. Oxidative stress in the heart rate affects myocardial function and improves atherosclerosis, whereas in type 2 diabetes, increased glucose inhibits insulin signaling and promotes oxidative damage that impairs glucose metabolism. Myocardium is particularly sensitive to oxidative damage and prevents heart repair, but oxidative stress in COPD and rheumatoid arthritis causes tissue destruction and inflammation, worsening lung and joint function. Oxidative stress in CKD also promotes kidney progression and fibrosis. Hyperglycemic oxidative stress connections exacerbate tissue damage and thus disrupt homeostasis and glucose tolerance. In general, oxidative stress contributes to important contributions to these disorders and there is a certain hope for antioxidant therapy, but the best treatment is the best treatment in the form of improvement, movement, and avoidance of environmental toxins. Recognition of this interaction is important to provide effective treatment and prevention strategies to combat oxidative stress and improve health outcomes.

References

1. Archuleta, T. L., Lemieux, A. M., Saengsirisuwan, V., Teachey, M. K., Lindborg, K. A., Kim, J. S., & Henriksen, E. J. (2009). Oxidant stress induced loss of IRS-1 and IRS-2 proteins in rat skeletal muscle: Role of p38 MAPK. *Free Radical Biology and Medicine, 47*, 1486–1493.
2. Aryaeian, N., Djalali, M., Shahram, F., Jazayeri, S. H., Chamari, M., & Nazari, S. A. (2011). Beta-carotene, vitamin E, MDA, glutathione reductase and arylesterase activity levels in patients with active rheumatoid arthritis. *Iranian Journal of Public Health, 40*, 102–109.
3. Barreiro, E., Peinado, V. I., Galdiz, J. B., Ferrer, E., Marin-Corral, J., Sanchez, F., Gea, J., & Barbera, J. A. (2010). Cigarette smoke-induced oxidative stress: A role in chronic obstructive pulmonary disease skeletal muscle dysfunction. *American Journal of Respiratory and Critical Care Medicine, 182*, 477–488.
4. Bartoli, M. L., Novelli, F., Costa, F., Malagrino, L., Melosini, L., Bacci, E., Cianchetti, S., Dente, F. L., Di Franco, A., Vagaggini, B., et al. (2011). Malondialdehyde in exhaled breath condensate as a marker of oxidative stress in different pulmonary diseases. *Mediators of Inflammation*, 891752.
5. Baylis, C. (2008). Nitric oxide deficiency in chronic kidney disease. *American Journal of Physiology. Renal Physiology, 294*, F1–F9.
6. Bierhaus, A., Schiekofer, S., Schwaninger, M., Andrassy, M., Humpert, P. M., Chen, J., Hong, M., Luther, T., Henle, T., Klöting, I., et al. (2001). Diabetes-associated sustained activation of the transcription factor nuclear factor-kappaB. *Diabetes, 50*, 2792–2808.
7. Biernacki, W. A., Kharitonov, S. A., & Barnes, P. J. (2003). Increased leukotriene B4 and 8-isoprostane in exhaled breath condensate of patients with exacerbations of COPD. *Thorax, 58*, 294–298.
8. Bigagli, E., & Lodovici, M. (2019). Circulating oxidative stress biomarkers in clinical studies on Type 2 diabetes and its complications. *Oxidative Medicine and Cellular Longevity, 2019*, 5953685.
9. Chen, J., Jing, J., Yu, S., Song, M., Tan, H., Cui, B., & Huang, L. (2016). Advanced glycation endproducts induce apoptosis of endothelial progenitor cells by activating receptor RAGE

and NADPH oxidase/JNK signaling axis. *American Journal of Translational Research, 8*, 2169–2178.
10. Choi, B. H., Kang, K. S., & Kwak, M. K. (2014). Effect of redox modulating NRF2 activators on chronic kidney disease. *Molecules, 19*(12727), 12759.
11. Chong, W. C., Shastri, M. D., & Eri, R. (2017). Endoplasmic reticulum stress and oxidative stress: A vicious nexus implicated in bowel disease pathophysiology. *International Journal of Molecular Sciences, 18*, 771.
12. Croia, C., Bursi, R., Sutera, D., Petrelli, F., Alunno, A., & Puxeddu, I. (2019). One year in review 2019: Pathogenesis of rheumatoid arthritis. *Clinical and Experimental Rheumatology, 37*, 347–357.
13. Daiber, A., Hahad, O., Andreadou, I., Steven, S., Daub, S., & Munzel, T. (2021). Redox-related biomarkers in human cardiovascular disease—Classical footprints and beyond. *Redox Biology, 42*, 101875.
14. Das, D. C., Jahan, I., Uddin, M. G., Hossain, M. M., Chowdhury, M. A. Z., Fardous, Z., Rahman, M. M., Kabir, A. K. M. H., Deb, S. R., Siddique, M. A. B., et al. (2021). Serum CRP, MDA, vitamin C, and trace elements in Bangladeshi patients with rheumatoid arthritis. *Biological Trace Element Research, 199*, 76–84.
15. Du, X., Matsumura, T., Edelstein, D., Rossetti, L., Zsengeller, Z., Szabo, C., & Brownlee, M. (2003). Inhibition of GAPDH activity by poly(ADP-ribose) polymerase activates three major pathways of hyperglycemic damage in endothelial cells. *The Journal of Clinical Investigation, 112*, 1049–1057.
16. Du, X., Edelstein, D., Obici, S., Higham, N., Zou, M. H., & Brownlee, M. (2006). Insulin resistance reduces arterial prostacyclin synthase and eNOSactivities by increasing endothelial fatty acid oxidation. *The Journal of Clinical Investigation, 116*, 1071–1080.
17. DuPont, J. J., Ramick, M. G., Farquhar, W. B., Townsend, R. R., & Edwards, D. G. (2014). NADPH oxidase-derived reactive oxygen species contribute to impaired cutaneous microvascular function in chronic kidney disease. *American Journal of Physiology Renal Physiology, 306*, F1499–F1506.
18. Filippin, L. I., Vercelino, R., Marroni, N. P., & Xavier, R. M. (2008). Redox signalling and the inflammatory response in rheumatoid arthritis. *Clinical & Experimental Immunology, 152*, 415–422.
19. Fortuño, A., Beloqui, O., San José, G., Moreno, M. U., Zalba, G., & Díez, J. (2005). Increased phagocytic nicotinamide adenine dinucleotide phosphate oxidase-dependent superoxide production in patients with early chronic kidney disease. *Kidney International Supplement, 61*(99), S71–S75.
20. Gao, X.-H., Qanungo, S., Pai, H. V., Starke, D. W., Steller, K. M., Fujioka, H., Lesnefsky, E. J., Kerner, J., Rosca, M. G., Hoppel, C. L., et al. (2013). Aging-dependent changes in rat heart mitochondrial glutaredoxins—Implications for redox regulation. *Redox Biology, 1*, 586–598.
21. Ge, Z. D., Lian, Q., Mao, X., & Xia, Z. (2019). Current status and challenges of NRF2 as a potential therapeutic target for diabetic cardiomyopathy. *International Heart Journal, 60*, 512–520.
22. Giacco, F., & Brownlee, M. (2010). Oxidative stress and diabetic complications. *Circulation Research, 107*, 1058–1070.
23. Gomes, M. J., Martinez, P. F., Pagan, L. U., Damatto, R. L., Cezar, M. D., Lima, A. R., Okoshi, K., & Okoshi, M. P. (2017). Skeletal muscle aging: Influence of oxidative stress and physical exercise. *Oncotarget, 8*, 20428–20440.
24. Gorin, Y., & Block, K. (2013). Nox4 and diabetic nephropathy: With a friend like this, who needs enemies? *Free Radical Biology & Medicine, 61*(130), 142.
25. Granata, S., Zaza, G., Simone, S., et al. (2009). Mitochondrial dysregulation and oxidative stress in patients with chronic kidney disease. *BMC Genomics, 10*, 388.
26. Hansson, G. K., & Libby, P. (2006). The immune response in atherosclerosis: A double-edged sword. *Nature Reviews Immunology, 6*, 508–519.
27. Hetz, C., & Papa, F. R. (2018). The unfolded protein response and cell fate control. *Molecular Cell, 69*, 169–181.

28. Holterman, C. E., Read, N. C., & Kennedy, C. R. (2015). Nox and renal disease. *Clinical Science (London), 128*, 465–481.
29. Ichinose, M., Sugiura, H., Yamagata, S., Koarai, A., & Shirato, K. (2000). Increase in reactive nitrogen species production in chronic obstructive pulmonary disease airways. *American Journal of Respiratory and Critical Care Medicine, 160*, 701–706.
30. Jia, G., DeMarco, V. G., & Sowers, J. R. (2016). Insulin resistance and hyperinsulinaemia in diabetic cardiomyopathy. *Nature Reviews Endocrinology, 12*, 144–153.
31. Jia, G., Hill, M. A., & Sowers, J. R. (2018). Diabetic cardiomyopathy: An update of mechanisms contributing to this clinical entity. *Circulation Research, 122*, 624–638.
32. Joseph, A. M., Adhihetty, P. J., & Leeuwenburgh, C. (2016). Beneficial effects of exercise on age-related mitochondrial dysfunction and oxidative stress in skeletal muscle. *The Journal of Physiology, 594*, 5105–5123.
33. Kalpakcioglu, B., & Senel, K. (2008). The interrelation of glutathione reductase, catalase, glutathione peroxidase, superoxide dismutase, and glucose-6-phosphate in the pathogenesis of rheumatoid arthritis. *Clinical Rheumatology, 27*, 141–145.
34. Karamouzis, I., Sarafidis, P. A., Karamouzis, M., et al. (2008). Increase in oxidative stress but not in antioxidant capacity with advancing stages of chronic kidney disease. *American Journal of Nephrology, 28*, 397–404.
35. Keane, K. N., Cruzat, V. F., Carlessi, R., de Bittencourt, P. I., & Newsholme, P. (2015). Molecular events linking oxidative stress and inflammation to insulin resistance and beta-cell dysfunction. *Oxidative Medicine and Cellular Longevity, 2015*, 181643.
36. Lenzen, S., Drinkgern, J., & Tiedge, M. (1996). Low antioxidant enzyme gene expression in pancreatic islets compared with various other mouse tissues. *Free Radical Biology and Medicine, 20*, 463–466.
37. Lesnefsky, E. J., Chen, Q., & Hoppel, C. L. (2016). Mitochondrial metabolism in aging heart. *Circulation Research, 118*, 1593–1611.
38. Lu, D., Ma, Y., Zhang, W., Bao, D., Dong, W., Lian, H., Huang, L., & Zhang, L. (2012). Knockdown of cytochrome P450 2E1 inhibits oxidative stress and apoptosis in the cTnT(R141W) dilated cardiomyopathy transgenic mice. *Hypertension, 60*, 81–89.
39. Lushchak, V. I. (2015). Free radicals, reactive oxygen species, oxidative stresses and their classifications. *The Ukrainian Biochemical Journal, 87*, 11–18.
40. Maiese, K. (2016). The bright side of reactive oxygen species: Lifespan extension without cellular demise. *Journal of Translational Science, 2*, 185–187.
41. Mukherjee, S., & Ghosh, A. (2020). Molecular mechanism of mitochondrial respiratory chain assembly and its relation to mitochondrial diseases. *Mitochondrion, 53*, 1–20.
42. Munzel, T., Gori, T., Keaney, J. F., Maack, C., & Daiber, A. (2015). Pathophysiological role of oxidative stress in systolic and diastolic heart failure and its therapeutic implications. *European Heart Journal, 36*, 2555–2564.
43. Muthuramu, I., Singh, N., Amin, R., & De Geest, B. (2015). Role of lipids and lipoproteins in myocardial biology and in the development of heart failure. *Journal of Clinical Lipidology, 10*, 329–342.
44. Newsholme, P., Cruzat, V. F., Keane, K. N., Carlessi, R., & de Bittencourt, P. I. (2016). Molecular mechanisms of ROS production and oxidative stress in diabetes. *Biochemical Journal, 473*, 4527–4550.
45. Onyango, A. N. (2018). Cellular stresses and stress responses in the pathogenesis of insulin resistance. *Oxidative Medicine and Cellular Longevity*, 4321714.
46. Papaconstantinou, J. (2019). The role of signaling pathways of inflammation and oxidative stress in development of senescence and aging phenotypes in cardiovascular disease. *Cells, 8*, 1383.
47. Paredi, P., Kharitonov, S. A., Leak, D., Ward, S., Cramer, D., & Barnes, P. J. (2000). Exhaled ethane, a marker of lipid peroxidation, is elevated in chronic obstructive pulmonary disease. *American Journal of Respiratory and Critical Care Medicine, 162*, 369–373.
48. Petersen, M. C., & Shulman, G. I. (2018). Mechanisms of insulin action and insulin resistance. *Physiological Reviews, 98*, 2133–2223.

49. Pinto, A. R., Ilinykh, A., Ivey, M. J., Kuwabara, J. T., D'Antoni, M. L., Debuque, R., Chandran, A., Wang, L., Arora, K., Rosenthal, N. A., et al. (2016). Revisiting cardiac cellular composition. *Circulation Research, 118*, 400–409.
50. Quinonez-Flores, C. M., Gonzalez-Chavez, S. A., DelRioNajera, D., Pacheco-Tena, C. (2016). Oxidative stress relevance in the pathogenesis of the rheumatoid arthritis: A systematic review. *BioMed Research International.*
51. Rains, J. L., & Jain, S. K. (2011). Oxidative stress, insulin signaling, and diabetes. *Free Radical Biology & Medicine, 50*, 567–575.
52. Ray, P. D., Huang, B.-W., & Tsuji, Y. (2012). Reactive oxygen species (ROS) homeostasis and redox regulation in cellular signaling. *Cellular Signalling, 24*, 981–990.
53. Ricciardolo, F. L., Caramori, G., Ito, K., Capelli, A., Brun, P., Abatangelo, G., Papi, A., Chung, K. F., Adcock, I., Barnes, P. J., et al. (2005). Nitrosative stress in the bronchial mucosa of severe chronic obstructive pulmonary disease. *The Journal of Allergy and Clinical Immunology, 116*, 1028–1035.
54. Rizvi, F., Preston, C. C., Emelyanova, L., Yousufuddin, M., Viqar, M., Dakwar, O., Ross, G. R., Faustino, R. S., Holmuhamedov, E. L., & Jahangir, A. (2021). Effects of aging on cardiac oxidative stress and transcriptional changes in pathways of reactive oxygen species generation and clearance. *Journal of the American Heart Association, 10*, e019948.
55. Roden, M., Price, T. B., Perseghin, G., Petersen, K. F., Rothman, D. L., Cline, G. W., & Shulman, G. I. (1996). Mechanism of free fatty acid-induced insulin resistance in humans. *The Journal of Clinical Investigation, 97*, 2859–2865.
56. Santos, C. X., Anilkumar, N., Zhang, M., Brewer, A. C., & Shah, A. M. (2011). Redox signaling in cardiac myocytes. *Free Radical Biology & Medicine, 50*, 777–793.
57. Seok, H., Lee, H., Lee, S., Ahn, S. H., Lee, H. S., Kim, G., Peak, J., Park, J., Cho, Y. K., Jeong, Y., et al. (2020). Position-specific oxidation of miR-1 encodes cardiac hypertrophy. *Nature, 584*, 279–285.
58. Sharma, R. B., & Alonso, L. C. (2014). Lipotoxicity in the pancreatic beta cell: Not just survival and function, but proliferation as well? *Current Diabetes Reports, 14*, 492.
59. Sies, H. (2021). Oxidative eustress: On constant alert for redox homeostasis. *Redox Biology, 41*, 101867.
60. Sies, H., & Jones, D. P. (2020). Reactive oxygen species (ROS) as pleiotropic physiological signalling agents. *Nature Reviews Molecular Cell Biology, 21*, 363–383.
61. Suvakov, S., Damjanovic, T., Stefanovic, A., et al. (2013). Glutathione S-trans ferase A1, M1, P1 and T1 null or low-activity genotypes are asso ciated with enhanced oxidative damage among haemodialysis patients. *Nephrology, Dialysis, Transplantation, 28*, 202–212.
62. Tanner, R. M., Brown, T. M., & Muntner, P. (2012). Epidemiology of obesity, the metabolic syndrome, and chronic kidney disease. *Current Hypertension Reports, 14*, 152–159.
63. Tsushima, K., Bugger, H., Wende, A. R., Soto, J., Jenson, G. A., Tor, A. R., McGlauflin, R., Kenny, H. C., Zhang, Y., Souvenir, R., et al. (2018). Mitochondrial reactive oxygen species in lipotoxic hearts induce post-translational modifications of AKAP121, DRP1, and OPA1 that promote mitochondrial fission. *Circulation Research, 122*, 58–73.
64. van der Pol, A., van Gilst, W. H., Voors, A. A., & van der Meer, P. (2019). Treating oxidative stress in heart failure: Past, present and future. *European Journal of Heart Failure, 21*(425), 435.
65. Veselinovic, M., Barudzic, N., Vuletic, M., Zivkovic, V., Tomic-Lucic, A., Djuric, D., & Jakovljevic, V. (2014). Oxidative stress in rheumatoid arthritis patients: Relationship to diseases activity. *Molecular and Cellular Biochemistry, 391*, 225–232.
66. Watson, J. D. (2014). Type 2 diabetes as a redox disease. *Lancet, 383*, 841–843.
67. Wongsurawat, T., Woo, C. C., Giannakakis, A., Lin, X. Y., Cheow, E. S. H., Lee, C. N., Richards, M., Sze, S. K., Nookaew, I., Kuznetsov, V. A., et al. (2018). Distinctive molecular signature and activated signaling pathways in aortic smooth muscle cells of patients with myocardial infarction. *Atherosclerosis, 271*, 237–244.
68. Yuan, T., Yang, T., Chen, H., Fu, D., Hu, Y., Wang, J., Yuan, Q., Yu, H., Xu, W., & Xie, X. (2019). New insights into oxidative stress and inflammation during diabetes mellitus-accelerated atherosclerosis. *Redox Biology, 20*, 247–260.

Antioxidant Agents and Their Potential Applications in Treating Various Chronic Diseases

H. Sowmya Lakshmi, S. Subhashini, D. Rajkumar, and R. Ramesh

Abstract Antioxidants had a significant attraction among biomedical researchers because of their unique property of neutralizing power with reactive oxygen species and their ability to mitigate oxidative stress. It plays a crucial role in the pathogenesis of numerous chronic diseases. Antioxidants are molecules that can be produced endogenously or exogenously by various supplementary diets by keeping safe the bimolecular such as lipids, DNA, and proteins from cellular damage, and they play a very crucial role in the development and prevention of various chronic diseases including cardiovascular, chronic kidney diseases, neurodegenerative disease, cancer, and inflammatory conditions. In the case of heart-related diseases, antioxidants and oxidative stress affect endothelial dysfunction and myocardial injury. Vitamins C, Vitamin E, and other antioxidants like polyphenols help reduce lipid peroxidation and improve vascular function. Research findings revealed that antioxidants enhance insulin sensitivity and help in reducing glucose-induced oxidative damage. Antioxidants like coenzyme Q10 and curcumin were found to have neuroprotective properties that affect the cross blood–brain barrier help in preventing neuroinflammation and preserve cognitive function. Similarly in cancer, the ROS contributes to carcinogenesis by stimulating DNA mutations, and the restricted use of antioxidants helps in preventing tumor production and multiplication through genomic stability and modulates signaling pathways. Antioxidants derived from fruits, plants, and vegetables possess strong anti-inflammatory effects and are found to have potential in treating autoimmune disorders and bone-related rheumatoid arthritis. Despite the promising preclinical and clinical findings, the healing application of the antioxidants remains very complex and sometimes very controversial due to the bioavailability, dosage level, interaction with medications, and the redox state of the target influence efficacy sometimes may lead to adverse effects. In this chapter, the potential applications of various antioxidants in the prevention and treatment of various chronic diseases,

H. Sowmya Lakshmi · D. Rajkumar · R. Ramesh (✉)
Department of Physics, Sacred Heart College (Autonomous), Tirupattur 635601, Tamil Nadu, India
e-mail: rramesh@shctpt.edu

S. Subhashini
School of Physiotherapy, Dhanalakshmi Srinivasan University, Tiruchirappalli 621112, Tamil Nadu, India

K. Kasinathan and A. Muthukumaran (eds.), *Recent Advances in Oxidative Stress Associated Chronic Diseases Volume 2*, https://doi.org/10.1007/978-981-95-3758-7_2

therapies with optimized delivery systems, elucidating precise mechanisms of action, and the ability of antioxidants to combat oxidative stress are discussed in detail.

Keywords Antioxidants · Oxidative stress · Chronic diseases · Therapeutic applications · Free radicals · Disease prevention

1 Introduction

It is essential to consume an ideal diet in order to ensure the well-being of humans and prevent associated diseases. The food that we consume in our day-to-day lives in our regular diet can be a natural tool of many health-promoting compounds including antioxidants [1]. Antioxidants help remove the toxins that are present in our bodies. In general, antioxidants are found naturally within our body itself such as glutathione, catalase, an important antioxidant, superoxide dismutase (SOD), and coenzyme. A significant role is played by phytochemicals in protecting the body and our cells and acting as a supporting system to repair damaged cells [2]. The human body continuously produces free radicals by using oxygen. The term free radicals are defined as unstable atoms or molecules. Free radicals are naturally created in our body due to cellular processes such as metabolism, exposure to environmental stressors like UV radiation, breathing in pollutants, consumption of junk foods, and alcohols, and lack of daily exercise and smoking. Production of the free radicals is also due to oxidative stress which means between the formation of free radicals and the elimination of them, a balanced equilibrium cannot be achieved body's capacity in equalizing them with oxidative stress [3–5]. As we know, free radicals are unpaired electrons. The healthy cells (*paired electrons*) in our brain, heart, and blood are damaged because of donating an electron to free radicals. Due to this conversion of healthy cells into unhealthy cells by free radicals, the antioxidants that are naturally present in our body will support unhealthy cells. But there are no sufficient antioxidants in our body as this process takes place constantly [6–8]. Thus, to enhance the antioxidants in our body we should lead a healthy lifestyle like consuming fresh fruits and vegetables that must contain antioxidants. Antioxidants are chemical substances that are found in our food very organically which can prevent or reduce the oxidative stress in our body [9, 10]. An immense source of naturally occurring antioxidants is found in plants and animals. Antioxidants are groups of carotenoids, flavonoids, melatonin, vitamin A, vitamin C, vitamin E (mainly α-tocopherol), phytochemicals such as beta-carotene and lycopene, and minerals such as selenium and zinc [11, 12]. WHO reports that chronic diseases are very prevalent for over 60% of all unfortunate deaths worldwide. The generation or synthesis of free radicals in the body due to oxidative stress can cause various chronic diseases such as diseases in the heart system, respiratory illness, oral diseases, arthritis, Chronic Obstructive Pulmonary Disease (COPD), Osteoporosis, depression, pregnancy women with chronic illness as shown in Fig. 1 [13–15]. Hence, to overcome these chronic diseases, antioxidant agents play a key role. The antioxidant can be measured by oxygen radical

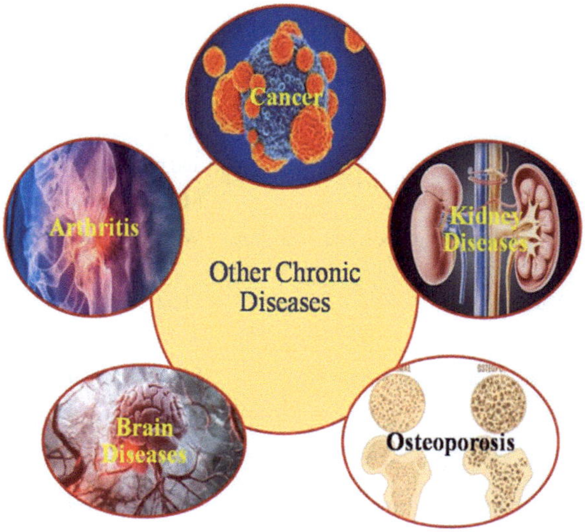

Fig. 1 Different characteristics are associated with chronic diseases

absorbance capacity (ORAC). The basic need for antioxidants for a human per day is 5000 (ORAC). For example, 1 g of turmeric contains 1000 ORAC, and 100 g of country fruit pomegranate contains 3000–5000 ORAC. To reduce oxidative stress, free radicals and antioxidants are compensated by taking antioxidants additionally through our regular diet apart from the antioxidants produced within our body as shown in Table 1. The benefits of taking antioxidant supplements are preventing various chronic diseases, enhancing the natural antioxidants that are within our body, helping in deducing oxidative stress, and building up our immunity power [16–18]. This chapter highlights the antioxidant agents and their potential applications in the treatment of various chronic diseases. The below figure shows the donation of electrons from the healthy cell to the free radical. Because of this reason, there is a loss of electrons in the healthy cell and becomes an unhealthy cell. Now, the antioxidants support the unhealthy cells to be stable by donating electrons to healthy cells [19–22].

2 Implementation of AI in Chronic Disease Monitoring and Management

Chronic diseases like cardiovascular disease, diabetes millets, Alzheimer's disease, kidney-related diseases, bone-associated rheumatoid arthritis disease, chronic obstructive pulmonary disease (COPD), hypertension problems asthma, and osteoarthritis are common worldwide [31–34]. Utilizing digital method therapeutics, a virtual mode of a platform for tracking disease symptoms, and artificial intelligence and risks can be diagnosed. Patients can take an active role in managing their health.

Table 1 In nature, antioxidants can be found in a wide variety of forms

Sources	Antioxidant compound	Ref.
Apple	Phenolics, flavonoids	[23]
Pecan nuts	Hydrobenzoinc compounds including gallic acid and ellagic acid	[24]
Brews from coffee	Melanoidins and polyphenols	[25]
Juice extract of grapes, red cabbages, legumes, etc.,	Water-soluble anthocyanins compounds	[26]
Walnut	Plant phenolics	[27]
Berry	Anthocyanins	[28]
Sesame coat waste	Sesamin sesamolin lignans	[29]
Tomato	Vitamin C and E, phenolics, lycopene flavonoids	[30]

This approach aims to improve health results reduce the costs in health care management and enhance the active involvement of the patient himself and adherence to effective plans of treatment, transforming long-term care through remote monitoring [35].

2.1 Symptom Analyzing and Risk Monitoring Development System

The successful implementation asymptomatic analyzing the risks involved assessment monitoring relies heavily on careful methodology in planning and data collection execution strategies. These systems are designed to gather a diverse array of health—related information efficiently. They can track various metrics, including levels of glucose in the blood, pressure, symptoms of asthma, regular dietary practices, and physical activity [36–38]. This extensive data collection enables the system to forecast health trends and develop a thorough understanding of each patient's health by synthesizing information from multiple sources. At this stage of system development, the integration of AI models is essential. The system analyzes real-time health data using machine learning algorithms, particularly neural systems. Machine learning algorithms identify subtle indicators that suggest deteriorating wellness or elevated disease risk and provide strategies for reducing these risks [39]. For example, if a diabetic patient shows consistently high blood sugar levels, the system may initiate automated interventions or provide tailored recommendations. A crucial component of the system is its alert mechanism, which notifies, for healthcare consumers and providers to be aware of important warning signs and symptoms, they need to receive education [40]. This alert system can quickly respond when health issues are about to worsen. The nature of the alerts can be based on symptom severity, ranging from immediate action requirements to lifestyle guidance and recommendations. Throughout the development phase, techniques for dataset augmentation are employed to improve the AI model's performance. These techniques may include generating samples, a wide range of statistical sources are used to enhance the model's flexibility and applicability, including ethnic minorities [41–45]. Integrated virtual care (IVC) offers a transformative environment for patients, caregivers, and healthcare professionals. This innovative approach is set to enhance the management an illness that has lasted for a long time [46]. The use of digital health, continuous data collection, and individualized care management will help improve health outcomes, and this method significantly improves the continuity of care for chronic health issues. Central to integrated virtual care is the well-coordinated two-way communication between team members providing clinical care to patients. These communication pathways empower patients to actively participate in their clinical care teams. Using these communication channels allows patients to provide relevant information, gain insight, and actively participate in teleconversations, which motivates them to take an active role in their medical care [47, 48]. Through platform interfaces that simplify content accumulation, sharing, and collaboration, users will have a more enjoyable experience. For healthcare professionals, it is essential that their systems are easily integrated. The interaction between healthcare workers and patients is comfortable and easy for everyone involved [49]. With this device, healthcare professionals can track multiple patients' concerns, address their concerns promptly, and respond to their medical plans appropriately. As a result of the collaboration, insight is gained into, and modifications are made to care for a specific patient, enhancing care coordination. With its health tracking capabilities, the integrated virtual care system enables continuous monitoring of key metrics of a patient's health. It has become possible to determine a patient's health improvement or decline with the use of AI-driven forecasting. The triage capability facilitates the identification of problematic cases as quickly as possible by healthcare professionals [50]. A clinician is alerted when an individual deviates significantly from his or her starting point. In addition to digital therapies, various other sources contribute data, including current health information and medication history.

2.2 Development of AI-Driven Digital Therapeutics

To tailor treatments effectively, the AI-driven digital therapeutics platform assesses each patient's goals, challenges, and cognitive patterns. This advancement offers avenues for personalized interventions. This program continuously analyzes patient information to create tailored treatments based on each patient's background, interests, and behavior [51]. Patients are empowered through personalized reminders and supportive messaging, especially when they encounter difficulties with medication adherence. The AI-enhanced digital therapy system utilizes cognitive behavioral strategies, incorporating elements such as enhancing positive behavior by identifying goals, developing skills, and reinforcing them. Users can set health objectives across various areas, including a balanced diet, a physically active lifestyle, and prescription medicines management [52]. In order to develop lasting attitudes, a network design that is optimal and monitored continuously is necessary. As a system enhances the accuracy of behavioral by utilizing enhancements, datasets can be enhanced techniques informed by behavioral insights. It integrates lifestyle, socioeconomic, and user-generated information alongside health records. By adopting a comprehensive approach, we can ensure that the AI model accurately interprets patient behavior, allowing us to provide customized solutions that go beyond traditional clinical assessments. The evolution of AI-facilitated digital therapies focuses on ongoing improvement, with system enhancements based on the effectiveness and user-friendliness of the system evaluations. AI models are built using patient records to ensure that treatments are relevant, cost-efficient, and customized according to the needs of each patient. It is vital to assess an individual's ability to self-care for prolonged illnesses to develop a plan that meets their needs. A careful assessment of the patient's health is conducted in this method [53–55].

3 Common Diagnostic Methods and Treatment for Heart Diseases

Most elderly people fail to diagnose heart failure, which is a well-known clinical syndrome. A detailed history of medical conditions is crucial, including not only the immediate manifestations of angina, but also signs or symptoms of primary conditions, such as arterial diseases, high blood pressure, or heart valve disease. Assessing the patient's past cardiac history, particularly myocardial infarction, is essential as shown in Fig. 2. Heart failure patients have left ventricular systolic dysfunction and diastolic dysfunction, which can be detected by Valsalva's maneuver during the physical examination [56]. Additionally, an electrocardiogram should be performed, as well as a radiographic examination of the chest. Diastolic and systolic dysfunction can be assessed using ultrasound echocardiography. In addition, patients with coronary artery disease who also suffer from severe chest pain might benefit from coronary angiography, as might those who have an electrocardiogram showing ischemia or myocardial infarction. A person with heart failure has difficulty breathing because their heart cannot pump enough blood to meet their body's needs. A high respiratory rate in the left ventricle can also make it impossible to sustain normal respiration [57].

3.1 Evaluation of Heart Disease

Prior to the development of noninvasive methods for ascertaining systolic and diastolic dysfunction, systolic symptoms were determined by Framingham, Duke, and Boston criteria, respectively. These criteria aim to help diagnose heart failure. Among them, the Boston criteria have demonstrated the best-combined sensitivity of 50%.

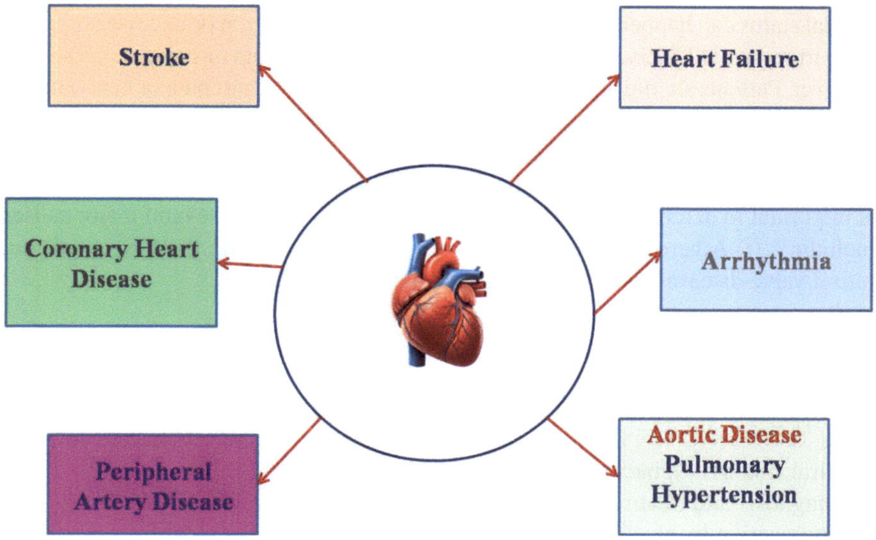

Fig. 2 Chronic diseases associated with the cardiovascular system

These criteria are particularly useful for identifying advanced or severe heart failure, and people with an abnormal cardiac output ratio are at risk of developing this condition in 20–40% of cases [58]. Early detection of heart failure is crucial for effectively treating the root causes and, in some cases, stopping further heart damage and worsening health. However, diagnosing it can be challenging since symptoms can range from none at all to severe issues like pulmonary edema and cardiogenic shock. Studies suggest that only about 50% of patients with heart failure are accurately diagnosed a first. To diagnose this condition more accurately, a structured approach is beneficial. The very first step in determining diabetes is to obtain a comprehensive medical history. It is important to inquire about symptoms such as breathlessness, coughing, frequent urination at night, and tiredness that may indicate heart failure [59]. Shortness of breath, a key symptom of heart failure, often worsens from being out of breath during activity to difficulty breathing while lying down, sudden nighttime breathlessness, and even at rest. A cough, typically occurring at night and not producing mucus, may happen alongside shortness of breath, especially during exertion or when lying flat. Frequent nighttime urination is another common sign of heart failure, resulting from increased blood flow to the kidneys when the patient is lying down [60]. General fatigue, due to reduced blood flow, and swelling in the legs that makes it hard to wear regular shoes are also common issues. As heart failure worsens, gastrointestinal symptoms like bloating, loss of appetite, and a feeling of fullness in the upper right abdomen may appear. In severe cases of long-term heart failure, patients may experience cardiac cachexia, which is severe weight loss in the intensities and higher levels of certain inflammatory substances. This condition can resemble the weight loss seen in advanced cancer patients. Confusion and changes in mental status can happen due to reduced blood flow to the brain or liver issues related to heart failure [61]. Liver problems in heart failure arise from ongoing congestion in the liver. Patients should also be questioned about any past chest pain or heart attacks; coronary artery disease is responsible for diminished left ventricular function in 75% of patients with heart failure. In contrast to other health issues, cardiovascular history provides a more reliable indication of heart failure. In order to prevent heart failure, it is important to determine the cause of hypertension. There are several major causes, including: (i) Atherosclerosis; (ii) Having high blood pressure; and (iii) Aortic and mitral valve disease are the most common heart valve diseases.

3.2 Additional Causes Encompass

- Viral (like HIV), bacterial, and parasitic infections.
- Diagnosis and treatment of pericardial diseases.
- Intoxication by alcohol.
- Inflammatory diseases of the connective tissue.
- An increase in heart rate.
- Heart disease caused by obstructive cardiomyopathy.
- Drugs (e.g., doxorubicin [Adriamycin]).
- Dystrophies, myotonic dystrophies, and Friedreich's ataxia are all neuromuscular disorders.
- Metabolic disorders, such as Pompe's disease and McArdle's disease (glycogen storage diseases).
- The presence of nutritional deficiencies such as beriberi and kwashiorkor.
- A phaeochromocytoma is an abnormal cell growth.
- A significant amount of radiation exposure.
- There is an increase in the number of patients with endomyocardial fibrosis.
- The disease of endomyocardium caused by eosinophils.
- The following diseases are considered cardiovascular risk factors: Heart failure with high output, intracardiac shunts, atrioventricular fistulas, beriberi, pregnancy, Paget's disease, hyperthyroidism, and anemia.
- During pregnancy, there is a risk for peripartum cardiomyopathy.
- Diabetic cardiomyopathy with dilated blood vessels.
- There are several types of intracardiac shunts, including atrioventricular fistulas, beriberi, pregnancy, Paget's disease, and hyperthyroidism.
- The drug cyclophosphamide [Cytoxan], as well as cocaine.
- There are several diseases that infiltrate the body, such as amyloidosis, sarcoidosis, and cancers.

3.3 Physical Examination

In order to diagnose heart failure, there must also be a thorough physical examination. Cachexia, cyanosis, and resting dyspnea should be evaluated by evaluating the patient's overall appearance. An accurate record of the patient's vital signs and rate of heartbeat must be kept. It is possible to determine whether blood pressure is high, normal, or low by measuring it [62]. A poor prognosis is associated with patients exhibiting a systolic blood pressure below 90 to 100 mm Hg in the absence of medication. Tachycardia can indicate heart failure, particularly in its decompensated phase, as the heart rate may increase as a compensatory mechanism to make sure your heart is active enough. A reduction in heartbeat rate during medical intervention can be used as an indirect indicator of treatment effectiveness. Additionally, a weak, thread pulse and pulses alternant are linked to diminish an assessment of the left ventricle's function. It is also important to check the individual for Cheyne–Stokes breathing symptoms, which can be a sign of chronic respiratory problems [63]. (i) *Jugular venous distention: t*o assess jugular venous distention, the patient is laid in a supine position with their upper body elevated 45 degrees above the horizontal angle. Venous distention is determined based on the peak of the internal jugular pulsation curves [64]. The position of the sternal angle is established by drawing an imaginary horizontal line from this point. If the difference between the sternal inclination and this line exceeds 4–5 cm, then venous pressure may be raised. When the left ventricle is inflated to its maximum capacity, increased jugular venous flow can serve as a specific (90%) but insensitive (30%) indicator. Despite its significance, the jugular venous distention measurement is surprisingly inconsistent [65]. (ii) *Point of maximal impulse and Pulmonary Examination:* in the left atrium, the maximum impulse (MI) occurs in the fifth intercostal space near the midclavicular axis. When the patient is seated, the physician employs their fingerprints to pinpoint this location. In cases of cardiomegaly, the cardiac impulse is often displaced laterally and inferiorly. Locating the PMI can sometimes prove challenging, resulting in a loss of sensitivity (66%). Although it is not perfect, it is still an effective tool for measuring heart volume (96% accurate). If the lungs are assessed accurately, drains and interpleural fusions might be evident [66]. It is possible that there are symptoms of congested lungs without showing any signs due to increased lymphatic drainage and adaptation in surrounding tissues. People with heart failure may misdiagnose asthma as pulmonary congestion due only to breathlessness. (iii) *Liver size and Hepatojugular reflux and Valsalva's maneuver:* in addition to measuring the liver's dimensions, the abdomen exam also includes a question regarding its condition. When the heart fails unilaterally, it leads to swelling in the veins which is known as hepatomegaly [67]. It is important to perform a hepatojugular recirculation test in a patient who is suffering from acute right heart failure. When performing HI assessment, the individual must lie on their back with their shoulders positioned at 45 degrees. Performing the Valsalva maneuver may lead to falsely positive results if the patient maintains an open airway and breathes normally. Patients with circulatory failure are rarely subjected to Valsalva's maneuver [68]. Despite this, this technique is relatively easy to perform while being highly accurate (91% specificity) and reliable (69% accuracy) when determining left ventricular dysfunction. A 15 mm Hg difference is made between the arterial blood pressure and systolic blood pressure during Valsalva's maneuver. Patient must not only listen to the physician through the brachial arteries, but also perform voluntary expirations through restricted airways. At the start of the straining phase (phase I), Korotkoff sounds are heard along with an increase in systolic blood pressure (phase I). As a result of maintaining the maneuver (phase II), the blood pressure in the heart decreases, and Korotkoff sounds cease to be heard. Phase III (releasing the maneuver) typically results in the blood pressure returning to normal [69]. A patient who does not experience overshoot after releasing the maneuver may be suffering from heart failure if there is no overshoot.

3.4 Diagnostic Challenges

Older adults are more likely to suffer from cardiovascular disease due to the unusual signs and symptoms they experience. Heart disease typically manifests in elderly people as anorexia, fatigue, and malaise. A common psychological condition is depression, as well as stress. It is generally true that older people experiencing physical discomfort reduce their activities, thereby essentially preventing symptoms from appearing [70]. Over time, exercise-induced symptoms that worsen with decreased activity can gradually lead to restlessness (NYHA class IV). As chronic heart disease progresses, interpreting physical findings becomes increasingly challenging. It is uncommon to have resting tachycardia, and assessing abnormal pulse contours can be difficult if there are concurrent respiratory diseases, which may change the pulse contours due to peripheral arteriosclerosis.

3.5 Diagnosis Tests—Electrocardiography and Chest Radiography

Every person suffering from cardiomyopathy must undergo an electrocardiogram (ECG). There are very few definite ECG characteristics that indicate heart failure, but atrioventricular and supraventricular arrhythmias are very common. Patients suffering from irreversible heart failure are more likely to develop the condition than those suffering from cardiomyopathy. A person with atrial fibrillation, tachycardia, or latent bundle branch block has a significantly worse prognosis [71]. It may indicate amyloidosis if the ECG shows abnormally low voltage and conductivity. In order to diagnose hypertension, chest radiographs are important. If viewed from a posterior–anterior perspective, cardiomegaly affects the ratio to the extent that it exceeds 0.50. When systolic insufficiency occurs in a person with normal ventricular size, it serves as a guide to how much systolism is abnormal. When viewed from the side, there is an insufficient gap between the heart and the chest which indicates an enlarged right ventricle. As a result of a high intravenous pressure, a reversal of blood flow from the lungs' base to their apices is often observed on radiographic examinations, indicating a linear density (Kerley's lines) indicative of interstitial densities, resulting in hazy patterns primarily in the pericardium. Chest radiographs can also diagnose heart failure-related pleural fluid accumulation [72].

3.6 Echocardiography and Angiography

A transthoracic two-dimensional echocardiography paired with a doppler flow study is strongly recommended for anyone with heart failure. The left ventricle can be assessed according to its size, mass, and function with this examination. It represents a noninvasive method for evaluating the systolic as well as diastolic function of the heart [73]. Because of faulty image quality or discrepancies between the clinical assessment and the two-dimensional echocardiography, this imaging modality is particularly helpful in providing conclusive results. Left ventricular ejection fraction and regional wall motion can be evaluated quantitatively with radionuclide angiography. While ectopic activity and atrial fibrillation can have a negative impact on its precision, measurements can still be problematic due to their presence. By using left ventricular angiography, the ejection fraction and left ventricular volume are evaluated, as well as valvular regurgitation or stenosis. Furthermore, assessing the diastolic and diastolic relaxation rates of the left ventricle can be helpful in diagnosing dysfunction [74]. It is also possible to measure ejection fraction and evaluate regional wall motion using MRI and computed tomography (CT) such as magnetic resonance imaging (MRI). There is, however, a limited application of these techniques in cardiac function assessment, and they are still considered inferior to echocardiography and angiography. In some cases, controlling coronary artery disease has an important role to play in preventing heart failure. If a patient with heart failure suffers from angina, coronary angiography and cardiac catheterization should be strongly recommended. If the patient has established coronary artery disease but does not have angina and is not suffering from angina, coronary arteriography or noninvasive testing methods (such as thallium stress tests or stress echocardiograms) should be considered, and if signs of ischemia occur, coronary arteriography should be performed. It is dependent on the likelihood that a patient has coronary artery disease whether ischemic heart disease should be explored thoroughly [75].

3.7 Antioxidant Vitamins and the Prevention of Coronary Heart Disease

A significant reduction in the risk for coronary heart disease (CHD) could be achieved by taking multivitamin supplements. According to epidemiological studies, those with a high intake of antioxidants in their diet are less likely to experience coronary heart disease. Several nutrients have been shown to help prevent coronary heart disease events, according to clinical trials. Carotenoids are not recommended for supplementation because they may have negative effects on the body [76]. In studies, 400 IU vitamin E and 500 mg vitamin C are found to be effective in reducing the risk coronary heart disease in patients who are at high risk. For patients with high coronary heart disease risks, vitamin C could be considered as a risk reduction intervention. It is confirmed that folate can reduce homocysteine levels, which is why it is recommended to take it every day as a supplement. However, recent studies suggest that antioxidant supplements may help lower coronary artery disease (CHD) risk. Vitamins C, E, folate, beta-carotene, ubiquinone (coenzyme Q10), and bioflavonoids are examples of antioxidants [77].

3.8 A Combination of Vitamin E, Vitamin C, and β-Carotene

Polyunsaturated fatty acids in cell membranes are prevented from peroxidizing through vitamin E. Tocopherol, the form of vitamin E with the greatest bioactivity, has the greatest accessibility. Vitamin E helps keep LDL from oxidizing because it is integrated into lipoproteins and cellular membranes [78]. The agent also blocks the activation of platelets and monocyte adhesion. Vitamin E can be found in vegetable and seed oils, wheat germ, meats, fish, fruits, and vegetables in smaller amounts. In terms of dietary reference intake (RDI), vitamin E is recommended at 30 μg per day. In a typical diet, it can be challenging to get enough vitamin E, and multivitamins typically contain 30–50 mg. Plasma primarily contains vitamin C as an antioxidant. Vitamin B2 protects LDL particles from free radical oxidation by scavenging free radicals in both the plasma and the cells. As well as regenerating active vitamin E, vitamin C induces cholesterol excretion [79]. Vasodilation is induced by this vitamin when the endothelium is activated, and monocyte adhesion is decreased when this vitamin is present. Supplementing with 1000 mg of vitamin C and 800 IU of vitamin E before eating a high-fat meal can reverse endothelial dysfunction and vasoconstriction. In addition to citrus fruits, strawberries, cantaloupe, tomatoes, cabbage, and leafy green vegetables, vitamins C can also be obtained from dietary sources. The good thing about vitamins C is that they are available in raw or supplement form since cooking can degrade them. Women who are pregnant or loco, smokers, and those who are recovering from wounds should take more vitamin C than the RDA of 60 mg. The exact role of carotenoids remains largely unknown, despite their abundance. β-carotene is turned into vitamin A via plasma and LDL and is transported in the body [80]. LDL oxidation is prevented by it, but it decreases uptake of oxidized LDL. Foods high in carotenoids include fruits, yellow and orange vegetables (like carrots and squash), and dark green vegetables (like spinach and broccoli). As of right now, there is no established RDA for -carotene.

4 Diagnostic and Management of Diabetes

In the beginning, diabetes was classified as juvenile-onset diabetes mellitus (now called type I diabetes mellitus) and adult-onset diabetes mellitus (now called type 2). In total, over 50 subcategories have been classified, each resulting from an independent pathogenic mechanism or associated with a different disease or syndrome [81]. In the field of diabetes diagnosis, physicians have evolved from identifying typical symptoms to detecting ambient hyperglycemia and later established criteria for determining diabetes based on elevated plasma glucose levels following an overnight fast and/or a glucose challenge (oral glucose tolerance test). A1c (glycated hemoglobin) is now being measured on a routine basis for diagnostic purposes. Globally and in the United States, diabetes prevalence has significantly increased, highlighting its significant social consequences [82]. Diabetes risks can be identified through contemporary testing methods as well as the risks associated with pregnancy and their fetuses. A majority of patients with type 1 diabetes have autoantibodies and T cells that react with islet cell antigens both before and after hyperglycemia develops. In this type of diabetes, cells in the pancreatic islet beta cells are targeted by the immune system [83]. Hypoglycemia and ketosis occur when approximately 80% of beta cells are compromised or destroyed. It results in osmotic diuresis, a condition characterized by frequent urination, excessive thirst, and loss of weight. Alternatively, type 2 diabetes is caused by a combination of insulin resistance and relative resistance, especially in people who are overweight, sedentary, or aging [84].

4.1 Management Diagnosis of Diabetes

When diabetes mellitus (DM) is not treated, severe complications may develop, making early detection critical to preventing these problems. A high level of blood glucose over an extended period of time, frequent urination, thirst, and hunger are the primary signs of diabetes. Prediabetes or diabetes can be diagnosed through a variety of biochemical tests [85]. Diabetes screening usually involves the HbA1c test and the oral glucose tolerance test (OGTT). A specific amount of sugar is consumed, and the body is tested to see how well it can absorb glucose. It is typically required that an individual suspected of having diabetes consume 75 g of glucose, and the plasma glucose level is assessed two hours after consumption. It is possible to diagnose diabetes when one's plasma glucose level exceeds 11.1 mmol/L. Diagnosing diabetes also requires measuring fasting plasma glucose levels, which are typically at least 7.0 mmol/L for individuals with diabetes [86]. After consuming 75 g of glucose, a person's blood glucose level should be at least 7.8 mmol/L. In addition, HbA1c is commonly used to diagnose diabetes since patients with type 2 diabetes mellitus (T2DM) typically carry 48 mmol/mol (6.5 DCCT%) glycosylated hemoglobin. In addition to random blood sugar monitoring, diabetic patients may also benefit from regular blood pressure monitoring. Prior to the discovery of diabetes mellitus (DM), diabetes was thought to be a lifelong illness; however, DM can be remitted. It is important to maintain a healthy body weight, change your diet, exercise regularly, monitor your lipid levels, and use the appropriate medication when needed in order to effectively manage diabetes [87]. Changing one's diet plays a crucial role in controlling diabetes. Choosing foods low in glycemic index, complex carbohydrates, proteins, and polyunsaturated fats (PUFAs) can help stabilize blood sugar. In addition to helping reduce obesity, moderate exercise lowers blood glucose levels by facilitating glucose transport into muscle tissues that is insulin independent. It is significantly more likely for diabetic women who have gestational diabetes to develop type 2 diabetes after childbirth. Postpartum T2DM can be reduced by approximately 40% in women who breastfeed for at least three months [88]. The effects of breastfeeding on postpartum glucose tolerance have been scientifically proven. During lactation, estrogen levels decrease, which contributes to a better sugar and lipid metabolism as well as more even distribution of fat. Consequently, women with GDM may be less likely to develop long-term T2DM if they experience prolonged lactation. When it comes to managing diabetes, medication is the last resort. A type 1 diabetic requires insulin, while a type 2 diabetic needs insulin in 25–30% of cases. It is more dangerous to have hypoglycemia than hyperglycemia if your insulin dose is incorrect. The risk of this can be reduced by carrying sugar or chocolate for emergencies in diabetic patients who are on insulin [89]. Diabetes patients with peripheral neuropathy may benefit from vibration therapy. Diabetes-associated metabolic syndrome and atherosclerosis can be lowered if a healthy lipid profile is maintained. The administration of antioxidant supplements has been proven to reduce plasma insulin levels as well as HbA1c levels. The antioxidant properties of vitamin C are known to prevent atherosclerosis and the oxidative damaged of LDL (low-density lipoprotein). In managing diabetes, it is crucial to prioritize physical activity.

4.2 Role of Antioxidant in the Treatment and Prevention of Diabetes Mellitus

There is currently no definitive cure for diabetes; however, the progression of its complications can be managed through medication, allowing for some degree of disease control. Among the most effective treatments for non-insulin-dependent diabetes mellitus are sulphonylureas and biguanides. Given that each pharmaceutical agent carries its own set of side effects, there has been a growing preference among both patients and healthcare providers for medications derived from natural herbal sources [90]. Metformin, derived from the herb Galega officinalis, stands out as the primary conventional medication utilized in the management of non-insulin-dependent diabetes. Recognizing that oxidative stress underlies many diabetes-related complications, researchers have extensively explored the antioxidant properties of various substances, particularly those of plant origin, to identify more effective agents for diabetes management. Numerous studies have demonstrated that antioxidants such as lycopene, retinol, α- and γ-tocopherol, β-cryptoxanthin, ascorbic acid, α- and β-carotene, lutein, and zeaxanthin, found in various plants, significantly mitigate diabetes complications [91]. One study indicated that the aqueous extract of Chrysobalanus icaco, known for its antioxidant properties, effectively lowered blood glucose levels. Additional research has confirmed and elaborated on the beneficial effects of antioxidant phytochemicals in reducing complications associated with chronic conditions, including diabetes, cardiovascular diseases, and obesity [92]. These phytochemicals enhance the activity of α-glucosidase and lipase due to their antioxidant effects, leading to lower glycemic levels, improved pancreatic function, and a synergistic effect with hypoglycemic medications, thereby contributing to better diabetes management. Overall, the findings from various studies suggest that the consumption of vegetables, fruits, and seeds is beneficial in preventing diabetes complications due to their antioxidant properties. Some potent antioxidants that contribute to the better management of diabetes include vitamin C, vitamin E, carotenoids, and polyphenolic compounds.

5 Diagnostic and Treatment for Pregnancy Women with Chronic Illness

Here, we are going to discuss about the diagnosis and treatment for pregnant women with hypertension. These hypertensions, preeclampsias, and eclampsias are all hypertensive conditions of pregnancy (HDP). In addition to chronic hypertension, these symptoms can also occur. The World Health Organization (WHO) defines pregnancy hypertension as elevated blood pressure (BP) over 140/90 mm Hg. There is a wide range of criteria for initiating antihypertensive therapy, however, which stems from the ambiguity surrounding maternity advantages of reducing BP and the possible risks to the fetus associated with medication-induced alterations in uteroplacental circulation and exposure to antihypertensive drugs in utero [93].

5.1 Blood Pressure Measurement

No matter how many months a patient has been pregnant, accurate blood pressure measurements are necessary for the diagnosis of hypertension and the commencement of treatment. Since mercury sphygmomanometers are limited in availability, aneroid devices often are used, even though their accuracy may be reduced [94]. Women with gestational hypertension and preeclampsia can benefit from oscillometric automated devices. Hypertension management in pregnancy is primarily based on office blood pressure readings in current guidelines; however, out-of-office blood pressure measurements are widely considered more accurate and reliable as a measure of cardiovascular morbidity and mortality in the general population. Blood pressure levels have been measured many times during pregnancy using self-measurement or ambulatory blood pressure monitoring, but there is not much data concerning appropriate out-of-office thresholds for detecting hypertensive disorders [95]. It is recommended that women with chronic or gestational hypertension, especially if the condition remains uncontrolled, use self-measured blood pressure, especially if the condition is not controlled by the American College of Obstetricians and Gynecologists (ACOG). Currently, there is no consistent difference in blood pressure readings during pregnancy between self-measured and office readings, suggesting that self-monitoring thresholds may be comparable to standard clinic thresholds for treating and diagnosing pregnancy; however, further research is needed to determine how appropriate methodologies and device validation can be applied [96].

5.2 *Nonsustained Hypertension and Secondary Hypertension*

Around 25% of nonpregnant adults experience white coat hypertension (elevated blood pressure in clinical settings). Approximately 4–30% of pregnant women are affected, according to estimates. A 24 h blood pressure assessment found that 32% of women with hypertension had white coat hypertension. However, only 8% were officially diagnosed with white coat hypertension. Meta-analysis of white coat hypertension found that females with high blood pressure had a higher risk for preeclampsia and poorer prenatal outcomes. Women who experienced chronic hypertension or gestational hypertension faced greater risks than those who did not. There has not been enough research on masked hypertension during pregnancy and its clinical implications [97]. It is necessary to monitor blood pressure continuously during pregnancy if an elevated blood pressure occurs temporarily during pregnancy. Blood pressure measurements, such as those taken by yourself, play an essential role in identifying temporary blood pressure elevations, such as masked hypertension and white coat hypertension. Blood pressure results taken four hours apart are used to determine whether hypertension is present in pregnancy [98]. In approximately 90% of cases of chronic hypertension diagnosed in women, it is a primary condition. In a smaller percentage of cases, mothers and their fetuses may suffer from secondary hypertension. It is advisable to evaluate this condition when the mother is under 35, suffers from severe or resistant hypertension, does not have a family history of hypertension, or exhibits laboratory signs of hypokalemia, elevated creatinine levels, or albuminuria in the early stages of pregnancy. Furthermore, women of reproductive age are becoming increasingly obese, and women suffering from obstructive sleep apnea may be at greater risk of developing secondary hypertension. There are no specific guidelines for managing obstructive sleep apnea during pregnancy. Therefore, pregnant women with this condition need to consult with a sleep medicine specialist to determine the appropriate diagnostic and therapeutic approach [99].

5.3 BP Goals for Treatment of Hypertension in Pregnancy

For nonpregnant individuals with stage 1 hypertension, the threshold is 130/80 mmHg, while for stage 2 hypertension, it is 140/90 mmHg, according to guidelines established by the American College of Cardiology and the American Heart Association. Diagnoses and treatments have increased because of this change. In general, research has shown that lowering blood pressure decreases cardiovascular disease risk [100]. It is noteworthy that most cardiovascular events occur in individuals whose blood pressure is 140–159/90–109 mmHg. In addition, even younger hypertensive patients may show early signs of vascular remodeling and endothelial dysfunction, particularly in smaller arteries and arterioles. If left untreated, large blood vessels become stiffer, and organ damage occurs. Hypertension is defined as a blood pressure more than 140/90 mmHg; however, specific treatment thresholds and targets vary according to guidelines. Low blood pressure thresholds are strongly advocated for several reasons [101]. It has been shown that a more proactive approach to hypertension during pregnancy can help prevent severe hypertension, as demonstrated both by a systematic review of randomized trials, and by the Control of Hypertension in Pregnancy Study (CHIPS), which reported a BP of 133/85 mmHg on average under strict control. Although preeclampsia rates were not significantly different among those with less stringent control and those with stringent control, those with less stringent control had higher rates of thrombocytopenia and elevated liver enzymes, which indicate a serious disease [102]. The results of this trial, as well as those from others, suggest that rigid control may reduce the chance of preterm births. A debate has existed over the significance of severe hypertension as an outcome; however, exploratory analyses of the CHIPS data, adjusted for the allocated groups and prognostic factors, indicated that severe hypertension is comparable to preeclampsia when it comes to adverse maternal and perinatal outcomes. The situation is particularly relevant for high-risk groups, including Black women, who are at a higher risk of complications associated with hypertension. It has been found that black women who are chronically hypertensive who receive antihypertensive treatment before 20 weeks of gestation and achieve a blood pressure level below 140/90 mmHg have a lower rate of superimposed preeclampsia and preterm deliveries before 35 weeks than those with high blood pressure. Preeclampsia rates are also lower when blood pressure is maintained lower during pregnancy, including episodes with severe features [103].

5.4 *Interaction and Regulation of Antioxidant Defense Mechanisms, Antioxidant Supplementation in Pregnancy-Induced Hypertension*

It is important to note that the effectiveness of antioxidant defense mechanisms depends on not only the content of the various elements, but also how they interact with one another and how they are regulated. In addition to nutritional status, lifestyle choices, and environmental exposures, antioxidant enzymes are affected by several factors. The body is capable of counteracting oxidative stress by consuming plenty of antioxidant-rich foods. Oxidative damage can be further worsened by negative factors such as smoking, pollution, and poor dietary practices [104]. A number of signaling pathways and transcription factors play critical roles in modulating antioxidant gene expression, including nuclear factor erythroid 2-related factor 2 (Nrf2). By activating Nrf2 and thereby increasing antioxidant enzyme and protein synthesis, the cell becomes more resistant to oxidative damage. PIH can cause complications such as preeclampsia and eclampsia during pregnancy, posing significant risks for maternal and fetal health. Considering the correlation between oxidative stress and PIH, antioxidant supplements could be used as a therapeutic approach to help ameliorate oxidative damage [105]. Supplemental antioxidants are discussed in this section in terms of what they are, what types of supplements they can be, what dosages they can be prescribed, and the clinical evidence supporting their use.

5.5 *Rationale for Antioxidant Supplementation, Dosage, and Timing of Supplementation*

The antioxidants in the body are important for both neutralizing reactive oxygen species (ROS) in the body as well as helping to restore redox balance in PIH. PIH is associated with mechanisms such as endothelial dysfunction and enhanced vascular resistance as a result of oxidative stress during pregnancy [106]. By adding antioxidants to a diet, blood pressure can be reduced, endothelial function can be improved, and reactive oxygen species (ROS) can be reduced. Nitric oxide (NO), one of the key mediators of vasodilation, may also be enhanced during pregnancy by antioxidant supplements. There are different types of antioxidants containing PIH which require different dosages and timings. The daily doses of vitamin C and vitamin E in clinical trials generally range from 500 to 1000 mg and between 100 and 400 mg. In a number of studies, it has been suggested that pregnant women can increase their antioxidant intake by starting it during pregnancy. Furthermore, some studies indicate that taking antioxidants early in pregnancy can offer increased health benefits, with some studies indicating a greater benefit during the first trimester. Optimal dosages and regimens for ensuring mother's and fetus' safety and effectiveness are still awaiting research [107].

6 Diagnostic and Treatment Method for osteoporosis

Osteoporosis is a chronic systemic disease that results in bone deterioration, bone microarchitecture deterioration, and higher fracture rates as a result of bone loss. It is possible for this condition to result in pain and other complications that can make it difficult for the patient to perform daily activities and may be associated with significant mortality and morbidity. In addition to the ongoing increase in life expectancy, age-related diseases are on the rise, especially senile osteoporosis, which has a substantial economic impact [108]. It is well known that osteoporosis can be treated with a variety of treatments. However, many people are becoming interested in complementary, alternative, and natural medicine as well as antioxidants. Considering oxidative stress and how it contributes to osteoporosis development as well as the potential benefits of natural antioxidants can help manage osteoporosis. Numerous studies indicate that oxidative stress decreases bone mineral density by adversely affecting the bone remodeling process. Additionally, inflammation related to oxidative stress appears to contribute to osteoporosis pathogenesis [109]. Endogenous physiological changes (such as metabolic changes, hormonal fluctuations, and aging) can cause oxidative stress, along with environmental influences (including nutrition and pollution), as well as pathological conditions caused by inflammatory cytokines or long-term pharmacological treatments. Superoxide anion, hydroxyl radicals, and hydrogen peroxide are often generated excessively when enzymes that produce reactive oxygen species are activated excessively, interspersed with the inhibition of antioxidant enzymes. Antioxidant levels are both declining exogenously and endogenously. It stands out that glutathione (GSH) and its precursors play a key role in bone metabolism. In cells (between 2 and 10 mM), this thiol antioxidant maintains the redox environment of the cell. Many detoxifying enzymes utilize GSH as a substrate, including glutathione peroxidase (GPX), which help convert GSH into GSSG by oxidizing it. Induction of oxidative stress (OS) is frequently accompanied by an increase in GSSG levels, which results in a reduction in GSH/GSSG ratios. GSH/GSSG ratios typically range between 200:1 and 200:3, making them important indicators of cellular redox status; decreases in this ratio indicate impaired mineralization. Cells maintain their redox equilibrium by synthesizing GSH, reducing GSSG, and consuming exogenous GSH. In addition, GSH modulates transcription factor activity and intracellular protein activity through glutathionylation reaction, which involves multiple signaling pathways [110]. There is evidence that low levels of antioxidants contribute to increased blood hydrogen peroxide concentrations and reduced bone density in postmenopausal women. Reactive oxygen species are in fact generated by enzymes activated by inflammatory bone processes as well as extended use of nonsteroidal anti-inflammatory drugs, contributing to osteoporosis. GSH levels are reduced in postmenopausal/senile osteoporosis and secondary osteoporosis, both of which are characterized by impaired intestinal absorption of antioxidants. Ovariectomy (OVX) rats show significant oxidative stress and significant reductions in antioxidant enzyme activity, including catalase, superoxide dismutase (SOD), and GPX, as well as lower GSH levels. Nonenzymatic antioxidant levels, which tend to be diminished in postmenopausal women, are sometimes restored by hormone replacement therapy [111]. Further, in postmenopausal osteoporosis, oxidative stress has been linked to increased superoxide production and reduced antioxidant enzyme synthesis as a result of NADPH oxidase activation. According to recent research, oxidative stress and inflammation are implicated in osteoporosis due to their disruption of calcium-phosphorus metabolism and alteration of mineralization processes, which ultimately leads to altered bone microarchitecture and fracture risk [112].

6.1 *Natural Antioxidants and Derivatives on Bone Metabolism, the Process of Mineralization, Treatment, and Prevention of Osteoporosis*

Several studies have examined how natural antioxidants can reduce bone resorption, improve remodeling processes, and stimulate bone mass. The molecular factors that regulate bone metabolism are negatively affected by oxidative stress (OS), which these antioxidants can prevent and/or inhibit. By inhibiting the osteoclast genesis mechanism and simultaneously stimulating osteoblast differentiation and osteogenic activity, the drugs facilitate the formation of bone matrix [113]. In recent years, studies have shown that antioxidant therapy has positive effects on bone health when applied to cells, animals, and select patient groups suffering from osteoporosis or inflammatory bone conditions. A large number of interventions involving antioxidants have been conducted on animals, which has demonstrated that consumption of antioxidant-rich foods is associated with bone protection, as evidenced by increased bone mineral content and bone mineral density (BMD). Osteoporosis and fragility fractures are managed pharmacologically based on severity, selecting anti-fracture medications based on calcium and vitamin D levels, as well as lifestyle changes. As of now, most osteoporosis treatments aim to inhibit osteoclast genesis and bone resorption to reduce bone loss; less frequently, treatments that facilitate bone formation are used. An oral or intravenous bisphosphonate, as well as a monoclonal antibody targeting RANKL, are used as antiresorption therapies for osteoporosis [114]. There are several anabolic treatments with anti-fracture efficacy, which are effective at both increasing bone mass and strength and enhancing bone formation. They include teriparatide (1–34), a synthetic form of parathyroid hormone, and abaloparatide, a synthetic analogue of PTHrP, which is not available everywhere. ROMOSOSUMAB, a monoclonal anti-sclerostin antibody, was recently approved as a new treatment for osteoporosis; it exhibits dual functions of increasing bone production and decreasing bone resorption, albeit to a lesser extent. There has been some research suggesting that antioxidants should be used in tandem with antiresorptive therapies, since they inhibit osteoclast activity without causing cell death [115]. In contrast to conventional antiresorptive medications, these antioxidants restore communication between osteoclasts and osteoblasts, facilitating the physiological remodeling process. Activating osteocytes and osteoblasts is modulated through signaling pathways activated by antioxidants, which scavenge reactive oxygen species (ROS). Consequently, antioxidants may be useful for enhancing management approaches for the prevention of fragility fractures and fostering the restoration of anabolic processes [116]. Primary osteoporosis can also cause fractures, with secondary osteoporosis often accompanied by inflammation and oxidative stress. In addition to providing some benefits for bone health when supplemented with isolated antioxidants, whole plant foods typically supply a higher level of antioxidant protection. Although the relationship between antioxidant properties in various nutrients and bone metabolism is not always clear, current research indicates they are fundamental. Although larger clinical trials are needed to substantiate definitive clinical guidelines [117], smaller trials are sufficient.

7 Diagnostic and Treatment for Depression

Depression in the elderly is a pathological condition rather than a typical response to the aging process. While most individuals manage the challenges of aging and many experience happiness and fulfillment, there exists a prevailing bias among healthcare professionals and society at large that accepts diminished functioning and increased symptoms in older adults Many elderly individuals tend to deny their depression, influenced by a cultural background that discourages the expression of emotions, which complicates the diagnostic process. Additionally, the presence of comorbid medical issues, a propensity for somatization, cognitive decline, and various life changes often involving loss, such as bereavement, retirement, or downsizing, further complicate the ability to accurately diagnose depression in this population. The importance of diagnosing depression cannot be overstated. This condition is associated with heightened rates of morbidity and mortality, placing additional burdens on family members, as well as social and healthcare systems [118]. Fortunately, depression often shows a positive response to treatment, particularly in the early stages, leading to an enhancement in quality of life and a potential decrease in mortality rates. Antidepressant therapies have proven to be effective, safe even in the presence of coexisting physical illnesses and cost-efficient when evaluated against the expenses associated with untreated depression [119].

- *Signs and symptoms*

  The physical manifestations of aging, including a hunched posture and wrinkled skin, can affect how elderly patients are perceived by others, often leading to assumptions of depression. On the other hand, symptoms associated with depression, such as reports of pain or fatigue, may be mistakenly linked to the aging process or to underlying medical conditions [120].
- *Depression and dysphoria*

A depressed mood or feelings of sadness may not be evident, as the patient may deny experiencing such emotions when asked. More commonly, they report difficulties in coping, along with symptoms of anxiety or irritability (Reference Shulman, 1989). These patients often express concerns regarding disturbances in sleep or appetite, a decrease in energy levels, and other physical symptoms; typically, they will recognize a diminished interest or enjoyment in their everyday activities [121].

- *Crying*

  In the current elderly population, it is uncommon for men to exhibit crying behavior, as many belong to the 'stiff upper lip' generation. Consequently, instances of crying may indicate emotional instability linked to cerebral ischemia [122].
- *Anxiety/agitation*

  Over 80% of elderly individuals suffering from depression display symptoms such as nervousness, irritability, or intrusive behavior, accompanied by varying levels of anxiety. This condition is often correlated with diminished self-esteem and a reduced capacity to cope, leading to sleep disturbances or restlessness that can result in fatigue. Irritability may strain interpersonal relationships, potentially causing essential support to be withdrawn. Feelings of worthlessness and perceptions of being a 'nuisance' or 'burden' may prevent individuals from seeking assistance, with their needs either denied by themselves or overlooked by caregivers [123].
- *Low energy, fatigue*

  Reports of fatigue are prevalent among the elderly, yet these complaints may be mistakenly attributed to physical ailments or sleep disturbances, resulting in missed diagnoses.
- *Physical retardation*

  A noticeable slowing of movements and, in more severe cases, a deceleration of thought processes can complicate daily living activities, lead to poor nutrition, and ultimately result in a cessation of eating and drinking. Diminished mobility may escalate to complete immobility, increasing the risk of pressure sores. In severe instances, this condition poses a life-threatening risk [124].
- *Cognitive deficits*

  Cognitive impairments are often observed in older adults with depression and can sometimes be severe enough to be classified as pseudodementia. Deficiencies in language and memory processing are relatively frequent and are not always attributable to cognitive slowing or lack of motivation. A depressed individual experiencing memory issues may show improvement with prompts, indicating a difficulty in accessing information rather than a complete loss of it. However, even mild cognitive deficits may not be entirely reversible [125].
- *Depression and dementia*

  Research within communities indicates that instances of depression are significantly more prevalent than those of dementia. Depression may serve as an initial symptom of dementia, particularly in cases of Alzheimer's disease. This form of depression frequently proves resistant to therapeutic interventions, and cognitive impairments are typically evident, albeit initially mild. Additionally, a low mood is often observed in the early phases of any dementia when the individual retains some level of insight [126].

- *Somatization*

  The concept of somatization is gaining greater acknowledgement in contemporary discussions. Lipowski characterized somatization as 'the inclination to perceive and express somatic distress and symptoms that lack corresponding pathological evidence, to attribute these experiences to physical ailments, and to seek medical assistance for them.' Somatization can result in iatrogenic conditions due to unnecessary diagnostic procedures and treatments, as well as the perpetuation of the underlying depressive disorder [127].
- *Suicidal ideation*

  The experience of feeling that life lacks value and harboring a desire to die can manifest even in the absence of a depressive mood. Various factors associated with the desire to end one's life include being unmarried, experiencing poor self-rated health, living with disabilities, enduring pain, having sensory impairments, and residing in a nursing home or similar facility. Transient suicidal thoughts are prevalent among the elderly, particularly those suffering from physical ailments or disabilities. The occurrence of suicide is not uncommon, with age and physical illness recognized as significant risk factors [128]. It is likely that official statistics underreport the true incidence of suicide in this demographic, as not all cases are overt; for instance, non-adherence to prescribed treatments for physical health issues may result in death classified as 'natural causes'.
- *Behavioral disturbances*

  A range of behavioral changes may emerge as indicators of depression in older adults, particularly among those who rely heavily on caregivers. These changes can manifest as refusal to eat, inappropriate urinary and fecal incontinence, episodes of screaming, and dramatic 'falls' where individuals intentionally drop to the ground. In nursing home settings, such behaviors may also include aggression toward caregivers, as well as biting, scratching, and recent conflicts with fellow residents [129].

7.1 Risk Factors Associated with the Development of Depression

The following factors are recognized as being associated with the onset of depression and can serve as indicators for identifying and focusing on high-risk populations:

- Chronic physical illness, particularly when it is painful or debilitating
- Feelings of loneliness (not necessarily linked to living alone)
- Recent loss or other significant life challenges
- Auditory impairments
- Personal or familial history of depression

- Dementia, whether in early stages or with preserved insight
- Certain medications (e.g., corticosteroids, major tranquillizers)
- Female gender (with a ratio of approximately 70:30 female to male).

Thus, the role of antioxidants in depression disorder is Vitamin E. It acts as a supportive system and serves multiple roles in human health, functioning as an antioxidant, anti-inflammatory agent, anti-cancer compound, and anti-atherogenic substance. It also influences enzymatic activities and regulates gene transcription. Furthermore, vitamin E is crucial for the central nervous system, with emerging evidence suggesting its potential in preventing and/or treating certain neurological disorders [130]. Notably, recent research has highlighted the involvement of vitamin E in the modulation of major depressive disorder (MDD). Various factors, such as inflammation and oxidative and nitrosative stress, have been associated with the pathophysiology of this condition.

8 Diagnostic and Treatment Method for chronic obstructive Pulmonary Disease (COPD)

Chronic obstructive pulmonary disease (COPD) poses a significant Infectious diseases challenge, impacting approximately more than 300 million individuals globally and ranking as among the world's leading causes of death. Furthermore, it incurs a substantial economic strain, with hospitalization costs related to disease exacerbations amounting to USD 18 billion annually based in the USA The primary pathophysiological characteristic of COPD is the irreversible obstruction of the airways, accompanied by a progressive decline in lung function. This decline is particularly pronounced in individuals with ongoing besides pollutants such as cigarette smoke and energy-related smoke, air pollution is also a major concern, and some patients may experience a less severe progression of lung function deterioration as shown in Fig. 3 [131–134]. Consequently, it is essential to prevent further decline in lung function among individuals, people who are at risk for developing COPD, but do not show symptoms of the disease developing the condition (pre-COPD). Additionally, the early identification of preclinical COPD is vital for patients exhibiting as evidenced by spirometry, and obstruction is present, especially those with minimal or no symptoms, to prevent significant deterioration in their condition, including lung function decline, symptom exacerbation, or acute episodes. To prevent the overtreatment of individuals in the early stages of COPD, also known as preclinical COPD, it is crucial to understand the variations in disease progression. Researchers are conducting studies to analyze lung function profiles of COPD patients and identify those at the highest risk of developing severe COPD [135].

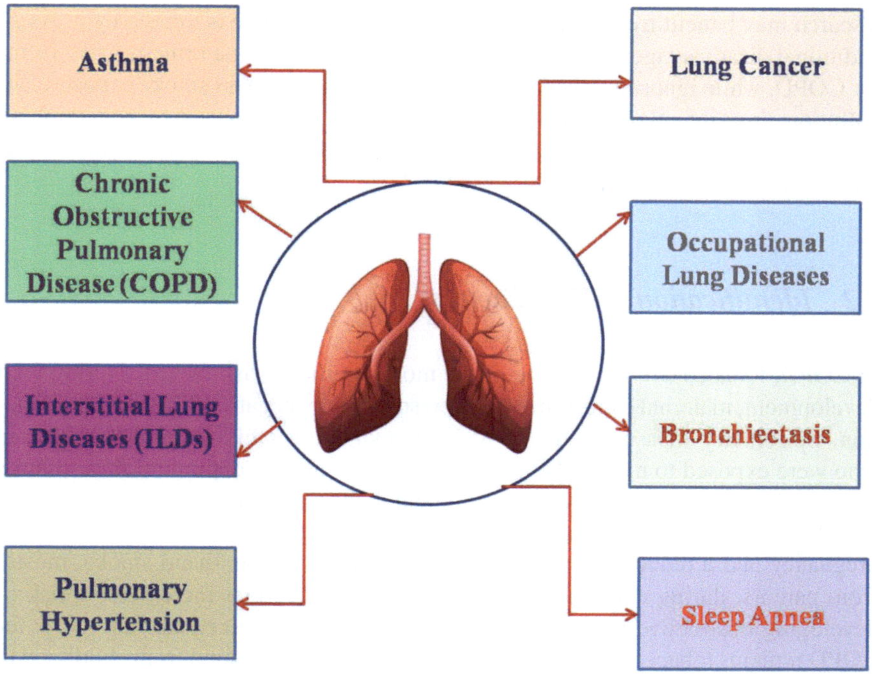

Fig. 3 Health problems relating to the lungs

8.1 Diagnosis of COPD

COPD should be considered as a longitudinal disease, rather than as an acute issue. Stages 1 and 2 are indicative of an earlier phase of COPD development, characterized by the absence of typical symptoms. There is an important distinction to be made between this and 'mild COPD,' which is usually understood as a cross-sectional assessment of mild spirometric airway obstruction following a short course of the disease [136]. There is insufficient evidence to accurately classify patients in the early stages of COPD, and therefore, many research groups refer to mild COPD as the early stages of COPD without clearly distinguishing between the two. Recently, Martinez et al. proposed a definition for early COPD, indicating that diagnosis should occur in individuals under 50 years of age with a smoking history of at least 10 pack-years, along with one or more of the following criteria: a forced expiratory volume in one second (FEV1) to forced vital capacity (FVC) ratio below the lower limit of normal; relevant computed tomography (CT) findings, such as visual emphysema, air trapping, or bronchial wall thickening; or evidence of a rapid decline in FEV1 ($\geq$ 60 mL/year). As a practical matter, this definition was formed based on the premise that younger patients are unlikely to have developed significant COPD symptoms [137–139]. Moreover, it focuses on individuals who have smoked for at least 10 pack-years, which is deemed the threshold for the pathogenesis of the disease. Academic research may benefit from this definition, but clinical practice is limited by it. As an additional disadvantage, it focuses primarily on smoking as the principal risk factor for COPD, while ignoring other critical factors such as biomass smoke exposure, air pollution, prior infections, and occupational hazards. COPD caused by non-smoking accounts for 25–45% of all cases, highlighting a critical gap in the definition that ignores this phenotype [140].

8.2 Identification of Risk Factors

In COPD, tobacco smoking is one of the most significant risk factors. During fetal development, maternal exposure to tobacco smoke may negatively impact lung function. Spirometry measurements of 8863 non-smoking children aged 8–12 years who were exposed to maternal smoking indicated that their expiratory flow volume decreased [141]. Furthermore, a longitudinal study which followed male subjects for 21 years suggested a significant relationship between maternal smoking during pregnancy and a reduction in FEV1. The presence of second-hand smoke, mostly from parents, during childhood and young adulthood further increases the risk of developing obstructive lung diseases later in life. The most critical risk factor for COPD remains tobacco smoking, evidence of which can be found from childhood to old age. Smoking is not the only risk factor associated with COPD, however. Childhood infections, respiratory conditions like asthma, exposure to biomass smoke, air pollution, and occupational hazards are also important but often overlooked [142]. A significant proportion of COPD cases are not caused by smoking, which is higher than what clinicians typically expect. It is also crucial for parents to ensure that their children do not have a history of respiratory infections or asthma during childhood. In children suffering from chronic respiratory infections, there are greater chances of developing chronic bronchitis, decreased lung function, thicker airway walls, and COPD exacerbations later in life. A two-fold explanation for these associations has been proposed by Hayden et al. Early childhood infections are thought to alter the airways, resulting in lung diseases in adulthood, or a preexisting lung condition may increase the risk of pneumonia in children [143–145].

8.3 Treatment for Early COPD

It is likely underreported that the prevalence of early chronic obstructive pulmonary disease (COPD) is large, and it has been connected to significant symptoms, increased exacerbation risk, lung function decline, and a diminished quality of life. Even though early COPD has received a great deal of attention, few treatments have been studied that address groups aligned with contemporary COPD definitions [146]. Prior studies involving populations with similar characteristics can help determine the most effective treatment strategies. In this section, earlier studies that investigated early COPD treatment, using various definitions, will be reviewed. A pre-COPD individual does not yet meet the criteria for an obstruction in the airway in a lung function test but is at risk for COPD. Smokers who have maintained normal lung function throughout their lives are at greater risk for developing clinically significant COPD [147]. The treatment of these individuals should begin as early as possible to slow or prevent disease progression. Overt COPD and pre-COPD can be prevented or mitigated by quitting smoking. When a smoker quits smoking, his or her lung function declines at a normal rate, as opposed to accelerating before quitting. An intervention program that addresses smoking cessation in middle-aged smokers has successfully halted the decline in FEV1 among COPD patients with mild to moderate airway obstruction, according to the Lung Health Study, a multicenter, randomized, controlled trial [148–150]. A subsequent study of smokers who successfully quit found a reduction in the rate of FEV1 decline by half at five years following their cessation.

8.4 Pulmonary Function Test and Other Physiological Assessments

PFTs are carried out to assess whether or not an individual has chronic obstructive pulmonary disease (COPD) and to determine the extent of the obstruction preventing airflow [151]. A decline in FEV1 exceeding 60 mL per year is a reliable indication of early COPD diagnosis, as outlined by Martinez et al. By roughly doubling the average change in FEV1 observed in nonsmokers, a decline of 25–30 mL per year was determined as the threshold. According to Washko et al. the average annual decline in FEV1 among those in the accelerated lung function decline group was 51 mL [152]. A further study conducted by Lange et al. examined lung function data from three extensive cohort studies, finding that after 22 years of obstruction without airway obstruction, subjects with COPD experienced a mean annual decline in FEV1 of 53 mL. Based on these results, it appears that individuals who experience rapid decline may be overlooked by the 60 mL per year threshold when they are at a higher risk of developing COPD in the future. Although FEV1 measurements performed longitudinally and sequentially are crucial to identifying those undergoing rapid decline, the cutoff level of FEV1 is yet to be thoroughly validated [153].

8.5 Role of Antioxidants in COPD

COPD is characterized by chronic inflammation and oxidative stress. In addition to oxidative stress affecting lung physiology and COPD progression in a profound way, oxidative stress causes antiproteases and surfactants to be inactivated, mucus to be produced excessively, membranes to be oxidized, alveolar epithelial cells to be damaged, extracellular matrix remodeling, and apoptosis to occur [154–157]. Therefore, managing COPD may be made easier by using antioxidants or enhancing the body's natural antioxidants. Antioxidant and/or anti-inflammatory compounds, including thiol molecules (such as glutathione and mucolytic agents like N-acetyl-L-cysteine and N-acystelyn), dietary polyphenols (notably curcumin-diferuloylmethane, a key component of turmeric), resveratrol (a flavonoid present in red wine), green tea (theophylline and epigallocatechin-3-gallate), ergothioneine (an inhibitor of xanthine and peroxynitrite), quercetin, erdosteine, and carbocysteine lysine salt, have been shown to modulate NF-kB activation, regulate genes involved in glutathione biosynthesis, and influence chromatin remodeling, thereby affecting inflammatory gene expression [158]. Additionally, specific spin traps such as α-phenyl-N-tert-butyl nitrone, catalytic antioxidants such as manganese (III) meso-tetrakis (N, N′-diethyl-1,3-imidazolium-2-yl) porphyrin (AEOL 10150 and AEOL 10113), and superoxide dismutase mimetic M40419 have demonstrated the ability to mitigate inflammatory responses induced by cigarette smoke in vivo. Since various oxidants, free radicals, and aldehydes are implicated in COPD's pathogenesis, using multiple antioxidants might be a useful therapeutic approach [159–161].

9 Telemedicine in Health Care

Advanced technologies combined with high-quality network services empower individuals to enhance healthcare delivery, making it accessible to an increasing number of people. Telemedicine represents a significant advancement that facilitates access to preventive care and supports long-term health outcomes [162]. This is especially beneficial for those who encounter financial or geographical barriers to obtaining quality treatment. Telehealth has the capacity to render healthcare more efficient, organized, and widely available [163]. Although research in this field is still developing, it is progressing rapidly. For instance, telephone-based care and telemonitoring of vital signs in patients with heart disease have been shown to reduce mortality and hospitalization rates while improving quality of life [164]. There are numerous compelling reasons for individuals to seek a diagnosis or treatment plan, as this can instill confidence in the quality of care they are receiving. Telemedicine is particularly advantageous for addressing mental health concerns, as it mitigates several obstacles that may hinder patients from accessing this essential form of treatment. Telemedicine allows patients to receive medical care at a time that is convenient for both the physician and the patient, while also ensuring safety [165]. This arrangement means that individuals do not need to take time off work or arrange for childcare. Visiting a doctor's office often involves close proximity to others, which can increase the risk of infection, particularly for those with chronic health conditions or compromised immune systems. By utilizing telemedicine, the risk of contracting infections in a clinical setting is significantly reduced. Additionally, telemedicine service providers may operate with lower overhead costs, enabling clinicians to increase their patient load and supplement their income [166]. When healthcare providers consult with patients online, they are less exposed to potential infections that patients may carry. Patients also benefit from the convenience of not having to travel to the office, wait for treatment, or risk exposure to infections in a hospital environment [167].

10 Conclusion

Biological systems and food systems generate various forms of free radicals. A range of cellular defense mechanisms effectively neutralize oxygen-derived free radicals that are generated by the normal metabolism of oxygen within cells. Yet, biological systems are frequently subjected to prooxidant conditions that produce excessive amounts of activated oxygen species, either because of excessive production by certain causative agents, or because regulators fail to function properly. It produces superoxide, hydroperoxy radicals, singlet oxygen, hydroxyl radicals, and hydrogen peroxide as activated oxygen forms. It is possible for transient metals, iron, copper, xenobiotic metabolites, and hyperbaric oxygen tension to trigger prooxidant states in biological systems. Both membrane and cytosolic lipid oxidation can occur when these active oxygen species interact, resulting in degenerative diseases, mutations, and cytotoxicity. Malonaldehyde and cholesterol oxidation products are formed in food systems when lipids are oxidized. There are a number of key causal agents involved in this disease, such as heat, photosensitization, transition metals, metalloproteins, and radiation. By oxidizing dietary lipids, food products become rancid, protein-damaged, and pigment-oxidized, thereby losing their sensory qualities, nutrients, and shelf life. Thus, the purpose of this chapter is to review various chronic diseases, as well as how antioxidants play a role in combating them. It is essential for prevention and treatment of various chronic diseases that antioxidants are available to humans.

References

1. Wagner, B. A., & Buettner, G. R. (1996). Burns CP, vitamin E slows the rate of free radical-mediated lipid peroxidation in cells. *Archives of Biochemistry and Biophysics, 334*, 261–267.
2. Du, J., Cullen, J. J., & Buettner, G. R. (2012). Ascorbic acid: Chemistry, biology and the treatment of cancer. *Biochimica et Biophysica Acta, 1826*, 443–457.
3. Rietjens I. M., Boersma M. G., Haanm Ld., Spenkelink B., Awad H. M., Cnubben N. H., et al. (2002). The pro-oxidant chemistry of the natural antioxidants vitamin C, vitamin E, carotenoids and flavonoids. *Environmental Toxicology and Pharmacology, 11*, 321–333.
4. Kang, M. J., Lee, S. S., & Koh, H. C. (2012). Prooxidant properties of ascorbic acid in the nigrostriatal dopaminergic system of C57BL/6 mice. *Toxicology, 294*, 1–8.
5. Heo, J. H., et al. (2013). The possible role of antioxidant vitamin C in Alzheimer's disease treatment and prevention. *American Journal of Alzheimer's Disease and Other Dementias, 28*, 120–125.

6. Putchala, M. C., Ramani, P., Sherlin, H. J., Premkumar, P., & Natesan, A. (2013). Ascorbic acid and its pro-oxidant activity as a therapy for tumours of oral cavity a systematic review. *Archives of Oral Biology, 58*, 563–574.
7. Bast, A., & Haenen, G. R. (2013). Ten misconceptions about antioxidants. *Trends in Pharmacological Sciences, 34*, 430–436.
8. Halliwell, B. (1996). Commentary: vitamin C: antioxidant or pro-oxidant in Vivo? *Free Radical Research, 25*, 439–454.
9. Halliwell B. (2006). Reactive species and antioxidants. Redox biology is a fundamental theme of aerobic life. *Plant Physiology, 141*, 312–322.
10. Halliwell, B. (1991). Drug antioxidant effects. *Drugs, 42*, 569–605.
11. Hegde, M. V., Patil, S., & Bhalerao, S. (2008). A philosophy for integration of ayurveda with modern medicine: A biochemist's perspective. *Current Science, 95*, 721–722.
12. Farnsworth, N. R., Akerele, O., Bingel, A. S., Soejarto, D. D., & Guo, Z. (1995). Medicinal plants in therapy. *Bulletin of the World Health Organization, 63*, 965–981.
13. Zhang, K., & Kaufman, R. J. (2008). From endoplasmic-reticulum stress to the inflammatory response. *Nature, 454*, 455–462.
14. Schrader, M., & Fahimi, H. D. (2006). Peroxisomes and oxidative stress. *Biochimica et Biophysica Acta, 63*, 1755–1766.
15. Rodrigo, R., Guichard, C., & Charles, R. (2007). Clinical pharmacology and therapeutic use of antioxidant vitamins. *Fundamental & Clinical Pharmacology, 21*, 111–127.
16. Kohen, R., & Nyska, A. (2002) Oxidation of biological systems: oxidative stress phenomena, antioxidants, redox reactions, and methods for their quantification. *Toxicologic Pathology, 30*, 620–650.
17. Prior, R. L., Wu, X., & Schaich, K. (2005). Standardized methods for the determination of antioxidant capacity and phenolics in foods and dietary supplements. *Journal of Agricultural and Food Chemistry., 53*, 4290–4302.
18. Ou, B., Woodill-Hampsch, M., & Prior, R. L. (2001). Development and validation of an improved oxygen radical absorbance capacity assay using fluorescein as the fluorescent probe. *Journal of Agriculture and Food Chemistry, 49*, 4619–4626.
19. Alam, M. N., Bristi, N. J., & Rafiquzzaman, M. (2013). Review on in vivo and in vitro methods evaluation of antioxidant activity. *Saudi Pharmaceutical Journal, 13*(21), 143–152.
20. Winkel-Shirley, B. (2002). Biosynthesis of flavonoids and effects of stress. *Current Opinion in Plant Biology, 5*, 218–223.
21. Nicholson, R. L., & Hammerschmidt, R. (1992). Phenolic compounds and their role in disease resistance. *Annual review of Phytopathology, 30*, 369–389.
22. Szalai, G., Kellos, T., Galiba, G., & Kocsy, G. (2009). Glutathione as an antioxidant and regulatory molecule in plants under abiotic stress conditions. *Journal of Plant Growth Regulation, 28*, 66–80.
23. Korkina, L. G. (2007). Phenylpropanoids as naturally occurring antioxidants: From plant defense to human health. *Cellular and Molecular Biology, 53*, 15–25.
24. Meyer, A. J., & Hell, R. (2005). Glutathione homeostasis and redox-regulation by sulfhydryl groups. *Photosynthesis Research, 86*, 435–457.
25. Lü, J. M., Lin, P. H., Yao, Q., & Chen, C. (2010). Chemical and molecular mechanisms of antioxidants: Experimental approaches and model systems. *Journal of Cellular and Molecular Medicine, 14*, 840–860.
26. del Río, L. A., Sandalio, L. M., Corpas, F. J., Palma, J. M., & Barroso, J. B. (2006). Reactive oxygen species, reactive nitrogen species in peroxisomes. Production, scavenging, and role in cell signaling. *Plant Physiology, 141*, 330–335.
27. Houmani, H., Rodríguez-Ruiz, M., Palma, J. M., Abdelly, C., & Corpas, F. J. (2016). Modulation of superoxide dismutase (SOD) isozymes by organ development and high long-term salinity in the halophyte Cakile maritima. *Protoplasma, 253*, 885–894.
28. Holzmeister, C., Gaupels, F., Geerlof, A., Sarioglu, H., Sattler, M., Durner, J., & Lindermayr, C. (2015). Differential inhibition of Arabidopsis superoxide dismutases by peroxynitrite-mediated tyrosine nitration. *Journal of Experimental Botany, 66*, 989–999.

29. Hemre, G. I., Lock, E. J., Olsvik, P. A., Hamre, K., Espe, M., Torstensen, B. E., Silva, J., Hansen, A. C., Waagbo, R., Johansen, J. S., Sanden, M., & Sissener, N. H. (2016). Atlantic salmon (*Salmo salar*) require increased dietary levels of B-vitamins when fed diets with high inclusion of plant based ingredients. *PeerJ, 4*, e2493.
30. Das, D., Kabir, M. E., Sarkar, S., Wann, S. B., Kalita, J., & Manna, P. (2022). Antidiabetic potential of soy protein/peptide: A therapeutic insight. *International Journal of Biological Macromolecules, 194*, 276–288.
31. Sheik Abdul, N., & Marnewick, J. L. (2022). Green rooibos extract attenuates high glucose induced oxidative stress in a human derived (HepG2) liver cell line. *South African Journal of Botany, 151*, 852–865.
32. Sambraus, F., Fjelldal, P. G., Remo, S. C., Hevroy, E. M., Nilsen, T. O., Thorsen, A., Hansen, T. J., & Waagbo, R. (2017). Water temperature and dietary histidine affect cataract formation in Atlantic salmon (*Salmo salar* L.) diploid and triploid yearling smolt. *The Journal of Fish Diseases, 40*, 1195–1212.
33. Parodi, J. (2019). Something old, something new and something used in alzheimer's; the idea of pore, ethanol and the use of oocytes to understand the disease. *Current Chemical Biology, 13*, 105–109.
34. Sambamoorthi, U., Tan, X., & Deb, A. (2015). Multiple chronic conditions and healthcarecosts among adults. *Expert Review Pharmacoeconomics Outcomes Res., 15*(5), 823–832.
35. Wu, L.-T., Zhu, H., & Ghitza, U. E. (2018). Multicomorbidity of chronic diseases andsubstance use disorders and their association with hospitalization: Results from electronic health records data. *Drug and Alcohol Dependence, 192*, 316–323.
36. Dennis, M., & Scott, C. K. (2007). Managing addiction as a chronic condition. *Addiction Science & Clinical Practice, 4*(1), 45–71.
37. Nesti, L., Pugliese, N. R., Sciuto, P., & Natali, A. (2020). Type 2 diabetes and reduced exercise tolerance: A review of the literature through an integrated physiology approach. *Cardiovascular Diabetology, 19*(1), 1–17.
38. Ding, J., Karp, J. E., & Emadi, A. (2017). Elevated lactate dehydrogenase (LDH) can be a markerof immune suppression in cancer: Interplay between hematologic and solid neoplasticclones and their microenvironments. *Cancer Biomarkers, 19*, 353–363.
39. Nelson, R. (2017). Telemedicine and telehealth: The potential toimprove rural access to care. *American Journal of Nursing, 117*(6), 17–18.
40. Di Meo, S., Reed, T. T., Venditti, P., & Victor, V. M. (2016). Role of ROS and RNS sources in physiological and pathological conditions. *Oxidative Medicine and Cellular Longevity*, 1245049.
41. De Bont, R., & van Larebeke, N. (2004). Endogenous DNA damage in humans: A review of quantitative data. *Mutagenesis, 19*, 169–185.
42. Davalli, P., Mitic, T., Caporali, A., Lauriola, A., & D'arca, D. (2016). ROS, cell senescence, and novel molecular mechanisms in aging and age-related diseases. *Oxidative Medicine and Cellular Longevity*, 3565127.
43. Danielson, S. R., & Andersen, J. K. (2008). Oxidative and nitrative proteinmodifications in Parkinson's disease. *Free Radical Biology & Medicine, 44*, 1787–1794.
44. Chen, X., Guo, C., & Kong, J. (2012). Oxidative stress in neurodegenerative diseases. *Neural Regeneration Research, 7*, 376–385.
45. Chen, X.-F., Wang, L., Wu, Y.-Z., Song, S.-Y., Min, H.-Y., & Yang, Y. (2018). Effect of puerarin in promoting fatty acid oxidation by increasing mitochondrial oxidative capacity and biogenesis in skeletal muscle in diabeticrats. *Nutrition & Diabetes, 8*, 1–13.
46. Cardoso, B. R., Hare, D. J., Bush, A. I., & Roberts, B. R. (2017). Glutathioneperoxidase 4: A new player in neurodegeneration? *Molecular Psychiatry, 22*, 328–335.
47. Cadet, J., & Wagner, J. R. (2013). DNA base damage by reactive oxygen species, oxidizing agents, and UV radiation. *Cold Spring Harbor Perspectives in Biology, 5*, a012559.
48. Cadet, J., Ravanat, J. L., Tavernaporro, M., Menoni, H., & Angelov, D. (2012). Oxidatively generated complex DNA damage: tandem and clustered lesions. *Cancer Letters, 327*, 5–15.

49. Cadet, J., Davies, K. J. A., Medeiros, M. H. G., Di Mascio, P., & Wagner, J. R. (2017). Formation and repair of oxidatively generated damage in cellular DNA. *Free Radical Biology and Medicine, 107*, 13–34.
50. Buj, R., & Aird, K. M. (2018). Deoxyribonucleotide triphosphate metabolism in cancer and metabolic disease. *Frontiers in Endocrinology, 9*, 177.
51. Barrera, G. (2012). Oxidative stress and lipid peroxidation products in cancer progression and therapy. *ISRN Oncology*, 137289–137289.
52. Aroor, A. R., Mandavia, C., Ren, J., Sowers, J. R., & Pulakat, L. (2012). Mitochondria and oxidative stress in the cardiorenal metabolic syndrome. *Cardiorenal Medicine, 2*, 87–109.
53. Antonioni, A., Fantini, C., Dimauro, I., & Caporossi, D. (2019). Redoxhomeostasis in sport: Do athletes really need antioxidant support? *Research in Sports Medicine, 27*, 147–165.
54. Gao, K., Shi, X., & Wang, W. (2017). The life-course impact of smoking on hypertension, myocardial infarction and respiratory diseases. *Science and Reports, 7*(1), 1–7.
55. Fernández-Solà, J. (2015). Cardiovascular risks and benefits of moderate and heavy alcohol consumption. *Nature Reviews Cardiology, 12*(10), 576–587.
56. Ezzati, M., & Riboli, E. (2013). Behavioral and dietary risk factors for non communicable diseases. *New England Journal of Medicine, 369*(10), 954–964.
57. Díaz-Gutiérrez, J., Ruiz-Canela, M., Gea, A., Fernández-Montero, A., & Martinez-González, M. Á. (2018). Association between a healthy lifestyle score and the risk of cardiovascular disease in the SUN cohort. *Revista Espanola de Cardiologia, 71*(12), 1001–1009.
58. Chomistek, A. K., Chiuve, S. E., Eliassen, A. H., Mukamal, K. J., Willett, W. C., & Rimm, E. B. (2015). Healthy lifestyle in the primordial prevention of cardiovascular disease amongyoung women. *Journal of the American College of Cardiology, 65*(1), 43–51.
59. Bes-Rastrollo, M., Perez Valdivieso, J., Sanchez-Villegas, A., Alonso, A., & Martinez-Gonzalez, M. (2005). Validation of self-reported weight and body mass index of the participants of a cohort of university graduates. *Rev Esp Obes, 3*(6), 352–358.
60. Barlow, C. E., LaMonte, M. J., FitzGerald, S. J., Kampert, J. B., Perrin, J. L., & Blair, S. N. (2006). Cardiorespiratory fitness is an independent predictor of hypertension incidence among initially normotensive healthy women. *American Journal of Epidemiology, 163*(2), 142–150.
61. Münzel, T., Camici, G. G., Maack, C., Bonetti, N. R., Fuster, V., & Kovacic, J. C. (2017). Impact of oxidative stress on the heart and vasculature. *Journal of the American College of Cardiology., 70*(2), 212–229.
62. Mozaffarian, D., Hao, T., Rimm, E. B., Willett, W. C., & Hu, F. B. (2011). Changes in diet and lifestyle and long-term weight gain in women and men. *New England Journal of Medicine, 364*, 2392–2404.
63. Lam, J., & Perales, F. (2018). Chronic illness and mental strain: The moderating role of marital status over the disease cycle. *Longit Life Course Study, 9*(3), 279–298.
64. Lee, C. H., Lee, M. C., Shu, C. C., Lim, C. S., Wang, J. Y., Lee, L. N., & Chao, K. M. (2013). Risk factors for pulmonary tuberculosis in patients with chronic obstructive airway disease in Taiwan: A nationwide cohort study. *BMC Infectious Diseases, 13*(1), 194.
65. Islam, F. M., Wu, J., Jansson, J., & Wilson, D. P. (2012). Relative risk of cardiovascular disease among people living with HIV: A systematic review and meta-analysis. *HIV Medicine, 13*(8), 453–468.
66. McPhail, S. M. (2016). Multi morbidity in chronic disease: Impact on health care resources and costs. *Risk Management and Healthcare Policy., 9*, 143.
67. Kingston, A., Robinson, L., Booth, H., Knapp, M., & Jagger, C. (2018). Projections of multi-morbidity in the older population in England to 2035: Estimates from the population ageing and care simulation (PACSim) model. *Age and Ageing, 47*(3), 374–380.
68. Bos-Touwen, I., Schuurmans, M., Monninkhof, E. M., Korpershoek, Y., Spruit-Bentvelzen, L., Ertugrul-van der Graaf, I., deWit, N., & Trappenburg, J. (2015). Patient and disease characteristics associated with activation for self-management in patients with diabetes, chronic obstructive pulmonary disease, chronic heart failure and chronic renal disease: A cross-sectional survey study. *PLoS ONE, 10*, e0126400.

69. Karami, N., Kazeminia, M., Karami, A., Salimi, Y., Ziapour, A., & Janjani, P. (2023). Global prevalence of depression, anxiety, and stress in cardiac patients: A systematic review and meta-analysis. *Journal of Affective Disorders, 324*, 175–189.
70. Rozario, S. S., & Masho, S. W. (2018). The associations between mental health status, hypertension, and hospital inpatient visits in women in the United States. *American Journal of Hypertension, 31*, 804–810.
71. Yusuf, L. (2016). Depression, anxiety and stress among female patients of infertility; A case control study. *Pakistan Journal of Medical Sciences, 32*, 1340.
72. Kruk, J., & Aboul-Enein, H. Y. (2017). Reactive oxygen and nitrogen species in carcinogenesis: Implications of oxidative stress on the progression and development of several cancer types. *Mini Reviews in Medicinal Chemistry., 17*, 904–919.
73. Khanna, H. D., Sinha, M. K., Khanna, S., & Tandon, R. (2008). Oxidativestress in hypertension: Association with antihypertensive treatment. *Indian Journal of Physiology and Pharmacology, 52*(3), 283–287.
74. Oguntibeju, O. O. (2019). Type 2 diabetes mellitus, oxidative stressand inflammation: Examining the links. *International Journalof Physiology, Pathophysiology and Pharmacology, 11*(3), 45–63.
75. Lee, I. M., Cook, N. R., Gaziano, J. M., Gordon, D., Ridker, P. M., Manson, J. A. E., Hennekens, C. H., & Buring, J. E. (2005). Vitamin E in the primary prevention of cardiovascular disease and cancer. The women's health study: A randomized controlled trial. *The Journal of the American Medical Association, 294*, 56–65.
76. Kim, S., Song, J., Ernst, P., Latimer, M. N., Ha, C. M., Goh, K. Y., Ma, W., Rajasekaran, N. S., Zhang, J., & Liu, X. (2020). MitoQ regulates redox-related noncoding RNAs to preserve mitochondrial network integrity in pressure-overload heart failure. *American Journal of Physiology. Heart and Circulatory Physiology, 318*, H682–H695.
77. Imamura, H., Yamaguchi, T., Nagayama, D., Saiki, A., Shirai, K., & Tatsuno, I. (2017). Resveratrol ameliorates arterial stiffness assessed by cardio-ankle vascular index in patients with type 2 diabetes mellitus. *International Heart Journal, 58*, 577–583.
78. Timmers, S., Konings, E., Bilet, L., Houtkooper, R. H., Van De Weijer, T., Goossens, G. H., Hoeks, J., Van Der Krieken, S., Ryu, D., & Kersten, S. (2011). Calorie restriction-like effects of 30 days of resveratrol supplementation on energy metabolism and metabolic profile in obese humans. *Cell Metabolism, 14*, 612–622.
79. Beaglehole, R., & Yach, D. (2003). Globalisation and the prevention and control of noncommunicable disease: The neglected chronic diseases of adults. *Lancet, 362*(9387), 903–908.
80. Tenney, L., McKenzie, L. M., Matus, B., Mueller, K., & Newman, L. S. (2019). Effect of an opioid management program for Colorado workers' compensation providers on adherence to treatment guidelines for chronic pain. *American Journal of Industrial Medicine, 62*(1), 21–29.
81. Price, D. D. (2000). Psychological and neural mechanisms of the affective dimension of pain. *Science, 288*(5472), 1769–1772.
82. Closs, S. J., & Briggs, M. (2002). Patients' verbal descriptions of pain and discomfort following orthopaedic surgery. *International Journal of Nursing Studies, 39*(5), 563–572.
83. Tang, N. K., & Crane, C. (2006). Suicidality in chronic pain: A review of the prevalence, risk factors and psychological links. *Psychological Medicine, 36*(5), 575–586.
84. Petrosky, E., Harpaz, R., Fowler, K. A., Bohm, M. K., Helmick, C. G., Yuan, K., & Betz, C. J. (2018). Chronic pain among suicide decedents, 2003 to 2014: Findings from the national violent death reporting system. *Annals of Internal Medicine, 169*(7), 448–455.
85. Long, C. J. (1981). The relationship between surgical outcome and MMPI profiles in chronic pain patients. *Journal of Clinical Psychology, 37*(4), 744–749.
86. Atluri, S., Akbik, H., & Sudarshan, G. (2012). Prevention of opioid abuse in chronic noncancer pain: an algorithmic, evidence based approach. *Pain Physician, 15*(3), ES177–ES189.
87. Copenhaver, D. J., Karvelas, N. B., & Fishman, S. M. (2017). Risk management for opioid prescribing in the treatment of patients with pain from cancer or terminal illness: Inadvertent oversight or taboo? *Anesthesia and Analgesia, 125*(5), 1610–1615.

88. Arnow, B. A., Hunkeler, E. M., Blasey, C. M., Lee, J., Constantino, M. J., Fireman, B., Kraemer, H. C., Dea, R., Robinson, R., & Hayward, C. (2006). Comorbid depression, chronic pain, and disability in primary care. *Psychosomatic Medicine, 68*(2), 262–268.
89. Fishbain, D. A., Cole, B., Lewis, J. E., & Gao, J. (2014). Does pain interfere with antidepressant depression treatment response and remission in patients with depression and pain? *An evidence-based structured review. Pain Med., 5*(9), 1522–1539.
90. Zhang, W., Moskowitz, R. W., Nuki, G., Abramson, S., Altman, R. D., Arden, N., Bierma-Zeinstra, S., Brandt, K. D., Croft, P., Doherty, M., Dougados, M., Hochberg, M., Hunter, D. J., Kwoh, K., Lohmander, L. S., & Tugwell, P. (2008). OARSI recommendations for the management of hip and knee osteoarthritis, Part II: OARSI evidence-based, expert consensus guidelines. *Osteoarthritis Cartilage, 6*(2), 137–162.
91. Bjordal, J. M., Ljunggren, A. E., Klovning, A., & Slørdal, L. (2004). Non-steroidal anti-inflammatory drugs, including cyclo-oxygenase-2 inhibitors, in osteoarthritic knee pain: Meta-analysis of randomised placebo controlled trials. *The BMJ, 329*(7478), 1317.
92. Santos, J., Alarcão, J., Fareleira, F., Vaz-Carneiro, A., & Costa, J. (2015). Tapentadol for chronic musculoskeletal pain in adults. *The Cochrane Database of Systematic Reviews, 5*, CD009923.
93. Krebs, E. E., Gravely, A., Nugent, S., Jensen, A. C., DeRonne, B., Goldsmith, E. S., Kroenke, K., Bair, M. J., & Noorbaloochi, S. (2018). Effect of opioid vs nonopioid medications on pain-related function in patients with chronic back pain or hip or knee osteoarthritis pain: The SPACE randomized clinical trial. *JAMA, 319*(9), 872–882.
94. Dublin, S., Walker, R. L., Shortreed, S. M., Ludman, E. J., Sherman, K. J., Hansen, R. N., Thakral, M., Saunders, K., Parchman, M. L., & Von Korff, M. (2019). Impact of initiatives to reduce prescription opioid risks on medically attended injuries in people using chronic opioid therapy. *Pharmacoepidemiology and Drug Safety, 28*(1), 90–96.
95. Els, C., Jackson, T. D., Kunyk, D., Lappi, V. G., Sonnenberg, B., Hagtvedt, R., Sharma, S., Kolahdooz, F., Straube, S. (2017). Adverse events associated with medium- and long-term use of opioids for chronic non-cancer pain: An overview of cochrane reviews. *The Cochrane Database of Systematic Reviews, 10*(10), CD012509.
96. Freynhagen, R., & Bennett, M. I. (2009). Diagnosis and management of neuropathic pain. *BMJ, 12*(339), b3002.
97. Gilron, I., Baron, R., & Jensen, T. (2015). Neuropathic pain: Principles of diagnosis and treatment. *Mayo Clinic Proceedings, 90*(4), 532–545.
98. Wiffen, P. J., Derry, S., Moore, R. A., & McQuay, H. J. (2011). Carbamazepine for acute and chronic pain in adults. *The Cochrane Database of Systematic Reviews, 19*(1), CD005451.
99. Padayatty, S. J., Sun, H., Wang, Y., Riordan, H. D., Hewitt, S. M., Katz, A., Wesley, R. A., & Levine, M. (2004). Vitamin C pharmacokinetics: Implications for oral and intravenous use. *Annals of Internal Medicine, 140*(7), 533–537.
100. Walsh, Z., Gonzalez, R., Crosby, K., Thiessen, M. S., Carroll, C., & Bonn-Miller, M. O. (2017). Medical cannabis and mental health: A guided systematic review. *Clinical Psychology Review, 51*, 15–29.
101. Borghetti, G., von Lewinski, D., Eaton, D. M., Sourij, H., Houser, S. R., & Wallner, M. (2018). Diabetic cardiomyopathy: Current and future therapies. Beyond glycemic control. *Frontiers in Physiology, 9.*
102. Lontchi-Yimagou, E., Sobngwi, E., Matsha, T. E., & Kengne, A. P. (2013). Diabetes mellitus and inflammation. *Current Diabetes Reports, 13*, 435–444.
103. Mirtschink, P., Jang, C., Arany, Z., & Krek, W. (2018). Fructose metabolism, cardio metabolic risk, and the epidemic of coronary artery disease. *European Heart Journal, 39*, 2497–2505.
104. Lawson, C., Vicencio, J. M., Yellon, D. M., & Davidson, S. M. (2016). Microvesicles and exosomes: New players in metabolic and cardiovascular disease. *Journal of Endocrinology, 228*, R57–R71.
105. Faria, A., & Persaud, S. J. (2017). Cardiac oxidative stress in diabetes: Mechanisms and therapeutic potential. *Pharmacology & Therapeutics, 172*, 50–62.

106. Kaludercic, N., & Di Lisa, F. (2020). Mitochondrial ROS formation in the pathogenesis of diabetic cardiomyopathy. *Frontiers in Cardiovascular Medicine, 7*.
107. Adameova, A., Shah, A. K., & Dhalla, N. S. (2020). Role of oxidative stress in the genesis of ventricular arrhythmias. *International Journal of Molecular Sciences, 21*, 4200.
108. Sharma, A., Fonarow, G. C., Butler, J., Ezekowitz, J. A., & Felker, G. M. (2016). Coenzyme Q10 and heart failure. *Circulation: Heart Failure, 9*, e002639.
109. Dhalla, N. S., Temsah, R. M., & Netticadan, T. (2000). Role of oxidative stress in cardiovascular diseases. *Journal of Hypertension, 18*, 655–673.
110. Fang, W., Wang, C., He, Y., Zhou, Y., Peng, X., & Liu, S. (2018). Resveratrol alleviates diabetic cardiomyopathy in rats by improving mitochondrial function through PGC-1α deacetylation. *Acta Pharmacologica Sinica, 39*, 59–73.
111. Fourny, N., Lan, C., Sérée, E., Bernard, M., & Desrois, M. (2019). Protective effect of resveratrol against ischemia-reperfusion injury via enhanced high energy compounds and eNOS-SIRT1 expression in Type 2 diabetic female rat heart. *Nutrients, 11*, 105.
112. Homan, E. A., Reyes, M. V., Hickey, K. T., & Morrow, J. P. (2019). Clinical overview of obesity and diabetes mellitus as risk factors for atrial fibrillation and sudden cardiac death. *Frontiers in Physiology, 9*.
113. Vona, R., Gambardella, L., Cittadini, C., Straface, E., & Pietraforte, D. (2019). Biomarkers of oxidative stress in metabolic syndrome and associated diseases. *Oxidative Medicine and Cellular Longevity*, 1–19.
114. Maharani, N., Kuwabara, M., & Hisatome, I. (2016). Hyperuricemia and atrial fibrillation. *International Heart Journal, 57*, 395–399.
115. Canton, M., Menazza, S., Sheeran, F. L., Polverino de Laureto, P., Di Lisa, F., & Pepe, S. (2011). Oxidation of myofibrillar proteins in human heart failure. *Journal of the American College of Cardiology, 57*, 300–309.
116. Ferenczyova, K., Kalocayova, B., & Bartekova, M. (2020). Potential implications of quercetin and its derivatives in cardioprotection. *International Journal of Molecular Sciences, 21*, 1585.
117. Cong, L., Su, Y., Wei, D., Qian, L., Xing, D., Pan, J., Chen, Y., & Huang, M. (2020). Catechin relieves hypoxia/reoxygenation-induced myocardial cell apoptosis via down-regulating lncRNA MIAT. *Journal of Cellular and Molecular Medicine, 24*, 2356–2368.
118. Paulis, G., & De Giorgio, G. (2022). Full regression of Peyronie's disease plaque following combined antioxidant treatment: A three-case report. *Antioxidants, 11*, 1661.
119. Zujko, M. E., Witkowska, A. M., Waśkiewicz, A., Piotrowski, W., & Terlikowska, K. M. (2015). Dietary antioxidant capacity of the patients with cardiovascular disease in a cross-sectional study. *Nutrition Journal, 14*, 26.
120. Sies, H. (2015). Oxidative stress: A concept in redox biology and medicine. *Redox Biology, 4*, 180–183.
121. Sharma, K. (2016). Obesity and diabetic kidney disease: Role of oxidant stress and redox balance. *Antioxidants & Redox Signaling, 25*, 208–216.
122. Huang, W.-J., Zhang, X., & Chen, W.-W. (2016). Role of oxidative stress in Alzheimer's disease. *Biomedical Reports, 4*, 519–522.
123. Prasad, K. N., & Bondy, S. C. (2018). Oxidative and inflammatory events in prion diseases: Can they be therapeutic targets? *Current Aging Science, 11*, 216–225.
124. Hosseini, A., & Abdollahi, M. (2013). Diabetic neuropathy and oxidative stress: Therapeutic perspectives. *Oxidative Medicine and Cellular Longevity*, 168039.
125. Vázquez, M. C., Balboa, E. Alvarez, A. R., & Zanlungo, S. (2012). Oxidative stress: A pathogenic mechanism for Niemann-pick Type C disease. *Oxidative Medicine and Cellular Longevity*, e205713.
126. Aoyama, K., Matsubara, K., & Kobayashi, S. (2006). Aging and oxidative stress in progressive supranuclear palsy. *European Journal of Neurology, 13*, 89–92.
127. Komsiiska, D. (2019). Oxidative stress and stroke: A review of upstream and downstream antioxidant therapeutic options. *Comparative Clinical Pathology, 28*, 915–926.
128. Ray, P. D., Huang, B. W., & Tsuji, Y. (2012). Reactive oxygen species (ROS) homeostasis and redox regulation in cellular signaling. *Cellular Signalling, 24*, 981–990.

129. Haigis, M. C., & Yankner, B. A. (2010). The aging stress response. *Molecular Cell, 40*, 333–344.
130. Arulselvan, P., Fard, M. T., Tan, W. S., Gothai, S., Fakurazi, S., Norhaizan, M. E., & Kumar, S. S. (2016). Role of antioxidants and natural products in Inflammation. *Oxidative Medicine and Cellular Longevity*, 5276130.
131. Fukai, T., & Ushio-Fukai, M. (2011). Superoxide dismutases: Role in redox signaling, vascular function, and diseases. *Antioxidants & Redox Signaling, 15*, 1583–1606.
132. Nimse, S. B., & Pal, D. (2015). Free radicals, natural antioxidants, and their reaction mechanisms. *RSC Advances, 5*, 27986–28006.
133. DeTure, M. A., & Dickson, D. W. (2019). The neuropathological diagnosis of Alzheimer's disease. *Molecular Neurodegeneration, 14*, 32.
134. Breijyeh, Z., & Karaman, R. (2020). Comprehensive review on Alzheimer's disease: Causes and treatment. *Molecules, 25*, 5789.
135. Buccellato, F. R., D'Anca, M., Fenoglio, C., Scarpini, E., & Galimberti, D. (2021). Role of oxidative damage in Alzheimer's disease and neurodegeneration: From pathogenic mechanisms to biomarker discovery. *Antioxidants., 10*, 1353.
136. Rahman, T., Hosen, I., Islam, M. M. T., & Shekhar, H. U. (2012). Oxidative stress and human health. *Advances in Bioscience and Biotechnology, 03*, 997–1019.
137. Butterfield, D. A., & Halliwell, B. (2019). Oxidative stress, dysfunctional glucose metabolism and Alzheimer disease. *Nature Reviews Neuroscience, 20*, 148–160.
138. Elfawy, H. A., & Das, B. (2018). Crosstalk between mitochondrial dysfunction, oxidative stress, and age related neurodegenerative disease: Etiologies and therapeutic strategies. *Life Sciences, 218*, 165–184.
139. Park, H. A., Hayden, M. M., Bannerman, S., Jansen, J., & Crowe-White, K. M. (2020). Anti-apoptotic effects of carotenoids in neurodegeneration. *Molecules, 25*, 3453.
140. Young, A., & Lowe, G. (2018). Carotenoids-antioxidant properties. *Antioxidants, 7*, 28.
141. Tan, B. L., & Norhaizan, M. E. (2019). Carotenoids: How effective are they to prevent age-related diseases? *Molecules, 24*, 1801.
142. Min, J., & Min, K. (2014). Serum lycopene, lutein and zeaxanthin, and the risk of Alzheimer's disease mortality in older adults. *Dementia and Geriatric Cognitive Disorders, 37*, 246–256.
143. Ashok, A., Singh, N., Chaudhary, S., Bellamkonda, V., Kritikos, A. E., Wise, A. S., Rana, N., McDonald, D., & Ayyagari, R. (2020). Retinal degeneration and Alzheimer's disease: An evolving link. *International Journal of Molecular Sciences, 21*, 7290.
144. Dhas, N., & Mehta, T. (2020). Cationic biopolymer functionalized nanoparticles encapsulating lutein to attenuate oxidative stress in effective treatment of Alzheimer's disease: A non-invasive approach. *International Journal of Pharmaceutics, 586*, 119553.
145. Boothby, L. A., & Doering, P. L. (2005). Vitamin C and vitamin E for Alzheimer's disease. *Annals of Pharmacotherapy, 39*, 2073–2080.
146. Donnelly, J., Appathurai, A., Yeoh, H.-L., Driscoll, K., & Faisal, W. (2022). Vitamin E in cancer treatment: A review of clinical applications in randomized control trials. *Nutrients, 14*, 4329.
147. Wołoszynowska-Fraser, M. U., Kouchmeshky, A., & McCaffery, P. (2020). Vitamin A and retinoic acid in cognition and cognitive disease. *Annual Review of Nutrition, 40*, 247–272.
148. Wimalawansa, S. J. (2019). Vitamin D deficiency: Effects on oxidative stress, epigenetics, gene regulation, and aging. *Biology, 8*, 30.
149. Hamid, M., Mansoor, S., Amber, S., & Zahid, S. (2022). A quantitative meta-analysis of vitamin C in the pathophysiology of Alzheimer's disease. *Frontiers in Aging Neuroscience, 14*, 970263.
150. Shahidi, F., & Ambigaipalan, P. (2015). Phenolics and polyphenolics in foods, beverages and spices: Antioxidant activity and health effects—A review. *Journal of Functional Foods, 18*, 820–897.
151. Spencer, J. P. E. (2007). The interactions of flavonoids within neuronal signalling pathways. *Genes & Nutrition, 2*, 257–273.

152. Murphy, M. P., & Smith, R. A. (2007). Targeting antioxidants to mitochondria by conjugation to lipophilic cations. *Annual Review of Pharmacology and Toxicology, 47*, 629–656.
153. Banks William, A., & Elizabeth, M. (2021). Rhea the blood–brain barrier, oxidative stress, and insulin resistance. *Antioxidants., 10*, 1695.
154. Baum, L., & Ng, A. (2004). Curcumin interaction with copper and iron suggests one possible mechanism of action in Alzheimer's disease animal models. *Journal of Alzheimer's Disease, 6*, 367–377.
155. Mathew, B. C., & Biju, R. S. (2008). Neuroprotective effects of garlic a review. *Libyan Journal of Medicine, 3*, 23–33.
156. Tamagno, E., & Guglielmotto, M. (2022). Estrogens still represent an attractive therapeutic approach for Alzheimer's disease. *Neural Regeneration Research, 17*, 93.
157. Qajari, N. M., & Khonakdar-Tarsi, A. (2021). Short communication silibin in and neurological diseases. *Journal of Neurobiology and Physiology, 3*, 8–9.
158. Tarabasz, D., & Kukula-Koch, W. (2020). Palmatine: A review of pharmacological properties and pharmacokinetics. *Phytotherapy Research, 34*, 33–50.
159. Imenshahidi, M., & Hosseinzadeh, H. (2020). Berberine neuroprotection and antioxidant activity. Oxidative stress diet. *Antioxidant Therapy in Neurologic Disease*, 199–216.
160. Alam, J. (2020). Vitamins: A nutritional intervention to modulate the Alzheimer's disease progression. *Nutritional Neuroscience*, 1–18.
161. Holohan, C., Van Schaeybroeck, S., Longley, D., & Johnston, P. (2013). *Nature Reviews Cancer, 13*, 714–726.
162. Sisubalan, N., Ramkumar, V., Pugazhendhi, A., Karthikeyan, C., Indira, K., Gopinath, K., Hameed, A., & Basha, M. (2018). *Environmental Science and Pollution Research, 25*, 10482–10492.
163. Zehiroglu, C., & Ozturk Sarikaya, S. B. (2019). The importance of antioxidants and place in today's scientific and technological studies. *Journal of food science and technology, 56*(11), 4757–4774.
164. Mason, S. A., Trewin, A. J., Parker, L., & Wadley, G. D. (2020). Antioxidant supplements and endurance exercise: Current evidence and mechanistic insights. *Redox biology, 35*, 101471.
165. Lobo, V., Patil, A., Phatak, A., & Chandra, N. (2010). Free radicals, antioxidants and functional foods: Impact on human health. *Pharmacognosy reviews, 4*(8), 118.
166. Hsieh, C., & Rajashekaraiah, V. (2021). Ferric reducing ability of plasma: A potential oxidative stress marker in stored plasma. *Acta Haematologica Polonica, 52*(1), 61–67.
167. Gulcin, İ. (2020). Antioxidants and antioxidant methods: An updated overview. *Archives of toxicology, 94*(3), 651–715.

Metabolism, Mechanisms of Antioxidants in Plants, Microbes and Animals

M. Mouhli, S. Subhashini, C. Thirupathy, and R. Ramesh

Abstract Antioxidants are decisive bio molecules that alleviate the damaging effects of reactive oxygen species (ROS) and reactive nitrogen species (RNS), which are incessantly formed as by-products of usual cellular metabolism. Organisms across all domains of life including plants, animals and microbes have evolved complex antioxidant systems to uphold redox homeostasis and guarantee survival under both normal and stress conditions. The metabolism of antioxidants comprises both enzymatic and non-enzymatic components. Enzymatic antioxidants comprise superoxide dismutase, catalase, glutathione peroxidase (GPx) and peroxiredoxins, which act synergistically to counteract various ROS. Non-enzymatic antioxidants including ascorbic acid, tocopherols, glutathione, flavonoids, carotenoids and polyphenols, function by directly scavenging free radicals or by regenerating oxidized antioxidant molecules. In plants, antioxidant systems are highly diversified in their components such as mitochondria, chloroplasts and peroxisomes, and plants undergo oxidative stress through various mechanisms including drought, salinity, photosynthesis and biotic and abiotic stress. All these mechanisms were controlled by the network of antioxidants and protect plant tissues. In the case of animals, the antioxidants help in oxidative damage through metabolic activity, ageing and some diseases that involve the Nrf2-ARE signalling pathway and it maintains the gene coding, etc. Similarly, many microorganisms like bacteria, fungi and archaea comprise antioxidant mechanisms in the case of host–pathogen interactions, and exposure to environmental conditions by synthesizing metabolites like mycothiol, ergothioneine and microbial carotenoids. In this chapter, the antioxidant metabolism and mechanisms in different organisms including plants, animals and microbes were discussed and provided deep insights into the applications in medicine, agriculture and industry which help in future to carry synthetic biology by enabling the enhanced antioxidant traits in probiotics, crops and therapeutics and helps in improving health-related issues.

M. Mouhli · C. Thirupathy · R. Ramesh (✉)
Department of Physics, Sacred Heart College (Autonomous), Tirupattur 635601, Tamil Nadu, India
e-mail: rramesh@shctpt.edu

S. Subhashini
School of Physiotherapy, Dhanalakshmi Srinivasan University, Tiruchirappalli 621112, Tamil Nadu, India

K. Kasinathan and A. Muthukumaran (eds.), *Recent Advances in Oxidative Stress Associated Chronic Diseases Volume 2*, https://doi.org/10.1007/978-981-95-3758-7_3

Keywords Non-enzymatic antioxidants · Host–pathogen interactions · Oxidized antioxidant molecules

1 Introduction

When a body cell undergoes a chemical process, food transforms into energy. Energy is essential for body function, including acting, conceiving and developing. During metabolism, a number of chemical mechanisms are influenced by the proteins in the body. The term mycotoxins refer to highly toxic substances originating from fungi mainly because there are multiple types of fungi in this species, including *Penicillium* (*Aspergillus*), *Fusarium* and *Penicilla* (*Penicillium*) [1]. They are classified as low molecular weight secondary metabolites. Mycotoxins can be either pathogenic or saprophytic and potentially can contaminate an extensive range of food products, both liquid and a solid food product as well as animal feed. This contagious disease poses significant risks to it is important to keep animals healthy as well as humans. Fungi are some types of plants can produce multiple types of mycotoxins [2]. The presence of these toxins has serious implications for the global economy and international trade. Ochratoxins are a significant group of mycotoxins that include ochratoxin A (OTA), B and C, with OTA being the most prevalent and harmful. OTA is commonly found in various foods and this substance is produced by many different types of fungi from the *Aspergillus* and *Penicillium genera*, particularly [3–5]. A variety of aspergillus species, such as *Aspergillus ochraceus*, *Aspergillus niger*, *Aspergillus carbonarius*, *Aspergillus westerdijkiae*, *Aspergillus malleus*, *Aspergillus alliaceus*, *Aspergillus auricomus*, *Aspergillus verrucosum*, as well as *Penicillium viridicatum* are commonly found in advanced environments. The environmental conditions that promote OTA production vary depending on the fungal genus. Ochratoxins can be harmful to the immune, nervous and renal systems in both humans and animals. They are also associated with DNA damage, carcinogenic effects and teratogenicity [6]. Among the species that produce Zaralenone (ZEA), the *Fusarium* genus produces it primarily. *Fusarium graminearum*, *Fusarium roseum*, *Fusarium culmorum*, *Fusarium equiseti*, *Fusarium cerealis*, *Fusarium verticillioides*, *Fusarium incarnatum*, *Fusarium crookwellense* and *Fusarium semitectum* [7]. ZEA is a naturally occurring oestrogen with significant implications as a health precaution for humans and animals, particularly in swine. There has been links to it hormonal imbalances and an increased risk of human breast cancer. In pregnant animals, ZEA in feed can reduce embryonic survival rates and can also lead to conditions such as vulvar swelling, reduced sperm quality, rectal prolapse and hormonal imbalances [8, 9]. ZEA consumed dairy cows excrete these substances into their milk. In general, fumonisins (FUM) which are predominantly produced by *Fusarium* species such as *F. verticillioides*, *Fusarium proliferatum* and *A. niger*, are divided into four categories, A, B, C and P. Among the six recognized fumonisins, including FA1, FA2 and FB1—FB4, FB1 is the most toxic and prevalent. While an analysis of the effects of fumonisins on human health is limited, they are known to cause severe damage to animal health, including liver and kidney cancer [10]. Fumonisins are linked to leukoencephalomalacia (ELEM) in horses, pulmonary oedema and hydrothorax in pigs; moreover, fish suffer damage to their livers and neural tissues.

There are a number of types of trichothecenes, chemically diverse there are a wide variety of mycotoxins produced by various fungal genera, including *Fusarium*, *Myrothecium*, *Trichoderma*, *Trichothecium*, *Cephalosporium*, *Verticillium* and *Stachybotrys* matrix. The compounds are divided into four groups (A, B, C and D: T-2 and HT-2 toxins appear in group A, while deoxynivalenol and nivalenol appear in group B. These mycotoxins can cause delay in growth, leukotoxicity, hyperhemotoxicity and septicemia. Deoxynivalenol damages the digestive system and reduces appetite, while nivalenol has severe effects are observed on bone tissue, intestine lining and lymphatic organs. Sterigmatocystin (STC) is a mycotoxin contains more than 50 fungi genera, primarily from the *Aspergillus* genus [11, 12]. It is associated with tumorigenicity and various cancers, including hepatocellular carcinomas. STC is highly toxic to animals, leading to severe liver damage, nephrotoxicity and, in extreme cases, death. Ergot alkaloids (EAs) are produced by *Claviceps* and *Epichloe* fungi and include over 50 compounds, such as ergometrine and ergotamine. These alkaloids are linked to neurotoxicity, weight loss and hormonal imbalances. Phomopsins, which are produced by Diaporthetoxica, are classified into five groups, with group A being the most toxic [13]. These mycotoxins are associated with nephrotoxicity, immunotoxicity and carcinogenicity, and they are responsible for a condition known as lupinosis in grazing animals. Citrinin is synthesized by species of *Aspergillus*, *Penicillium* and *Monascus*. Besides being harmful to stomachs and autoimmune systems, citrinin may also cause DNA damage [14]. Alternaria mycotoxins, produced by Alternaria species, encompass over 70 different toxins that significantly impact both human and animal health [15]. They exhibit phytotoxic, genotoxic and mutagenic properties, particularly affecting mice, chickens and dogs. Mycotoxin contamination causes significant economic losses, often requiring the destruction of contaminated crops like maize, wheat, rye, barley and oats. Losses due to mycotoxins have been reported in over 100 countries, with financial damages amounting to billions annually [16]. Efforts to regulate mycotoxin levels in food and feed are evolving worldwide, with the European Union and other nations imposing stringent limits to mitigate their impact on health and trade. Mycotoxins disrupt cellular homeostasis in effect, it boosts the generation of free radicals and reactive oxygen species (ROS). While low to moderate levels of ROS play essential physiological roles, elevated levels can result during oxidative reactions, proteins, fats and DNA are all affected. Antioxidant defence mechanisms, which can be classified as enzymatic or non-enzymatic, work to counteract these harmful effects [17]. However, exposure to mycotoxins like aflatoxins can impair these antioxidant defences, leading to increase the result of oxidative damage, particularly to the hepatic and renal systems. The ability of plants to adapt to pathogen attacks and the strategies developed through co-evolution have been crucial, analysing livestock farming strategies from a historical perspective. Foreign stimuli are capable of causing plants to become susceptible or resistant, pathogens is determined by the recognition and signalling pathways that occurs pathogen's interaction with the host [18–20].

The presence of microbes or pathogens molecular patterns (MAMPs/PAMPs) and avirulence (Avr) gene products correspond to specific cell receptors in the host in biological organisms and work alongside resistance (*R*) a gene's function initiate a signalling in cascades [21]. This cascade leads, in addition to producing toxic oxygen radicals, phytoalexins and antimicrobial proteins, activating a wide array of plant defence mechanisms effective against various pathogens. In plants, ROS are generated through a transport mechanism for electrons, chains microsomes and chromoplasts are also affected by photoreabsorption [22]. It is thought that anapoplastic ROS can be formed at the cell cytoplasm by reactions that involve homologous oxidases (RBOH) including NADPH oxidase and pulmonary oxalate oxidase (PAO), which eliminates spermidine, and oxalate oxidase [23]. As a result of the growth of plants, a large amount of ROS is generated that is dependable and abundant, are hydrogen peroxide (H_2O_2), superoxide (O_2^-), hydroxyl radical (OH) and singlet oxygen [24]. These ROS can rapidly interconvert, providing diverse functional variability. Of these, hydrogen peroxide is stability is the best and is transported across membranes via aquaporins. The physiological it is important to note that aspects of ROS development depend on the concentration of ROS. At lower concentrations, ROS serve as signalling molecules, whereas excessive accumulation can lead to oxidative Injuries that damage RNA, DNA, macromolecular structures and membranes, eventually resulting in apoptosis [25]. An equilibration involving calcification channels, NADPH oxidases (NOX) and fluxes of calcium, in response to stress, free radicals are generated, which initiate ROS currents, which serve as long-distance communication mechanisms. By triggering the mitochondrial burst oxidase homolog D (RBOHD) polypeptide, a self-perpetuating cycle of ROS generation is triggered [26]. As a result of this process, a lone cell produces more ROS, which is then transmitted to its surroundings to enhance ROS production. There is recent research that emphasizes the unique role of the ROS wave in the coordination of a variety of stress reaction mechanisms, across the plant and influencing systemic stomatal responses. This process involves, increasing ROS levels in endoplasmic reticulum, facilitated by RBOHD, thereby playing a significant role in plant-wide stress responses [27].

2 Defence System Against ROS Production and Accumulation

Plant cells have been identified as possessing ROS immunity since the end of the twentieth century, in conjunction with their numerous functions and elements. Recent studies have revealed that ROS's protective effects range beyond exclusively scavenging mechanisms, including enzyme-dependent and enzyme—independent systems, which are both influenced by environmental stresses [28]. As a result of oxidative stress, ROS are produced by enzymes and non-enzymatic mechanisms during normal cellular metabolism. In order to combat free radical damage, enzymatic defence systems are employed, such as ascorbate peroxidase (APX), catalase (CAT), superoxide dismutase (SOD) and glutathione peroxidase (GPX). The non-enzymatic system is composed of glutathione, ascorbic acid, phenolic compounds and tocopherols. Despite the complexity of the plant's defence system, ROS are neutralized by enzymes as well as by non-enzymatic agents [29]. As well as ROS production, ROS scavenging occurs in many different types of organelles, including peroxisomes, mitochondria and chloroplasts, thus indicating that these organelles cooperate well during this process. Generally, plants maintain a balance between producing ROS and scavenging other compounds. During stress, this equilibrium can be disrupted, which results in increased ROS levels and subsequent oxidative stress on cells [30]. A plant's natural defences have evolved to combat ROS levels, which are evidently increasing with increasing evolution. Photosynthesis periods of abiotic stress are characterized by high levels of ROS within plants, which are reduced by the ROS defence system. Throughout history, plants have evolved sophisticated and complex defence mechanisms that effectively manage ROS production and development (Table 1).

Table 1 Mechanism and generation of ROS in plants and its detoxification

S. No.	Type of ROS	Production	Detoxification	Ref.
1	Superperoxide radical	Mitochondrial respiration, Electron transport, Superoxide dismutases, Xanthine oxidoreductase	Superperoxide dismutases	[31]
2	Alkoxyl radical	Through Polyunsaturated fatty acid metabolism	Non-enzymatic Scavengers	[32, 33]
3	Peroxyl radical	Polyunsaturated fatty acid metabolism	Non-enzymatic Scavengers	[34]
4	Singlet Oxygen	Generated through nonsensitive reaction of O_2 generation	O_2 is quickly occupied in numerous oxidation reactions, Polyamines	[35]
5	Hydroxyl radical	Generated through Fenton and Haber–Weiss reaction Thymidine catabolism Aconitase production	OH has very small half-life and is readily involved in other reactions with polyamines	[36]
6	Hydrogen peroxide (H_2O_2)	Generated via Polyamine catabolic reaction, Thymidine Catabolism NADPH reaction Lysyl, Dihydroorotate dehydrogenase, CYP cycle	Glutathione Peroxidases Catalase Peroxiredoxins Glutaredoxins Polyamines	[37]
7	Hydroperoxyl radical	Protonated form of O_2	Non-enzymatic scavengers	[38]

3 Antioxidant Defence Efficiency and Stress Tolerance in Plants

The efficiency of antioxidant defence mechanisms varies significantly among different plant species, genotypes and the types and frequency of stress conditions. El-Badri et al. studies have shown that mutants tolerant to salt exhibit lower concentrations of ROS and malondialdehyde (MDA) when antioxidants are activated, like superoxide dismutase (SOD), peroxidase (POD), catalase (CAT) and ascorbate peroxidase (APX). It has been shown that these enzymes are capable of scavenging ROS in the presence of salinity stress [39]. There were several key factors contributing to stress tolerance, including amino acids, sucrose, flavonoid metabolism, which increased when exposed to salt stress. The tolerant cultivar exhibited higher antioxidant enzyme activities and increased metabolite accumulation, resulting in significant improvements in tolerance to salinity. Developing high-salt tolerance rapeseed cultivars can be simplified with the help of these findings [40]. As part of their study, Begum et al. have examined the antioxidant activity in soybeans when exposed to drought or salinity stresses (polyethene glycol 5, 10 and 15%), as well as combinations of both, especially during the germinating stage. A study by the authors found that oxidative stress increased antioxidant enzyme activities, such as SOD, POD, CAT and APX. These effects can be observed, in particular, with cultivars PI31 and PI90 when grown in conditions of salinity and drought. A study concluded that soybeans accumulate antioxidant enzyme activity and secondary metabolites, which are important for stress tolerance and sprouting [41]. It is possible to produce ROS in plants in both a beneficial and detrimental manner. However, the specific mechanisms by which ROS reduce stress are still unclear. Sachdev et al. reviewed current knowledge regarding the sites of ROS production, signalling mechanisms, effects and management strategies under stress. They also highlighted the role of modern advancements, modelling of crops, molecular biology and genomics are some of the areas in which microbial priming is done, in mitigating oxidative stress and redirecting ROS into signalling pathways [42]. Cross-tolerance, which is an adaptive response to abiotic stress, is closely linked to ROS signalling. Rani et al. Cold acclimatization was reported to have been achieved significantly increased both enzymatic (SOD, CAT, APX and glutathione reductase) and non-enzymatic (ascorbic acid and glutathione) antioxidant activities in tolerant chickpea genotypes, resulting in reduced leaf damage from low-temperature-induced cold stress. This information is valuable for enhancing cold tolerance in crops [43].

Stress and oxidative stress are regulated by phytohormones, and these hormones also directly impact growth of plants. Plants have been shown to be more resilient to a variety of stresses when bacteria such as *Bacillus*, *Pseudomonas*, *Enterobacter*, *Micrococcus* and *Lysobacter* are used as probiotics. Qadir et al. Acinetobacter Bouvet P1, a rhizobacteria that produces phytohormones, stimulated sunflower growth when stressed with chromium (Cr^{+6}). The sunflower's antioxidant defence were strengthened, antioxidant enzyme activity was increased (CAT, APX, SOD, POD) and antioxidant components were enhanced (flavonoids, polyphenols, prostaglandins, glutamine) [44]. In addition, Fatma et al. It was found that wheat grown at 42 °C was significantly more likely to increase antioxidant enzyme activity, decrease hydrogen peroxide (H_2O_2) concentration and increase photosynthesis when treated exogenously with methyl jasmonate (MeJA). It has been shown that nitric oxide (NO), hydrogen sulphide (H_2S) and hydrogen peroxide (H_2O_2) are important molecules for antioxidant protection. Stress-resistance is enhanced in species that are adaptable, such as plants and halophytes [45]. NO is a signalling molecule involved in stress resistance. According to Hasanuzzaman et al., the antioxidant defence controlled by NO are critical to salt tolerance in the halophyte Kandeliaobovata. As a result of sodium nitroprusside (SNP) treatment, salt-stressed plants were stimulated to produce endogenous NO, inhibit ion toxicity, enhance nutrient uptake and produce more proline. CAT, APX, monodehydroascorbate reductase (MDHAR) and Dehydroascorbate reductase (DHAR) were also expressed more highly after treatment with single nucleotide polymorphism (SNP).

4 The Metabolic Process of Ferulic Acid (FA) in the Animal Organism

Plenty of investigations have been conducted on the bioavailability and fate of phenolic acids after digestion. During the transit of stomach contents to the small intestine and then through the large intestine, metabolizable phenolic acids may react with glucuronic acid before being absorbed by the intestinal cell walls. Meanwhile, upon entering the large intestine, phenolic acids are still bound, where bacteria and microbial esterases further metabolize and degrade themselves [46]. Esterified FA was only released, in the barley matrix, through enzymatic action, and this primarily happened during intestinal digestion as opposed to gastric digestion, according to the findings of in vitro simulated digestion trials. Trypsin, lipase and pancreatic amylase are among the pancreatic α-tic enzymes ligands and interactions that FA has with, which could be another reason, during the digestive process of the intestine's, esterified FA is more concentrated [47]. FA can interact with hydrophobic groups at the site of activity and alpha-amylases and glucosidase at the site of activity in addition to hydrophobic interactions, according to molecular docking studies. In particular, the amino acid Gly 334 and the carbonyl groups C3 and C4 form a mono-hydrogen bond with α-amylase, demonstrating competitive inhibition.

5 Antioxidants in Plants

An inoculation of *Bacillus* sp. was performed on ZK MD. Drought moderately affects dry weight of Cachystachyapachystachya plants. In this study, two PGPB species (*Azospirillumbrasilense* and *Bacillus* sp.) were inoculated to investigate their antioxidant capacities in drought-exposed *Cecropia pachystachya* and *Cariniana estrellensis* vines. There is no research in the literature to show that plant growth-promoting bacteria (PGPB) increase antioxidant activity in crops [48]. However, this has yet to be demonstrated in Neotropical trees. The biometrical parameters of the seedlings as well as the amount of non-enzymatic antioxidants. The process involves activating enzymes that fight free radicals' leaves were assessed. The number of non-enzymatic antioxidant molecules, further assessment is made of the activities of antioxidant enzymes were reduced in both tree species by dryness. *A. brasilense* inoculation mostly impacted the enzymatic and non-enzymatic pathways for *C. pachystachya*, increasing the activities of superoxide dismutase and ascorbate peroxidase (APX) and favourably influencing the concentration of non-enzymatic antioxidant chemicals [49]. *A. brasilense* inoculation increased APX activity in *C. estrellensis*. However, compared to non-inoculated plants under drought, *A. brasilense* and *Bacillus* sp. inoculation had a stronger impact. Due to the non-enzymatic mechanism of both microbes, secondary metabolites were accumulated at greater levels (such as synaptic acid, gallic acid, rutin and chlorogenic acid). PGPB enhanced biometrical indicators of drought tolerance for both plants, every leaf is given its area-to-surface ratio and its surface area. Our findings show that PGPB increased drought tolerance in drought-stressed Neotropical plants by inducing antioxidant processes. As a result, PGPB inoculation offers a biotechnological substitute to increase forestry initiatives' productivity [50].

6 Antioxidant in Plant Source

Numerous naturally occurring substances with antioxidant properties are present in plants, categorized into three main groups: terpenoids, polyphenols (which include flavonoids, phenolic acids, stilbenes and lignans) and vitamins (specifically C and E vitamins). These vitamins are particularly abundant in various plant-based foods. Notable sources of these vitamins include fruits from the Rosaceae family (such as blackberries, strawberries and sour cherries), the Empetraceae family (cowberries), the Ericaceae family (blueberries), the Asteraceae family (sunflower seeds) and the Punicaceae family (pomegranates). Vegetables rich in vitamins C and E encompass tomatoes, lettuce, leeks, broccoli, Brussels sprouts, green cabbage and kohlrabi [51]. The vitamins derived from plants are considered primary antioxidant agents. The microbiota plays a crucial role in the breakdown of polyphenols, leading to the production of various metabolites. These polyphenols are therefore more readily available to the body thanks to the microbiome, resulting in metabolites that positively influence the microbiota by decreasing harmful bacteria and increasing beneficial strains [52]. Studies have demonstrated that phytochemicals improve bacteria in the stomachs of pigs and poultry, thus reducing their stress levels. An in vitro study has been conducted that polyphenols possess significant antioxidant capabilities, including the chelation of metal ions, the removal of (ROS), and the stimulation of antioxidant enzyme transcription. However, due to their poor absorption, polyphenols undergo xenobiotic degradation, necessitating a comparison there are a variety of antioxidants in these foods, including glutathione, albumins, vitamin C, vitamin E and uric acid. Additionally, research has indicated that plant polyphenols can lower ROS levels in animals suffering from inflammatory intestinal disorders [53]. Pigs that were fed grape pomace exhibited improved growth performance, and their final products displayed altered fatty acid profiles, which are relevant for human nutrition. It is also important to note that, despite their beneficial effects, polyphenols can exhibit pro-oxidative properties and form complexes with metals, thereby reducing their bioavailability [54].

6.1 Adapting to Drought Stress in Plants: Encompassing Morphological, Anatomical and Biochemical Alterations

Chemical and molecular changes ultimately disrupt plant functioning. Drought conditions adversely affect plant performance at various developmental stages. Insufficient water supply can significantly hinder germination rates and the growth of young plants. As plants mature, drought primarily impacts their water relations, leading to considerable disturbances in overall. In plants, metabolic processes are regulated at the cellular, enzymatic and physiological levels, which are influenced by the intensity and duration of the stress. Water scarcity modifies numerous plant processes, with a primary consequence being the reduction of photosynthetic activity, ultimately affecting plant yield. Under drought stress, oxidative stress occurs directly. The direct generation of substances within plants serves as a primary factor influencing their responses, leading to damage to the cell membrane. This damage disrupts membrane integrity and a physiological and biochemical change, which ultimately leads to acute metabolic disorders and a decline in plant productivity [55].

6.2 Drought Stress and Its Impact on Plant Development

Stress caused by drought is widely recognized as a significant factor that impacts different aspects of growing plants and their evolution. It is critical that seeds germinate, that plants remain healthy, and that coleoptiles remain long, all of which are essential for the progress of plants [56]. The germination of seeds, when they begin to grow, is especially susceptible to drought conditions. Observable changes have occurred in the survival of many plant species, such as corn, wheat and sorghum. Plants suffering from water deficiency in the growing stage show visible symptoms including leaf wilting, shortened height and irregular bud and flower formation. Furthermore, drought causes nutrients to be trapped in the soil, resulting in a reduction of stem length. Moreover, it is evident that the growing length of *Lathyrus sativus* also decreases under water-deficient conditions. Root structures are strengthened by plants to draw water deeper into the soil when water is scarce. Various root characteristics, including width and spacing of roots, as well as longer lateral roots, are influenced by them. They are also essential to determining water availability [57]. In the rhizosphere, roots interact symbiotically with microorganisms to improve micronutrient uptake and water absorption, both of which can influence plant growth. Roots of *Crocus sativus* L grow longer when drought stresses them. It is important for a plant's growth to have a strong root system during its early stages. Plants are better able to produce biomass when their roots are longer, since they retain more water and nutrients. There is a substantial decline in plant biomass that occurs when water deficits occur. As a result, root-to-shoot ratios increase, but often in conjunction with an increase in root-to-shoot ratios as shown in Fig. 1 [58]. Leaf's are the most important component of a plant, as they are the site of most of the plant's photosynthesis. A reduction in the leaf coverage of *Andrographis paniculate* was caused by water stress. The primary factor driving plant development is photosynthesis, which requires sufficient leaf area to ensure healthy leaf growth. Reduced leaves, which results in reduced photosynthesis, subsequently decreasing crop yield, are the results of water stress [59]. When leaves are stressed by water, their surface area appears to be reduced in *Petroselinum crispum* L. *Stevia rabaudiana* is perhaps adapting to balance root water absorption. Plants with smaller leaf areas are more drought-resistant, since they lose less moisture through transpiration due to a smaller leaf area. The decrease in leaf area is caused by a reduction in cell division, which reduces the turgidity of the leaves. The turgor pressure in guard cells and stomatal closure decrease when soil moisture declines. In drought-stricken conditions, leaf senescence is accelerated prematurely.

6.3 Drought Stress and Photosynthetic Activity

Plants whose water levels are inadequate are unable to perform photosynthesis. Factors such as diminished leaf area, increased, inhibition of stomatal closure and the resulting decrease in leaf cooling through evapotranspiration contribute to a number of factors can damage the photosynthetic machinery, such as osmotic stress [60]. The decline in photosynthetic activity during drought conditions is primarily linked to reduced CO_2 conductance through stomata and limitations within the mesophyll. Additionally, the diminished capacity for stomatal movement further exacerbates the decrease in photosynthesis. This decline is largely driven by a reduction in CO_2 uptake, which negatively impacts, besides decreasing rubisco activity, it reduces the functions of nitrate reductase, sucrose phosphate synthase and produces urea ribulose

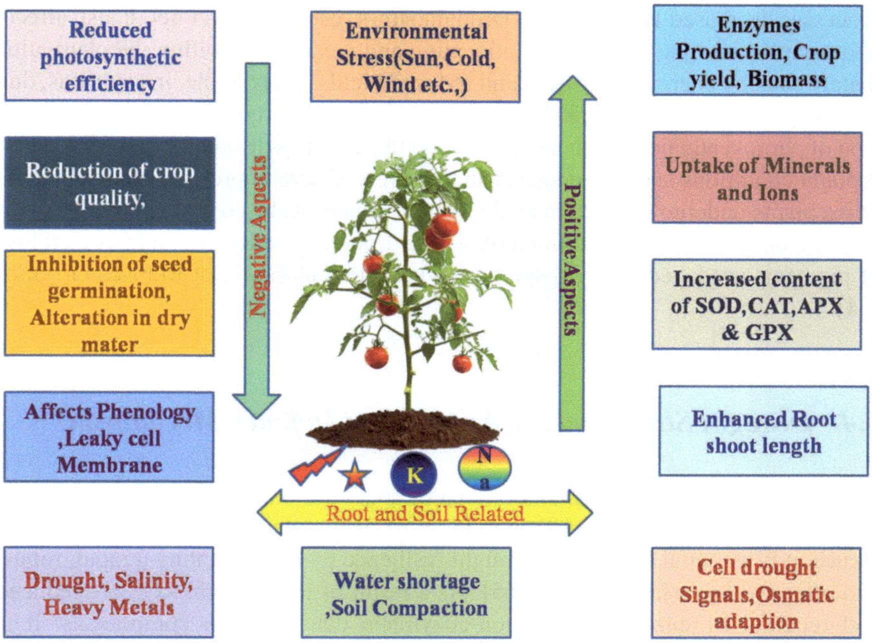

Fig. 1 Analysis of ROS's impact on plants

bisphosphate (RuBP). Notably, CO_2 enrichment has been shown to mitigate many of the initial metabolic and transcriptomic responses in maize that are associated with drought stress. Furthermore, water deficit leads to a reduction in leaf area per shoot, resulting in changes to canopy structure, this can have an impact on gas exchange and water flow, vegetative growth and the development of sinks, such as fruits or grains [61]. For instance, in grapevines, this can influence berry sugar concentration, while in maize, prolonged water stress has been linked to decreased kernel number and reduced 100-kernel dry weight. Chlorophyll content contributes to the success of photosynthesis and is significantly affected by water scarcity, serving an important indicator of photo-oxidation it also degrades chlorophyll. During drought stress, vegetable plants produce less chlorophyll in their leaves, so measurements are taken of their chlorophyll *a* to *b* ratios [62]. This stress leads to a reduction in photosynthetic activity, a decrease in chlorophyll levels. Plant water balance disruptions reduced photochemical efficiency of photosystem II, and alterations in stomatal behaviour are all contributing factors, all contributing to reduced productivity [63]. A primary factor in the reduction of chlorophyll during drought is the increase in reactive oxygen species, such as O_2^- and H_2O_2, which cause lipid peroxidation and subsequent chlorophyll degradation. The decline in growth and yield across various water-stressed plant species conditions is frequently linked to diminished photosynthetic performance and impaired chlorophyll content. Water scarcity disrupts, in photosynthetic processes, chlorophyll plays a vital role, leading to activation of photosynthetic processes decreased *Vigna mungo*. Drought stress results in lower net, it also affects photosynthesis along with carbon allocation and metabolism within the plant, ultimately causing energy dissipation and reduced yield. For example, in soybeans, due to drought stress, physiological metabolic disorders are reduced through suppression of photosynthetic growth and disruption of carbon cycle balances. Additionally, drought conditions lead to a decrease in the levels of several proteins involved in the Olive plants undergo the Calvin cycle, which reduces Rubisco and glutathione, two antioxidant compounds. Additionally, severe droughts can lead to an increased loss of pigments produced by photosynthesis due to the inhibition of Rubisco enzymes [64].

6.4 Drought Stress and the Antioxidant Defence Mechanism

Reactive oxygen species (ROS) are a major source of damage to plants caused by drought conditions and are mitigated by most of the plant's defence mechanisms. Hence, it is essential to enhance drought resilience by implementing a rapid, robust and effective antioxidant system [65]. As a result of this system, ROS are reduced and repaired by enzymatic and non-enzymatic detoxification components. It is through the enhancement of the antioxidant system that ROS are scavenged, thereby preventing electrolyte leakage and lipid peroxidation, thus maintaining organelle and membrane vitality and structural integrity. It is widely acknowledged that drought conditions lead to oxidative stress through the production of ROS, such as superoxide ($O_2\cdot^-$), hydroxyl radicals ($OH\cdot$), singlet oxygen and hydrogen peroxide (H_2O_2). The balance between ROS production and the activity of antioxidant enzymes determines the cellular redox state, influencing either ROS management or cellular damage and death when ROS levels surpass physiological thresholds [66]. Numerous investigations conducted under water-limited conditions have reported increased activities of key antioxidant enzymes, including catalase (CAT), superoxide dismutase (SOD), peroxidase (POD) and ascorbate peroxidase (APX). Typically, a species, variety or genotype that resists drought has a higher antioxidant enzyme activity compared to their non-tolerant counterparts, underscoring their critical role in drought resistance, particularly in regulating the production and movement of H_2O_2 and O_2^- within leaf tissues. The generation of $O_2\cdot^-$ and H_2O_2 is managed by the actions of SOD, peroxidase (POX) and catalase (CAT), with their activity notably heightened in drought-resistant potato genotypes. Additionally, ascorbate peroxidase (APX) functions as a significant ROS scavenger, H_2O_2 can be converted to water by using ascorbate as a substrate. Its activity increases under stress [67]. APX activity changes are more likely to be noticed in leaves than in fibrous roots, since APX mainly resides in chloroplasts and cytoplasm, where it plays a critical role in scavenging H_2O_2.

6.5 Drought Stress and Secondary Metabolite Production

Plants generate secondary metabolites as a means of adapting to different environmental stresses. In addition to temperature, light exposure and nutrient availability, environmental conditions affect the production of these secondary metabolites. In situations where water supplies are scarce, secondary metabolites are more likely to be produced. This is because biomass development decreases and CO_2 is diverted to carbon-based secondary metabolites instead of being assimilated. Consequently, excess sugars do not inhibit photosynthesis as a feedback mechanism [68]. Plants with a water deficit display significant increases in phenolic acids. The Iranian *Origanum vulgare* species includes two native subspecies, namely subsp. In addition to Gracile, there is a subspecies called Caryophyllene which is markedly above average under water-limiting conditions. The presence of drought increases oleanolic acid and betulin levels in *Betula platyphylla*, while drought increases glycyrrhizin production in *Glycyrrhiza glabra.* When drought stresses are applied to *Cynodondactylon* L., Tifton-85 also exhibits an increase in lignin content [69]. Moreover, Labisia-pumila's medicinal properties are markedly improved under high water deficits due to its flavonoid content. Cyanogenic glucosides are produced in greater concentrations when *Phaseolus lunatus* is drought tolerant. A decrease in essential oil content is observed in *Lavandula latifolia* and *Salvia sclarea* within the *Lamiaceae* family; however, under drought conditions, *Mentha piperita, Salvia lavandulifolia, Thymus capitatus* and *Thymus mastichina*'s essential oil levels increase, because of a reduction in leaf area resulting in an increase in oil gland density. Furthermore, phenolics and flavonoids persist in Achillea species when drought stress is applied, whereas flavonoids decrease in *Achillea pachycephala* as phenolic acids increase [70].

6.6 Drought Stress and Nutritional Minerals

Water scarcity typically diminishes the overall availability of soil nutrients, impairs the translocation of nutrients within roots, the ions are ultimately reduced through this process concentration in various plant tissues. Under conditions of water deficit, the uptake of potassium (K) by plants is notably decreased [71]. This reduction in potassium is linked to diminished mobility of K, transpiration rate is lowered and root membrane transporters are less effective. Drought-stressed *Malus hupehensis* plants also exhibited a decrease in potassium levels. Among *Triticum durum* genotypes, the resistant varieties displayed the highest potassium content, while the susceptible ones had elevated sodium (Na) levels. Water deficit conditions inhibited the genes responsible for K transporters; Calcium sensors called calcineurin B are responsible for activating the inner K channels, while the protein kinase CIPK23 controls the outer K channels. Grapevine roots inhibit this K channel, while leaves activate it [72]. *N* levels in *Mentha piperita, Salvia lavandulifolia, Salvia sclarea* and *Thymus capitatus* remained unchanged under drought stress. Meanwhile, leaf phosphorus (P) levels decreased across the board for all species except *S. scrarea*, which retained its concentration. Photosynthesis was found to decline and leaf senescence to occur due to nitrogen deficiency. During water deficit conditions, Ocimumgratissimum leaf phosphorus and potassium levels significantly decreased, while Thymus diagenesis showed an increase in potassium. In addition, plants with limited water availability showed decreased potassium levels. A decrease in potassium occurs primarily in leaves where water scarcity disrupts stomatal movement and lowers the turgidity of cells. Additionally, soybeans were accumulating more manganese (Mn), molybdenum (Mo), phosphorus (P), potassium (K), copper (Cu), calcium (Ca) and zinc (Zn) due to drought conditions [73].

7 Techniques for Improving Crop Tolerance by Using Plant Tolerance Mechanisms to Increase Crop Resilience to Drought Stress

There is still debate on the exact meanings of drought resistance and tolerance. To improve crop water use efficiency (WUE), there is a possibility the phrase "*plants that reduce water consumption*" refers in plants, water is used for growth and development primarily. In economics, WUE refers to the economic return on investment, generated per unit of water used, and there is no conclusive evidence linking it to drought tolerance [74]. On the other hand, the generally agreed—upon definition of drought resistance refers to an organism's ability to endure or flourish in water-scarce environments, which may entail strategies like drought recovery, dehydration avoidance or dehydration tolerance. The progressive decrease in the amount of water present in plant tissues is known as dehydration. Since there is already thorough research on the subject, it is outside the purview of this review to distinguish tolerable droughts and resistant droughts because it can be very complex. Therefore, drought-resistant plants will be called "*tolerant*" without distinguishing between drought tolerance and drought resistance, in the following sections of this review. Changes at the molecular, metabolic and morphological levels are all part of plant drought tolerance. The ability of a plant to withstand difficult drought circumstances depends on the existence of one or more tolerance factors. Practically speaking, a thorough comprehension, there are many of these processes that can be applied to pinpointing crops, genotypes, etc. or varieties that are less susceptible to water constraints [75]. To identify some of these qualities as possible criteria in the following sections, we explain the physiological, biochemical and molecular mechanisms by which plants with drought—tolerant characteristics can withstand drought conditions better.

8 Antioxidants in Animals

8.1 The Impact of Mycotoxins on the Antioxidant Levels in Swine and Avian Species

Mycotoxins are recognized as significant contaminants in animal feed, posing potential risks to livestock health. The intensive farming of poultry and swine increases the likelihood of mycotoxin exposure due to their high intake of cereals and oilseeds, which are often contaminated [76]. These toxins can adversely affect various organs, including the gastrointestinal tract, liver and immune system, ultimately leading to decreased productivity. While individual mycotoxins can be detrimental, their combined presence may result in heightened toxicity due to synergistic effects. Common mycotoxins found in animal feeds Aflatoxin A, ochratoxin A, zearalenone, fumonisins, deoxynivalenol and T-2 toxin are among the most common toxins. As a result of mycotoxins' toxicological effects, oxidative stress (OS) is induced and free radicals are produced. Damage to DNA, proteins and lipids can occur when free radicals increase and antioxidative systems are compromised [77]. Typically, cells maintain a balanced production of oxygen-free radicals and antioxidants. However, external factors can exacerbate oxidative stress and the overproduction of free radicals, disrupting cellular homeostasis. When cells are oxidized due to excessive reactive oxygen species (ROS), such as hydroxyl radicals, per-hydroxyl radicals, superoxide anions and reactive nitrogen species (RNS), such as nitric oxide, oxidative stress occurs. As a result of an excess of ROS, DNA, proteins and membrane lipids can be oxidized [78]. As a result of lipid peroxidation, mycotoxin toxicity has significant implications, particularly for cell death. It is unclear whether mycotoxins cause lipid peroxidation by causing free radicals to form or by impairing antioxidant defences.

8.2 The Transformation of Mycotoxins and Their Occurrence in Animal Products

By contaminating feed, mycotoxins or their metabolites are transferred into animals, which is a major concern for consumers and the market. A variety of absorption and metabolic processes affect how these toxins are deposited in farm animals' tissues. Since mycotoxins and their metabolites are extensively metabolized by the liver, mycotoxins accumulate in the muscle tissues of animals in low concentrations, frequently under detection thresholds [79]. Muscles and fat tissues contain lower concentrations of mycotoxins and their metabolites than blood, kidneys and liver. Therefore, these organ meats should be consumed with caution. Mycotoxins can be absorbed by humans through meat products when they are aged or cooked by methods such as dry-curing, or when they are used in spices contaminated with mycotoxins, such as nutmeg, peppers, coriander and paprika. Curing, fermenting or ripening meats are commonly contaminated with mould species from the genera Penicillium and Aspergillus, increasing their organoleptic quality [80]. In spite of this, mycotoxins can be generated on these substrates by toxigenic fungi. Also, eggs have been investigated for their ability to transmit mycotoxins. Researchers have found it is either impossible to detect residues of aflatoxins in eggs from hens fed contaminated diets, or they are measured to be five times lower than EU maximum residue limits (MRLs). Similarly, aflatoxins have been detected in eggs and chicken meat from Pakistan, yet their concentrations remained below the aforementioned MRL, with the highest levels detected in live birds. Overall, there is a notable multi-exposure risk for humans to mycotoxins through the consumption of milk, with AFM1 being the hydroxylated derivative of AFB1 [81].

8.3 *Antioxidants Derived from Animal Sources*

Oxidants such as ovotransferrin and phosvitin are found in egg white proteins. Ionic iron is chelated to prevent metal-catalyzed lipid oxidation. Ovotransferrin exhibits a unique mechanism for self-cleaving when it interacts with metal ions. It is well established that many peptides produced from proteins through enzymatic hydrolysis or various physicochemical methods are antioxidants. Several factors affect how effective these peptides are as antioxidants, including their amino acid sequences, compositions and lengths. A diet rich in meat, milk, eggs and fish product peptides, as well as their by-products, including head, skin, fin, intestine and blood can provide such antioxidant peptides [82]. These bioactive peptides are derived from various proteins and are often used in traditional medicine to achieve specific physiological effects. Normally, peptides are synthesized through enzyme hydrolysis, which is responsible for their non-toxicity, easy excretion and degradability. It has been proven both in vitro and in vivo that bioactive peptides have physiological benefits. Processes, such as fermentation, generate a large number of bioactive peptides from animal sources [83]. A microbial enzyme breaks down polypeptides into smaller peptides after being hydrolysed into polypeptides. Polypeptides can be converted to dipeptides, tripeptides or even individual amino acids by the action of exopeptidase enzymes, all which function as antioxidants. In addition to providing high quality proteins, milk and its by-products, such as whey, have been used for decades to produce bioactive peptides. Endogenous proteases present in milk are responsible for breaking down milk proteins and producing peptides with the amino acid sequence VLPVPQK that have antioxidant properties [84].

8.4 The Impact of Polyphenols on Animal Health

There was no definitive evidence that polyphenols had an impact on animal growth when incorporated into animal feed. As a result of low digestion enzyme secretions, increased protein excretion and diminished digestibility, metabolism may be adversely affected. Weaners whose diets include spent hops, a natural polyphenol source, demonstrate a decrease in body weight, decreased nutritional efficiency and decreased digestibility of overall proteins and cellulose [85]. Although this supplement did not affect overall performance, the gain-to-feed ratio was more favourable in the experimental group than the control group (638 g/kg versus 579 g/kg). When compared to both the positive and negative control groups (887.6 g; 918.7 g), chickens who were fed Moringa oleifera, a plant that contains quercetin and kaempferol, gained significantly more weight at 21 days (low: 928 g; median: 932.5 g; high: 954.6 g). Furthermore, feed conversion improved (1.47; 1.44; 1.45) compared with the positive control group with a conversion rate of 1.53. Additionally, adding 0.2% plant extracts from *Chelidonium majus*, *Lonicera japonica* and *Saposhnikovia divaricata* to the chicken diet produced significant weight increases (1949 g, 1930 g, 1930 g versus 1845 g) and daily gains of approximately 3 g compared to the control group [86]. These plant extracts contained flavonoids, tannins, phenolic compounds, saponins, terpenoids and essential oils. A heat stressed chicken (32–40 °C) that was fed a diet enhanced with dried peppermint and ginkgo biloba (a source of flavonoids) also showed an increase in body weight and a decrease in feed conversion when compared to birds fed individual herbs or vitamin C supplements. However, when chickens were fed grape seed extract (7.2 g/kg diet) for 21 days, body weight decreased significantly [87].

8.5 The Impact of Silymarin on Quail

During a toxicity study of Japanese quails exposed to carbon tetrachloride, two doses of silymarin (0 and 1 ml/kg) were administered, along with 2 dosages of carbon tetrachloride (0 and 1 ml/kg). Blood serum protein levels decreased after 0 ml/kg of silymarin was administered in conjunction with 1 ml/kg of carbon tetrachloride [88]. As a contrast, combining 1 mg of silymarin with 0 mg of carbon tetrachloride increased serum protein levels (p 0.05). In a 35 day study of silymarin, total antibodies and IgG concentrations tended to be elevated ($p = 0.05$). In contrast, carbon tetrachloride reduced IgG levels on day 35 (p 0.05), but did not significantly affect other parameters of the study. After the rearing period, no difference was observed in the levels of IgG, IgM or total antibodies. In dogs treated with silymarin, the bursa Fabricius and spleen weights decreased, whereas the thymus weight increased. However, carbon tetrachloride treatment resulted in increased spleen and thymus weights (p 0.05). A higher percentage of white blood cells was present in birds treated with silymarin (p 0.05). As a result of studying silymarin in Japanese quail during oxidative stress, this study found that its administration could positively influence their humoral immune response [89]. Researchers examined the effect of silymarin on oxidative stress induced by carbon tetrachloride in Japanese quails, measuring the tibia bone and blood parameters before and after treatment. As part of the treatments, (1) a standard diet was served, (2) silymarin was added to the control diet at 1 ml/kg body weight, (3) carbon tetrachloride was added to the control diet at 1 ml/kg body weight and (4) a combined treatment group received silymarin and carbon tetrachloride at 1 ml/kg body weight, in addition to the previous treatment groups. As part of the breeding period, silymarin was administered every third day via gavage, while carbon tetrachloride was injected intraperitoneally every third day [89]. A significant increase in serum calcium and vitamin D3 levels was found with silymarin, while a significant reduction in serum glucose levels was found (*p* 0.05). As a result of the combination of carbon tetrachloride and silymarin, cholesterol and alkaline phosphatase levels were the lowest (*p* 0.05). In the silymarin group, high-density lipoprotein cholesterol (HDL-C) was lowest and in the carbon tetrachloride group, triglycerides were lowest (*p* 0.01). The effect of silymarin on tibial properties is statistically significant (p 0.05) due to its thickened inner layer and increased percentage of tibia ash [90]. Also, both silymarin and carbon tetrachloride treatment groups displayed the lowest weights and the largest tibial circumferences (*p* 0.05). A significantly lower tibial bone density was also observed after exposure to carbon tetrachloride (*p* 0.05). Japanese quail were subjected to a separate study to assess whether silymarin might reduce oxidative stress induced by carbon tetrachloride. During this study, silymarin (0 and 1 ml/kg of body weight) and carbon tetrachloride (0 and 1 ml/kg of body weight) were administered at different levels [91]. According to the findings, silymarin reduced bilirubin levels, malondialdehyde levels, liver enzymes ALT and ALP and blood cholesterol levels, including total cholesterol, low-density lipoprotein (LDL) and triglycerides. Additionally, it significantly decreased blood serum levels of albumin, total protein, superoxide dismutase (SOD), total antioxidants and glutathione peroxidase (GPx). To further investigate the influence of silymarin on the weight of lymphatic organs and serum protein in Japanese quail experiencing oxidative stress from carbon tetrachloride, an experiment was designed using a factorial arrangement of 2×2, where two levels of silymarin (0 and 1 ml/kg of body weight) and two levels of carbon tetrachloride (0 and 1 ml/kg of body weight) were administered. As a result of silymarin treatment, total protein levels significantly changed (*p* 0.05). A significant decrease in the thymus weight was also observed, whereas an increase in the spleen weight (*p* 0.05) was noted with carbon tetrachloride [92]. According to these findings, silymarin may modulate the immune system of Japanese quails when they are exposed to oxidative stress induced by carbon tetrachloride. As part of another experiment, silymarin was tested on Japanese quail poisoned with carbon tetrachloride (1 ml/kg body weight) for effects on serum lipid concentrations. We used a 1 day-old Japanese quail in a factorially designed study with a completely randomized design. It was shown that silymarin reduced triglycerides, total cholesterol and LDL cholesterol (*p* 0.05), whereas carbon tetrachloride increased triglycerides ($p = 0.001$). This study utilized 480 one-day-old Japanese quail as the experimental subjects. A two-by-two factorial design was used to administer four treatments and four replications to the Japanese quail. Controlled and silymarin treatments (100 mg versus 1 mg/kg body weight administered via gavage) were administered, as well as carbon tetrachloride (1 mg/kg body weight administered intraperitoneally). As part of the combined silymarin and carbon tetrachloride treatment (100 mg of silymarin per kilogram in the blood and 1 mg of carbon tetrachloride per kilogram in the blood), each child received 100 mg of silymarin. One ml/kg of silymarin without carbon tetrachloride significantly increased devilli length and the ratio of devilli length to crypt depth after treatment with silymarin. As a result of the use of silymarin, intestinal length and weight were significantly increased (p 0.05). The experimental treatments had no significant effect on the muscle layer thickness or the crypt depth in the jejunum, however [93]. There was a significant increase in villi length, villi width and villi length to crypt depth among birds treated with silymarin. In a study using silymarin and zero carbon tetrachloride, villi length and villi-crypt depth ratios were significantly improved. Aflatoxin-fed chickens receiving milk thistle experienced improved blood parameters than those fed just aflatoxin, which is due to silymarin's powerful antioxidant properties. As a result, intestinal cholesterol absorption may be reduced, blood sugar levels may be reduced and liver metabolism may be improved. Milk thistle's high raw fibre content in the diet contributes to a reduction in blood parameters, which leads to an increase in bile and bile acids excreted as well as natural steroids secreted, which in turn reduces cholesterol levels [94].

9 Antioxidants in Microbes

9.1 Classification and Characteristics of Microbial Extracellular Polymeric Substances (EPSs)

EPSs (exopolysaccharides) are substances produced by microorganisms that come in various chemical compositions. Hemopolysaccharides and homopolysaccharides are the most common types of EPS (Fig. 2).

9.2 Homopolysaccharides Are Polysaccharides Composed of a Single Type of Monosaccharide Unit

An homopolysaccharide is formed by repeating units of only one monosaccharide. In yeast, there are three types of macro soluble sugars: α-d-glucans, β-d-glucans and fructans. Dextran and mutans belong to the group of compounds known as α-d-glucans. The homopolysaccharide mutan differs from dextran in that it cannot

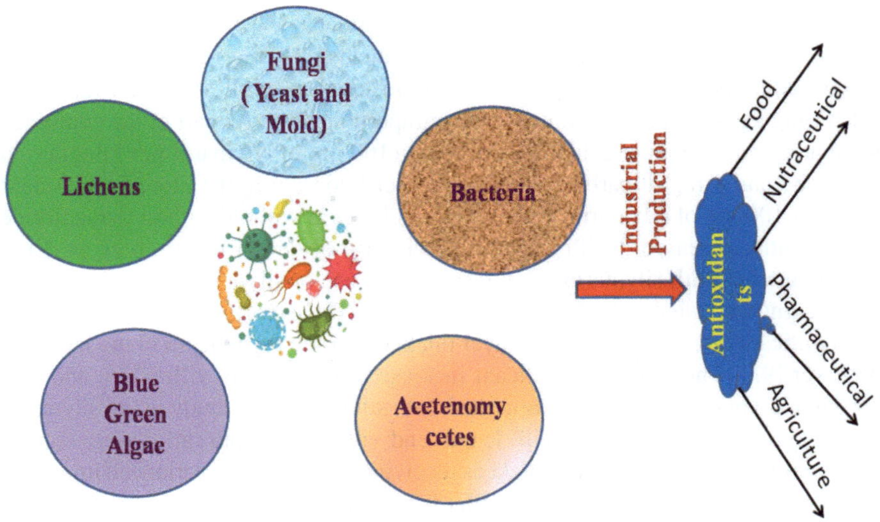

Fig. 2 Review of microbiological production and its uses

be dissolved in water. *Leuconostoc bacterium, Lactobacillus bacterium, Limoslactobacillus bacterium, Weissella bacterium, Streptococcus bacterium, Lentilactobacillus bacteria* and *Latilactobacillus bacterium* produce dextran in different quantities [95]. In addition to *Streptococcus downei, Streptococcus mutans, Streptococcus salivarius* and *Limosilactobacillusreuteri*, there are many types of bacteria that produce mutants. In addition to alternan and reuteran, there are other types of α-d-glucans that are soluble in water. A variety of bacteria produce alternan, including *Leuconostocmesenteroides, Leuconostocacidum* and *Streptococcus salivarius*, while *Limosilactobacillusreuteri* is also known to produce reuteran. Solvent—soluble curdlan, cellulose and salecan are unbranched glucans. Curdlan is produced by *Agrobacterium* sp. ATCC 31,749, *Cellulomonas flavigena, Alcaligenes faecalis, Rhizobium meliloti* and *Agrobacterium radiobacter*. Cellulose is produced by Acetobacter, Pseudomonas, Agrobacterium and Rhizobium. Species of Agrobacterium. HoPS can be synthesized by ZX09 from salecan [96]. A fructan is a molecule that is water-soluble, such as levan, inulin and fructooligosaccharides. Among the strains of Limosilactobacillusreuteri and L. mesenteroides that produce levan are *Streptococcus mutans, Bacillus subtilis* and *Streptococcus salivarius*. As for *Streptococcus mutans, Limosilactobacillusreuteri, Leuconostoccitreum* and *Lactobacillus johnsonii*, they are capable of making inulin as well. A culture of *Lactobacillus reuteri* 121 contains fructooligosaccharides as well [97].

9.3 Heteropolysaccharides

Polysaccharides (HePSs) are complex because they contain more than one type of monosaccharide. It is possible for polysaccharides to contain pentose, hexose, as well as N-acetylated monosaccharides and non-carbohydrate components, which can be branched or unbranched. HePSs usually have a molecular weight of 10^4–10^6 Da. Magellan, diutan and well all have l-rhamnopyranose, d-glucopyranuronic acid and d-glucopyranose as their backbones [98]. The gellan structure is linear, whereas the wellan and diutan structures are branched. Water-soluble polysaccharides have been documented in Sphingomonas and Pseudomonas strains. HePS containing d-glucopyranose and d-galactopyranose, such as kefiran, is another branched HePS. A range of *Lactobacillus* species can produce this water-soluble polysaccharide, such as *A. kefir*, *L. parakefir*, *L. kefiranofaciens*, *L. kefirgranum* and *L. delbrueckii* subsp. bulgaricus. In terms of its backbone, xanthan has the same shape as cellulose, while it has branched structures composed of mannopyranose and glucopyranuronic acid. By producing *Xanthomonas campestris* and *Sphingomonaspaucimobilis*, xanthan becomes a water-soluble exopolysaccharide. In addition to being water-soluble, d-glucuronic acid and d-N-acetylglucosamine are also present in hyaluronic acid [99]. The bacteria produce this natural antibiotic in the body such as *Streptococcus equi*, *Streptococcus influenzae*, *Streptococcus pneumoniae*, *Pasteurella multocida* and *Cryptococcus neoformans*. The linear HePS alginate, composed of d-mannopyranuronic acid and l-guluronic acid subunits, is also produced by *Pseudomonas aeruginosa* and *Azotobacter*.

9.4 Recent Developments in the Biological Functions of Microbial Exopolysaccharides

In addition to exhibiting considerable diversity, exopolysaccharides from microorganisms are multifunctional carbohydrates that can provide substantial health benefits. In recent years, researchers have demonstrated that microbial EPS's have remarkable health benefits, because of their distinct characteristics compared with other synthetic polysaccharides. At present, microorganisms contribute a significant amount to commercially available EPS's. EPS's from microorganisms have several advantages, including customizable chemical composition and structure, which explains why they are in high demand in pharmaceutical and medical industries [100]. Microbial EPSs might offer a viable alternative to conventional chemicals such as antibiotics, antioxidants, anticancer agents, antivirals and antifungals. In addition to being non-toxic, biocompatible, thermally stable and biodegradable, microbial EPSs are biocompatible and biocompatible. By applying antibacterial or antifungal EPSs, the antagonistic activity of normal flora against pathogens may be maintained. Antibiotic treatments may diminish this activity. Additionally, microbial EPSs may provide safer cancer therapy options than synthetic anticancer treatments since their adverse effects are reduced, and they stimulate the immune system [101]. Aside from contributing to the alleviation of oxidative stress, EPSs are also considered promising eco-friendly alternatives to synthetic antioxidants since they help to scavenge a wide range of free radicals, inhibit lipid peroxidation, chelate metal ions and enhance both enzymatic and non-enzymatic antioxidant functions. There is a significant impact on the functional activities of EPSs the physicochemical properties, such as the molecular weight, the degree of branching, the composition of the monosaccharides, the glycosidic linkages, the electric charge and the functional groups. Identifying the link between structure and function may lead to improved bioactivity by modifying identified EPSs [102].

10 Natural Source Antioxidants

Antioxidants are crucial components of dietary supplements that aim to promote health and prevent serious disease. Scientists are increasingly interested in discovering natural antioxidant alternatives because of their carcinogenic and toxic properties. As a result of these explorations, vitamin A, C and E, have been recognized as effective antioxidants. Natural antioxidants are found to reduce the risk of cardiovascular diseases in vegetables, fruits, tea and wine, according to epidemiological research [103]. Natural antioxidants are increasingly discussed in relation to human health and nutrition due to their potentially beneficial properties. Vitamin C, vitamin E, tocopherol, beta-carotene and other phenols in diet are antioxidant phytochemicals that provide protection against free radicals. There is a wide range of variability in the antioxidant efficacy of these phytochemicals, influenced by the reactive groups present in them. Among the antioxidant properties of vitamin C are its high concentration of reactive hydroxyl groups, unlike that of vitamin E and A. In addition to enhancing health, antioxidants derived from plants and fungi can assist in managing a variety of diseases [104]. As a natural antioxidant source, mushrooms, medicinal herbs, spices, vegetables and fruits are among the most documented species. In particular, phenolic compounds inhibit lipid peroxidation, reverse hydroperoxide degradation and chelate metal ions. They are excellent hydrogen donors. As antioxidants, phenolic compounds that originate in nature are generally superior to synthetic compounds. There are several fungi known to produce gallic acid, ferulic acid and ellagic acid, so fermentation by microbes may also prove to be a useful means of producing these beneficial compounds [105].

11 Microbial—Derived Antioxidants

Innovative biotechnology relies on the abundance of novel chemicals produced by microbial diversity. Actinomycetes produce 42% of the identified microbial secondary metabolites, fungi produce a similar amount (42%) and Eubacteria produce 16%. In recent years, microbial metabolites have been used to manufacture lifesaving pharmaceuticals, including amphotericin, erythromycin, penicillin, streptomycin, tetracyclines and vancomycin, to treat bacterial and fungal infections, cancer, transplant rejection, high cholesterol (lovastatin and mevastatin) and cancer rejection (bleomycin, doxorubicin, daunorubicin and mitomycin). In addition to this, some of these primary and secondary metabolites are also antioxidants [106].

11.1 Microbial Populations in Red Raspberry Enzymes

Among the bacterial phyla identified from all samples were Firmicutes, Actinobacteria, Verrucomicrobia, Gemmatimonadetes, Cyanobacteria, Bacteroidetes, Chloroflexi, Planctomycetes, Proteobacteria and Acidobacteria. As fermentation progressed, Firmicutes grew significantly, and the other phyla were far less prevalent. During the opening week of fermentation, the proportion of Firmicutes in M samples reached 40%, then declined slightly before stabilizing around 30% over the remainder of the process. In contrast, after 60 days of fermentation, Firmicutes proportions increased in the S samples to 26.1%, up from 11.7% in the first week. After the raw materials are mixed, the bacteria community is evolving until the ripening process begins, during which the bacterial composition diversifies and stabilizes until the process is concluded [107]. Several distinct phyla of fungal species were identified based on the sequence analysis. All samples were dominated by Ascomycota, except for fresh samples and S samples fermented for more than 15 days, which contained significant amounts of fungi from an unknown phylum. In fermented alcoholic beverages, Ascomycota dominate the fungi community. As well as Basidiomycota and Mortierellomycota contributing to the community, other phyla also might have influenced spontaneous fermentation. Analysing the genus-level composition of bacteria and fungi associated with enzyme fermentation provided further insight into their dynamics. It was found that 1,014 bacterial genera were identified during fermentation, with the top 30 genera highlighted. By using the universal 16S rRNA primers, we were able to detect and amplify the Rubus hybrid cultivar gene while analysing bacterial diversity in the F samples and early fermented S samples. Further, Lactobacillus dominated early fermentation stages in the M group due to the incorporation of Lactobacillus plantarum starter cultures [108]. As part of food fermentation, lactobacillus produces several antimicrobial metabolites, including organic acids that contribute to amino acid metabolism. As fermentation progressed, the ratio of Lactobacillus to Rhodococcus gradually increased, though there were fluctuations in both genera' proportions throughout. According to speculation, these fluctuations were caused by changes in pH and alcohol content caused by fermentation. As well as Acinetobacter, other notable genera present in the samples included Akkermansia, Streptococcus and Prevotella. There were 1032 fungal genera identified during fermentation. Around 43.8% of the F group consists of the genus Cladosporium. This genus produces benzofuran derivatives and isocoumarin compounds through fermentation, which have notable antibacterial properties. Also, Cladosporium is an enzyme contributor that contributes aroma and taste to enzymes and inhibits the growth of harmful microbes during fermentation. The M group's fungi were predominantly from the genus Kodamaea, primarily due to the inoculation of K. ohmeri WP-1 as starters. A common source of Kodamaeaohmeri is naturally fermented foods such as bacon, pickles and cheese, which possess the typical fermentation flora and are capable of degrading biogenic amines. Despite 90 days of fermentation, Kodamaea remained the predominant fungus in M90, followed by Fusarium, Mycosphaerella and Aspergillus. S7 and S15 samples also contained a notable presence of Saccharomyces. When Saccharomyces autolyzes, it releases various substances, such as sugars, proteins, amino acids organic acids, vitamins and inorganic salts. These substances serve as carbon and nitrogen sources for fermentation, thereby facilitating the growth of other microorganisms. Fermented products are often enhanced by Saccharomyces, according to previous studies. A reduction in the proportion of the original primary genera was observed as fermentation progressed, while an increase in the proportion of Fusarium and Mortierella took place. It is believed that Fusarium promotes the circulation of biological substances within fermenting environments by breaking down cellulose and degrading organic matter [109].

11.2 Technologies and Challenges in the Production of Microbial Pigments

As a food additive and colouring agent, microbial pigments are produced from algae, fungi and bacteria. To produce industrial goods, these pigments must be extracted. In a few research studies, it has been shown that optimal pigment-producing microbes can utilize a diverse range of carbon and nitrogen sources, as well as being tolerant of changes in pH, temperature and mineral content, while also yielding a substantial colour output. Coloured microbes can be obtained from a variety of environmental locations after which they can be cultivated and purified. Recent reviews suggest that using agro-industrial residues to cultivate these microbes might be an effective and inexpensive alternative to synthetic growth media. In addition to high hydrostatic pressure (HHP), pulse electric field (PEF), sonication-assisted extraction, gamma irradiation, enzymatic extraction and membrane technology, there are also other advanced techniques for pigment extraction available today. Genetic engineering has been observed to enhance microbial pigment production, contrary to conventional chemists' scaling-up methods. In addition to temperature, pH, carbon source, fermentation type, mineral content, nitrogen source, moisture level and aeration rate, other important factors impact microbial pigment production. Microorganisms must be non-pathogenic and non-toxic, capable of utilizing a wide range of carbon and nitrogen sources, as well as producing an attractive colour effect, and must be tolerant of salt concentrations, temperatures and pH. To minimize the number of steps involved in the extraction process [110], the process should be streamlined.

11.3 Recent Advancements in Technologies to Produce Microbial Pigments

Statistical methods are used to design a number of factors that affect microbe pigment production, such as the medium composition, stimulators and optimal culture conditions [110]. It has been demonstrated that response surface methodology (RSM) and artificial neural networks (ANN) can be used to enhance the synthesis of microbial pigment. An assessment of medium component impacts on pigment production is conducted using the Plackett–Burman design, followed by the optimization of significant factors through Central Composite Design, with regression analysis used to analyse their responses. As a result, large-scale production processes can be developed using this methodology. Using immobilized strains may additionally promote cell proliferation and enhance metabolic activity, which can help optimize pigment production. An increase in prodigiosin yield has been demonstrated by using a solid carrier in the fermentation of the compound. Furthermore, osmotic pressure and elevated temperatures can enhance pigment production among microbes due to their adaptability through limited growth [111].

11.4 Challenges

Despite manufacturers' best efforts, the extraction of highly concentrated pigments in a purified form remains challenging. Extraction and purification of intracellular pigments require further cell disruption, resulting in longer downstream processing times. Moreover, environmental stressors such as UV exposure or environmental stress compromise microbial pigment stability. Microbes hold the key to increasing pigment yields while maintaining cost-effective manufacturing processes [112]. This is the industry's greatest challenge. Extracting and purifying pigments presents many obstacles and limitations. It is possible to create undesirable isomers by using high temperatures and solvents for long periods of time. Because enzymes are expensive, they raise concerns as an alternative. Additionally, the FDA does not approve some microbial pigments, including Monascus pigments, which are azaphilone pigments mixed. Using these pigments in the food industry has been controversial due to their contents of mycotoxin citrinin and other toxic substances [113].

11.5 The Antioxidant Characteristics of Microbial Pigments and Their Advantages

Free radicals are captured by antioxidants, which makes them unique. Besides neutralizing photosensitizers and interacting with singlet oxygen, they eliminate peroxy radicals as well. Free radicals are implicated in a wide range of chronic illnesses such as cancer, diabetes, cardiovascular disease and autoimmune disorders [114]. Increasing the use of microbial pigments in food products has demonstrated beneficial effects as antioxidants. Biochemical processes have contributed to carotenoids, naphthoquinone and violacein's antioxidant properties. As well, xanthomonadin inhibits photodynamic lipid peroxidation in liposomes, which is another example of antioxidant activity. A research study has shown that vitelacein enhances mucosal defence mechanisms in gastric ulcers, and thus is beneficial in the prevention of oxidative damage [115]. Atherosclerosis, cataracts, age-related macular degeneration and multiple sclerosis can be prevented with the help of these pigments. Xanthophyll, lutein and zeaxanthin, which are antioxidants found in corn, spinach and kale, are important in the prevention of AMD, a leading cause of blindness in humans. An antioxidant that is beneficial in the prevention of disease, in enhancing the immune system, in preventing cataracts, and in combating Helicobacter pylori, astaxanthin is a powerful antioxidant [116]. Cardiovascular diseases can be mitigated, the immune system strengthened, sunburn can be prevented and certain types of cancer can be inhibited by carotenoids. In a recent study, researchers found that carotenoids isolated from M. maritimus and M. maritimus possess UV-protective, antioxidant and antibacterial properties. A similar mechanism is used by lycopene to produce oxidized low-density lipoprotein cholesterol that prevents arteriosclerosis and coronary disease. It has been well documented that prodigiosin acts as an anticancer pigment in microbial cultures [117].

12 Key Microbial Pigments in the Food Industry

- *β-Carotene*

  Various hues of yellow, orange and red are associated with carotenoids. In their production, Serratia, Micrococcus, Mycobacterium, Agrobacterium, Sulfolobus and Streptomyces play a role. For carotenoid production, agro-industrial by-products such as cheese whey and sugarcane molasses have been gaining popularity, along with glucose syrup, peat hydrolysate and cellobiose [118]. In addition to serving as an antioxidant, acarotene has been shown to be beneficial in treating certain diseases. Studies indicate that Blakesleatrispora, Mucor circinelloides and Phycomycesblakesleeanus produce most of this fungus. Vegetable oils, orange beverages, margarine and microencapsulated bead lets commonly contain this compound [119].
- *Arprink Red*

  In soil, Penicillium oxalicum is isolated from Arprink red to produce the red pigment. As part of its Codex Alimentarius Commission, arprink red levels in various foods have been established [120].
- *Riboflavin*

  Some microbial species produce riboflavin, which is a yellow, water-soluble vitamin. In commercially produced fermentations, Ashbya gossypii is used, which is a high yielding, genetically stable ascomycete, as well as filamentous fungi like

Candida famata and the bacteria Bacillus subtilis. Overproduction of riboflavin can be categorized into weak, moderate and strong overproduction. Candida flaleri, Saccharomyces cerevisiae and Bacillus subtilis are examples of strong producers. Among its many uses, riboflavin is found in cereals, pastas, sauces, dairy products, juices, energy drinks and juice drinks as an additive [121].

- *Monascus Pigments*

  An Ascomycete family, Monascaceae, produces an ascomycete pigment called monascus. As far as the food industry is concerned, the main strains used are M. plurosus, M. sanguineus, M. sanguineus and M. ruber flora. A food product's colour is influenced by these pigments [122].

12.1 Microbial Synthesis of Polyphenolic Compounds

A polyphenol is a natural antioxidant that consists of phenolic acids, flavonoids and stilbenes. In nature, there are thousands of unique varieties of polyphenols. The phenylpropanoid pathway produces most of these compounds as secondary metabolites. It can be difficult for traditional extraction methods to isolate and purify polyphenols because they are often conjugated with one another or with organic acids and carbohydrates. As a result of genetically enhanced microbial hosts like *Escherichia coli* and *Saccharomyces cerevisiae* reconfiguring specific biosynthetic pathways, various polyphenols have been heterologously produced, some of which have already demonstrated large-scale potential [123].

12.2 Microbial Synthesis of Carotenoids

Lycopene and beta-carotene are examples of carotenoids that plants and microorganisms produce naturally. An isopentenyl diphosphate precursor (IPP) is converted to its isomer dimethylallyl diphosphate (DMAPP) through condensation reactions [124]. In most eukaryotic organisms, the mevalonate (MVA) pathway produces IPP and DMAPP, and the 2-C-methyl-D-erythritol 4-phosphate (MEP) pathway generates DMAPP. In the early stages of heterologous carotenoids production, yeast was used, but the best success has been achieved through metabolic engineering of *E. coli*, leading to significant yields of carotenoinds [125]. It is noteworthy that *E. columbachia* contains both the MEP pathway and the prenyl-diphosphate pathway in natural form. Phytoene synthase (crtB) and phytoene desaturase (crtI) from Erwinia are both sufficient for producing lycopene at low levels. The introduction of a lycopene β-cyclase (LYC-β) further extended the metabolic pathway to produce β-carotene. There is an enzyme that is responsible for limiting the rate of prenyl-diphosphate synthesis: geranylgeranyl diphosphate synthase (GGPP). GGPP and DMAPP levels are increased when both native enzymes as well as exogenous isoenzymes are overexpressed or incorporated. IPP and DMAPP levels were also increased when the enzymes involved in the MEP pathway were overexpressed [126]. A variety of innovative biotechnological methods have been employed for the production of carotenoids besides the conventional optimization strategies mentioned above. A notable method involved modifying carbon flux towards a heterologous pathway in response to cellular metabolic conditions by implementing an artificial dynamic regulatory circuit. Through this method, lycopene levels were increased from negligible levels to 150 mg/L, and adverse metabolic effects were greatly reduced [127]. In addition to gene knockouts, other techniques were employed. Using a systematic computational search coupled with random transposon mutagenesis, multiple targets were identified, leading to efficient triple knockout strains that produced lycopene in high levels at 18 mg/g dry weight. Using a shotgun library, positive lycopene-producing genes were identified in *E. Coli.* Multiplex automated genome engineering (MAGE) was developed to increase the efficiency of lycopene production by generating targeted genomic diversity [128]. The recent advancement of optimizing carbon sources with the addition of sugars and amino acids led to the production of 1.35 g/L of lycopene and 2.47 g/L of β-carotene. A number of other carotenoids, such as zeaxanthin, astaxanthin and lutein, have been produced heterologously by expanding the existing platforms, as discussed in several other reviews [129].

12.3 Microbial Synthesis of Antioxidant Vitamins

Known for its antioxidant properties, coenzyme Q10 plays an important role in the transport of electrons within cells and is a vitamin-like ubiquinone [130]. Fermentative methods were first used with mutant strains of natural producers to produce its biosynthesis, which could reach 350 mg/L concentrations. By expanding the MVA/MEP pathway to synthesize decaprenyl diphosphate (DPP), microorganisms can enhance their biosynthesis by identifying native enzymes [131]. This can then be enhanced by expressing ubiquinone-related genes (ubiA, D, B, G, H, E, F) for the synthesis of coenzyme Q10. Agrobacterium tumefaciens contributed decaprenyl diphosphate synthase gene (ddsA gene) along with the heterologous MVA pathway from Streptococcus pneumoniae and Ralstoniaeutropha to the production of DPP. A combination of this approach and the presence of 4-hydroxybenzoate (4HB) resulted in coenzyme Q10 production at a yield of 2.4 mg/g DCW in E. Thermophilus DH5a was used as a model organism. A metabolic engineering method for the synthesis of vitamin E (tocopherols) has not been described, although synthesis of vitamin C (ascorbate) has been demonstrated in yeast metabolically engineered with glucose [132].

13 Conclusion

The metabolism and mechanism of antioxidants in plants, animals and microbes are very complex to understand by simply emphasizing the widespread significance of these biomolecules in maintaining cellular health through various chemical reactions by the generation of reactive oxygen species and thereby preventing the plants from various external and internal stress factors and continued existence beneath stress. Each group of organisms has developed convoluted antioxidant systems both enzymatic and non-enzymatic to neutralize reactive oxygen and nitrogen species, ensuring redox homeostasis. In plants, antioxidants are key to stress resilience and defence responses, in animals, they protect against cellular damage and support physiological functions and in microbes, they aid in survival in hostile environments. A comprehensive understanding of these antioxidant systems not only deepens our knowledge of biological stress responses but also opens avenues for future advancements in healthcare, crop improvement and industrial biotechnology.

References

1. Kulczyński, B., Sidor, A., & Gramza-Michałowska, A. (2019). Characteristics of selected antioxidative and bioactive compounds in meat and animal origin products. *Antioxidants, 8*, 335.
2. Zeb, A. (2020). Concept, mechanism, and applications of phenolic antioxidants in foods. *Journal of Food Biochemistry, 44*, e13394.
3. Abeyrathne, E. D. N. S., Nam, K. C., & Ahn, D. U. (2021). Analytical methods for lipid oxidation and antioxidant capacity in food systems—A review. *Antioxidants, 1*(10), 1587.
4. Orian, M., & Escriche, I. (2015). Antioxidants: Characterization, natural sources, extraction, and analysis. *Food Research International, 74*, 10–36.
5. Alshkh, N., de Camargo, A. D., & Shahidi, F. (2015). Phenolics of selected lentil cultivars: Antioxidant activities and inhibition of low-density lipoprotein and DNA damage. *Journal of Functional Foods, 18*, 1022–1038.
6. Single, R. K., Dubey, A. K., Garg, A., Sharma, R. K., & Fiorino, M. (2019). Natural polyphenols: Chemical classification, definition of classes, subcategories, and structure. *Journal of AOAC International, 102*, 1397–1400.
7. Pandey, K. B., & Rizvi, S. I. (2020). Resveratrol may protect plasma proteins from oxidation under conditions of oxidative stress in vitro. *Journal of the Brazilian Chemical Society, 21*, 909–913.
8. Nagapan, T. S., Ghazali, A. R., Basri, D. F., & Lim, W. N. (2018). Photoprotective effect of stilbenes and its derivatives against ultraviolet radiation-induced skin disorders. *Biomedical and Pharmacology Journal, 11*, 1199–1208.
9. Rosalina, R., & Weerapreeyakul, N. (2021). An Insight into sesamolin: Physicochemical properties, pharmacological activities, and future research prospects. *Molecules, 26*, 5849.
10. Moreira, A. S. P., Nunes, F. M., Domingues, M. R., & Coimbra, M. A. (2012). Coffee melanoidins: Structures, mechanisms of formation and potential health impacts. *Food & Function, 3*, 903–915.
11. Oswell, N. J., Thippareddi, H., & Pegg, R. B. (2017). Practical use of natural antioxidants in meat products in the US: A review. *Meat Science, 145*, 469–479.
12. Huang, X., & Ahn, D. U. (2019). Lipid oxidation and its implications to meat quality and human health. *Food Science and Biotechnology, 28*, 1275–1285.

13. Park, Y. W., & Nam, M. S. (2015). Bioactive peptides in milk and dairy products: A review. *Korean Journal for Food Science of Animal Resources, 35*, 831–840.
14. Albenzio, M., Santillo, A., Caroprese, M., della Malva, A., & Marino, R. (2017). Bioactive peptides in animal food proteins. *Foods, 6*, 35.
15. Barreca, D. (2021). Mechanisms of plant antioxidants actions. *Plants, 10*, 35.
16. Bast, A., & Haenen, G. R. (2013). Ten misconceptions about antioxidants. *Trends in Pharmacological, 34*, 430–436.
17. Gulcin, İ. (2020). Antioxidants and antioxidant methods: An updated overview. *Archives of toxicology, 94*(3), 651–715.
18. Lobo, V., Patil, A., Phatak, A., & Chandra, N. (2010). Free radicals, antioxidants and functional foods: Impact on human health. *Pharmacognosy reviews, 4*(8), 118.
19. Hsieh, C., & Rajashekaraiah, V. (2021). Ferric reducing ability of plasma: A potential oxidative stress marker in stored plasma. *Acta Haematologica Polonica, 52*(1), 61–67.
20. Mason, S. A., Trewin, A. J., Parker, L., & Wadley, G. D. (2020). Antioxidant supplements and endurance exercise: Current evidence and mechanistic insights. *Redox biology, 35*, 101471.
21. Bose, J., Pottosin, I., Shabala, S. S. S., Palmgren, M. G., & Shabala, S. (2011). Calcium efflux systems in stresssignaling and adaptation in plants. *Frontiers in Plant Science, 85*, 1–17.
22. Flowers, T. J., & Colmer, T. D. (2008). Salinity tolerance in halophytes. *New Phytologist, 179*, 945–963.
23. Kronzucker, H. J., & Britto, D. T. (2011). Sodium transport in plants: A critical review. *New Phytologist, 189*, 54–81.
24. Manishankar, P., Wang, N., Koster, P., Alatar, A. A., & Kudla, J. (2018). Calcium signaling during salt stress and in the regulation of ion homeostasis. *Journal of Experimental Botany, 69*, 4215–4226.
25. Nouri, H., Borujeni, S. C., Nirola, R., Hassanli, A., Beecham, S., Alaghmand, S., & Mulcahy, D. (2017). Application of green remediation on soil salinity treatment: A review on halophytoremediation. *Process Safety and Environmental Protection, 107*, 94–107.
26. Apel, K., & Hirt, H. (2004). Reactive oxygen species; metabolism, oxidative stress and signal transduction. *Annual Review of Plant Biology, 55*, 373–399.
27. Arora, A., Sairam, R. K., & Srivastava, G. C. (2002). Oxidative stress and antioxidative system in plants. *Current Science, 82*, 1227–1273.
28. Baxter, A., Mittler, R., & Suzuki, N. (2014). ROS as key players in plant stress signalling. *Journal ofExperimental Botany, 65*, 1229–1240.
29. Bhattacharjee, S. (2014). Membrane lipid peroxidation and its conflict of interest: The two faces of oxidativestress. *Current Science, 107*, 1811–1823.
30. Bienert, G. P., & Chaumont, F. (2014). Aquaporin-facilitated transmembrane diffusion of hydrogen peroxide. *Biochemical et Biophysical Acta, 1840*, 1596–1604.
31. Brunetti, C., Di Ferdinando, M., Fini, A., Pollastri, S., & Tattini, M. (2013). Flavonoids as antioxidants anddevelopmental regulators: Relative significance in plants and humans. *International Journal of Molecular Sciences, 14*, 3540–3555.
32. Choudhury, F. K., Rivero, R. M., Blumwald, E., & Mittler, R. (2017). Reactive oxygen species, abiotic stress andstress combination. *Plant Journal, 90*, 856–867.
33. Chung, W. H. (2017). Unraveling new functions of superoxide dismutase using yeast model system: Beyond itsconventional role in superoxide radical scavenging. *Journal of Microbiology, 55*, 409–416.
34. Corpas, F. J., Barroso, J. B., Palma, J. M., & Rodriguez-Ruiz, M. (2017). Plant peroxisomes: A nitro-oxidativecocktail. *Redox Biology, 11*, 535–542.
35. Couto, N., Wood, J., & Barber, J. (2016). The role of glutathione reductase and related enzymes on cellular redoxhomoeostasis network. *Free Radical Biology and Medicine, 95*, 27–42.
36. Das, K., & Roychoudhury, A. (2014). Reactive oxygen species (ROS) and response of antioxidants as ROSscavengersduring environmental stress in plants. *Frontiers in Environmental Science, 2*, 53.
37. del Rio, L. A. (2015). ROS and RNS in plant physiology: An overview. *Journal of Expermiental Botany, 66*, 2827–2837.

38. Das, P., Nutan, K. K., Singla-Pareek, S. L., & Pareek, A. (2015). Oxidative environment and redox homeostasisin plants: Dissecting out significant contribution of major cellular organelles. *Frontiers in Environmental Science, 2*, 70.
39. De Gara, L., & Foyer, C. H. (2017). Ying and Yang interplay between reactive oxygen and reactive nitrogenspecies controls cell functions. *Plant Cell and Environment, 40*, 459–461.
40. del Rio, L. A., & Lopez-Huertas, E. (2016). ROS generation in peroxisomes and its role in cell signalling. *Plantand Cell Physiology, 57*, 1364–1376.
41. El-Beltagi, H., & Mohamed, H. I. (2013). Reactive oxygen species, lipid peroxidation and antioxidative defensemechanism. *Notulae Botanicae Horti Agrobotanici Cluj-Napoca, 41*, 44–57.
42. Foyer, C. H., & Noctor, G. (2016). Stress-triggered redox signalling: What's in pROSpect? *Plant Cell and Environment, 39*, 951–964.
43. Gharechachi, J., Sharifi, G., Komatsu, S., & Salekdeh, G. H. (2016). Proteomic analysis of crop plants under lowtemperature: A review of cold responsive proteins. *Agricultural Proteomics, 2*, 97–127.
44. Gligorovski, S., Strekowski, R., Barbati, S., & Vione, D. (2015). Environmental implications of hydroxyl radicals (·OH). *Chemical Reviews, 115*, 13051–13092.
45. Lokhandwala, A., & Hoeksema, J. D. (2019). Priming by arbuscular mycorrhizal fungi of plant antioxidant enzyme production: A meta-analysis. *The Annual Review of Plant Biology, 2*, 1069–1084.
46. Gholinezhad, E., Darvishzadeh, R., Moghaddam, S. S., & Popovic- Djordjevic, J. (2020). Effect of mycorrhizal inoculation in reducing water stress in sesame (*Sesamum indicum* L.): The assessment of agrobiochemical traits and enzymatic antioxidant activity. *Agricultural Water Management, 238*, 106234.
47. Langeroodi, A. R., Osipitan, O. A., Radicetti, E., & Mancinelli, R. (2020). To what extent arbuscular mycorrhiza can protect chicory (*Cichorium intybus* L.) against drought stress. *Scientia Horticulturae, 263*, 109109.
48. Maheshwari, R., & Dubey, R. S. (2019). Nickel-induced oxidative stress and the role of antioxidant defence in rice seedlings. *Plant Growth Regulation, 59*, 37–49.
49. Grondin, A., Mauleon, R., Vadez, V., & Henry, A. (2016). Root aquaporins contribute to whole plant water fluxes under drought stress in rice (*Oryza sativa* L.). *Plant, Cell & Environment, 39*, 347–365.
50. He, Y., Fang, J., Xu, W., & Shi, P. (2022). Substantial increase of compound droughts and heat waves in wheat growing seasons worldwide. *International Journal of Climatology, 42*, 5038–5054.
51. Prabha, C., Sharadamma, N., & Nagesh, B. R. (2019). Arbuscular mycorrhizal influence on oxidative stress in French bean (*Phaseolus vulgaris*) under drought conditions. *GSC Biological and Pharmaceutical Sciences, 7*, 21–29.
52. Triantafyllou, K., Chang, C., & Pimentel, M. (2014). Methanogens, methane, and gastrointestinal motility. *Journal of Neurogastroenterology and Motility, 20*, 31–40.
53. Martinez-Guryn, K., Leone, V., & Chang, E. B. (2019). Regional diversity of the gastrointestinal microbiome. *Cell Host & Microbe, 26*, 314–324.
54. Kanehisa, M., & Susumu, G. (2000). KEGG: Kyoto Encyclopedia of genes and genomes. *Nucleic Acids Research, 28*, 27–30.
55. Dąbrowska, I., Grzędzicka, J., Niedzielska, A., & Witkowska-Piłaszewicz, O. (2023). Impactof chlorogenic acid on peripheral blood mononuclear cell proliferation, oxidativestress, and inflammatory responses in racehorses during exercise. *Antioxidants, 12*, 1924.
56. Duarte, M. E., Stahl, C. H., & Kim, S. W. (2023). Intestinal damages by F18+ *Escherichia coli* andits amelioration with an antibacterial bacitracin fed to nursery pigs. *Antioxidants, 12*, 1040.
57. Pastorelli, G., Benamri, R., Faustini, M., De Bellis, R., Serra, V., Turin, L., Haumont, M., Durand, P., Bianchessi, L., & Prost-Camus, E. (2023). Partial replacement of synthetic vitamin E by polyphenols in post-weaning piglets. *Antioxidants, 12*, 1752.

58. Karageorgou, A., Tsafou, M., Goliomytis, M., Hager-Theodorides, A., Politi, K., & Simitzis, P. (2023). Effect of dietary supplementation with a mixture of natural antioxidantson milk yield, composition, oxidation stability and udder health in dairy ewes. *Antioxidants, 12*, 1571.
59. Craig, A., Muns, R., Gordon, A., & Magowan, E. (2023). Extended nursing and/or increased starter diet allowances for low weaning weightpigs. *Asian-Australasian Journal of Animal Sciences, 33*, 1301–1309.
60. Guerbette, T., Boudry, G., & Lan, A. (2022). Mitochondrial function in intestinal epithelium homeostasis and modulation in diet-inducedobesity. *Molecular Metabolism, 63*, 101546.
61. Valera-Alberni, M., & Canto, C. (2018). Mitochondrial stress management: A dynamic journey. *Cell Stress, 2*, 253–274.
62. Tang, X., Liu, X., & Fang, R. (2021). Role of epidermal growth factor in regulation of intestinal health of weaned piglets. *Chinese Journal of Animal Nutrition, 33*, 614–621.
63. Tuller, J., De, S., & Jerry, D. (2014). Dietary influence of fucoidan supplementation on growth of lates calcarifer. *Aquaculture Research, 45*, 749–754.
64. Sivagnanavelmurugan, M., Thaddaeus, B., Palavesam, A., & Immanuel, G. (2014). Dietary effect of Sargassum wightii fucoidan to enhancegrowth, prophenoloxidase gene expression of Penaeus monodon and immune resistance to Vibrio parahaemolyticus. *Fish Shellfish. Immun., 39*, 439–449.
65. Kumar, S., Wang, M., Liu, Y., Zhu, Z., Fahad, S., Qayyum, A., & Zhu, G. (2022). Vanadium stress alters sweet potato (*Ipomoea batatas* L.) growth, ROS accumulation, antioxidant defense system, stomatal traits, and vanadium uptake. *Antioxidants, 11*, 2407.
66. Rishi, G., & Subramaniam, V. (2017). The liver in regulation of iron homeostasis. *The American Journal of Physiology-Gastrointestinal, 313*, 157–165.
67. Vercellino, I., & Sazanov, L. (2021). The assembly, regulation and function of the mitochondrial respiratory chain. *Nature Reviews Molecular Cell Biology, 23*, 141–161.
68. Robinson, M. W., Harmon, C., & O'Farrelly, C. (2016). Liver immunology and its role in inflammation and homeostasis. *Cellular & Molecular Immunology, 13*, 267–276.
69. Mooli, R. G. R., Mukhi, D., & Ramakrishnan, S. K. (2022). Oxidative stress and redox signaling in the pathophysiology of liver diseases. *Comprehensive Physiology, 12*, 3167–3192.
70. Surai, P. F. (2015). Silymarin as a natural antioxidant: An overview of the current evidence and perspectives. *Antioxidants, 4*, 204–247.
71. Shanmugam, S., Park, J. H., Cho, S., & Kim, I. H. (2022). Silymarin seed extract supplementation enhances the growth performance, meatquality, and nutrients digestibility, and reduces gas emission in broilers. *Animal Bioscience, 35*, 1215–1222.
72. Baradaran, A., Samadi, F., Ramezanpour, S. S., & Yousefdoust, S. (2019). Hepatoprotective effects of silymarin on CCl4-induced hepaticdamage in broiler chickens model. *Toxicology Reports, 6*, 788–794.
73. Abhilash, P. A., Harikrishnan, R., & Indira, M. (2019). Ascorbic acid is superior to silymarin in the recovery of ethanol-induced inflammatoryreactions in hepatocytes of guinea pigs. *Journal of Physiology and Biochemistry, 69*, 785–798.
74. Sharma, S., Le Guillou, D., & Chen, J. Y. (2023). Cellular stress in the pathogenesis of nonalcoholic steatohepatitis and liver fibrosis. *Nature Reviews Gastroenterology & Hepatology, 20*, 662–678.
75. Nunnari, J., & Suomalainen, A. (2012). Mitochondria. *Sickness and in Health Cell, 148*, 1145–1159.
76. Bellance, N., Lestienne, P., & Rossignol, R. (2009). Mitochondria: From bioenergetics to the metabolic regulation of carcinogenesis. *Frontiers in Bioscience-Landmark, 14*, 4015–4034.
77. Schinwald, M., Creutzinger, K., Keunen, A., Winder, C., Haley, D., & Renaud, D. (2022). Predictors of diarrhea, mortality, and weight gainin male dairy calves. *Journal of Dairy Science, 105*, 5296–5309.
78. Jouany, J. P., & Morgavi, D. (2007). Use of 'natural' products as alternatives to antibiotic feed additives in ruminant production. *Animal, 1*, 1443–1466.
79. Caprarulo, V., Giromini, C., & Rossi, L. (2020). Chestnut and quebracho tannins in pig nutrition: The effects on performance and intestinalhealth. *Animal, 15*, 100064.

80. Gourlay, G., & Constabel, C. P. (2019). Condensed tannins are inducible antioxidants and protect hybrid poplar against oxidative stress. *Tree Physiology, 39*, 345–355.
81. Alshaibani, D., Zhang, R., & Wu, V. C. (2017). Antibacterial characteristics and activity of *Vaccinium macrocarpon* proanthocyanidinsagainst diarrheagenic *Escherichia coli*. *Journal of Functional Foods, 39*, 133–138.
82. Lipinski, K., Mazur, M., Antoszkiewicz, Z., & Purwin, C. (2017). Polyphenols in monogastric nutrition. *Annals of Animal Science, 17*, 41–58.
83. Serra, V., Salvatori, G., & Pastorelli, G. (2021). Dietary polyphenol supplementation in food producing animals: Effects on the quality of derived products. *Animals, 11*, 401.
84. Heim, K. E., Tagliaferro, A. R., & Bobilya, J. D. (2002). Flavonoid antioxidants: Chemistry, metabolism and structure-activity Relationships. *Journal of Nutritional Biochemistry, 13*, 572–584.
85. Campbell, J. M., Crenshaw, J. D., & Polo, J. (2013). The biological stress of early weaned piglets. *Journal of Animal Science and Biotechnology, 4*, 19.
86. Aziza, A. E., Quezada, N., & Cherian, G. (2010). Antioxidative effect of dietary camelina meal in fresh, stored, or cooked broiler chicken meat. *Poultry Science, 89*, 2711–2718.
87. Rymer, C., & Given, D. I. (2010). Effects of vitamin E and fish oil inclusion in broiler diets on meat fatty acid composition and on the flavor of a composite sample of breast meat. *Journal of the Science of Food and Agriculture, 90*, 1628–1633.
88. Halliev, B. (2006). Reactive species and antioxidants. Redox biology is a fundamental theme of aerobic life. *Plant Physiology, 141*, 312–322.
89. Ramos Tovar, E., & Muriel, P. (2020). Free radicals, antioxidants, nuclear factor-E2-related factor-2 and liver damage. *Journal of Applied Toxicology, 40*, 151–168.
90. Lee, J., Giordano, S., & Zhang, J. (2012). Autophagy, mitochondria and oxidative stress: Cross-talk and redox signalling. *Biochemical Journal, 441*, 523–540.
91. Ramos-Tovar, E., & Muriel, P. (2020). Molecular mechanisms that link oxidative stress, inflammation, and fibrosis in the liver. *Antioxidants, 9*, 1279.
92. Majima, H. J., Indo, H. P., Suenaga, S., Matsui, H., Yen, H. C., & Ozawa, T. (2011). Mitochondria as possible pharmaceutical targets for theeffects of vitamin E and its homologues in oxidative stressrelated diseases. *Current Pharmaceutical Design, 17*, 2190–2195.
93. Snezhkina, A. V., Kurdymon, O. L., Savvateeva, M. V., Melnikova, N. V., Krasnov, G. S., Dmitriev, A. A. (2019). ROS generation and antioxidant defense systems in normal and malignant cells. *Oxidative Medicine and Cellular Longevity*, 6175804.
94. Gessner, D. K., Ringseis, R., & Eder, K. (2017). Potential of plant polyphenols to combat oxidative stress and inflammatory processes infarm animals. *Journal of Animal Physiololgy & Animal Nutrition, 101*, 605–628.
95. Adams, L., Franco, M. C., & Estevez, A. G. (2015). Reactive nitrogen species in cellular signaling. *Experimental Biology and Medicine, 240*, 711–717.
96. Phaniendra, A., Jestadi, D. B., & Periyasamy, L. (2015). Free radicals: Properties, sources, targets, and their implication in various diseases. *Indian Journal of Clinical Biochemistry, 30*, 11–26.
97. Venza, I., Venza, M., Visalli, M., Lentini, G., Teti, D., d'Alcontres, F. S. (2021). ROS as regulators of cellular processes in melanoma. *Oxidative Medicine and Cellular Longevity*, 1–19.
98. Amrit, B. K., Ponnampalam, E. N., Macwan, S., Wu, H., Aziz, A., Muir, S., Dunshea, F. R., & Suleria, H. A. R. (2023). Comprehensive screeningand characterization of polyphenol compounds from pasture grasses used for livestock production under temperate region. *Animal Feed Science and Technology, 300*, 115657.
99. Duda-Chodak, A., & Tar, T. (2023). Possible side effects of polyphenols and their interactions with medicines. *Molecules, 28*, 2536.
100. Asfaw, D. E. (2022). Antioxidant properties of phenolic compounds to manage oxidative stress—A review. *Journal of Advances in Agronomy and Crop Science, 1*, 1–16.
101. Jiamboonsri, P., Pithayanukul, P., Bavovada, R., Leanpolchareanchai, J., Gao, S., & Hu, M. (2016). In vitro glucuronidation of methylgallate and pentagalloyl glucopyranose by liver microsomes. *Drug Metabolism and Pharmacokinetics, 31*, 292–303.

102. Belkaid, Y., & Hand, W. T. (2014). Role of the microbiota in immunity and inflammation. *Cell, 157*, 121–141.
103. Rastmanesh, R. (2011). High polyphenol, low probiotic diet forweight loss because of intestinal microbiota interaction. *Chemico-Biological Interactions, 189*, 1–8.
104. Gasaly, N., & Gottelan, M. (2022). Interference of dietary polyphenols with potentially toxic amino acid metabolites derived from the colonic microbiota. *Amino Acids, 54*, 311–324.
105. Hachimura, S., Totsuka, M., & Hosono, A. (2018). Immunomodulation by food: Impact on gut immunity and immune cell function. *Bioscience, Biotechnology, and Biochemistry, 82*, 584–599.
106. Shimizu, M. (2017). Multifunctions of dietary polyphenols in the regulation of intestinal inflammation. *Journal of Food and Drug Analysis, 25*, 93–99.
107. Pereira, A. (2016). Plant abiotic stress challenges from the changing environment. *Frontiers in Plant Science, 7*, 1123.
108. Etsuo, N. (2016). Oxidative Stress and antioxidants: Distress or eustress? *Archives of Biochemistry and Biophysics, 595*, 19–24.
109. Chen, Y., & Junger, W. G. (2012). Measurement of oxidative burst in neutrophils. *Methods in Molecular Biology, 844*, 115–124.
110. Obata, T., & Fernie, R. (2012). The use of metabolomics to dissect plant responses to abiotic stress. *Cellular and Molecular Life Sciences, 69*, 3225–3243.
111. Forni, C., Duca, D., & Glick, B. B. (2017). Mechanisms of plant response to salt and drought stress and their alteration by rhizobacteria. *Plant and Soil, 410*, 335–356.
112. Sarker, U., & Oba, S. (2018). Catalase, superoxide dismutase and ascorbate-glutathione cycle enzymes confer drought tolerance of Amaranthus tricolor. *Science and Reports, 8*, 16496.
113. Zahedi, A., On, V., Phandthong, R., Chaili, A., Remark, G., Bhanu, B., & Talbot, P. (2018). Deep analysis of mitochondria and cell health using machine learning. *Science and Reports, 8*, 16354.
114. Javadov, S., Chapa-Dubocq, X., & Makarov, V. (2018). Different approaches to modeling analysis of mitochondrial swelling. *Mitochondrion, 38*, 58–70.
115. Singh, R., Kumar, M., Mittal, A., & Mehta, P. K. (2017). Microbial metabolites in nutrition, healthcare and agriculture. *3 Biotech, 7*, 1–14.
116. Chandra, P., Sharma, R. K., & Arora, D. S. (2020). Antioxidant compounds from microbial sources: A review. *Food Research International, 129*, 108849.
117. Feng, T., Wang, J. (2020). Oxidative stress tolerance and antioxidant capacity of lactic acid bacteria as probiotic: A systematic review. *Gut Microbes, 12*.
118. Mitra, A. K. (2020). Antioxidants: A masterpiece of mother nature to prevent illness. *Journal of Chemical Reviews, 2*, 243–256.
119. Ayyanna, R., Ankaiah, D., & Arul, V. (2018). Anti-inflammatory and antioxidant properties of probiotic Bacterium *Lactobacillus mucosae* AN1 and *Lactobacillus fermentum* SNR1 in Wistar Albino Rats. *Frontiers in Microbiology, 9*, 3063.
120. Koller, M., Muhr, A., & Braunegg, G. (2014). Microalgae as versatile cellular factories for valued products. *Algal Research, 6*, 52–63.
121. Chaki, M., Begara-Morales, J. C., & Barroso, J. B. (2020). Oxidative stress in plants. *Antioxidants, 9*, 481.
122. Raja, V., Majeed, U., Kang, H., Andrabi, K. I., & John, R. (2017). Abiotic stress: Interplay between ROS, hormones and MAPKs. *Environmental and Experimental Botany, 137*, 142–157.
123. Nahar, K., Rhaman, M. S., Parvin, K., Bardhan, K., Marques, D. N., Garcia-Caparros, P., & Hasanuzzaman, M. (2022). Arsenic-induced oxidative stress and antioxidant defense in plants. *Stresses, 2*, 179–209.
124. Sachdev, S., Ansari, S. A., Ansari, M. I., Fujita, M., & Hasanuzzaman, M. (2021). Abiotic stress and reactive oxygen species: Generation, signaling, and defense mechanisms. *Antioxidants, 10*, 277.
125. Rillig, M. C., Wright, S. F., Nichols, K. A., Schmidt, W. F., & Torn, M. S. (2001). Large contribution of arbuscular mycorrhizal fungi to soil carbon pools in tropical forest soils. *Plant and Soil, 233*, 167–177.

126. Mandal, S. M., Chakraborty, D., & Dey, S. (2010). Phenolic acids act as signaling molecules in plant-microbe symbioses. *Plant Signaling & Behavior, 5*, 359–368.
127. Lenoir, I., Fontaine, J., & Sahraoui, A. L. (2016). Arbuscular mycorrhizal fungal responses to abiotic stresses: A review. *Phytochemistry, 123*, 4–15.
128. Millar, N. S., & Bennett, A. E. (2016). Stressed out symbiotes: Hypotheses for the influence of abiotic stress on arbuscular mycorrhizal fungi. *Oecologia, 182*, 625–641.
129. Omura, T. (2006). Mitochondrial P450s. *Chemico-Biological Interactions, 163*(1–2), 86–93.
130. Storz, P. (2005). Reactive oxygen species in tumor progression. *Frontiers in Bioscience, 10*(1–3), 1881–1896.
131. Brand, M. D. (2010). The sites and topology of mitochondrial superoxide production. *Experimental Gerontology, 45*(7–8), 466–472.
132. Mari, M., Morales, A., Colell, A., Garcia-Ruiz, C., & Fernandez-Checa, J. C. (2009). Mitochondrial glutathione, a key survival antioxidant. *Antioxidants & Redox Signaling, 11*(11), 2685–2700.

An Overview of the Health Benefits and Risks of Substance Use

S. Srivaishnavi, S. Subhashini, and R. Ramesh

Abstract Substances in general are both legal and illegal and it has very many effects both positive and negative on human health. Some of the substances like alcohol, tobacco products, cannabis, prescription medications and some of the illicit drugs also have therapeutic benefits. Substances with moderate misuse or disorder may provide some health benefits cannabidiol (CBD) has become one of the therapeutic agents in chronic pain, anxiety and inflammatory diseases. Low-level usage of caffeine is linked with increased mental awareness and helps in preventing neurogenerative diseases. Despite the potential benefits of using substances, it also has some severe consequences in terms of mental health and as well as physical health. For instance, excessive usage of addiction to alcohol may lead to severe heart problems, liver diseases and a variety of cancers among regular users. Similarly, tobacco use leads to stroke and lung cancer. Cannabis is one of the substances that may cause impaired memory and dysfunction of cognition and create long-term mental health. Overall, the individual is not only going to be affected because of the use of substances, but the societal impact among the youngsters in this current scenario is a challenging one. Substance abuse has adverse effects on individuals, and families, creating violence, social marginalization, sexual harassment, etc. In this chapter, overall details of the substances, natural and synthetic substances, their medications and their misuse, and the benefits and risks of using substances were discussed in depth.

Keywords Substance use · Health benefits · Addiction · Drug abuse · Risk factors · Preventive strategies

S. Srivaishnavi · R. Ramesh (✉)
Department of Physics, Sacred Heart College (Autonomous), Tirupattur 635601, Tamil Nadu, India
e-mail: rramesh@shctpt.edu

S. Subhashini
School of Physiotherapy, Dhanalakshmi Srinivasan University, Tiruchirappalli 621112, Tamil Nadu, India

K. Kasinathan and A. Muthukumaran (eds.), *Recent Advances in Oxidative Stress Associated Chronic Diseases Volume 2*, https://doi.org/10.1007/978-981-95-3758-7_4

1 Introduction

In recent times, there has been a spread of addiction diseases and nowadays it has taken a viral increase and is considered to be a threat to a future generation which is termed as '*substance use*' or simply known as '*drug abuse*'. As we know, adolescence age is a very crucial phase in every individual life where we get into good and bad habits or get into work which will take to lot of loss and gain which will unexpectedly turn our life's table. Regarding this, substance use is a major problem in adults [1–3]. Substances which is obtained from plants, animals or any medical complex product that always enjoyed great popularity drugs present in cosmetics, care products, etc., and the exact substance like drugs, are said as substance is a complex matter that has a specific composition of some materials of some specific structure, pharmacological effects and potential to affect the balance of the hormones which is meant for some of the feeling in the human being [4–8]. The use of psychoactive substances, these substance which are responsible for altering the brain chemistry which leads to changes in mood, perception and behaviour for depression—taking alcohol is one of the ways that keep our subconscious mind out of our control, another drug benzodiazepines which is responsible to slow down the activity of the brain and helps to get rid of depression [9]. There is nothing wrong with having the substance, but one must take in the limited amount, overtaking or continuous taking of the substance over a period leads to many risks or even leads to death sometimes. So, substances can be of any type and can be added to any material like foods, tablets, perfumes, etc., which makes us addicted to it with or without our knowledge [10–12]. Coffee has caffeine substance which makes people addicted to coffee, but it has some health benefits and some risks for instance, we know that taking caffeine can improve the skin but there is a risk of overtaking that can lead to heart disease and weight gain. According to the World Health Organization (WHO), estimated that 270 million people get used to illicit drugs from 2019 and 35 million people will suffer from this substance use disorders [13–16]. This substance use can have devastating situations, overdose and long-term use which leads to cardio diseases and mental health disorders, respectively. It also affects socioeconomic problems, which include poverty, homelessness and criminal justice involvement etc. And as if there a good and bad in everything we same way there are more risks and benefits of the usage of substances and they are categorized on public health strategies as adding one of the drugs as an essential thing in every medicine they prepare for certain diseases in limited amount, policy-making like regulation in taking with harm reduction approaches, taking in the form of pills with prescribed addictive medications, respectively, [17–19]. The substances are classified based on the nature of the products, drugs' texture and purpose in different types. Initially, a concise summary of the extensive and intricate relationships humans has formed with substances is presented, highlighting the advantages and dangers associated with both medical and nonmedical applications of these substances [20–22]. This chapter review aims to give a comprehensive yet limited summary by concentrating on various substances (along with their pharmacological ancestors and successors) that have arguably, had a significant presence

(*Example., Opium, cannabis, alcohol, cocaine*). To understand these substances and the advantages and dangers of their long-term use [23].

2 Natural Substances: Risks and Benefits

Natural substances are those that occur in nature or are by-products of natural processes. According to European Union regulation concerning the Registration, Evaluation, Authorization, and Restriction of Chemicals (REACH) [Article 3(39)], a naturally occurring substance is defined as one that is either unprocessed or processed only through manual methods, mechanical means or gravitational techniques [24]. The processing methods include techniques such as dissolution in water, flotation, extraction with water, filtration, steam distillation and heating the residue to remove water to obtain the raw material, among others. As shown in Fig. 1, natural substances can be classified according to their appearance, texture and field use. Natural substances are also used in the field of medicine due to their pharmaceutical and therapeutic properties processed by it. Some of the natural substances are turmeric, honey, ginger, etc., as they possess unique benefits in both modern and traditional ways [25–27]. Let us discuss the risks and benefits of some of the natural substances that are helpful in traditional and modern ways in medicinal and various field.

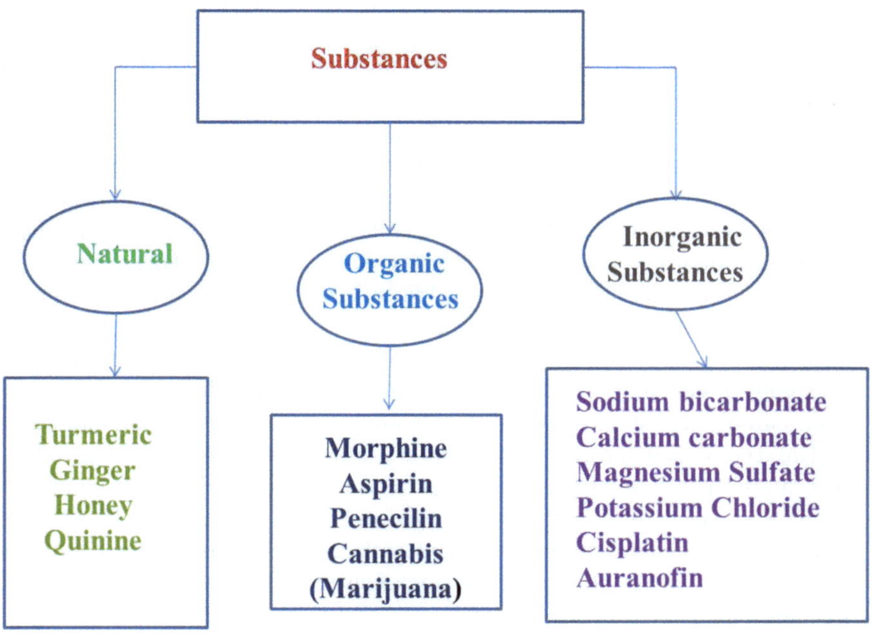

Fig. 1 There are several categories in which this substance can be classified

2.1 *Turmeric—The Golden Healer*

Turmeric is a famous and easily available antibiotic substance that can be used in medicinal ways and can be used in cooking [28]. It is a bright yellow substance that can be called a spice; it is a staple in traditional medicine, especially in Ayurveda and instant medicine for wounds also. This is celebrated as a powerful anti-inflammatory, antioxidant and antibiotic agent, respectively.

- *Active Compound*

  The active compound responsible for the turmeric to possess medicinal properties is '*Curcumin*'. It is a polyphenolic compound that has significant properties in some therapeutic areas [29]. This has been considered for its anti-inflammatory effects in conditions such as arthritis, cardio cardio-related diseases and responsible actions on the skin it cures acne and pimples and helps in the removal of facial hairs etc. For instance, according to Rizaldy et al., this compound presence has exhibited its potential as a natural and good alternative to pain relief in the treatment of disease Arthritis pain [30, 31].
- *Anti-cancer Agent*

  This compound has anti-cancer characteristics. It hinders cancer cell proliferation by disrupting the cell cycle, triggering apoptosis which destroys the cancer cells and decreases angiogenesis. As per the study by done by Eric et al., cancer disease cause mortality to one-fifth of the generation globally. Epigenetic and genetic changes culmination in apoptosis, severe tumour growth, angiogenesis and metastasis are the real driving force of the cancer [32].
- *Gastrointestinal Effects*

  Turmeric offers various protective benefits to the digestive system. It actively prevents the development of ulcers triggered by factors such as stress, alcohol, indomethacin, risperidone and pyloric ligation by enhancing gastric wall mucus in rats exposed to these digestive challenges. Additionally, it alleviates intestinal spasms and boosts the secretion of bicarbonate, gastrin, secretin and pancreatic enzymes [33–35]. A phase II open-label trial involving 25 patients with endoscopically confirmed gastric ulcers, who were administered 600 mg of powdered turmeric five times a day, revealed that 48 per cent achieved complete healing. No negative side effects or blood anomalies were observed. Curcumin demonstrated a reduction in mucosal damage in mice with induced colitis. When 0 mg/kg of curcumin was administered for ten days before the colitis using 1,4,6 trinitrobenzene sulfonic acid, it led to a notable decrease in diarrhoea, neutrophil infiltration and lipid peroxidation within colonic tissue. Furthermore, all inflammation indicators showed a decline and symptoms experienced enhancement [36]. In rat models of induced pancreatitis, curcumin managed to lessen inflammation. In instances of pancreatitis induced by cerulean or ethanol, curcumin also suppressed inflammatory mediators, resulting in a reduction of disease severity as assessed through histology and pancreatic evaluations.
- *Alzheimer's and Turmeric*

Alzheimer's condition is when aged mice burned with extensive plaque formations akin to those found in Alzheimer's condition were given curcumin, and there was a notable decrease in plaque accumulation [37]. It mitigated oxidative harm and reversed the amyloid anomalies in a transgenic mouse model of Alzheimer's condition. Curcumin's robust antioxidant and anti-inflammatory qualities also alleviated symptoms associated with inflammation and oxidation in Alzheimer's condition.

- *Antioxidant Activity*
  Curcumin has demonstrated remarkable efficacy as a scavenger of reactive oxygen species. Its antioxidant properties rival those of vitamins C and E. It can safeguard lipids or haemoglobin from oxidative damage [38]. Furthermore, it can greatly impede the production of highly reactive oxygen species like H_2O_2, anions of superoxides and radicals of nitrides generated by the macrophage's activation. Its derivatives, bismethoxycurcumin, also exhibit antioxidant properties. Pretreatment with curcumin has been found to alleviate ischemia-induced oxidative stress and alterations in the heart. A laboratory study investigating the effects of curcumin on an inducible stress protein showed increased cellular resistance to oxidative harm [39].

2.2 *Ginger-The Spicy Medicinal Root and Its Effects*

Ginger is another remarkable natural remedy that has been utilized for centuries for its healing properties. It is a part of the *Zingiberaceae* family, and the biological name is *Zingiberaceae officinale.* Ginger has a rich history in traditional medicine systems in a worldwide range [40]. It is recognized for alleviating nausea, enhancing digestive problems and delivering anti-inflammatory benefits as well. The active compounds present in ginger are gingerol and shogaol. Gingerol contributes to the distinctive spicy flavour of ginger and is also noted for anti-inflammatory, antioxidant and anti-cancer effects. Shogaol, which emerges in dried ginger or cooked ones also helps in boosting anti-inflammatory and pain-relieving properties. It exhibits a good anti-inflammatory property; it offers benefits for conditions like osteoarthritis and rheumatoid arthritis. A study presented by Altman and Marcussen and his team found that ginger extract significantly helps in relieving pain and stiffness in osteoarthritis patients and participants who consumed ginger extract reported a marked reduction in symptoms and improved health. Ginger is widely as it provides good results in digestion and alleviates nausea [41]. A study by Mansour et al., explains that ginger could reduce the symptoms and control the frequency of nausea and vomiting, particularly during chemotherapy treatments. Additionally, ginger has been shown to improve gastrointestinal motility, making it an effective remedy for indigestion and bloating, respectively [10].

2.3 Honey—The Natural Sweetener with Healing Properties

Honey is one of the products available naturally, which was collected by the honeybees (*Apis mellifera*). They gather nectar a type of chemical solution that contains sugars and many other compounds synthesized from flowers, productions from plants or waste from plant-sucking insect organisms and convert these into honey, which shows anti-microbial, antibiotic and anti-inflammatory characteristics [42]. It has a variety of nutrients, vitamins, organic acids, minerals and vitamins and it also aids in reducing bloating. Honey diminishes the functions of cyclooxygenase1 and cyclooxygenase2, thereby exhibiting anti-inflammatory properties and shows immunomodulatory effects. Additionally, the consumption of diluted natural honey has been found to lower the levels of compounds including prostaglandin E_2, $F_{2\alpha}$ and B_2 thromboxane in the plasma cells of the individuals. A study demonstrated that the anti-inflammatory properties of the honey were comparable to those of the reference drug prednisolone [43]. Moreover, honey possesses an anti-inflammatory action devoid of negative side effects such as hindering responses, inhibiting tissue growth and causing gastric ulcers among many individuals. Honey exhibits a purifying effect on injuries, promotes cellular repair and diminishes swelling, and infused pads serve as nonsticking wound coverings. Clinical studies have demonstrated that honey dressings yielded superior improvements (for instance, compared to dressings with amniotic membranes, silver sulfadiazine and boiled potato skin) and promoted quicker healing with a reduced degree of contraction and scarring [44]. Studies on the anti-microbial characteristics and healing effects of the honey utilize a blend of multi-floral processed honey that was diluted in a water (1:1) ratio for over 5 days, which comparatively decreased the dental plaque in healthy subjects, however, about 0.2% chlorohexidine solution along with honey showed high effectiveness [45]. Other research explored the impact of Iranian honey on vulvovaginal candidiasis and revealed a notable decrease in the inflammation level, discharge and itching was completely cured in 8 days of complete treatment with 70% honey cream. Researchers found that although 1% of clotrimazole was highly effective, honey would act as a practical substitution for the treatment of vulvovaginal candidiasis. Availability in nature and very low cost, the honey was used in research comprising of young children who were suffering from a disorder called pyomyositis abscesses, were treated with gauze that was soaked in the raw honey cream and similarly, a medical solution that was applied on the affected wounds for 21 days, results revealed that the application of honey had drastically cleared the wounds and showed improvement in healing thereby reduced the time duration in hospital stay [46]. Another study demonstrated that the application of 30% of Iranian honey cream on the episiotomy wound in women completely healed and reduced discharge in women who had not given birth before. Numerous research efforts have examined the impact of dressings on foot ulcers in diabetics using natural honey had been reported and, in the study, it was observed that notable development in the wound healing and recovery duration time in the group treated with Beri honey versus the control group received a standard saline dressing [47]. Moghazy in his research findings shared comparable results, observing notable enhancements in the healing and recovery time of diabetic foot ulcers treated with Egyptian honey dressings and not including the control group in their findings. Shukrimi discovered no significant differences in clen nonsterile honey versus povidone-iodine. Although honey from different origins exhibits anti-microbial properties, they do not exceed the effectiveness of pharmaceutical treatments like chlorohexidine, clotrimazole or povidone-iodine [48].

3 Organic Substance as Drugs for Pharmacology: Risks and Benefits

The discovery of pure drug substances evolved from plant preparation of the plants that possess the medical properties. The identification of pure drugs emerged from the study of ethnobotany extracts became or termed medicines and they paved the way for the great changes in pharmacology, i.e. in pharmaceutical industries to elucidate and medicate the diseases, respectively [49]. The organic substances are aspirin, penicillin, morphine and cannabis. Organic compounds extracted from flora and fauna have played a fundamental role in shaping contemporary medicine. Among these, morphine and cannabis are particularly notable due to their powerful medicinal properties. Utilized for centuries, these substances have predominantly been employed for alleviating pain and serving various therapeutic needs. Recent research has enhanced our comprehension of these natural materials, uncovering both their healing capabilities and the hazards of their misuse. This chapter delves into the past, pharmacodynamics, health advantages and scientific inquiry surrounding morphine and cannabis, emphasizing their importance within the medical domain [50].

3.1 Aspirin—Pain Reliever

Aspirin is a natural drug that is found in willow bark and combined with salicylic acid which is called as a derivative of the natural drugs. It is a nonsteroidal, anti-inflammatory drug; it works with the inhibition of some enzymes called PG endoperoxide or Cyclooxygenase (COX), respectively. Aspirin drug works is carried by enzymes the effect of this drug on the stomach, in the human body [51]. The so-called protective role of PGs in averting gastric erosions and ulcers primarily arises from the endogenously synthesized prostacyclin and PGE2, which diminish gastric acid production and have a direct dilatory effect on the gastric mucosa. Beyond these principal functions, proteinoids enhance the release of thick mucus and gastric fluid, along with duodenal bicarbonate [52]. The increased mucus serves as a safeguarding physical barrier while preserving an alkaline milieu for the neutralization of excess acid. In most organisms, including humans, most of the protective PGs are formed by COX1, although minor amounts of COX2 are also found in the stomachs of rats.

Thus, it remains puzzling why mice lacking the COX1 gene do not exhibit spontaneous gastric bleeding or erosions. It is possible that during pregnancy, compensatory protective mechanisms such as nitric oxide synthesis assume the function of the PGs.

3.2 Morphine and Its Effects

Morphine was isolated from Opium poppy in 1806 by Freidrich Serturner. The discovery of morphine led to a development towards dose-controlled medicine for pain. It has been used as a short-term and long-term pain reliever for many years in the field of medicine and it works with the receptors of the whole-body system and is a small molecule that enables the opiate receptors and their subtypes very easily [53]. Morphine primarily exerts its effects by attaching to specific opioid receptors within the central nervous system, particularly the mu-opioid receptor. This attachment hampers the release of neurotransmitters responsible for signalling pain, ultimately resulting in analgesia (*pain relief*). Beyond alleviating pain, morphine fosters sensations of euphoria, which adds to its potential for addiction. Morphine finds application in various clinical environments, particularly for immediate pain relief after surgeries, pain stemming from cancer and discomfort resulting from trauma or injury. It may also be utilized in managing persistent conditions, such as end-stage cancer, when alternative treatments prove ineffective [54]. A study demonstrated that morphine, when given appropriately, is exceptionally effective in alleviating severe pain and enhances the quality of life for patients battling advanced cancer. Nevertheless, the administration of morphine necessitates vigilant monitoring due to the associated risks of dependency, tolerance and addiction. Despite its efficacy, morphine carries notable disadvantages. The most significant risks entail dependency, tolerance and addiction. The body develops tolerance to morphine, necessitating larger doses to achieve the same degree of pain relief. This increase raises the likelihood of overdose and fatality. Moreover, prolonged use of morphine may result in physical dependence, with withdrawal symptoms that consist of anxiety, perspiration, nausea and muscle discomfort when the medication is reduced or discontinued. The opioid crisis, especially in nations like the United States, has highlighted the perils associated with the widespread use of opioids, including morphine [55]. A study in 2016 indicated that opioid-related fatalities, inclusive of those involving morphine, have emerged as a crucial public health concern, with an increasing number of deaths linked to prescription opioids. It was found that taking morphine with an overdose can cause breathing problems, colour changes in the skin to bluish and in the nails. If an individual uses morphine for the long term, it affects the adrenal glands thereby disturbing the process of making cortisol and this syndrome is called adrenal insufficiency. It also produces low blood pressure and makes an individual feel confused. A state called increased sensitivity to the pain symptoms will be caused while taking morphine [56]. It is also advised this morphine cannot be used with other drugs or medicines like vitamins, minerals or herbal supplements, etc., it interacts with those things and creates a very severe effect, especially with medicines like tizanidine, Klonopin, Ativan, serotonin, sumatriptan, Butrans, etc.

3.3 Cannabis—The Medicinal Herb

Cannabis, also known as marijuana, is derived from the *Cannabis sativa* plant. It is established more than ten thousand in both Taiwan and Europe. It originated in Afghanistan and was grown in Europe and Arabia, like various nations around the globe. It is utilized in the production of rope, textiles and fibres, as well as in the manufacture of paper in China and Japan during the second and fifth centuries. During the seventeenth and eighteenth centuries, Europe utilized this and documented that this substance shows antibiotic and analgesic properties and is acknowledged as a sedative agent [57]. In India, this is referred to as bhang and ganja. Long-term use of this substance may elevate the likelihood of mental deterioration and significantly raise the risk of developing mouth cancer or other cancer-related illnesses. A report revealed that usage of cannabis flowers can be utilized for peripheral neuropathy pain when it is administered through an inhalation process. Studies on HIV/AIDS-affected individuals investigated using cannabinoids as an effective solution for weight loss and Anorexia associated with HIV/AIDS involving 255 patients. Also, one of the studies reported that taking cannabinoids orally can be very effective for the multiple sclerosis symptoms.

3.4 Medical Uses and Benefits

Cannabis has been exhibited to be an effective drug when used in a limited amount in some of the conditions which includes chronic pain, multiple sclerosis glaucoma and nausea which is related to chemotherapy treatment [58]. A review says that cannabinoids will have a great effect on such medical conditions as chronic pain, in neuropathic pain and they also help in removing the symptoms of depression, anxiety and post-traumatic stress disorder problems. It also explored its potential active participation as a reducing agent in the reduction of opioid use. A study states cannabis was legal, and it does a significant reduction in opioid overdose deaths it also serves as a safer alternative for pain management.

3.5 Risks and Side Effects

Cannabis is not advisable in a prescribed manner if it is extended to an extent, it can cause effects such as impaired memory, concentration and motor coordination [59].

Long-term use of cannabis usage in high doses mental health issues, including an increased risk of suffering from cancer, psychosis schizophrenia, etc.

4 Inorganic Substances

Inorganic compounds are some of the chemical compounds which is mixed in a certain amount with some drug materials and produced as supplements in the medicinal field and industries side also. As there are a lot of inorganic substances there are a few important substances that are added to every medicine or material they prepare [60]. Among these, sodium bicarbonate, calcium carbonate, magnesium sulphate and potassium chloride are important substances with various benefits as well as risks. These materials are used widely in the market, especially for therapeutic benefits, in food production and on the industrial side, respectively. Many inorganic substances find very many applications in the field of medicine. Lithium is one of the compounds that can be used as a neurological agent in the form of Lithium carbonate and the Li^+ ions can block the inositol production recycling process and interact with Magnesium of the cells to form MgLi-inosine monophosphate and act as a specific inhibitor of the GSK_3 compounds. Metal-derived compounds can bin the DNA of biomolecules and can be used to kill cancer cells. Cisplatin is one of the famous metal-based drugs that was reported as an effective medicine for testicular cancer [61]. Gold-based inorganic substances were prescribed for rheumatoid arthritis disease.

4.1 Sodium Bicarbonate ($NaHCO_3$)—A Multifunctional Compound

Sodium Bicarbonate which is commonly known as baking soda is the most common and most used inorganic substances. It is represented as a white powder which is highly solubility in water. It is used in applications like fermenting ground rice flour, household cleaning and medical interventions [62]. It has chemical properties such as the ability to neutralize acids, rendering it an asset across various fields. When sodium bicarbonate is taken it dissociates from Sodium (Na^+) ions and bicarbonate (HCO_3^-) absorbs the hydrogen ions and gets converted into carbonic acid which aids in the conversion of water into carbon dioxide for the excretion from the lungs. The main impact of sodium bicarbonate is increasing the plasma bicarbonate levels and in turn, buffers the excess hydrogen concentration. Studies revealed that taking sodium bicarbonate orally helps athletes in their performance in their sports activities also found to be effective against the toxicity created by the drugs that affect the sodium channel [63]. Sodium bicarbonate is classified as a very weak type of base and can produce carbon dioxide gas, some salt and water when reacts with some specific acids. This property is effective as an antacid, neutralizing excess to relieve symptoms of heartburn, etc., sodium bicarbonate helps to restore the body's pH balance [64]. For instance, Ione de Brito-Ashurst et al., investigated sodium bicarbonate to slow the progression of chronic kidney disease and has given good results. It improves nutritional status among patients with CKD.

4.2 Calcium Carbonate—A Key Mineral in Medicine and Industry

The chemical compound that naturally occurs in limestone, marble and chalk. It is a white odourless powder substance which used in medical and industrial fields due to the availability of the substance. This is very essential in healthcare which maintains bone health [65]. Calcium carbonate acts as an acid agent in stomach digestion problems and this helps in bone healthcare especially. Less intake of calcium leads to bone-related problem conditions like osteoporosis. The supplementation of calcium carbonate is recommended for persons who are aged and lack calcium content in their bodies. In this study, the intake of calcium supplementation indicates a sustained beneficial effect and says that daily intake of some 800–1000 mg for about a few months can improve the calcium deficiency in the body [66]. In addition to its role in bone health, calcium carbonate is used in medical settings to treat problems in dental products, toothpaste dental fillers, etc., and that helps to treat low blood calcium deficient problems. Calcium carbonate plays a significant role in the construction and manufacturing field. It is used as a filler in products such as paints, coatings, paints and plastics. It is used in the production of cement and lime [67].

4.3 Magnesium Sulphate—A Compound with Multiple Functions

Magnesium sulphate is one of the chemical compounds that is available in the form of powder, solution and foam with the combination of magnesium, oxygen and Sulphate. It is commonly used in agricultural fields as fertilizers, in medical treatments and industrial sides. It is an essential mineral as calcium for the human body both magnesium and sulphate serve as necessary ingredients for various biological processes which happen in the human body. Magnesium sulphate is a multifunctional compound that plays a vital role in medicinal uses. It is commonly used in the prevention of obstetrics and treat seizures in women with preeclampsia it is a condition that causes trouble with high blood pressure condition during pregnancy it helps to relax the blood vessels and in a reduction in risk of seizures [68]. This magnesium sulphate is widely utilized for preventing eclamptic seizures. The efficacy of magnesium sulphate is observed through various mechanisms, including peripheral and cerebral vasodilation, protection of the blood–brain barrier and as an anticonvulsant. Although the exact mechanisms of action are not fully understood, the role of magnesium sulphate in preventing eclampsia is probably influenced by multiple factors. Magnesium sulphate might function as a vasodilator, influencing either the peripheral or cerebral vasculature, to lower peripheral vascular resistance and/or alleviate vasoconstriction. Moreover, magnesium sulphate might also safeguard the blood–brain barrier and reduce the formation of cerebral oedema, or it could operate via a central anticonvulsant effect [69].

4.4 Potassium Chloride—Essential Electrolyte Balance

Potassium chloride is a chemical compound made up of potassium and chlorine. It plays a crucial role in sustaining the body's electrolyte equilibrium. Potassium is vital for optimal nerve and muscle functioning. In both medical and industrial applications, it helps in many purposes' chloride is mainly employed in healthcare to manage or avert hypokalaemia (*deficient potassium levels*) in individuals. Potassium is a key electrolyte that assists in maintaining fluid equilibrium, muscle performanc and nerve conduction. Inadequate potassium can result in muscle weakness, erratic heart rhythms and in extreme cases, paralysis or fatality [70]. A study reported in '*The New England Journal of Medicine*' examined the application of potassium chloride in preventing hypokalemia among patients undergoing diuretic treatment for hypertension. The research demonstrated that potassium chloride supplementation effectively thwarted potassium depletion and aided in preserving electrolyte balance. Potassium chloride is additionally utilized in intravenous (IV) solutions to sustain electrolyte balance during surgical procedures or for patients unable to ingest food or fluids [71]. It is critical in emergency care environments where rapid correction of fluid and electrolyte discrepancies is necessary. Beyond its medicinal uses, potassium chloride serves various industrial purposes. It is a fundamental component in fertilizers, as potassium is an essential nutrient for plant development. Potassium chloride is also incorporated in salt production, the synthesis of numerous chemicals and acts as a de-icing agent on roads during winter.

5 Controlled Substances

Parliament has long acknowledged that while numerous substances can have positive qualities, some are especially harmful to individuals and to society due to their significant potential for abuse [72]. In London, the rules and regulations of drug use started with the Arsenic Act of 1852 and the passage of the Poisons and Pharmacy Act in 1908 which began in the parliament and addressed the drug abuse by lowering the sales of opium drugs in the country. However, subsequent administrations have tackled the growing issue of drug addiction in a fragmented way, implementing isolated laws to solve specific issues. For instance, a concerning rise in substance abuse among soldiers during World War I resulted in the establishment of the Defence of the Realm regulations in 1917, which restricted drug distribution to the troops without any prescription. In the 1960s, many individuals used amphetamines illegally and that resulted in the formulation of the new act the creation of the Drug Act of 1964 commenced that it is illegal to use or possess amphetamines without the doctor's prescription [73]. By the early 1970s, the necessity for comprehensive measures to tackle drug abuse was acknowledged, leading to the creation of the Misuse of Drugs Act 1971, which identified all detrimental or harmful substances that would be regulated by a control system was adapted to prevent the misuse of drugs and then formulated the Misuse of drugs act 1971 along with associated regulations and applied the same rules all over the countries in the United Kingdom [74].

6 Different Schedules of Usage of Controlled Substances

The providers must be familiar with the rules guidelines and laws associated with the schedule of the drugs, based on their basis, need and purpose of the drug usage and the side effects and the adverse risks of using such scheduled drugs. In the US, they follow the strict regulations by the state and federal laws which guide the manufacturers as well as the distributors [75]. Misuse of scheduled drugs results in many health-associated problems and makes an individual an addictive person while using the substances. Based on the seriousness of the state an act was enacted in 1970 in the United States about drug abuse prevention and control and it comprises the categorizing and regulating the usage of drugs according to their purpose of usage and chemical content, synthesizing and manufacturing, distribution and exporting it and selling [76]. According to the act, there are five different schedules were established and categorized to regulate their production and distribution by getting a license from the Drug Enforcement Administration body by the pharmacists. This license consists of a specific number allocated to the distributor and the provider and tracked during the time of substance prescription. This schedule of drugs has several parameters including their medical properties, risks associated when addiction and the harmful effects. The drugs were scheduled from the range Schedule I to V based on their level of risk or addiction and the abusability level.

6.1 Schedule I (CD Licence)

This category of schedule has the highest potential of abuse and causes severe disorder these substances are not prescribed for any medical usage and can only be supplied possessed or administrated by any individual who possesses a licence for it that was granted by the home office [77]. Such licenses are allocated for research or extraordinary circumstances. Medical professionals cannot prescribe these substances, nor can pharmacists dispense them for medicinal purposes and these drugs are illicit and called as street drugs. Some of the drugs in this category schedule-I included cannabis, LSD, raw opium, lysergic acid diethylamide, MDMA, methaqualone, heroin, mescaline, GHB, Psilocybin, Khat, Bath salts and marijuana. Misuse of these scheduled drugs by an individual can create many health issues based on their body condition which includes an increase in pulse rate, lowering the blood pressure, increased appetite, tiredness and bloodshot eyes and taking higher doses can give a type of hallucinations [78].

6.2 Schedule II (CD)

This category of drugs possesses a reduced potential than schedule-I drugs and has a higher risk for psychological and physical dependence and the possibility of misuse and use disorder. It was generally prescribed to individuals for the treatment of pain, insomnia, anxiety and ADHD problems. Those days these drugs were prescribed by using only paper prescription and now it is in the electronic form. Schedule II drugs include hydromorphone, secobarbital, oxycodone, fentanyl, opiates, cocaine, methadone, meperidine, dextroamphetamine, pentobarbital and methylphenidate [79]. These substances require secure storage and must be kept in a locked container, typically in a dedicated controlled drug cabinet or sanctioned safe, accessible only to the individual legally in possession of the controlled drug or an authorized associate. A doctor, dentist or qualified practitioner is permitted to administer the drug to a patient, as well as any person acting under the guidance of an accredited practitioner [80]. A log must be maintained for Schedule 2 controlled substances, adhering to the specific regulations in place. Disposal of these scheduled drugs needs proper authorization and the witness of the individual while disposing.

6.3 Schedule III (CD No Register)

This category of scheduled drugs has a lower level of potential for misuse and disorder than Schedule I and II. These substances were mainly used in medications for pain control, used as anaesthesia or as appetite suppression drugs. Some examples of this category are ketamine, anabolic steroids, benzphetamine and phendimetrazine these substances have the property of stimulants and other drugs possess a decreased risk of misuse and are not subjected to any protected storage requirements and kept on open shelves of a pharmacy excluding flunitrazepam, temazepam, buprenorphine and diethylpropion, which need to be stored in locked controlled drug containers [81]. Schedule III drugs follow the same specific prescribing protocols as those in Schedule II. There is no legal obligation to document these drugs and their transactions in a controlled register. Additionally, destruction regulations do not come into force unless these substances are personally produced in the manufacturing sector and the invoices can be kept for a period of minimum of two years.

6.4 Schedule IV Drugs

The items listed in this category of Schedule-IV do not require any secure storage, and the obligations for destruction solely pertain to importers, exporters and producers. There are no specific regulations for issuing prescriptions for controlled substances, and maintaining logs of controlled substances is not mandatory; nevertheless, it is recommended to document instances of manufacturing such substances or licenced importation or exportation [82]. Section of this schedule pertains to controlled substances called benzodiazepines, encompassing the majority of benzodiazepines along with eight other compounds like fecamfamin and mesocarb. Possessing these medications alone without a legitimate prescription is deemed an offence. Authorized possession is allowed for professionals and pharmacists carrying out their responsibilities in a professional role. Part II of this category addresses anabolic steroids, which include various androgenic and anabolic steroid compounds like clenbuterol, testosterone and other growth-enhancing hormones [83]. A home license is mandatory for importing and exporting this category of substances and it is a part of the medicinal items for personal use by an individual. Some of the substances that are very familiar in this schedule are clonazepam which has been used for panic disorder and seizure disorders, diazepam is another drug used for anxiety and muscle spasms. Zolpidem can be used as a sedative drug for the treatment of insomnia. To treat sleep disorders and narcolepsy modafinil has been used. Similarly, the following drugs are midazolam, alprazolam, temazepam, carisoprodol, triazolam and tramadol. Prescription for these substances may be verbally using a phone, paper prescription or through EPCS and refilling is allowed up to a maximum of five times in six months from the date of issue [84].

6.5 Schedule V Drugs

This category of substances is found to be the very least one that can be misused. It has only a very low level of psychological and physical dependence. Some of the drugs that fall into this schedule are atropine, pregabalin, codeine and ezogabine. Some of these had been used in the cough preparations with a minimum percentage of 200 mg of codeine in 100 ml per 10 gm for cough suppression. A small dose of non-injectable preparations can be obtained at pharmacies without a prescription, but once acquired, it is illegal to hand them over to another individual [85]. Many preparations fall into this category including well-known cough syrups and pain relievers. There are no limitations regarding the import, export, possession, administration or disposal of these preparations, nor do secure storage regulations apply. A practitioner, pharmacist or an individual possessing a suitable license is permitted to generate or produce such substances and the invoices for this category of drugs can be maintained for two years.

7 Therapeutic Applications of Regulated Substances

Although regulated substances carry the potential misuse and dependence, numerous compounds also provide notable healing advantages. Medications listed in Schedule II to v are very crucial for addressing a variety of health issues, ranging from pain relief to anxiety disorders. In this section, numerous applications of regulated substances were discussed in detail.

7.1 Pain Relief

A prevalent therapeutic application of regulated substances is in alleviating intense pain. Opioids categorized under Schedule II including morphine, oxycodone and fentanyl, prove highly effective in managing both acute and chronic pain conditions [86]. These compounds operate by combining with opioid receptors situated in the brain and spinal cord, obstructing pain signals and providing comfort. An investigation conducted in the year 2017 assessed the effectiveness of opioid treatments in chronic pain management. The findings indicated that while opioids are beneficial for immediate pain relief, their prolonged use involves risks of addiction and overdose. This study has been pivotal in comprehending the opioid epidemic and reconciling the necessity for pain relief with the potential dangers of dependence. A major public health challenge linked to controlled substances is the opioid epidemic, which has been intensified by the rampant usage of prescribed analgesics [87]. Opioids such as oxycodone, hydrocodone and fentanyl, categorized as Schedule II substances, have caused a spike in addiction, overdose fatalities and social repercussions. The Centres for Disease Control and Prevention (CDC) has indicated that over 500,000 individuals have succumbed to opioid overdoses in the United States since 1999. Research published in 2019 disclosed that between 2006 and 2015, there was a notable rise in opioid prescriptions and opioid-related overdose fatalities. The study indicates that excessive prescribing of opioids and the ensuing misuse of these drugs significantly contributed to the epidemic. In reaction to the opioid crisis, regulatory bodies have implemented measures to restrict access to prescribed opioids and promote the option of alternative pain management techniques. These strategies. These strategies encompass stricter prescribing protocols, increased utilization of prescription drug monitoring programs (PDMPs), and the encouragement of nonopioid pain relief options [88].

7.2 Sedation and Anaesthesia

Some of the controlled substances that are allowed to be used or injected into the human body to unconscious when undergoing any operations are termed sedative substances. A most common and necessary sedative is called as anaesthesia and the drugs used such as propofol and ketamine will come under schedule III, which is on the surgical side [89]. These are induced to relax the muscles they operate, ensuring the patients who undergo surgeries are not in pain. The study entitled Ketamine as a Novel Antidepressant: From Synapse to Behaviour by J.W. Murrough and his team of researchers highlights the use of ketamine in anaesthesia and its potential towards treating depression and providing rapid relief of depression symptoms, respectively [90].

7.3 Addiction Substances

Addiction is a multifaceted and enduring medical disorder that impacts millions globally. It is defined as the uncontrollable consumption of substances, regardless of detrimental effects. Substance addiction can include a broad array of items, such as alcohol, prescription medications, illegal drugs and even activities like gambling [91]. This chapter delves into the essence of substance addiction, the fundamental factors that contribute to it, its influence on society, and the contemporary methods for treatment and recovery as shown in Fig. 2. Addiction is also known as substance use disorder (SUB) it is a medical condition that develops a psychological and physical dependence on a substance. This makes an individual who takes a substance regularly and gets addicted to that without knowing to have uncontrollable cravings for such substances. If they don't get that instantly they behave very aggressively and control until taken.

7.4 Categories of Frequently Misused Substances

Dependency can arise from various substances, encompassing both legal and illicit drugs, alcoholic beverages and even specific behaviours such as gambling or overeating. Among the most frequently misused substances are discussed as follows. (i) *Beverages:* Alcohol is one of the most prevalent and socially embraced substances, nevertheless, overindulgence can result in alcohol use disorder (AUD), characterized by the inability to regulate alcohol consumption and the emergence of withdrawal symptoms upon decreasing or ceasing intake [92]. (ii) *Narcotics:* Narcotics, which include prescribed pain management medications like oxycodone, morphine and heroin, rank amongst the most habit-forming substances. Opioid dependency is especially alarming due to the threat of overdose and the ongoing opioid epidemic contributing to numerous fatalities worldwide [93]. (iii) *Stimulants:* Stimulants are substances such as cocaine, and methamphetamine and prescribed stimulants like Adderall are notably addictive. They activate the brain's reward pathways and may lead to intense physical and psychological reliance [94]. (iv) *Marijuana:* Although marijuana is frequently deemed less addictive compared to other substances, certain

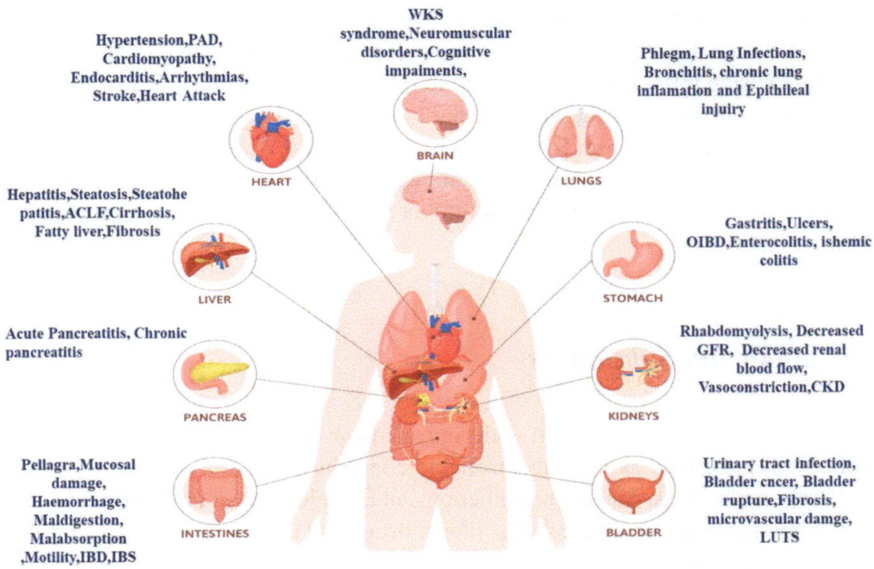

Fig. 2 Incidence of substance abuse-related diseases in humans

individuals can cultivate a dependency, particularly with the rising prevalence of high-potency cannabis products. (v) *Benzodiazepines:* Drugs such as diazepam (Valium) and alprazolam (Xanax) are indicated for anxiety and sleep disturbances. However, their tranquillizing effects can foster dependence, especially when taken for extended periods or misused [95].

7.5 Biological Factors, Effects of the Dependency on Individuals and the Society

Hereditary elements significantly influence addiction. Research indicates that people with a familiar background of addiction have a higher likelihood of developing substance use disorders [96]. A study published in '*Nature Neuroscience*' in 2012 pinpointed various genes linked it addiction risk which is particularly associated with the dopamine system which is crucial for processing rewards and motivation. The repercussions of dependency reach far beyond the individual, impacting families, communities and society at large. Dependency can trigger financial difficulties, strained relationships, unemployment and legal challenges. In extreme cases, dependency can lead to unlawful activities, such as drug trafficking thievery, or violent crimes [97]. Additionally, dependency significantly contributes to public health concerns, with individuals hooked on substances facing a heightened risk for infectious diseases. The financial burden of dependency is mind-boggling. As per a report from the National Institute on Drug Abuse (2017), the overall expense of substance misuse in the United States, encompassing healthcare expenses, lost productivity and crime-related costs, surpasses $740 billion each year. This statistically underscores the extensive implications of dependency for society.

7.6 The Addiction Effect on Neurology

Remarkable strides in neuroscience have enhanced our comprehension of the impacts that substances have on the brain, leading to addiction, which has resulted in the acknowledgement that addiction is a long-term brain disorder requiring treatment akin to any other health issue. Concurrently, our deepening insight into the neurobiological mechanisms linked to susceptibility to drug use and Substance Use Disorders (SUD), including the repercussions of facing detrimental environmental factors, is aiding in reshaping our perspectives on customized prevention strategies for vulnerable individuals. Nonetheless, considerable effort remains to encapsulate the intricacy of how drugs and other rewards affect our brains, to grasp how they interact, and how they ultimately drive behaviour [98]. A study demonstrated that the use of substances is propelled by the rewarding pharmacological effects they offer and is shaped by genetic, developmental and psychological influences that determine drug availability, societal norms and the presence or absence of supportive social structures. The reinforcing impact of substances is largely contingent on dopamine transmission in the nucleus accumbent. Chronic exposure to these substances initiates glutamatergic-driven neuroadaptions in dopaminergic striato-thalamo-cortical circuits (*primarily in prefrontal regions such as the orbitofrontal cortex and anterior cingulate cortex*) and limbic systems (*Including the amygdala and hippocampus*), which can lead to addiction in susceptible individuals [99]. Concurrently, alterations in the extended amygdala evoke negative emotional conditions that drive continued substance use as a means of temporary relief. Paradoxically, in those with addiction, the act of consuming the substance is linked to a diminished dopamine surge in the brain's reward areas, potentially fuelling a need to use drugs to bridge the gap between the anticipated reward, conditioned by drug-related cues and the actual reward experienced as shown in Fig. 3. Together, these phenomena lead to increased motivation to pursue the substance (*enhanced by dopamine increases linked to drug-related cues*) alongside compromised prefrontal self-regulation that promotes compulsive substance consumption amid emotions and heightened awareness of substance craving. Therapeutic interventions designed to counteract the neuroadaptions exhibit potential as effective strategies for treating addiction.

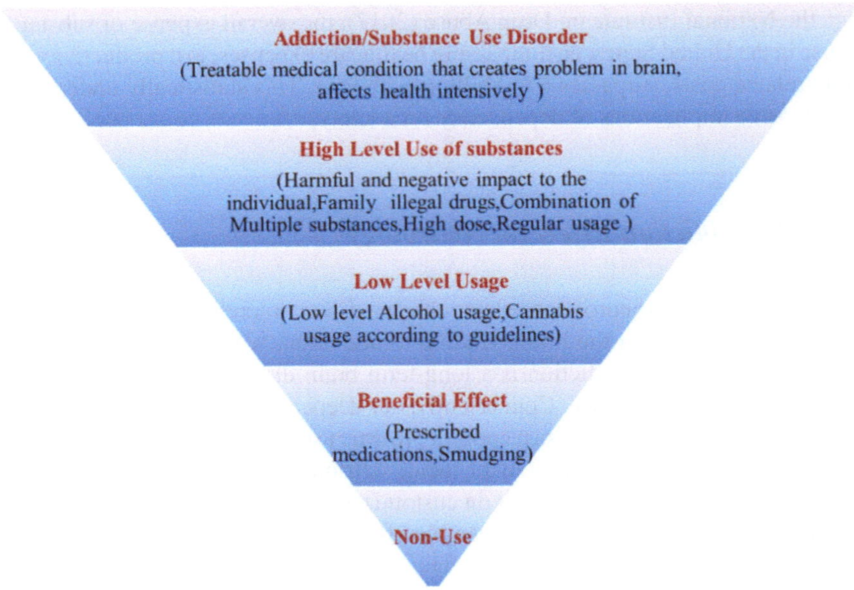

Fig. 3 Level of risks of using substances by an individual

8 Level of Risks and Addiction

Addiction is a multifaceted, intricate condition that affects millions around the globe. The notion of addiction isn't simply binary, with obvious Yes or No; rather, it resides on a continuum. This continuum fluctuates based on the substance in question, the individual's genetic composition, environmental factors, as well as the psychological and behavioural dynamics involved [100]. Within this continuum, it is vital to recognize the varying degrees of risks that different substances pose for addiction and how particular behaviours, situations or substances can heighten the chances of developing an addiction. The elements that affect the transition from sporadic use to dependency encompass the substance's physical and mental impacts, the frequency of consumption, the individual's susceptibility to addiction and external circumstantial factors such as stress or trauma.

8.1 Low-Risk Substances

Low-risk substances present a comparatively small risk for addiction. These substances typically do not elicit strong physical cravings or withdrawal symptoms. Although they may become habit-forming, they are less prone to result in severe dependence [101].

- *Caffeine*

  One of the most widely consumed psychoactive agents globally, caffeine exerts mild stimulating effects. While users may develop tolerance and withdrawal could trigger headaches and irritability, the potential for addiction remains relatively minimal. The risk of dependence is more associated with habitual usage rather than physical reliance. It ranks as the most widely consumed substance globally [102]. While ingesting low to moderate levels of caffeine is typically harmless, a growing body of clinical research indicates that certain individuals may become reliant on this substance and struggle to cut back despite being aware of persistent health issues linked to continued use. Consequently, the WHO and several healthcare professionals acknowledge caffeine dependence as a clinical condition. In this, we encapsulate the existing studies on a biological basis for caffeine dependence and a systematic analysis of the prevalence of caffeine dependence and the frequency of clinically significant signs of distress and functional impairment among regular caffeine consumers. Here the diagnostic criteria for caffeine use disorder is a topic designated for further exploration in the diagnostic and statistical. In the manual of mental disorders (5th edition) a proposal for a research framework to steer future clinical, epidemiological and genetic studies regarding caffeine dependence [103]. Various controlled laboratory studies reveal that caffeine induces behavioural and psychological effects akin to those of other addictive substances. Additionally, several recent clinical investigations suggest that caffeine dependence constitutes a clinically relevant disorder impacting a substantial segment of caffeine consumers. However, further research is essential to ascertain the reliability, validity and prevalence of this significant health concern.
- *Nicotine (at lower doses)*

  Although nicotine is highly addictive, it can be ingested at moderate levels without resulting in profound dependence, especially when consumed intermittently. However, prolonged use can culminate in addiction and health issues, as observed in the millions of smokers worldwide [104].

8.2 Moderate—Risk Substances

Substances classified as moderate risk possess a greater likelihood of leading to addiction because of their psychoactive traits and the increased possibility of developing tolerance and experiencing withdrawal symptoms [105]. These substances are frequently legal and can be socially embraced, which fosters their widespread consumption.

- *Alcohol*

Alcohol is commonly consumed for leisure purposes but presents a moderate addiction risk, particularly when ingested in large quantities. Extended and excessive drinking can result in tolerance, physical dependence and withdrawal symptoms such as tremors, anxiety and seizures. Long-term alcohol misuse is also associated with liver disorders, neurological impairments and mental health challenges [106].

- *Cannabis*

  The potential for cannabis addiction remains a topic of discussion, yet studies reveal that approximately 9% of users develop cannabis use disorder (CUD), with this figure rising for those who begin using it as teenagers or consume it daily. Cannabis dependence is mainly psychological, although some individuals might experience withdrawal symptoms like irritability, insomnia and reduced appetite [107]. A research study conducted by Vander Pol and his team in 2013 suggests that cannabis dependence correlates with various adverse psychological outcomes and can manifest in those who use cannabis regularly and for extended durations. The findings indicated that cannabis use disorder was associated with heightened risks for anxiety and depression.

8.3 Substances of Significantly (High) Risk

Substances of significant risk are those that possess a considerable likelihood of leading to addiction. These materials may precipitate physical dependence and a myriad of severe withdrawal reactions. Individuals who engage with these substances face a markedly elevated risk of developing comprehensive addiction, and the resultant physical, psychological and societal repercussions can be catastrophic [108].

- **Opioids**

  Opioids, which include both prescribed (*such as oxycodone and morphine*) and illegal (*like heroin and fentanyl*) variants, rank among the most habit-forming substances. They function by adhering to opioid receptors located in the brain and spinal cord, eliciting pain alleviation and profound sensations of euphoria [109]. The high tendency for addiction stems from the overwhelming pleasurable experiences, which cause a swift escalation of tolerance. Opioid dependency is frequently linked with withdrawal phenomena, encompassing intense pain, nausea, vomiting and in severe instances, fatality. Cocaine and methamphetamines are potent stimulants with a substantial propensity for misuse and addiction. Both substances induce a considerable release of dopamine in the brain, resulting in strong feelings of euphoria, heightened energy and increased self-assurance. Over time, regular usage can lead to significant psychological dependence, with users often pursuing the substance to stave off the feelings of dysphoria and exhaustion that arise as the effects diminish [70].

9 Transition from Risk to Dependency

The transition from casual use to severe dependency differs for each individual, but it typically involves the following phases.

(a) *Exploration:* Infrequent usage, often spurred by social pressures or a sense of curiosity.
(b) *Social Consumption:* Substances are ingested regularly for enjoyment, yet the person still exercises control.
(c) *Sensitization:* The individual starts requiring greater quantities of the substance to obtain the sought-after experience.
(d) *Reliance:* Both physical and mental reliance emerge, resulting in withdrawal symptoms when the substance is not consumed.
(e) *Dependency:* The individual has relinquished control over their substance use, and it becomes the central aspect of their existence.

10 Assessment Tools for Addiction

Addiction is an intricate and multifaceted state that impacts the physical, mental and social dimensions of a person's life. Accurate assessment is essential for recognizing and diagnosing addiction, formulating treatment plans and evaluating progress [110]. Addiction evaluations not only assist healthcare practitioners in comprehending the intensity and nature of substance use or addictive behaviour but also aid in customizing interventions that fit the unique requirements of the individual. Addiction evaluation involves examining an individual's substance consumption, behavioural patterns and the repercussions of their use on physical, emotional and social health [18]. This examination usually consists of collecting information from various sources, including clinical interviews, standardized questionnaires and objective indicators, to develop a thorough insight into the individual's addiction.

11 A Comprehensive Assessment Empowers Healthcare Experts to

(a) *Determine the existence of addiction:* This encompasses diagnosing substance use disorders and assessing the severity of the addiction.
(b) *Comprehend the context:* Recognizing environmental, genetic, psychological and social influences that contribute to the individual's substance use.
(c) *Steer treatment planning:* Assessments can direct the most suitable treatment strategy, which may involve therapy, medication and support assistance.

(d) *Track progress:* Regular assessments facilitate monitoring the individual's reaction to treatment and modifying strategies as necessary. Successful addiction evaluations play a crucial part in guaranteeing that patients receive customized, evidence-based interventions, which is vital for enhancing long-term results.

- *Biological Factors*
  Biological examinations deliver factual information regarding the existence of substances within an individual's body. These examinations are utilized alongside other evaluative tools to confirm substance use, oversee sobriety and identify relapses. Although they are unable to gauge the intensity of addiction, they yield crucial insights into a person's recent substance consumption [34].
- *Urine Drug Analysers*
  These analysers are frequently employed in healthcare environments to check for the existence of illegal drugs, prescribed medications or alcohol. They are commonly utilized in addiction recovery programs to supervise compliance with recovery protocols and spot relapses [35].
- *Blood Analyses*
  Blood analyses can be utilized to evaluate the presence of alcohol or particular drugs within the body. For instance, blood alcohol concentration (BAC) testing is applied to determine immediate intoxication levels.
- *Hair Strand Testing*
  Hair strand examinations are frequently used for extended monitoring as they can identify drug use over several months. They are less often applied in initial evaluations but may prove beneficial for tracking prolonged recovery. Biological examinations offer objectivity and accuracy but are constrained in their ability to assess the wider context of addiction, such as emotional and psychological dependency [36].
- *Observation and Behaviour Evaluations*
  Observation and behaviour evaluations entail clinicians scrutinizing an individual's actions. And interactions to uncover evidence of substance use addictive behaviours. These evaluations might involve noting alterations in an individual's physical condition, mood, actions or social interactions [49]. Clinicians might also apply '*Behavioural Evaluation tools*' which gauge behaviours like impulse management and risk-taking commonly associated with addiction.

12 Conclusion

An overview of substance use is a complex and tremendous issue that carries both positive and negative sides in one individual life. Some of the natural substances are harmful and beneficial for the health of mankind like turmeric which cures wounds and gives a heavenly feel around us, but if the use is more than limited, it may cause some allergies to some persons. Some organic substances which a derivative of some natural substances named as aspirin and penicillin has a lot of benefits it is used for many purposes and save countless lives as well. It is very essential to approach substance use with respect, precautionary measures and most importantly a deep understanding is needed to be known about the purpose risks and benefits of a substance. The responsibility of one who uses substances is a critical, aspect of maintaining public health, individual well-being, etc. By knowing the purposes, risks and benefits and awareness of substance use we can provide the needed and necessary substances with more benefits and fewer risks.

References

1. Scalbert, A., & Williamson, G. (2000). Dietary intake and bioavailability of polyphenols. *The Journal of Nutrition, 130*, 2073S-2085S.
2. Bast, A., & Haenen, G. R. (2013). Ten misconceptions about antioxidants. *Trends in Pharmacological Sciences, 34*, 430–436.
3. Halliwell, B. (1991). Drug antioxidant effects. *Drugs, 42*, 569–605.
4. Farnsworth, N. R., Akerele, O., Bingel, A. S., Soejarto, D. D., & Guo, Z. (1995). Medicinal plants in therapy. *Bulletin of the World Health Organization, 63*, 965–981.
5. Thoms, S., Grønborg, S., & Gärtner, J. (2009). Organelle interplay in peroxisomal disorders. *Trends in Molecular Medicine, 15*, 293–302.
6. Rodrigo, R., Guichard, C., & Charles, R. (2007). Clinical pharmacology and therapeutic use of antioxidant vitamins. *Fundamental & Clinical Pharmacology, 21*, 111–127.
7. Kohen, R., & Nyska, A. (2002). Oxidation of biological systems: oxidative stress phenomena, antioxidants, redox reactions, and methods for their quantification. *Toxicologic Pathology, 30*, 620.
8. Prior, R. L., Wu, X., & Schaich, K. (2005). Standardized methods for the determination of antioxidant capacity and phenolics in foods and dietary supplements. *Journal of Agricultural and Food Chemistry, 53*, 4290–4302.
9. Espinoza, A., San, M. A., López-Climent, M., Ruiz-Lara, S., Gómez-Cadenas, A., & Casaretto, J. A. (2013). Engineered drought-induced biosynthesis of α-tocopherol alleviates stress-induced leaf damage in tobacco. *Journal of Plant Physiology, 170*, 1285–1294.
10. Biswas, A., Amarajeewa, M., Senapati, S., Sahu, M., & Maiti, P. (2018). Sustained release of herbal drugs using biodegradable scaffold for faster wound healing and better patient compliance. *Nanomedicine, 14*, 2131–2141.
11. Kandola, A., Lewis, G., Osborn, D. P. J., Stubbs, B., & Hayes, J. F. (2020). Depressive symptoms and objectively measured physical activity and sedentary behaviour throughout adolescence: a prospective cohort study. *The Lancet Psychiatry, 7*, e262–e271.
12. Darke, S., Kelly, E., & Ross, J. (2004). Drug driving among injecting drug users in Sydney, Australia: Prevalence, risk factors and risk perceptions. *Addiction, 99*, 175–185.
13. Bergen, S. E., Gardner, C. O., Aggen, S. H., & Kendler, K. S. (2008). Socioeconomic status and social support following illicit drug use: Causal pathways or common liability? *Twin Research and Human Genetics, 11*, 266–274.
14. Berridge, V. (2012). The rise, fall, and revival of recovery in drug policy. *The Lancet, 379*(9810), 22–23.
15. Tiffany, S. T., Friedman, L., Greenfield, S. F., Hasin, D. S., & Jackson, R. (2012). Beyond drug use: A systematic consideration of other outcomes in evaluations of treatments for substance use disorders. *Addiction, 107*(4), 709–718.
16. Heather, N. (2017). Q: Is addiction a brain disease or a moral failing? A: Neither. *Neuroethics, 10*(1), 115–124.
17. Wu, L.-T., Zhu, H., & Ghitza, U. E. (2018). Multicomorbidity of chronic diseases and substance use disorders and their association with hospitalization: Results from electronic health records data. *Drug and Alcohol Dependence, 192*, 316–323.

18. Dennis, M., & Scott, C. K. (2007). Managing addiction as a chronic condition. *Addiction Science & Clinical Practice, 4*(1), 45–71.
19. McLellan, A. T., Lewis, D. C., O'Brien, C. P., & Kleber, H. D. (2000). Drug dependence, achronic medical illness: Implications for treatment, insurance, and outcomes evaluation. *JAMA, 284*(13), 1689–1695.
20. Galea, S., Rudenstine, S., & Vlahov, D. (2005). Drug use, misuse, and the urban environment. *Drug and Alcohol Review, 24*(2), 127–136.
21. Koob, G. F., & Volkow, N. D. (2010). Neurocircuitry of addiction. *Neuropsychopharmacology, 5*(1), 217–238.
22. Hyman, S. E., Malenka, R. C., & Nestler, E. J. (2006). Neural mechanisms of addiction: The role of reward-related learning and memory. *The Annual Review of Neuroscience, 29*, 565–598.
23. Goldman, D., Oroszi, G., & Ducci, F. (2005). The genetics of addictions: Uncovering the genes. *Nature Reviews Genetics, 6*(7), 521–532.
24. Kendler, K. S., Karkowski, L. M., Neale, M. C., & Prescott, C. A. (2000). Illicit psychoactive substance use, heavy use, abuse, and dependence in a US population-based sample of male twins. *Archives of General Psychiatry, 57*(3), 261–269.
25. Arria, A. M., & McLellan, A. T. (2012). Evolution of concept, but not action, in addiction treatment. *Substance Use and Misuse, 47*, 1041–1048.
26. Heyman, G. M. (2013). Quitting drugs: quantitative and qualitative features. *Annual Review of Clinical Psychology, 9*, 29–59.
27. Kingston, D. G. (2011). Modern natural products drug discovery and its relevance to biodiversity conservation. *Journal of Natural Products, 74*, 496–511.
28. Dias, D. A., Urban, S., & Roessner, U. (2012). A historical overview of natural products in drug discovery. *Metabolites, 2*, 303–336.
29. Cragg, G. M., & Newman, D. J. (2013). Natural products: A continuing source of novel drug leads. *Biochimica et Biophysica Acta, 1830*, 3670–3695.
30. Cragg, G. M., & Newman, D. J. (2005). Biodiversity: A continuing source of novel drug leads. *Pure and Applied Chemistry, 77*, 7–24.
31. Hawk, K., & D'Onofrio, G. (2018). Emergency department screening and interventions for substance use disorders. *Addiction Science & Clinical Practice, 13*, 18.
32. Le Roux, C., Tang, Y., & Drexler, K. (2016). Alcohol and opioid use disorder in older adults: neglected and treatable illnesses. *Current Psychiatry Reports, 18*, 87.
33. Fairman, K. A., & Early, N. K. (2020). Treatment needs and service utilization in older U.S. adults evidencing high-risk substance use. *Journal of Aging and Health, 32*(10), 1363–1375.
34. Cardoso, B. R., Hare, D. J., Bush, A. I., & Roberts, B. R. (2017). Glutathioneperoxidase 4: A new player in neurodegeneration? *Molecular Psychiatry, 22*, 328–335.
35. Gao, K., Shi, X., & Wang, W. (2017). The life-course impact of smoking on hypertension, myocardial infarction and respiratory diseases. *Science and Reports, 7*(1), 1–7.
36. Fernández-Solà, J. (2015). Cardiovascular risks and benefits of moderate and heavy alcohol consumption. *Nature Reviews Cardiology, 12*(10), 576–587.
37. Fan, A. Z., Li, Y., Elam-Evans, L. D., & Balluz, L. (2013). Drinking pattern and blood pressure among non-hypertensive current drinkers: Findings from 1999–2004 national health and nutrition examination survey. *Clinical Epidemiology, 5*(1), 21–27.
38. Carlos, S., de Irala, J., Hanley, M., & Martínez-González, M. (2014). The use of expensive technologies instead of simple, sound and effective lifestyle interventions: aperpetual delusion. *Journal of Epidemiology and Community Health, 68*(9), 897–904.
39. Pruchno, R. A., Wilson-Genderson, M., & Heid, A. R. (2016). Multiple chronic condition combinations and depression in community-dwelling older adults. *Journals of Gerontology Series A: Biomedical Sciences and Medical Sciences, 71*(7), 910–915.
40. Jayawant, S. S., & Balkrishnan, R. (2005). The controversy surrounding OxyContin abuse: Issues and solutions. *Therapeutics and Clinical Risk Management, 1*(2), 77–82.
41. D'Aunno, T., Park, S. E., & Pollack, H. A. (2019). Evidence-based treatment for opioid use disorders: A national study of methadone dose levels, 2011–2017. *Journal of Substance Abuse Treatment, 96*, 18–22.

42. Thomas, C. P., Ritter, G. A., Harris, A. H. S., Garnick, D. W., Freedman, K. I., & Herbert, B. (2018). Applying American society of addiction medicine performance measures in commercial health insurance and services data. *Journal of Addiction Medicine, 12*(4), 287–294.
43. Rigg, K. K., March, S. J., & Inciardi, J. A. (2010). Prescription drug abuse & diversion: Role of the pain clinic. *Journal of Drug Issues, 40*(3), 681–702.
44. O'Donnell, J., Gladden, R. M., Mattson, C. L., Hunter, C. T., & Davis, N. L. (2020). Vital signs: Characteristics of drug overdose deaths involving opioids and stimulants—24 States and the District of Columbia, January–June 2019. *MMWR Morbidity and Mortality Weekly Report, 69*(35), 1189–1197.
45. Price, D. D. (2000). Psychological and neural mechanisms of the affective dimension of pain. *Science, 288*(5472), 1769–1772.
46. Closs, S. J., & Briggs, M. (2002). Patients' verbal descriptions of pain and discomfort following orthopaedic surgery. *International Journal of Nursing Studies, 39*(5), 563–572.
47. Tang, N. K., & Crane, C. (2006). Suicidality in chronic pain: A review of the prevalence, risk factors and psychological links. *Psychological Medicine, 36*(5), 575–586.
48. Lee, C. S., Kim, D., Park, S. Y., Lee, S. C., Kim, Y. C., & Moon, J. Y. (2019). Usefulness of the Korean version of the CAGE-adapted to include drugs combined with clinical predictors to screen for opioid-related aberrant behavior. *Anesthesia and Analgesia, 129*(3), 864–873.
49. Atluri, S., Akbik, H., & Sudarshan, G. (2012). Prevention of opioid abuse in chronic non-cancer pain: An algorithmic, evidence based approach. *Pain Physician, 15*(3), ES177–ES189.
50. Copenhaver, D. J., Karvelas, N. B., & Fishman, S. M. (2017). Risk management for opioid prescribing in the treatment of patients with pain from cancer or terminal illness: Inadvertent oversight or taboo? *Anesthesia and Analgesia, 125*(5), 1610–1615.
51. Arnow, B. A., Hunkeler, E. M., Blasey, C. M., Lee, J., Constantino, M. J., Fireman, B., Kraemer, H. C., Dea, R., Robinson, R., & Hayward, C. (2006). Comorbid depression, chronic pain, and disability in primary care. *Psychosomatic Medicine, 68*(2), 262–268.
52. Fishbain, D. A., Cole, B., Lewis, J. E., & Gao, J. (2014). Does pain interfere with antidepressant depression treatment response and remission in patients with depression and pain? An evidence-based structured review. *Pain Medications, 5*(9), 1522–1539.
53. Rasmussen-Barr, E., Held, U., Grooten, W. J. A., Roelofs, P. D. D. M., Koes, B. W., van Tulder, M. W., Wertli, M. M. (2017). Nonsteroidal anti-inflammatory drugs for sciatica: An . *Spine (Phila Pa 1976), 42*(8), 586–594.
54. McNicol, E. D., Midbari, A., Eisenberg, E. (2013). Opioids for neuropathic pain. *The Cochrane Database of Systematic Reviews, 8*, CD006146.
55. Von Korff, M., Kolodny, A., Deyo, R. A., & Chou, R. (2011). Long-term opioid therapy reconsidered. *Annals of Internal Medicine, 155*(5), 325–328.
56. Frieden, T. R., & Houry, D. (2016). Reducing the risks of relief-the CDC opioid-prescribing guideline. *New England Journal of Medicine, 374*(16), 1501–1504.
57. Dublin, S., Walker, R. L., Shortreed, S. M., Ludman, E. J., Sherman, K. J., Hansen, R. N., Thakral, M., Saunders, K., Parchman, M. L., & Von Korff, M. (2019). Impact of initiatives to reduce prescription opioid risks on medically attended injuries in people using chronic opioid therapy. *Pharmacoepidemiology and Drug Safety, 28*(1), 90–96.
58. Freynhagen, R., & Bennett, M. I. (2009). Diagnosis and management of neuropathic pain. *BMJ, 12*(339), b3002.
59. Cowan, J., & Devine, C. (2008). Food, eating and weight concerns of men in recovery from substance addiction. *Appetite, 50*(1), 33–42.
60. Cowan, J., & Devine, C. (2013). Diet and body composition outcomes of an environmental and educational intervention among men in treatment for substance addiction. *Journal of Nutrition Education and Behavior, 45*(2), 154–158.
61. Haque, M. M., Bhuiyan, M. R., Chowdhury, S., Islam, K. (2014). Nutritional status and dietary intake pattern of male drug addicts undergoing rehabilitation. *Journal of Nutritional Health & Food Engineering* 1(1).

62. Ross, L. J., Wilson, M., Banks, M., Rezannah, F., & Daglish, M. (2012). Prevalence of malnutrition and nutritional risk factors in patients undergoing alcohol and drug treatment. *Nutrition, 28*, 738–743.
63. Anderson, E. E., & DuBois, J. M. (2007). The need for evidence-based research ethics: A review of the substance abuse literature. *Drug and Alcohol Dependence., 86*(2), 95–105.
64. Molfenter, T., Boyle, M., Holloway, D., Zwick, J. (2015). Trends in telemedicine use in addiction treatment. *Addiction Science & Clinical Practice, 10*(14).
65. Bapat, S., Thornton, J. D., & Fleming, M. L. (2019). Changes in analgesic prescribing among Texas advanced practice providers after rescheduling of hydrocodone combination products. *Value in Health, 22*(2), S257–S258.
66. Sutherland, R., Peacock, A., Nielsen, S., Bruno, R. (2020). Alprazolam use among a sample of Australians who inject drugs: Trends up to six years post regulatory changes. *International Journal of Drug Policy, 79.*
67. Arigo, D., Jake-Schoffman, D. E., Wolin, K., Beckjord, E., Hekler, E. B., & Pagoto, S. L. (2019). The history and future of digital health in the field of behavioral medicine. *Journal of Behavioral Medicine, 42*, 67–83.
68. Farzam, K., Faizy, R.M., Saadabadi, A. (2024). Stimulants. In *StatPearls* (Internet), StatPearls Publishing.
69. Grant, B. F., Saha, T. D., & Ruan, W. J. (2016). Epidemiology of DSM-5 drug use disorder: Results from the national epidemiologic survey on alcohol and related conditions-III. *JAMA Psychiatry., 73*, 39–47.
70. Walsh, Z., Gonzalez, R., Crosby, K., Thiessen, M. S., Carroll, C., & Bonn-Miller, M. O. (2017). Medical cannabis and mental health: A guided systematic review. *Clinical Psychology Review, 51*, 15–29.
71. Meier, M. H., Hall, W., & Caspi, A. (2016). Which adolescents develop persistent substance dependence in adulthood? Using population representative longitudinal data to inform universal risk assessment. *Psychological Medicine, 46*, 877–889.
72. Dong, X., Deng, J., & Rashidian, S. (2021). Identifying risk of opioid use disorder for patients taking opioid medications with deep learning. *Journal of the American Medical Informatics Association, 28*, 1683–1693.
73. Barenholtz, E., Fitzgerald, N. D., & Hahn, W. E. (2020). Machine-learning approaches to substance-abuse research: Emerging trends and their implications. *Current Opinion in Psychiatry, 33*, 334–342.
74. Hatsukami, D. K., Benowitz, N. L., Rennard, S. I., Oncken, C., & Hecht, S. S. (2006). Biomarkers to assess the utility of potential reduced exposure to Bacco products. *Nicotine & Tobacco Research, 8*(4), 600–622.
75. Cunha-Oliveira, T., Cristina Rego, A., & Oliveira, C. R. (2015). Oxidative stress and drugs of abuse. *Mini-Reviews in Organic Chemistry, 12*(6), 1–12.
76. Turnbull, D. K., & Shepherd, D. B. (2003). Post-dural puncture headache: Pathogenesis, prevention and treatment. *British Journal of Anaesthesia, 91*(5), 718–729.
77. Ford, J. A., & Rigg, K. K. (2015). Racial/Ethnic differences in factors that place adolescents at risk for prescription opioid misuse. *Prevention Science, 16*, 633–641.
78. Ford, J. A., & Blumenstein, L. (2013). Self-Control and substance use among college students. *Journal of Drug Issues, 43*, 56–68.
79. Austin, A. E., Shanahan, M. E., & Zvara, B. J. (2018). Association of child abuse and prescription opioid use in early adulthood. *Addictive Behaviors, 76*, 265–269.
80. Greydanus, D. E., & Patel, D. R. (2003). Substance abuse in adolescents: A complex conundrum for the clinician. *Pediatric Clinics of North America, 50*, 1179–1223.
81. Whitesell, M., Bachand, A., Peel, J., Brown, M. (2013). Familial, social, and individual factors contributing to risk for adolescent substance use. *Journal of Addiction*, 579310.
82. Steinhoff, A., Bechtiger, L., Ribeaud, D., Eisner, M. P., Quednow, B. B., & Shanahan, L. (2021). Polysubstance use in early adulthood: Patterns and developmental precursors in an urban cohort. *Frontiers in Behavioral Neuroscience, 15*, 797473.

83. Afuseh, E., Pike, C. A., & Oruche, U. M. (2020). Individualized approach to primary prevention of substance use disorder: Age-related risks. *Substance Abuse Treatment, Prevention, and Policy, 15*, 58.
84. Jadidi, N., Nakhaee, N. (2014). Etiology of drug abuse: a narrative analysis. *Journal of Addiction*, 352835.
85. Santucci, K. (2012). Psychiatric disease and drug abuse. *Current Opinion in Pediatrics, 24*, 233–237.
86. Al-Eitan, L. N., Rababa'h, D. M., & Alghamdi, M. A. (2021). Genetic susceptibility of opioid receptor genes polymorphism to drug addiction: A candidate-gene association study. *BMC Psychiatry, 21*, 5.
87. Sansgiry, S. S., Bhansali, A. H., Bapat, S. S., & Xu, Q. (2017). Abuse of over-the-counter medicines: a pharmacist's perspective. *Integrated Pharmacy Research and Practice, 6*, 1–6.
88. Morales, A. M., Jones, S. A., Kliamovich, D., Harman, G., & Nagel, B. J. (2020). Identifying early risk factors for addiction later in life: A review of prospective longitudinal studies. *Current Addiction Reports, 7*, 89–98.
89. Mills, R., Kisely, S., Alati, R., Strathearn, L., & Najman, J. M. (2017). Child maltreatment and cannabis use in young adulthood: A birth cohort study. *Addiction, 112*, 494–501.
90. Habib, A. A., & Mitsumoto, H. (2011). Emerging drugs for amyotrophic lateral sclerosis. *Expert Opinion on Emerging Drugs, 16*, 537–558.
91. Carim-Todd, L., Mitchell, S. H., & Oken, B. S. (2013). Mind-body practices: an alternative, drug free treatment for smoking cessation? A systematic review of the literature. *Drug and Alcohol Dependence, 132*(3), 399–410.
92. Large, M., & Nielssen, O. (2017). Daily use of high-potency cannabis is associated with an increased risk of admission and more intervention after first-episode psychosis. *Evidence-Based Mental Health, 20*(2), 58.
93. Freeman, T. P., & Winstock, A. R. (2015). Examining the profile of high-potency cannabis and its association with severity of cannabis dependence. *Psychological Medicine, 45*(15), 3181–3189.
94. Schulte, M. T., Hser, Y. I. (2013). Substance use and associated health conditions throughout the lifespan. *Public Health Reviews, 35*(2).
95. Marin, S., Heshmatian, E., Nadrian, H., Fakhari, A., & Mohammadpoorasl, A. (2019). Associations between optimism, tobacco smoking and substance abuse among Iranian high school students. *Health Promotion Perspective, 9*(4), 279–284.
96. Schinke, S., Schwinn, T., Hopkins, J., & Wahlstrom, L. (2016). Drug abuse risk and protective factors among Hispanic adolescents. *Preventive Medicine Reports, 3*, 185–188.
97. Wilson, J. D., Vo, H., Matson, P., Adger, H., Barnett, G., & Fishman, M. (2017). Trait mindfulness and progression to injection use in youth with opioid addiction. *Substance Use and Misuse, 52*(11), 1486–1493.
98. Osborne, V., Serdarevic, M., Striley, C. W., Nixon, S. J., Winterstein, A. G., & Cottler, L. B. (2020). Age of first use of prescription opioids and prescription opioid non-medicaluse among older adolescents. *Substance Use Misuse., 55*(14), 2420–2427.
99. Biederman, J., Faraone, S. V., Monuteaux, M. C., & Feighner, J. A. (2000). Patterns of alcohol and drug use in adolescents can be predicted by parental substance use disorders. *Pediatrics, 106*(4), 792–797.
100. Whitesell, M., Bachand, A., Peel, J., Brown, M. (2018). Familial, social, and individual factors contributing to risk for adolescent substance use. *Journal of Addiction*.
101. Lynskey, M. T., Heath, A. C., Bucholz, K. K., Slutske, W. S., Madden, P. A. F., Nelson, E. C., Statham, D. J., & Martin, N. G. (2003). Escalation of drug use in early onset cannabis users vs co-twin controls. *JAMA, 289*, 427–433.
102. Squeglia, L. M., Jacobus, J., & Tapert, S. F. (2009). The influence of substance use on adolescent brain development. *Clinical Neuroscience Society ENCS, 40*(1), 31–38.
103. Verebey, K., & Gold, M. S. (1988). From coca leaves to crack: The effects of dose and routes of administration in abuse liability. *Psychiatric Annals, 18*, 513–520.

104. Hatsukami, D. K., & Fischman, M. W. (1996). Crack cocaine and cocaine hydrochloride: Are the differences myth or reality. *JAMA, 276*, 1580–1588.
105. Krohn, M. D., Lizotte, A. J., & Perez, C. M. (1997). The interrelationship between substance use and precocious transitions to adult statuses. *Journal of Health and Social Behavior, 38*(1), 87–103.
106. Gogtay, N., Giedd, J. N., Lusk, L., Hayashi, K. M., Greenstein, D., Vaituzis, A. C., Nugent, T. F., 3rd., Herman, D. H., Clasen, L. S., Toga, A. W., Rapoport, J. L., & Thompson, P. M. (2004). Dynamic mapping of human cortical development during childhood through early adulthood. *Proceedings of the National Academy of Sciences, 101*(21), 8174–8179.
107. Moss, R., & Munt, B. (2003). Injection drug use and right sided endocarditis. *Heart, 89*(5), 577–581.
108. Kelly, T. M., & Daley, D. C. (2013). Integrated treatment of substance use and psychiatric disorders. *Social Work in Public Health, 28*, 388–406.
109. Ross, S., & Peselow, E. (2012). Co-occurring psychotic and addictive disorders: Neurobiology and diagnosis. *Clinical Neuropharmacology, 35*(5), 235–243.
110. Das, D., Kabir, M. E., Sarkar, S., Wann, S. B., Kalita, J., & Manna, P. (2022). Antidiabetic potential of soy protein/peptide: A therapeutic insight. *International Journal of Biological Macromolecules, 194*, 276–288.

In Vitro and In Vivo Evaluation of Antioxidant Capacity

Rhea Biju Luke, J. S. Reshma, Vaishnavi Ajith, Swayam Gupta, Madhuri Patel, Janarthanan Sivasankar, Sharine Priscilla, Sahabudeen Sheik Mohideen, and S. Beer Mohamed

Abstract Various chronic diseases like cardiovascular, neurological or metabolic illnesses are linked to oxidative imbalance that arises from the presence and effects of free radicals or reactive oxygen species. Antioxidants neutralize the highly reactive oxygen species, thereby diminishing oxidative stress and minimizing cellular damage caused by these free radicals. This chapter analyzes the primary methodologies used for evaluating antioxidant capacity both in vitro and in vivo, outlining their applications and limitations. In vitro assays, such as oxygen radical absorbance capacity (ORAC), 2,2-diphenyl-1-picrylhydrazyl (DPPH), ferric reducing antioxidant power (FRAP), 2,2'-azino-bis(3-ethylbenzothiazoline-6-sulfonic acid) (ABTS), cupric reducing antioxidant capacity (CUPRAC) and other electrochemical methods, provide controlled, mechanism-based insights into radical scavenging, metal chelation, as well as redox potential. In vivo models, including the superoxide dismutase (SOD) activity, glycine, cysteine, and glutamate (GSH) assay, enzyme-linked immunosorbent assay (ELISA), liquid chromatography-mass spectrometry (LC–MS), and imaging techniques such as electron paramagnetic resonance (EPR) and fluorescent probes, indicate physiological complexity, bioavailability, metabolism, and systemic distribution. This chapter provides a comprehensive assessment of antioxidant evaluation methods, highlighting the need for increased multifaceted approaches integrating the mechanism-based findings in vitro along with the biochemical and systemic interactions in vivo, thereby improving the applicational relevance of antioxidant research.

R. Biju Luke · J. S. Reshma · V. Ajith · S. Gupta · M. Patel · J. Sivasankar · S. Priscilla · S. Sheik Mohideen
Department of Biotechnology, School of Bioengineering, College of Engineering and Technology, SRM Institute of Science and Technology, Kattankulathur, Chengalpattu District 603203, Tamil Nadu, India
e-mail: sahabuds@srmist.edu.in

S. Beer Mohamed (✉)
Department of Materials Science, School of Technology, Central University of Tamil Nadu, Thiruvarur, Tamil Nadu, India
e-mail: peermoha@gmail.com

K. Kasinathan and A. Muthukumaran (eds.), *Recent Advances in Oxidative Stress Associated Chronic Diseases Volume 2*, https://doi.org/10.1007/978-981-95-3758-7_6

Keywords Antioxidant capacity · Oxidative imbalance · Reactive oxygen species · Antioxidants · Bioavailability

1 Introduction

Oxidative stress is defined by the disproportion between the generation of surplus free radicals or redox species (ROS) and the organism's capacity to neutralize them using antioxidants [1]. Free radicals including those such as superoxide (O_2^-) or hydrogen peroxide (H_2O_2) are highly reactive and induce damage to cellular biomolecules like protein and lipids by mechanisms such as lipid peroxidation, protein oxidation or DNA damage. This in turn implicates the healthy functioning of the body and leads to the development of various ailments, including cardiovascular conditions, neurodegenerative disorders and cancer. Some notable diseases or conditions linked to oxidative stress include Alzheimer's disease, Parkinson's disease, hypertension, heart failure, and atherosclerosis among others. Antioxidants such as superoxide dismutase (SOD), catalase (CAT), ascorbic acid, on the other hand refers to molecules that neutralize the free radicals thereby inhibiting or preventing cellular damage caused by these redox species in the biological cells. They are essential in maintaining the heath and proper functioning of an organism [2]. Thus, accurately evaluating the antioxidant capacity is essential as it helps to assess the impacts of various compounds as well as potential uses for health benefits in developing therapeutic measures. This chapter presents a comprehensive overview into the current methodologies and techniques employed in assessing antioxidant capacity, both in vitro assays such as FRAP, DPPH, ORAC and more as well as the in vivo assays and techniques such as ELISA, microdialysis, electron paramagnetic resonance imaging (EPR), HPLC discussing their respective principles, advantages, limitations and applications.

2 In Vitro Evaluation of Antioxidant Capacity

In vitro methods offer a controlled and efficient means of assessing the antioxidant capacity of compounds. These techniques evaluate antioxidant activity by free radical scavenging, metal ion chelating, and lipid peroxidation inhibition and have two mechanisms: Hydrogen atom transfer (HAT) method [3] and electron transfer (ET) method. In vitro tests measures metal chelation, oxidative process inhibition, and radical scavenging to assess antioxidant activity. These techniques promote the prospective use of antioxidants in pharmaceutical and nutraceutical applications by assisting in the determination of the efficacy and underlined processes [3]. Table 1 lists a few cell lines that are routinely employed for investigating the antioxidant capacity.

Table 1 Different cell lines used in cell–based bioassay for antioxidant activity analysis

Cell line	Origin
Caco-2	Colon adenocarcinoma
HepG2	Liver carcinoma
MCF-7	Breast cancer
AGS	Gastric adenocarcinoma
L-929	Murine subcutaneous connective tissue
RBC and PMN cells	Blood
Int 407	Embryonic, intestine
MKN-45	Gastric adenocarcinoma
HT29	Colon-rectal adenocarcinoma

2.1 Mechanism

2.1.1 Hydrogen Atom Transfer Method

Hydrogen atom transfer method measures the antioxidant capacity of a molecule with its potential to scavenge free radicals or the redox species by donating hydrogen atoms. The assay usually consists of three key components: an antioxidant, a probe molecule (*fluorescent or colored substrate*), and a synthetic radical generator. The antioxidant competes with the substrate or probe molecule for the radicals and decreases the signal intensity. This change in the signal can be correlated with the antioxidant capacity as the presence of antioxidants prevents or slows down the degradation of the probe molecule. The radical generator generates free radicals usually by thermal or photochemical methods which react with the substrate to produce a detectable signal. The solvent, pH, and the bond dissociation enthalpy of the group donating hydrogen are the main factors that determine this process. The hydrogen atom donated interrupts the propagation phase of radical- driven chain reactions.

$$RH + R\bullet \rightarrow R\bullet + H\bullet \tag{1}$$

$$R\bullet + O2 \rightarrow ROO\bullet \tag{2}$$

$$ROO\bullet + RH \rightarrow ROOH + R\bullet \tag{3}$$

$$ROO\bullet + AH \rightarrow ROOH + A\bullet \tag{4}$$

The A• formed is a stable product which ends the reaction. The results are obtained in the form of a kinetic curve which plots the change in signal over time. The curve exhibits a lag phase, a slow decay rate or both depending on the antioxidant used. This assay is especially used for measuring the radical chain-breaking antioxidant activity of molecules and helps in developing effective antioxidants for pharmaceutical and food applications [3].

2.1.2 Electron Transfer Method

The ability of antioxidants to transfer one electron to decrease any molecule is measured using the electron transfer method. It has two components: antioxidant and an oxidant probe. The oxidant probe scavenges the electron from antioxidant and prevents it from cell damage, causing the color change of the probe. The chain reaction of oxidation ends once the free radical is stabilized. The degree of color change is directly proportional to the concentration of antioxidants. The change in absorbance vs concentration of antioxidant is plotted to get a linear curve and slope of the curve is measured to find the antioxidant's reducing capacity in Trolox equivalence or Gallic acid equivalent [3]. This method depends on the proton-reducing and ionization capacity of the relative functional group and therefore is dependent on pH. Ionization potential falls and electron-donating capacity with deprotonation rises as pH increases. Due to their time-consuming nature, electron transfer methods often measure the product's relative percentage drop rather than its kinetics or overall antioxidant capability. Because charged species stabilize in solvents, the ET process is solvent dependent. Some of the most commonly used assays are:

$$O_x + e^- \rightarrow \text{colored compound} \tag{5}$$

$$O_x = \text{oxidant probe} \tag{6}$$

$$e^- = \text{electron donated by the antioxidant} \tag{7}$$

$$\text{Colored compound} = \text{reduced from of the oxidant} \tag{8}$$

2.2 *Oxygen Radical Absorbance Capacity Assay (ORAC)*

Oxygen radical absorbance capacity assay is used to measure the oxidative degradation of fluorescent molecules when combined with an oxygen radical initiator. As the oxidative degradation increases, the intensity of fluorescence decreases but when the antioxidants are present the fluorescence is protected and thus the intensity is preserved across time. Thus, the longer the fluorescence is maintained, the less is the degradation by the oxygen radicals. **β**-phycoerythrin was initially used as the indicator which was then replaced by fluorescein [4]. It is based on the principle of fluorescence quenching by the reactive oxygen compounds and the ability of antioxidants to prevent this quenching. [2,2'-Azobis(2-amidinopropane) dihydrochloride] (AAPH), usually used as the free radical generator, produces alkyl radicals by thermal decomposition which produces peroxyl radicals when reacting with the molecular oxygen. These peroxyl radicals reduce the hydrogen from fluorescein, leading to the formation of fluorescein radicals. This damages the fluorophore, disrupting its electronic structure and reducing the fluorescence intensity. The antioxidant donates hydrogen atoms to peroxyl radicals and prevents it from reacting with fluorescein, thereby breaking the chain reaction of oxidative damage. The fluorescence intensity is measured at 485–520 nm and compared with the control which is associated with the antioxidant capacity. The effectiveness is quantified by measuring the area under the curve of the fluorescence decay vs time graph, represented as Trolox equivalents.

$$\text{AAPH} \rightarrow 2\text{R}^{\cdot} \tag{9}$$

$$\text{R}^{\cdot} + \text{O2} \rightarrow \text{ROO}^{\cdot} \tag{10}$$

$$\text{ROO}^{\cdot} + \text{Fluorescein} \rightarrow \text{Oxidized fluorescein} + \text{Other products} \tag{11}$$

$$\text{ROO}^{\cdot} + \text{AH} \rightarrow \text{ROOH} + \text{A}^{\cdot} \tag{12}$$

2.2.1 Reagents

AAPH is made at a concentration of 79.65 mM (pH 7.4). The first and second stock solution of fluorescein is made by dissolving it in phosphate buffer. A Trolox dilution series (50–6.25 mM) serves as the standard, while phosphate buffer is used as a blank.

2.2.2 Procedure

- 400 μl of fluorescein and 40 μl of sample was pipetted into first well.
- Fluorescence reading is taken for the time 0.
- After 4 cycles, 150 μl of AAPH is added to the respective wells with the time interval of 90's.
- The fluorescence reading is taken up to 35 cycles and is used for calculating the area under the curve and Trolox equivalents [4].

2.3 CUPric Reducing Antioxidant Capacity (CUPRAC) Assay

The CUPRAC assay measures the potential of antioxidants to convert cupric ions (Cu^{2+}) to cuprous ion (Cu^{+}) when neocuproine (Nc) is present. It forms a stable and colored product ($Cu(Nc\ )_2^{+}$) which can be measured spectrophotometrically.

$$Cu(Nu)_2^{2+} + AH \rightarrow [Cu(Nc)_2]^{+} + A + H^{+} \quad (13)$$

The Cu(II)-neocuproine acts as an electron acceptor and the antioxidant reduces Cu(II) to Cu(I), resulting in the orange-colored complex. The intensity of the color is measured at the absorbance 450 nm [5]. The CUPRAC chromophore, $[Cu(Nc)_2]$ is univalent, charged and is soluble in both organic and aqueous solvents which enables us to determine the lipophilic and hydrophilic antioxidants [6].

2.3.1 Reagents

$CuCl_2$ solution (0.01 mM) is prepared by dissolving $CuCl_2$ in water and diluting to 250 mL. Ammonium acetate buffer (pH 7.0, 1.0 M) is made by dissolving NH_4 Ac in water and diluting to 250 mL. Neocuproine (Nc) solution (0.0075 mM) is prepared by dissolving it in 96% ethanol and diluting to 25 mL with ethanol. Trolox (0.01 mM) is prepared in 96% ethanol.

2.3.2 Procedure

- Depending on the type of sample, the CUPRAC method was used in four consecutive steps: normal (N), incubated (I), hydrolyzed (H), and hydrolyzed & incubated (H&I) versions of the tests
- *Normal sample*: 1 mL of each of the buffer solutions of Cu (II), Nc, and NH_4Ac was added to a test tube. To obtain the final volume of 4.1 mL, H_2O and the antioxidant sample (or standard) solution were added to the original mixture. After 0.5 h, the absorbance at 450 nm was measured against a blank.
- *Incubated sample*: The combination is prepared as in the case of normal sample and is kept for incubation in water bath for 20 min at the temperature 50 °C. They were cooled down and absorbance at 450 nm was measured.
- *Hydrolyzed Sample*: A weighed amount of polyphenol was dissolved in 50% methanol to make a 1 mM solution. Hydrochloric acid was added to reach 1.2 M final concentration, then the solution was refluxed at 80 °C for 2 h. After cooling and neutralizing with 1 M NaOH, the hydrolysis product was analyzed using the normal sample measurement.
- *Hydrolyzed and Incubated sample*: The neutralized hydrosylate was kept for incubation at 50 °C in a water bath for 20 min and absorbance was measured at 450 nm after it cools down.

- By dividing the molar absorptivity of the sample by that of Trolox under similar conditions, the CUPRAC coefficients of phenolic compounds with linear calibration curves passing through the origin were determined [7].

2.4 DPPH Assay

The 2,2-diphenyl-1-picrylhydrazyl (DPPH) radical scavenging assay measures the anti-radical properties of various compounds. It is one of the standard assays for determining the antioxidant capacity of pure compounds and evaluates the ability of antioxidant molecules to scavenge free radicals. DPPH is an organic nitrogen radical and appears purple, absorbing at 515 nm in methanol. The assay is rooted on the principle that when DPPH accepts hydrogen atom from the antioxidant, it reduces to DPP H_2, resulting in the color change from purple to yellow. By observing the decrease in absorbance at 515–528 nm until the absorbance becomes consistent in organic media, the lowering capacity of antioxidants toward DPPH can be calculated. The percentage of DPPH can be calculated using the formula [3].

$$\%\mathrm{DPPH}_{\mathrm{rem}} = 100 \times [\mathrm{DPPH}]_{\mathrm{rem}}/[\mathrm{DPPH}]_T = 0 \quad (14)$$

By removing a hydrogen atom from the antioxidants, the nitrogen atom's single electron in DPPH is reduced to its associated hydrazine. When the antioxidant is mixed with the solution, the violet color disappears, due to the formation of the DPPH radical. The dislocation of the unpaired electron over the whole molecule gives DPPH a stable free radical. This ensures that, unlike many other free radicals, the DPPH radical won't dimerize. As a result of this electron delocalization, the molecule also takes on a dark purple color and a maximum absorption of the ethanol solution at 517 nm. As the electron pairs are removed from the DPPH radical, absorbance diminishes. The stoichiometry of the electron number determines the resulting color flare [8].

2.4.1 Reagents

DPPH stock solution (0.2 mM) was prepared by dissolving it in methanol in an Eppendorf tube. Centrifuge and transfer to 500-ml volumetric flask. Rinse with methanol (MeOH) several times.

2.4.2 Protocol

- Add to the 96-wells microplate the following reagents: a column with DPPH only, a column for each sample only and a column with sample plus DPH in equal volumes.

- Perform double dilutions and add 100 μl of DPPH (0.2 mM) solution to each well in the plate.
- Reactions were performed in triplicates and incubated at 30 °C for 30 min in the dark, with tubes wrapped in foil.
- Absorbance at 517 nm was recorded.

2.5 Ferric Reducing Antioxidant Power (FRAP) Assay

A standard method for assessing the reducing potential of antioxidants in biological and food samples is the ferric reducing antioxidant power (FRAP) assay. This assay is based on the ability of antioxidants to reduce the ferric-tripyridyl triazine (Fe^{3+}-TPTZ) complex to its ferrous form (Fe^{2+}-TPTZ) at acidic pH, leading to a distinct blue color. The intensity of this color, measured at 593 nm, is directly proportional to the total reducing capacity of the sample. Due to its simplicity, speed, and reproducibility, the FRAP assay is extensively used in antioxidant research. However, it does not distinguish between different antioxidant mechanisms, such as radical scavenging, and cannot measure thiol-containing antioxidants like glutathione [9]. The FRAP assay functions by utilizing an excess of Fe^{3+}, ensuring that the limiting factor in the reaction is the antioxidant potential of the sample. However, one limitation of this method is that it measures only the reducing capacity, which does not always correlate with true antioxidant activity in biological systems. Despite this concern, the FRAP assay remains a valuable tool for assessing antioxidant potential, especially when used alongside other complementary assays. To address certain limitations, the ferric reducing ability of plasma with stabilized ascorbate (FRASC) modification was developed to allow the specific and concurrent assessment of ascorbic acid and total reducing power in the biological sample [9].

2.5.1 Reagents

- 10 mM TPTZ
- Solution A: 40 mM HCl, 20 nM $FeCl_3$, and O.1 M acetate buffer (pH 3.6) in a ratio of 1:1:10.
- The FRAP reagent is prepared by mixing TPTZ in solution A.

2.5.2 Protocol

The mixture is incubated at 37 °C for 10 min to ensure stability. Then, 30 μL of the sample is combined with 200 μL of the FRAP reagent, and the volume is adjusted to 1 mL with distilled water. After a 10-min incubation, the sample is centrifuged at 10,000 × g, and the absorbance of the supernatant is measured at 593 nm against a blank prepared using a saline solution instead of plasma [10]. This rapid and cost-effective assay provides valuable insights into the antioxidant properties of various biological and food samples while maintaining reproducibility and ease of use.

2.6 *GSH Assay*

As an intracellular key antioxidant, glutathione (GSH) is important in ensuring that there is cellular detoxification, metabolic control, and preservation of redox balance. GSH, which is a tripeptide comprised of glutamate, cysteine, and glycine, is essential in averting cellular oxidative stress and neutralizing reactive species of oxygen (ROS) as well as fostering various enzymatic activities. Also, it takes part in the transport of some amino acids especially in the functions of the kidney, and its deficiency is linked with many clinical disorders including cataract formations [11]. Thus, GSH assessment is crucial within the plane of clinical drug design and development, diagnosis, and forensic biomedical science. Most of the time, GSH measurement is still done by using traditional methods of spectrophotometry, mainly because of ease, as was the case with Ellman's method devised in 1959. The capacity to decrease the disulfide bond of 5,5'-dithiobis-2-nitrobenzoic acid (DTNB) or Ellman's reagent results in a yellow 2-nitro-5-thiobenzoate (TNB) anion that absorbs at 412 nm, which is the basis for the GSH assay. The extent of GSH present in the sample is proportionate to the level of color change. It correlates directly with the concentration of GSH present in the sample. The assay employs protein precipitation using trichloroacetic acid (TCA) and ethylenediaminetetraacetic acid (EDTA) to guarantee precise measurement by eliminating interfering proteins [11] technique is effective; however, it cannot tell the difference between reduced and oxidized GSH (GSSG).

2.6.1 Reagents

- 0.1 M phosphate buffer
- TCA
- 1 mM EDTA
- Ellman's Reagent (0.1 mM DTNB in 0.3 M phosphate buffer containing 1% sodium citrate)

2.6.2 Protocol

The procedure initiates with the homogenization of tissue in 0.1 M phosphate buffer (pH 7.4), subsequently incorporating an equal volume of 20% TCA that includes 1 mM EDTA to facilitate protein precipitation. Following a 5-min incubation period, the mixture undergoes centrifugation at 2000 rpm for a duration of 10 min. The supernatant (200 μL) is combined with 1.8 mL of Ellman's reagent, and the total reaction volume is set to 2 mL. Upon completion, the absorbance is recorded at 412 nm relative to a blank, and the concentration of GSH is ascertained utilizing a standard curve [11]. This assay offers a swift, economical, and dependable approach for evaluating antioxidant capacity; however, it might necessitate additional techniques for a more thorough examination of the redox state in biological systems. When conducting in vivo studies, the same assay can be carried out in order to evaluate the antioxidant capacity of a living cell.

2.7 TLC Bioautography Assay

Thin-layer chromatography (TLC) bioautography is a technique used for isolation, localization, and evaluation of the active compound; it imparts antibacterial activity. It helps in investigation of complex antimicrobial effects in vitro and in vivo. TLC bioautography, after separating the compounds on a TLC plate, the plate is overlaid with a bioindicator such as a microorganism or enzyme such as *Escherichia coli (E. coli), Staphylococcus aureus*, *Candida albicans*, or *Aspergillus niger*. These microorganisms are sensitive to various bioactive compounds and can form inhibition zones on the TLC plate when exposed to antimicrobial or antifungal substances. After incubation, the areas where bioactive compounds (e.g., antimicrobial or enzyme-inhibiting compounds) are present will show a clear zone of inhibition or other biological effects. This indicates the presence of active compounds. It joins bioautography with high levels accuracy with TLC's practical, affordable, and user-friendly characteristics [12]. This TLC separation technique mainly comprises of determination of biological activity. In this method, materials are coated on surface by the adsorption materials where the stationary phase helps in mixture separation. By this method, the separation of different components is achieved [12].

2.7.1 Protocol

In TLC bioautography, first we prepare a TLC plate coated with a thin layer of adsorbent material such as silica—gel. On the same plate, a small amount of test sample is plated which is a plant of microbial extract. A chamber containing suitable solvent, which acts as a mobile phase is taken for placing the plate in it. Due to the interaction of the mixture component with the stationary phase, the solvent rises by capillary action. After this, residual solvent is removed from the plate by simply drying. To assess the biological activity, several bioautography method such as direct, contact, or agar-overlay bioautography are used. These involves the exposing of the TLC plate to the microorganisms or other biological systems. Now, under appropriate conditions, the setup is incubated for microbial growth. Zones where clear spots are visible, are zones where growth is inhibited, indicating the presence of biologically active compounds, which can further be identified by their location and R_f values [15].

3 In Vivo Evaluation of Antioxidant Capacity

In contrast to in vitro models, in vivo methods capture real-time metabolic and physiological interactions to evaluate antioxidant effectiveness in live systems. Many biochemical and molecular tests are used to assess antioxidant capability in vivo. Superoxide dismutase (SOD) tests evaluate superoxide radical—fighting ability [13]. HPLC and LC–MS are frequently used to identify and quantify antioxidant chemicals and their metabolites in plasma, tissues, and urine [14]. These approaches determine bioavailability, metabolic transformations, and systemic depression. These in vivo methods provide a solid foundation for studying antioxidant efficiency at enzymatic genetic and metabolic levels, enabling the creation of oxidative stress-induced illness treatments.

3.1 SOD Assay

The superoxide dismutase (SOD) test is the measurement of SOD activity, a crucial antioxidant enzyme that drives superoxide radical dismutation toward oxygen and hydrogen peroxide, so protecting cells from oxidative damage [15, 19].

3.1.1 Mechanism

Superoxide dismutase (SOD) is an important antioxidant enzyme that helps reduce oxidative stress. It functions by converting superoxide anions ($O_2^{-}\bullet$) into molecular oxygen (O_2) and hydrogen peroxide, equation [15]:

$$O_2^- + O_2^- + 2H^+ \rightarrow O_2 + H_2O_2 \tag{15}$$

In experiments involving mice, SOD assays are usually performed after induction of oxidative stress by drugs like lipopolysaccharide (LPS), carbon tetrachloride (CCl_4), or doxorubicin [17]. After treatment, tissues like liver, kidney, brain, or blood are excised and homogenized under cold conditions to maintain the enzyme [18]. The most frequently used method for the measurement of SOD activity in tissue homogenates is inhibition of superoxide-dependent reduction of nitro blue tetrazolium (NBT) [19]. Superoxide radicals generated by the xanthine—xanthine oxidase system in this process lower NBT to blue pigment formazan. SOD degrades superoxide, therefore interfering with NBT; the degree of inhibition is exactly correlated with SOD activity [20]. Alternatively, the pyrogallol autoxidation assay, SOD inhibits pyrogallol auto-oxidation by scavenging superoxide radicals, and absorbance changes at 420 nm are quantitated [20]. SOD activity is generally quantified as units per milligram of protein. Protein concentration is quantitated using the Bradford or Lowry assay [21]. In vivo SOD assays provide information on oxidative stress-related diseases such as neurodegeneration, inflammation, and cancer [22, 23] and are commonly employed in drug efficacy and antioxidant research.

3.2 MDA and TBARS Assay

The malondialdehyde (MDA) and thiobarbituric acid reactive substances (TBARS) assay is frequently used in vivo for the assessment of lipid oxidation, a marker of oxidative imbalance. MDA reacts with Thiobarbituric acid to form a colored complex measurable spectrophotometrically [24–26].

3.2.1 Mechanism

Malondialdehyde (MDA) is a reactive aldehyde created through the lipid oxidation of polyunsaturated fatty acids and is a reliable biomarker for oxidative imbalance in vivo [25]. Thio barbituric acid reactive substances (TBARS) test is a common way to measure MDA levels in biological samples, like tissue mixtures from mice that have experienced oxidative damage [27]. In this test, tissue samples (such as liver, brain, or kidney) are homogenized in ice-cold buffer, generally KCl or phosphate buffer, and centrifuged. The supernatant obtained is reacted with Thiobarbituric acid (TBA) in acidic medium and then heated to 95–100 °C. MDA in the sample combines with two molecules of TBA to create a pink color, known as the MDA-TBA complex, which can be quantified at wavelengths of 532–535 nm [27, 28]. In mice, oxidative stress is generally induced by toxicants (e.g., CCl_4, doxorubicin, or LPS), which initiate lipid peroxidation in cellular membranes. The extent of TBARS produced is proportional to the degree of lipid peroxidation and represents an indirect measurement of reactive oxygen species (ROS) activity [29]. Protein is usually determined with the Lowry or Bradford procedure to correct for TBARS amounts [30, 31]. Although the TBARS assay is straightforward and inexpensive, it is not very specific because other aldehydes and reducing compounds will also react with TBA [32]. However, it is still a commonly accepted technique for measuring lipid peroxidation in vivo models of oxidative stress [33].

3.3 ELISA Assay

Enzyme-linked immunosorbent assay (ELISA) is a sensitive immunoassay technique that is utilized for the detection and quantification of specific proteins, cytokines, hormones, or antigens in body fluids. ELISA is commonly applied in vivo mouse models for assessing molecular responses to experimental treatment or disease states by analyzing biomarkers in blood serum, plasma, or tissue homogenates [34, 38].

3.3.1 Mechanism

In mouse models, ELISA is widely applied to evaluate molecular responses to experimental treatments or disease conditions by measuring biomarkers in blood serum, plasma, or tissue homogenate [34, 39]. In a typical indirect or sandwich ELISA, a capture antibody specific to the target antigen (e.g., TNF-α, IL-6, insulin) is pre-coated on a microtiter plate. Mouse serum or tissue lysate is then added, allowing the antigen to bind to the immobilized antibody [37]. After washing, a biotinylated or enzyme-conjugated secondary antibody binds to the captured antigen. Upon the addition of a chromogenic or chemiluminescent substrate (such as TMB for HRP-conjugated antibodies), an enzymatic reaction occurs, producing a measurable signal proportional to the antigen concentration [38]. Sample collection from mice involves blood withdrawal (via retro-orbital, tail vein, or cardiac puncture) or tissue homogenization in lysis buffer with protease inhibitors. ELISA is then used to quantify in vivo changes in immune response, hormone levels, or oxidative stress markers [39, 40]. The assay's specificity is determined by the antibody pair used, and its sensitivity enables detection in the picogram range [41]. It is crucial to standardize the protocol using proper controls and reference standards to ensure accurate interpretation [42].

3.4 In Vivo Microdialysis

Microdialysis is a notable in vivo sampling technique that has both pre-clinical and clinical applications. This technique exhibits its usefulness in assays of extracellular tissue fluid and is widely used in many applications such as in the neuroscience studies as well as in pharmacology and drug discovery. It is a versatile technique as it allows the continuous measurements of the unbound analytes and their concentration in the extracellular tissue fluid while causing limited tissue trauma. Analytes refers to the chemical substance or component being measured in the sample through the analytical procedure. In Microdialysis, analytes assessed and measured can encompass a range substances such as oxidative stress markers (e.g., Isoprostanes), neurotransmitters (e.g., acetylcholine), amino acids (e.g., glutamate, GABA), as well as biogenic amines (e.g., dopamine, norepinephrine, serotonin) [43].

3.4.1 Principle and Mechanism of Microdialysis

Microdialysis uses the dialysis principle, which pertains to the movement of molecules across a concentration gradient by means of using a semipermeable probe in this technique. This technique setup employs the use of a small, thin, semipermeable dialysis probe. Liquid perfusion of the probe allows smaller molecular weight components to diffuse in both directions according to their concentration gradient. A stereotaxic apparatus is used for the precise placement into the tissue. This probe is

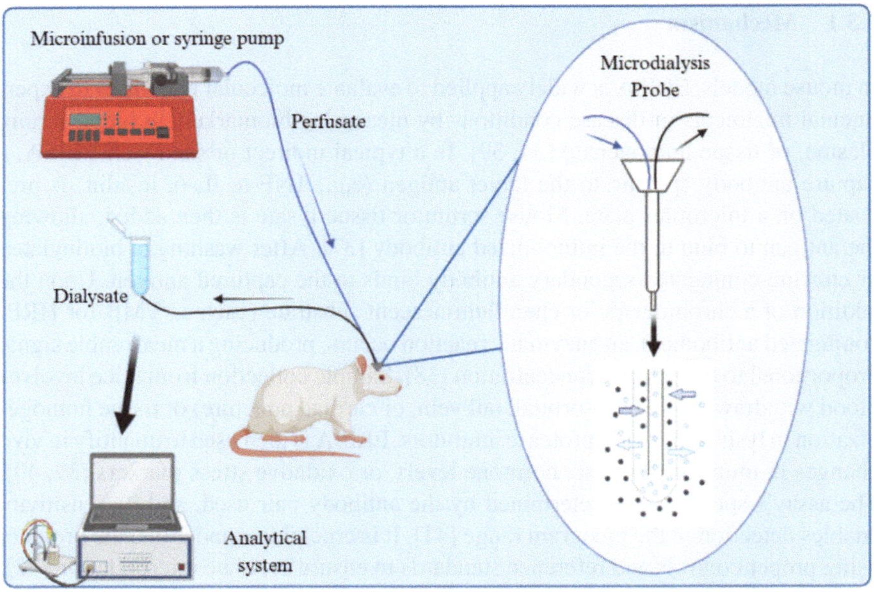

Fig. 1 Schematic Microdialysis setup for in vivo sampling in rat

inserted into the tissue of interest and allows for an exchange between the extracellular fluid and the perfusion solution (perfusate).

A micro infusion pump is used to deliver the perfusate at controlled rates with the perfusate being saline solutions or artificial cerebrospinal fluid (aCSF) that closely resemble the composition ions in the adjacent extracellular tissue fluid. This enables the free unbound substances, and small molecules present in the extracellular fluid, including oxidative stress markers, antioxidants like ascorbic acid or neurotransmitters, to passively diffuse across the probe's membrane in accordance with the concentration gradient. This is then collected as the dialysate over specified intervals as shown in Fig. 1. This collected dialysate is analyzed and quantified with the help of techniques such as capillary electrophoresis (CE) or HPLC alongside associated detectors. It can also be integrated, like with electrochemical detection in HPLC-EC (48) or Microdialysis—liquid chromatography (MD-LC), Microdialysis—capillary electrophoresis/microchip electrophoresis (MD–CE/ME), and could also be used in combination with electrochemical biosensors [44].

3.4.2 Uses of Microdialysis

Microdialysis can be applied in various studies, like for measuring ROS levels in vivo in various rodent tissues such as hydrogen peroxide and superoxide levels [49]; for detecting lipid peroxidation products and markers such as malondialdehyde (MDA) for the evaluation of antioxidant interventions in live animal models [45] or for measuring the antioxidant capacity of chemical compounds such as ascorbic acid [46].

3.5 Electron Paramagnetic Resonance (EPR) Imaging

It is a non-invasive imaging technique that establishes the localization of paramagnetic species utilized as contrast agents and presents their physical and chemical information [47]. It is similar to nuclear magnetic resonance and can be performed directly on living organisms to assess the oxidative stress and antioxidant capacity in real time. This technique provides insights into the redox status in living organisms which is necessary in monitoring oxidative damage and disease states in several neurological conditions including Parkinson's disease, and Alzheimer's disease.

3.5.1 Principle and Mechanism of EPR Imaging

It is an application of the spectroscopic based method which involves assessing the absorbed amount of microwave radiation by stable free radicals (paramagnetic substances) when subjected to an externally induced magnetic field. The unpaired electrons give rise to a magnetic moment due to an intrinsic angular momentum or spin. These magnetic moments align in parallel or antiparallel fashion when placed in a magnetic field which results in distinct energy levels (Zeeman effect). There is an energy difference between these levels which further leads to absorption of microwaves when the incident microwave radiation is comparable with the energy difference between the spin states of an unpaired electron [48]. Detection of this absorption of radiation is carried out which forms the basis for EPR spectroscopy. EPR Imaging consists of combining this traditional EPR spectroscopy with spatial resolution to map the EPR signal in three dimensions. This is done by inducing a magnetic field gradient which allows for the localization of detected signals. To execute this technique, exogenous spin probes such as nitroxides or trityl radicals are introduced into the system where on interaction with ROS alter EPR signal forming detectable paramagnetic species which help to spatially map the oxidative stress levels and assess the antioxidant capacity. The resulting EPR signals are collected, spatially encoded and analyzed to determine the presence and concentration of these species. The intensity and spatial distribution of the EPR signals help to quantify the ROS levels and are then processed using computational algorithms to reconstruct images that present the distribution of the paramagnetic species in the sample, which allows for visualization of areas with oxidative stress and other biochemical parameters.

3.5.2 Uses of EPR Imaging

EPR Imaging provides 2D or 3D images which depict the free radical distribution, redox status, and oxygen concentration in tissues. It is widely used in studies focused on tumor microenvironment and cancer research [49–51]; brain redox monitoring for brain disorders, including stroke, Alzheimer's disease and more [52]; in murine models of renal ischemia—reperfusion injury [57] among many of its other applications. It can also be coupled with different techniques like integrating with use of quantum sensors for the tracking of oxidative stress and total redox status in cells, helping in the assessment of the antioxidant capacity or properties of compounds [54].

3.6 Fluorescence Probes

Fluorescence Probes has become an indispensable tool in in vivo research as it provides a non-invasive means for real-time monitoring of the redox status within living organisms. It also has various other benefits such as higher sensitivity and specificity [55].

3.6.1 Principle and Mechanism of Fluorescent Probes

Fluorescence probes typically consist of fluorophore, a linker and a recognition group and uses chemical reactions or weak molecular reactions to detect its target compounds such as ROS [55]. They are also used in assessing redox imbalances, involving indications such as lipid oxidation or oxidative damage of protein as well as ROS detection. The probe is introduced via appropriate methods such as injection or ingestion, into the living organism including zebrafish, mice, rats or rabbits. The fluorophore probe undergoes a chemical transformation which leads to observance of a measurable change in fluorescence intensity or wavelength. The probes such as 2′7′-dichlorodihydrofluorescein diacetate ($DCFH_2$-DA) (57), MitoSOX Red [57], HyPer [58], Dihydroethidium (DHE) can be specifically designed and chosen to respond to ROS such as superoxide anions (O_2^-) or hydrogen peroxide (H_2O_2) to produce fluorescent signals.

The fluorescence changes over time are monitored and utilized to indicate the presence of and concentration of ROS as shown in Fig. 2. The fluorescence signals are then quantified and assessed which allow the selective detection of specific ROS as well as in the determination of oxidative stress levels which are utilized to estimate the antioxidant capacity of the system to be studied [59, 60].

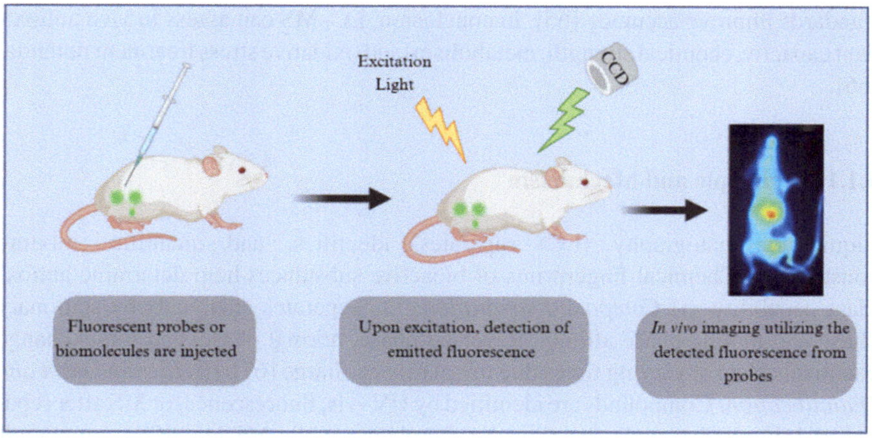

Fig. 2 Schematic representation of in vivo fluorescence imaging using fluorescent probes

3.6.2 Uses of Fluorescent Probe and Its Imaging

This tool is extensively used in animal models such as zebrafish, mice, and rats to assess the in vivo studies for conditions such as neurodegenerative disorders like Alzheimer's disease (AD) [61] for visualizing oxidative injuries such as in mouse kidney, in cancer or cardiovascular disease models.

4 Antioxidant Quantification Techniques

4.1 Liquid Chromatography (LC–MS) in Antioxidant Quantification

In vivo investigations reveal how antioxidant chemicals are absorbed, metabolized, and function in living animals, which in vitro testing cannot [14]. Plasma, serum, urine, and tissues are commonly tested for antioxidant chemicals and their metabolites using liquid chromatography-mass spectrometry (LC–MS). In in vivo antioxidant research, LC–MS has many benefits: Multiple antioxidants and their metabolites can be detected simultaneously to determine body presence and distribution [63]. It measures lipid peroxidation products, protein oxidation indicators, and DNA damage to assess antioxidant therapy [64]. It along with stable isotope tagging can follow antioxidant metabolism, bioavailability, and tissue targeting. It gives accurate data despite biological matrix complexity and low analyte concentration. Well designed to study methodologies assure reliable results, whereas isotope dilution and internal standards improve accuracy [65]. In conclusion, LC–MS can assess in vivo antioxidant capacity, chemical strength, metabolism, and oxidative stress treatment potential [66].

4.1.1 Principle and Mechanism

Liquid chromatography (LC) separates, identifies, and quantifies mixture constituents. Chemical fingerprints of bioactive substances help determine antioxidant capability. *(i) Compound separation:* LC separates chemicals by stationary phase and mobile phase affinity. Reversed-phase, normal phase, and ion-exchange chemicals elute at varying times due to polarity or charge [67]. *(ii) Identification and Quantification:* Compounds are identified by UV–vis, fluorescence, or MS after separation. UV–vis detected substances by absorbance while MS identified compounds by mass-to-charge ratio. Peak areas are quantified by comparing them to standard calibration curves (65). To determine the main contributors to antioxidant activity in plant extracts, LC data relates to antioxidant tests such as DPPH, ABTS and FRAP [68], Embelin, quercetin, and petunidin discovered by LC–MS are the main antioxidants in Oxalis corniculata [69].

4.1.2 Use of LC–MS

LC–MS is used to identify the phytochemical constituents present in the methanolic extract in vivo *O.corniculata* (71). Here are some examples of how LC–MS was used in this study. *(i) Identification of Phytochemicals:* LC–MS was used to identify compounds like embelin, phloridzin, quercetin, and petunidin in the extracts by comparing their mass-to-charge ratio (m/z) and retention times with those in the Metabolite and Chemical Entity Database (METLIN) database [64]. The identification involves matching the experimental m/z values of the unknown compounds with those theoretical m/z values of known compounds in databases such as METLIN [14]. High-resolution mass spectrometry provides accurate mass measurements, which are essential for the identification of compounds with analogous structures. *(ii) Comparative Analysis:* LC–MS was utilized to compare the phytochemical content of in vivo and in vitro extracts, thus highlighting differences in presence and relative abundance of different compounds like bilirubin, octanal, lecanoric acid, methyclothiazide, and epirubicin were detected in the in vivo extracts but not in the in vitro extracts. The comparative analysis highlights the role of growth conditions on the formation of secondary metabolites in O. corniculata. Relative abundance of a compound can be calculated from peak area noticed in the LC–MS chromatogram.

4.2 High-Performance Liquid Chromatography—Mass Spectrometry (HPLC—MS) Antioxidant Quantification

A common method for identifying and describing many antioxidant components in intricate biological matrices is high-performance liquid chromatography-mass spectrometry (HPLC–MS), which provides information on the compound's stability, bioavailability, and action [14]. With the use of improved chromatographic settings and assays like FRAP to measure antioxidant capacity, it has been applied to plant extracts such as Sonchus brachyotus DC [70]. To isolate important bioactive compounds that fight oxidative stress. HPLC–MS improves standardization and quality control when used in conjunction with chemometric techniques, guaranteeing consistent antioxidant activity in herbal formulations for pharmacological and therapeutic applications [65].

4.2.1 Principle and Mechanism

High-performance liquid chromatography (HPLC) is a method of mixture separation, identification, and quantitation [14]. It estimates antioxidant capacity on the basis of the following principles. HPLC is used to separate compounds by distributing them between a stationary phase (solid or liquid on a solid support) and a mobile phase (liquid solvent) [71]. The mobile phase is permitted to pass through the stationary phase, which is bundled within a column. Compounds interact with the phases in different ways depending on their characteristics such as polarity and size, leading to varying migration rates and resulting in separation [14]. Identification and quantification of separated compounds are made possible by HPLC with a diode array detector (DAD) (76). The DAD detects UV–visible absorbance as compound elute. The individual UV–vis spectra with reference standards [71]. The area of chromatogram peak is linearly related to the concentration of the compound, hence allowing accurate quantification [14]. HPLC identifies and quantifies antioxidant compounds and correlates them to the overall antioxidant activity of an extract. An awareness of concentration-activity relationship allows researchers to optimize extraction and purification for enhanced antioxidant capacity in natural products [70]. Although HPLC is utilized to detect and quantify antioxidant substances, it cannot measure the mechanism of action directly [72]. But by detecting the antioxidant substances, scientists can not only extrapolate the probable mechanisms from what is known of the properties of those substances but also provides an entire chemical profile to detect compounds that are responsible for antioxidant activity.

4.2.2 Use of HPLC Methods

HPLC for determination of antioxidant capacity [72] applied to HPLC–DAD to identify and describe phenolic compounds in Commiphora leptophloeos leaf extracts.

HPLC is an important analytical tool in antioxidant research based on the ability to separate and detect. It has some applications such as; HPLC–DAD identified compounds like rutin, vitexin, and quercetin di glycosides comparing UV absorption spectra and retention times to reference standards [72]. The technique employs the distinctive spectral fingerprints of molecules for identification [14]. Profiling the ethanol extract (EE), the ethanol extract was very active as an antioxidant and was also profiled by HPLC–DAD for its phenolic content [72]. Quantitative analysis also gave the concentration of each compound, giving insight into the composition and potential bioactivity of the extract [71].

5 Limitations

There are few limitations of various assays which have been discussed in terms of bioavailability and metabolism as results may not correlate with the antioxidant activity and are also sensitive to an experimental condition. In both in vivo and in vitro assays, a standardized protocol should be followed. Present study says that isolation, purification, molecular structure, and identification of the active compounds should be looked into more while performing several assays for an evaluation of an antioxidant capacity [73]. Some assays are time-consuming and complex to perform. Some assays depend on the food samples that is not always commercially available. While comparison between some assay at a fixed end point antioxidant capacity is obtained, which further shows differences. Also, some assays need only an organic media to get dissolved. Ionic equilibrium is influenced by addition of acidic and basic impurities. Radical quenching cannot be detected by FRAP [14]. GSH cannot be measured by FRAP, also some complexes can react with the antioxidants. Also, particularly thiols and proteins, this causes a serious serum underestimation. Hence, a complete reaction is not represented by the determination of antioxidant capacity.

6 Antioxidant Bioavailability and Bioaccessibility

Bioaccessibility is a compound that is available in the gut during digestion, which is released from the food matrix, which is further directed through modals such as gastrointestinal modals. On the other side, metabolism, proper distribution and compounds with several biological activities inside the body also states for bioavailability. Mostly animal or human models are used for the in vivo studies, which helps in determining the several roles in living organisms such as physiological roles and effects toward antioxidants. In vitro offers a more accurate and holistic understanding which provides an integrating and initial screening of important compounds and helps in the studies of their respective functional potential, development of effective nutraceuticals and helps guiding them and lastly the therapeutic strategies [14]. Some of the major factors that are affecting bioavailability and bioaccessibility are digestion, absorption, metabolism, and food matrix. Antioxidants bioavailability and bioaccessibility are critical in defining their efficiency both in vivo and in vitro. It has been noted that there is a difference in the behavior of antioxidants when subjected to in vitro and in vivo systems. Free radical scavenging and cellular uptake assays are considered in vitro, demonstrating the ability to achieve antioxidant potential due to the food matrix influences compound stability and release [76]. However, in vivo, studies show that post digestion, absorption, and metabolism, the actual bioavailability of antioxidants is far lesser than expected because of enzymatic changes and interactions with other dietary constituents [77]. The variations between these models explain the fundamental complexity of antioxidant activity where in vitro analyses tend to overvalue biological activity while in vivo realities tend to belittle it, as seen in the case of overzealous biological effects vis-a-vis the understated effects of in vitro experiments [78]. It has been proven that distinct formulations of foods can alter the way and degree to which antioxidants are absorbed, reinforcing the concept of metabolism and digestion as focal components for assessing efficacy [76].

7 Perspectives for the Future

Advancements in the biotechnology, nanotechnology, and personalized medicine is expected to redefine the future of in vitro and in vivo studies of antioxidant capacity. Use of the 3D organoids and organ-on-a-chip systems, mimicking human tissue responses will enable physiologically relevant assessment of antioxidants. Nanoantioxidants have been developed over the past decades, which are more sensitive and less cytotoxic with better cellular targeting. Much more research has to be done on the beneficial aspects and side effects of these nanoantioxidants for its safe use in vivo applications especially for long-term treatment [79]. The future research on nanoantioxidants should focus on the reaction of living species to this treatment, development of cheaper and easy fabrication methods, improving biocompatibility, biodegradability, reproducibility, and overall effectiveness of nanoantioxidants and lastly assessing any possible hazard or risk that might be associated with the large-scale production process [80]. The emergence of biomass-derived carbon dots as antioxidants remains a subject that requires further investigation. There are several obstacles to overcome before they are used in the practical applications. The main drawback is the difficulty in separating the carbon dots from biomass. A low production yield, high performing precursors for large-scale production are some other disadvantages [81]. Additionally, AI-predictive models for optimizing the selection of antioxidant treatment and biosensors for continuous assessment and early intervention have been developed. The integration of genomics, metabolomics, and proteomics will facilitate the identification of individual oxidative stress biomarkers for tailored antioxidant therapies. The standardization of antioxidant evaluation in pharmaceuticals and nutraceuticals will bridge the gap between the laboratory research and clinical application.

8 Conclusion

Antioxidant capacity is crucial to understanding how antioxidants reduce oxidative stress and related illness. In vitro tests including DPPH, FRAP, ORAC, and ABTS provide quick and controlled free radical scavenging and reduction potential evaluations, providing crucial antioxidant activity insights. In vivo methods are needed to adequately comprehend physiological circumstances because these methods may not entirely represent them. ELISA, EPR Imaging, microdialysis, and fluorescent probes help provide real-time monitoring and molecular study of oxidative stress and biological antioxidant responses. HPLC and LC–MS can detect and quantify antioxidant molecules and oxidative stress indicators with high sensitivity, improving accuracy. In vitro experiments are simple and efficient, while in vivo methods reveal antioxidant effectiveness, defense's biochemical and genetic pathways. These approaches provide a complete review that connects test results to clinical applications.

References

1. Rotariu, D., Babes, E. E., Tit, D. M., Moisi, M., Bustea, C., Stoicescu, M., et al. (2022). Oxidative stress—Complex pathological issues concerning the hallmark of cardiovascular and metabolic disorders. *Biomedicine and Pharmacotherapy, 1*(152), Article 113238.
2. Munteanu, I. G., Apetrei, C. (2021). Analytical methods used in determining antioxidant activity: A review. *International Journal Molecular Science, 22*(7)
3. Karadag, A., Ozcelik, B., Saner, S. (2009). Review of methods to determine antioxidant capacities. *Food Anal Methods [Internet], 2*(1), 41–60 [cited 2025 Apr 12]. Available from: https://doi.org/10.1007/s12161-008-9067-7
4. Armstrong, D. (2010). Editor. Advanced protocols in oxidative stress II, 594. [cited 2025 Apr 12]. Available from: https://doi.org/10.1007/978-1-60761-411-1
5. Apak, R., Güçlü, K., Özyürek, M., Çelik, S. E. (2008). Mechanism of antioxidant capacity assays and the CUPRAC (cupric ion reducing antioxidant capacity) assay. *Microchimica Acta [Internet], 160*(4), 413–9 [cited 2025 Mar 30]. Available from: https://doi.org/10.1007/s00604-007-0777-0
6. Özyürek, M., Güçlü, K., Tütem, E., Bakan, K. S., Erçağ, E., Esin Çelik, S., et al. (2025). A comprehensive review of CUPRAC methodology. *Analytical Methods [Internet], 3*(11), 2439–53 [cited 2025 Mar 24]. Available from: https://pubs.rsc.org/en/content/articlehtml/2011/ay/c1ay05320e
7. Apak, R., Güçlü, K., Özyürek, M., & Karademir, S. E. (2004). Novel total antioxidant capacity index for dietary polyphenols and vitamins C and E, using their cupric ion reducing capability in the presence of neocuproine: CUPRAC method. *Journal of Agriculture and Food Chemistry, 52*(26), 7970–7981.
8. Gulcin, İ., Alwasel, S. H. (2023). DPPH radical scavenging assay. *Processes 2023, 11*(8), 2248 [Internet] [cited 2025 Mar 31]. Available from: https://www.mdpi.com/2227-9717/11/8/2248/htm
9. Singh, S., Singh, R. P. (2008). In vitro methods of assay of antioxidants: An overview. *Food Reviews International [Internet], 24*(4), 392–415 [cited 2025 Mar 28]. Available from: https://doi.org/10.1080/87559120802304269
10. Pohanka, M., Bandouchova, H., Sobotka, J., Sedlackova, J., Soukupova, I., Pikula, J. (2009). Ferric reducing antioxidant power and square wave voltammetry for assay of low molecular weight antioxidants in blood plasma: Performance and comparison of methods. *Sensors, 9*(11),

9094–103 [cited 2025 Mar 28]. Available from: https://www.mdpi.com/1424-8220/9/11/9094/htm
11. Alam, M. N., Bristi, N. J., & Rafiquzzaman, M. (2013). Review on in vivo and in vitro methods evaluation of antioxidant activity. *Saudi Pharmaceutical Journal., 21*(2), 143–152.
12. Wang, M., Zhang, Y., Wang, R., Wang, Z., Yang, B., Kuang, H. (2021). An evolving technology that integrates classical methods with continuous technological developments: Thin-Layer chromatography bioautography. *Molecules, 26*(15), 4647 [cited 2025 Mar 29]. Available from: https://pmc.ncbi.nlm.nih.gov/articles/PMC8347725/
13. Van Raamsdonk, J. M., Hekimi, S. (2012). Superoxide dismutase is dispensable for normal animal lifespan. *Proceeding of National Academy Science USA, 109*(15), 5785–90 [cited 2025 Apr 4]. Available from: https://www.pnas.org/doi/abs/https://doi.org/10.1073/pnas.1116158109
14. Meija, J., Coplen, T. B., Berglund, M., Brand, W. A., De Bièvre, P., Gröning, M., et al. (2013). Atomic weights of the elements 2013 (IUPAC Technical Report). *Pure and Applied Chemistry. Walter de Gruyter GmbH, 88*, 265–91
15. McCord, J. M., Fridovich, I. (1969). Superoxide dismutase: An enzymic function for erythrocuprein (Hemocuprein). *Journal of Biological Chemistry, 244*(22), 6049–55 [cited 2025 Apr 10]. Available from: https://linkinghub.elsevier.com/retrieve/pii/S0021925818635045
16. Marklund, S., Marklund, G. (1974). Involvement of the superoxide anion radical in the autoxidation of pyrogallol and a convenient assay for superoxide dismutase. *European Journal Biochemical, 47*(3), 469–74 [cited 2025 Apr 10]. Available from: https://pubmed.ncbi.nlm.nih.gov/4215654/
17. Zhang, C., Bruins, M. E., Yang, Z. Q., Liu, S. T., & Rao, P. F. (2016). A new formula to calculate activity of superoxide dismutase in indirect assays. *Analytical Biochemistry, 15*(503), 65–67.
18. Aebi, H. (1984). Catalase in vitro. *Methods Enzymol, 105*(C), 121–6
19. Sun, Y., Oberley, L. W., Li, Y. (1988). A simple method for clinical assay of superoxide dismutase. *Clinical Chemistry, 34*(3), 497–500 [cited 2025 Apr 10]. Available from: https://doi.org/10.1093/clinchem/34.3.497
20. Marklund, S. L. (1985). Superoxide dismutase isoenzymes in tissues and plasma from New Zealand black mice, nude mice and normal BALB/c mice. *Mutation Research, 148*(1–2), 129–34 [cited 2025 Apr 10]. Available from: https://pubmed.ncbi.nlm.nih.gov/3969077/
21. Bradford, M. M. (1976). A rapid and sensitive method for the quantitation of microgram quantities of protein utilizing the principle of protein-dye binding. *Anal Biochemical, 72*(1–2), 248–54 [cited 2025 Apr 10]. Available from: https://pubmed.ncbi.nlm.nih.gov/942051/
22. Halliwell, B., Gutteridge, J. M. C. (2015). Free radicals in biology and medicine. *Free Radicals in Biology and Medicine*
23. Kakkar, P. M., Das, B., Viswanathan, P. N. (1984). A modified spectrophotometric assay of superoxide dismutase. *IJBB, 21*(2), 130–2 [cited 2025 Apr 10]. Available from: http://nopr.niscpr.res.in/handle/123456789/19932
24. Ohkawa, H., Ohishi, N., Yagi, K. (1979). Assay for lipid peroxides in animal tissues by thiobarbituric acid reaction. *Anal Biochemical, 95*(2), 351–8 [cited 2025 Apr 10]. Available from: https://pubmed.ncbi.nlm.nih.gov/36810/
25. Draper, H. H., Hadley, M. (1990). Malondialdehyde determination as index of lipid peroxidation. *Methods Enzymol, 186*(C), 421–31[cited 2025 Apr 10]. Available from: https://pubmed.ncbi.nlm.nih.gov/2233309/
26. Esterbauer, H., Cheeseman, K. H. (1990). Determination of aldehydic lipid peroxidation products: Malonaldehyde and 4-hydroxynonenal. *Methods Enzymol, 186*(C), 407–21
27. Yagi, K. (1987). Lipid peroxides and human diseases. *Chemistry and Physics of Lipids, 45*(2–4), 337–351.
28. Esterbauer, H., Cheeseman, K. H. (1990). Determination of aldehydic lipid peroxidation products: malonaldehyde and 4-hydroxynonenal. *Methods Enzymol, 186*(C), 407–21 [cited 2025 Apr 12]. Available from: https://pubmed.ncbi.nlm.nih.gov/2233308/
29. Janero, D. R. (1990). Malondialdehyde and thiobarbituric acid-reactivity as diagnostic indices of lipid peroxidation and peroxidative tissue injury. *Free Radical Biology Medicine, 9*(6), 515–40 [cited 2025 Apr 10]. Available from: https://pubmed.ncbi.nlm.nih.gov/2079232/

30. Lowry, O. H., Rosebrough, N. J., Farr, A. L., & Randall, R. J. (1951). Protein measurement with the folin phenol reagent. *Journal of Biological Chemistry., 193*(1), 265–275.
31. Bradford, M. M. (1976). A rapid and sensitive method for the quantitation of microgram quantities of protein utilizing the principle of protein-dye binding. *Analytical Biochemistry, 72*(1–2), 248–254.
32. Halliwell, B., Chirico, S. (1993). Lipid peroxidation: its mechanism, measurement, and significance. *The American Journal Clinical Nutrition, 57*(5 Suppl) [cited 2025 Apr 10]. Available from: https://pubmed.ncbi.nlm.nih.gov/8475889/
33. Del Rio, D., Stewart, A. J., & Pellegrini, N. (2005). A review of recent studies on malondialdehyde as toxic molecule and biological marker of oxidative stress. *Nutrition, Metabolism and Cardiovascular Diseases., 15*(4), 316–328.
34. Engvall, E., Perlmann, P. (1971). Enzyme-linked immunosorbent assay (ELISA). *Quantitative assay of immunoglobulin G. Immunochemistry, 8*(9), 871–4 [cited 2025 Apr 10]. Available from: https://pubmed.ncbi.nlm.nih.gov/5135623/
35. Voller, A., Bartlett, A., & Bidwell, D. E. (1976). Enzyme immunoassays for parasitic diseases. *Transactions of the Royal Society of Tropical Medicine and Hygiene, 70*(2), 98–106.
36. Voller, A., Bartlett, A., Bidwell, D. E. (1978). Enzyme immunoassays with special reference to ELISA techniques. *Journal Clinical Pathology, 31*(6), 507–20 [cited 2025 Apr 10]. Available from: https://pubmed.ncbi.nlm.nih.gov/78929/
37. Lequin, R. M. (2005). Enzyme immunoassay (EIA)/enzyme-linked immunosorbent assay (ELISA). *Clinical Chemistry, 51*(12), 2415–8 [cited 2025 Apr 10]. Available from: https://pubmed.ncbi.nlm.nih.gov/16179424/
38. Crowther, J. R. (2000). The ELISA guidebook. *Methods Molecule Biology, 149* [cited 2025 Apr 10]. Available from: https://pubmed.ncbi.nlm.nih.gov/11028258/
39. Bradley, B. P., Gossop, M., Phillips, G. T., Legarda, J. J. (1987). The development of an opiate withdrawal scale (OWS). *British Journal Addiction, 82*(10), 1139–42 [cited 2025 Apr 10]. Available from: https://doi.org/10.1111/j.1360-0443.1987.tb03294.x
40. Zhang, W. W., Ghosh, A. K., Mohamath, R., Whittle, J., Picone, A., Lypaczewski, P., et al. (2018). Development of a sandwich ELISA to detect Leishmania 40S ribosomal protein S12 antigen from blood samples of visceral leishmaniasis patients. *BMC Infection Disease, 18*(1)
41. Schuurs, A. H. W. M., Van Weemen, B. K. (1980). Enzyme-immunoassay: a powerful analytical tool. *Journal Immunoassay, 1*(2), 229–49 [cited 2025 Apr 10]. Available from: https://pubmed.ncbi.nlm.nih.gov/6785317/
42. Findlay, J. W. A., Smith, W. C., Lee, J. W., Nordblom, G. D., Das, I., Desilva, B. S., et al. (2000). Validation of immunoassays for bioanalysis: A pharmaceutical industry perspective. *Journal Pharmacy Biomedical Analysis, 21*(6), 1249–73 [cited 2025 Apr 10]. Available from: https://pubmed.ncbi.nlm.nih.gov/10708409/
43. Darvesh, A. S., Carroll, R. T., Geldenhuys, W. J., Gudelsky, G. A., Klein, J., Meshul, C. K., et al. (2011). In vivo brain microdialysis: advances in neuro psycho pharmacology and drug discovery. *Expert Opinion Drug Discover, 6*(2), 109–27 [cited 2025 Mar 11]. Available from: https://doi.org/10.1517/17460441.2011.547189
44. Xu, X., Zuo, Y., Chen, S., Hatami, A., Gu, H. (2024). Advancements in brain research: The in vivo/in vitro electrochemical detection of neurochemicals. *Biosensors, 14*(3), 125 [cited 2025 Mar 19]. Available from: https://www.mdpi.com/2079-6374/14/3/125/htm
45. Cooley, J. C., Lunte, C. E. (2011). Detection of malondialdehyde in vivo using microdialysis sampling with CE-fluorescence. *Electrophoresis, 32*(21), 2994–9 [cited 2025 Apr 2]. Available from: https://doi.org/10.1002/elps.201100143
46. Pretti, L., Bazzu, G., Serra, P. A., & Nieddu, G. (2014). A novel method for the determination of ascorbic acid and antioxidant capacity in Opuntia ficus indica using in vivo microdialysis. *Food Chemistry, 15*(147), 131–137.
47. Takakusagi, Y., Kobayashi, R., Saito, K., Kishimoto, S., Krishna, M. C., Murugesan, R., et al. (2023). EPR and related magnetic resonance imaging techniques in cancer research. *Metabolites, 13*(1), 69 [cited 2025 Apr 2]. Available from: https://www.mdpi.com/2218-1989/13/1/69/htm

48. Lohan, S. B., Lauer, A. C., Arndt, S., Friedrich, A., Tscherch, K., Haag, S. F., et al. (2015). Determination of the antioxidant status of the skin by in vivo-electron paramagnetic resonance (EPR) spectroscopy. *Cosmetics, 2*(3), 286–301 [cited 2025 Apr 3]. Available from: https://www.mdpi.com/2079-9284/2/3/286/htm
49. Khramtsov, V. V. (2018). In vivo molecular electron paramagnetic resonance-based spectroscopy and imaging of tumor microenvironment and redox using functional paramagnetic probes, 28(15), 1365–77. https://home.liebertpub.com/ars [cited 2025 Apr 3]. Available from: https://doi.org/10.1089/ars.2017.7329
50. Takakusagi, Y., Kobayashi, R., Saito, K., Kishimoto, S., Krishna, M. C., Murugesan, R., et al. (2023). EPR and related magnetic resonance imaging techniques in cancer research. *Metabolites, 13*(1), 69 [cited 2025 Apr 3]. Available from: https://www.mdpi.com/2218-1989/13/1/69/htm
51. Kao, J., Eaton, S. S., Krzykawska-Serda, M., Maggio, M. C., & Halpern, H. J. (2017). Imaging thiol redox status in murine tumors in vivo with rapid-scan electron paramagnetic resonance. *Journal of Magnetic Resonance., 1*(276), 31–36.
52. Fujii, H. G., Emoto, M. C., Sato-Akaba, H. (2019). Brain redox imaging using in vivo electron paramagnetic resonance imaging and nitroxide imaging probes. *Magnetochemistry, 5*(1), 11 [cited 2025 Apr 3]. Available from: https://www.mdpi.com/2312-7481/5/1/11/htm
53. Hirayama, A., Nagase, S., Ueda, A., Oteki, T., Takada, K., Obara, M., et al. (2005). In vivo imaging of oxidative stress in ischemia-reperfusion renal injury using electron paramagnetic resonance. *American Journal of Physiology Renal Physiology, 288*(357–3), 597–603 [cited 2025 Apr 3]. Available from: https://doi.org/10.1152/ajprenal.00020.2004
54. Lazarova, D., Semkova, S., Zlateva, G., Tatsuya, H., Aoki, I., Bakalova, R. (2025). Quantum sensors to track total redox-status and oxidative stress in cells and tissues using electron-paramagnetic resonance, magnetic resonance imaging, and optical imaging. *Anal Chemistry, 93*(5), 2828–37 [cited 2025 Apr 3]. Available from: https://doi.org/10.1021/acs.analchem.0c04116
55. Geng, Y., Wang, Z., Zhou, J., Zhu, M., Liu, J., James, T. D. (2023). Recent progress in the development of fluorescent probes for imaging pathological oxidative stress. *Chemical Society Reviews, 52*(11), 3873–926 [cited 2025 Apr 3]. Available from: https://pubs.rsc.org/en/content/articlehtml/2023/cs/d2cs00172a
56. Reiniers, M. J., Van Golen, R. F., Bonnet, S., Broekgaarden, M., Van Gulik, T. M., Egmond, M. R., et al. (2017). Preparation and practical applications of 2′,7′-Dichlorodihydrofluorescein in redox assays. *Anal Chemical, 89*(7), 3853–7 [cited 2025 Apr 9]. Available from: https://doi.org/10.1021/acs.analchem.7b00043
57. Čapek, J., Roušar, T. (2021). Detection of oxidative stress induced by nanomaterials in cells—The roles of reactive oxygen species and glutathione. *Molecules, 26*(16), 4710 [cited 2025 Apr 9]. Available from: https://www.mdpi.com/1420-3049/26/16/4710/htm
58. Bilan, D. S., Belousov, V. V. (2018). In vivo imaging of hydrogen peroxide with HyPer probes, 29(6), 569–84. https://home.liebertpub.com/ars [cited 2025 Apr 9]. Available from: https://doi.org/10.1089/ars.2018.7540
59. Ye, S., Hu, J. J., Zhao, Q. A., Yang, D. (2020). Fluorescent probes for in vitro and in vivo quantification of hydrogen peroxide. *Chemical Science, 11*(44), 11989–97 [cited 2025 Apr 3]. Available from: https://pubs.rsc.org/en/content/articlehtml/2020/sc/d0sc04888g
60. Wang, H., Wang, X., Li, P., Dong, M., Yao, S. Q., Tang, B. (2021). Fluorescent probes for visualizing ROS-associated proteins in disease. *Chemical Science, 12*(35), 11620–46 [cited 2025 Apr 3]. Available from: https://pubs.rsc.org/en/content/articlehtml/2021/sc/d1sc02165f
61. Lv, J., Li, H., Gao, J., Dong, N., Shi, W., Ma, H., et al. (2025). A dual-functional fluorescence probe for simultaneous in vivo imaging of Aβ aggregates and hydrogen peroxide in the brain of mice with Alzheimer's Disease. CCS Chemistry, 1–12 [cited 2025 Apr 3]. Available from: https://doi.org/10.31635/ccschem.024.202404927
62. Lv, Y., Cheng, D., Su, D., Chen, M., Yin, B. C., Yuan, L., et al. (2018). Visualization of oxidative injury in the mouse kidney using selective superoxide anion fluorescent probes. Chemical Science, *9*(39), 7606–13 [cited 2025 Apr 3]. Available from: https://pubs.rsc.org/en/content/articlehtml/2018/sc/c8sc03308k

63. Valko, M., Leibfritz, D., Moncol, J., Cronin, M. T. D., Mazur, M., Telser, J. (2007). Free radicals and antioxidants in normal physiological functions and human disease. *International Journal of Biochemistry and Cell Biology, 39*, 44–84
64. Chandrasekara, A., & Shahidi, F. (2011). Determination of antioxidant activity in free and hydrolyzed fractions of millet grains and characterization of their phenolic profiles by HPLC-DAD-ESI-MSn. *J Funct Foods., 3*(3), 144–158.
65. Halliwell, B. (2007). Oxidative stress and cancer: Have we moved forward? *Biochemical Journal,* 401, 1–11
66. Lobo, V., Patil, A., Phatak, A., Chandra, N. (2010). Free radicals, antioxidants and functional foods: Impact on human health. *Pharmacognosy Reviews, 4*, 118–26
67. El Rayess, Y., Nehme, N., Azzi-Achkouty, S., & Julien, S. G. (2024). *Wine phenolic compounds: Chemistry, functionality and health benefits* (Vol. 13). Multidisciplinary Digital Publishing Institute (MDPI).
68. Heim, K. E., Tagliaferro, A. R., & Bobilya, D. J. (2002). Flavonoid antioxidants: Chemistry, metabolism and structure-activity relationships. *Journal of Nutritional Biochemistry, 13*(10), 572–584.
69. Swami, D., Swami, D. R., Malpathak, N. P. (2018). Exploring in-vivo and in-vitro oxalis corniculata l. for phytochemicals using non-targeted LC-MS approach and its antioxidant capacity. *International Journal Pharmacy Science Research, 9*(10), 4151. Available from: https://doi.org/10.13040/IJPSR.0975-8232.9
70. Rice-Evans, C. A., Miller, N. J., Paganga, G. (1996). Structure-antioxidant activity relationships of flavonoids and phenolic acids. *Free Radical Biology and Medicine, 20*
71. Meyer, V. R. Practical high-performance liquid chromatography, Fifth edition
72. Cordeiro, M. L. D. S., Ribeiro, A. R. C., Melo, L. F. M. D., Silva, L, F, D., Fidelis. G. P., Silva, L. M. P., et al. (2001). Antioxidant activities of commiphora leptophloeos (Mart.) J. B. Gillett) (Burseraceae) Leaf extracts using in vitro and in vivo assays. Oxid Med Cell Longev. 2021
73. Yang, Z., Man, J., Liu, H., Wu, D., Gu, Q., Zhang, H., et al. (2025). Study on the In vitro and in vivo antioxidant activity and potential mechanism of Polygonum Viviparum L. Antioxidants, *14*(1), 41 [cited 2025 Mar 25]. Available from: https://www.mdpi.com/2076-3921/14/1/41/htm
74. Karadag, A., Ozcelik, B., & Saner, S. (2009). Review of methods to determine antioxidant capacities. *Food Analytical Methods, 2*(1), 41–60.
75. Toydemir, G., Gultekin Subasi, B., Hall, R. D., Beekwilder, J., Boyacioglu, D., Capanoglu, E. (2022). Effect of food processing on antioxidants, their bioavailability and potential relevance to human health. *Food Chem X, 14*, 100334
76. Francisco, J., Horta, A., Pedrosa, R., Afonso, C., Cardoso, C., Bandarra, N. M., et al., Bioaccessibility of antioxidants and fatty acids from Fucus spiralis. Foods, 9(4), 440 [cited 2025 Apr 4]. Available from: https://www.mdpi.com/2304-8158/9/4/440/htm
77. Holst, B., & Williamson, G. (2008). Nutrients and phytochemicals: From bioavailability to bioefficacy beyond antioxidants. *Current Opinion in Biotechnology, 19*(2), 73–82.
78. Schulz, M., Biluca, F. C., Gonzaga, L. V., Borges, G. da S. C., Vitali, L., Micke, G. A., et al. (2017). Bioaccessibility of bioactive compounds and antioxidant potential of Juçara fruits (Euterpe edulis Martius) subjected to in vitro gastrointestinal digestion. *Food Chemical, 228*, 447–54
79. Khalil, I., Yehye, W. A., Etxeberria, A. E., Alhadi, A. A., Dezfooli, S. M., Julkapli, N. B. M., et al. (2020). Nanoantioxidants: Recent trends in antioxidant delivery applications. *Antioxidants, 9*(1), 24 [cited 2025 Apr 3]. Available from: https://www.mdpi.com/2076-3921/9/1/24/htm
80. Khan, I., Hameed, H., Younas, K., Zaman, M., Faheem, S., Majeed, I., et al. (2025) Unlocking the power of antioxidant nanoparticles: Insights into classification, formulation, characterization, and biomedical applications. *Bionanoscience, 15*(1), 1–29 [cited 2025 Apr 3]. Available from: https://link.springer.com/article/https://doi.org/10.1007/s12668-024-01761-x
81. Sharma, N., Sharma, A., Lee, H. J. (2025). The antioxidant properties of green carbon dots: A review. *Environmental Chemistry Letters,* pp 1–49 [cited 2025 Apr 3]. Available from: https://doi.org/10.1007/s10311-025-01831-w

Recent Advances in Oxidative Stress Associated Chronic Diseases

Sana Firdous, Meenakshi Prakash, Shaikh Zaahid Iqbal, Yash Banabakode, R. Deepika, Rebeccas Thomas, T. Aravinth Karthik, Sharine Priscilla, S. Beer Mohamed, and Sahabudeen Sheik Mohideen

Abstract Reactive oxygen species (ROS), reactive nitrogen species (RNS), and reactive sulfur species (RSS)—collectively referred to as RONSS—are essential redox-active molecules that are involved in cellular communication and the initiation of oxidative stress in pathological circumstances. This chapter discusses the various roles of RONSS in modulating redox equilibrium and influencing disease processes in numerous chronic diseases. Although substantial research has been conducted on ROS and RNS separately, the combined functions with RSS in disease-specific pathophysiology and therapeutic potential are yet insufficiently investigated. The chapter delineates the molecular origins, interactions, and subsequent effects of RONSS in neurological illnesses, cardiovascular disorders, diabetes, cancer, and muscular dysfunctions. We emphasize mechanisms that include dysfunction in the mitochondria, free radical damage to nucleic acids and proteins, and inflammatory activation. Recent advancements in redox imaging, antioxidant modulation, and molecular targeting are examined as promising strategies for comprehending and addressing RONSS-related illness progression.

Keywords Oxidative stress · Redox signaling · Apoptosis · Tissue damage · Mitochondrial dysfunction

---

S. Firdous · M. Prakash · S. Z. Iqbal · Y. Banabakode · R. Deepika · R. Thomas · T. Aravinth Karthik · S. Priscilla · S. Sheik Mohideen
Department of Biotechnology, School of Bioengineering, College of Engineering and Technology, SRM Institute of Science and Technology, Kattankulathur, Chengalpattu District 603203, Tamil Nadu, India
e-mail: sahabuds@srmist.edu.in

S. Beer Mohamed (✉)
Department of Materials Science, School of Technology, Central University of Tamil Nadu, Thiruvarur, Tamil Nadu, India
e-mail: peermoha@gmail.com

K. Kasinathan and A. Muthukumaran (eds.), *Recent Advances in Oxidative Stress Associated Chronic Diseases Volume 2*, https://doi.org/10.1007/978-981-95-3758-7_7

1 Introduction

Reactive oxygen species (ROS), reactive nitrogen species (RNS), and reactive sulfur species (RSS) regulate numerous physiological and pathological processes. These reactive molecules, derived from oxygen, nitrogen, and sulfur, are crucial for immune functions, metabolism, and cellular communication. Excessive production disrupts redox balance, contributing to aging, neurological diseases, cancer, cardiovascular dysfunction, and inflammation [1, 2]. Understanding their chemical properties and biological roles is vital due to their dual role as both signaling molecules and potential destructive agents. ROS, including hydrogen peroxide (H_2O_2), superoxide anion (O_2•$^-$), and hydroxyl radicals (•OH), are mainly produced by mitochondrial respiration, NADPH oxidases, and peroxidases [3]. Superoxide is converted to H_2O_2 by superoxide dismutase (SOD) and can generate highly reactive hydroxyl radicals via Fenton chemistry [4]. RNS, such as nitric oxide (NO•) and peroxynitrite ($ONOO^-$), are produced by nitric oxide synthases (NOS), including inducible, endothelial (eNOS), and neuronal NOS (nNOS). Excess NO• can lead to $ONOO^-$, causing DNA, lipid, and protein damage, which contributes to inflammation and neurodegeneration [4]. RSS, including H_2S, persulfides (RSSH), and polysulfides (S_x^{2-}), unlike ROS and RNS, contribute to reversible thiol modifications that alter protein functions and cellular signaling [5]. RONS are crucial for apoptosis, immune defense, vasodilation, and cell signaling [6]. NO• influences memory, learning, and neural plasticity [1], H_2S impacts mitochondrial respiration, inflammation, and antioxidant defense by sulfhydrating cysteine residues, which alters metabolic pathways and stress tolerance (7). However, an imbalance in RONSS leads to oxidative, nitrosative, and sulfhydrative stress, contributing to cancer, cardiovascular disease, and dementia. Excess RNS causes nitrosative stress, resulting in S-nitrosylation and protein nitration, impairing enzyme function and aggravating inflammatory and autoimmune diseases [4]. Sulfhydrative stress from sulfur metabolism failure reduces mitochondrial efficiency, driving metabolic disorders and cardiovascular disease. Antioxidant mechanisms, both enzymatic (e.g., catalase, GPx) and non-enzymatic (e.g., glutathione, vitamins C and E), maintain redox balance and reduce oxidative and nitrosative damage [2, 8]. Redox repair systems, such as thioredoxin and glutaredoxin, restore oxidized proteins and prevent permanent damage [2]. Defensive mechanisms specific to RSS include interactions between sulfur metabolites and ROS/RNS to control damage and regulate protein function [7]. By understanding the molecular origins and effects of these species, we can develop treatments and preventive measures targeting their responses.

2 Sources of RONSS

2.1 Endogenous Sources

2.1.1 ROS Sources

Mitochondria are the primary source of cellular ROS, mainly through the electron transport chain at Complex I (NADH dehydrogenase) and Complex III (cytochrome bc1 complex). ROS form via forward (FET) and reverse electron transfer (RET), with RET—driven by $FADH_2$-linked substrates like succinate—producing high ROS levels at Complex I. Beyond mitochondria, NADPH oxidases (NOX) are key ROS sources: NOX2 generates superoxide during immune oxidative bursts, while NOX4 mediates redox signaling in fibroblasts and vascular cells. Peroxisomes also produce ROS during fatty acid oxidation via acyl-CoA and urate oxidases, generating hydrogen peroxide, which catalase detoxifies. Xanthine oxidase (XO) produces superoxide and hydrogen peroxide during hypoxanthine and xanthine metabolism, especially in ischemia–reperfusion and inflammation. Mitochondrial ROS can activate NOX enzymes, and NOX-derived ROS impair mitochondria, creating a feed-forward loop of oxidative stress linked to cardiovascular disease and neuroinflammation [9, 10].

2.1.2 RNS Sources

Nitric oxide (NO•) is a reactive nitrogen species (RNS) crucial for vascular health, immune regulation, and neurotransmission. It is synthesized by three nitric oxide synthase (NOS) isoforms: endothelial (eNOS), neuronal (nNOS), and inducible (iNOS). eNOS uses L-arginine with cofactors like tetrahydrobiopterin (BH4), NADPH, and flavins. nNOS, present in neurons and muscle, regulates contraction and synaptic activity, while iNOS, induced during inflammation, produces large NO• amounts for immune defense. NOS uncoupling from BH4 or L-arginine deficiency leads to superoxide ($O_2^{\bullet -}$) instead of NO•, forming peroxynitrite ($ONOO^-$), a potent oxidant. The NO•/ROS balance in vasculature dictates signaling vs. oxidative stress. Xanthine oxidase (XO) also generates superoxide and H_2O_2; under stress, XO and NO• form $ONOO^-$, impairing endothelium. Mitochondrial NO• inhibits cytochrome c oxidase, reducing ATP synthesis. Excess NO• reacts with superoxide to form $ONOO^-$, damaging mitochondria and triggering apoptosis. NADPH oxidases (NOX2, NOX4) also reduce NO• bioavailability, promoting nitrosative stress. Tight regulation of NO• and ROS is essential for cellular health [11, 12].

2.1.3 RSS Sources

Reactive sulfur species (RSS), including H_2S, persulfides (RSSH), and polysulfides (RS_nH, $n \geq 2$), are involved in redox regulation, signaling, and antioxidant functions in cells. Key endogenous RSS producers are cystathionine β-synthase (CBS), cystathionine γ-lyase (CSE), and 3-mercaptopyruvate sulfurtransferase (3-MST), which generate H_2S and derivatives from sulfur-containing amino acids like cysteine and homocysteine. CBS and CSE convert cystine to cysteine persulfide (CysSSH), which produces glutathione persulfide (GSSH) and other RSS in cells, tissue, and plasma [13]. H_2S is oxidized in mitochondria by sulfide quinone oxidoreductase (SQR) to form persulfides, which are further processed into thiosulfate by sulfur dioxygenase and rhodanese Polysulfides, including GSSH, act as potent antioxidants, scavenging hydrogen peroxide (H_2O_2) more effectively than glutathione (GSH) [14]. RSS also regulate protein function via sulfhydration, a post-translational modification influencing enzyme activity and redox signaling [13].

2.2 *Exogenous Sources*

2.2.1 ROS Sources

Environmental contaminants including heavy metals (e.g., arsenic, cadmium, and lead) and airborne particulate matter (PM2.5 and PM10) constitute the main external sources of ROS. These poisons either directly generate free radicals straight-forward or deplete cellular antioxidants, therefore generating oxidative stress [15]. *Chemicals and Drugs:* Many drugs—including chemotherapeutic medications (like cisplatin and doxorubicin)—generate ROS as part of their mechanism of action. For instance, doxorubicin generates semiquinone radicals combining with oxygen to produce superoxide anions. This ROS generation affects healthy organs as well as cancer cells, thereby producing side effects including cardiotoxicity [16].

2.2.2 RNS Sources

Major extrinsic causes of RNS are processed foods like cured meats and drinking water contaminated with nitrates and nitrites. First in the body, nitrates are converted to nitrites, which under an acidic environment in the stomach can be further changed to -NO. This path can improve cardiovascular health, but excessive nitrate/nitrite intake can lead to carcinogenic nitrosamines to emerge [17]. As major contributors of air pollution, nitrogen oxides (NOx), which consist of nitric oxide (-NO) and nitrogen dioxide (-NO_2) can with superoxide in the lungs produce peroxynitrite, which causes both asthma and chronic obstructive pulmonary disease [18].

2.2.3 RSS Sources

Foods heavy in sulfur-containing amino acids—including garlic, onions, and cruciferous vegetables—help to produce RSS. Metabolized to produce H_2S and other RSS in the body, these foods contain organosulfur compounds such diallyl disulfide and allicin. Studies on dietary RSS have exposed anti-inflammatory, antioxidant, and cardioprotective effects [19]. RSS accumulates in the body through exhaust from industrial setting like an oil refinery or wastewater treatment plant surrounded by hydrogen sulfide gas. Low levels of H_2S provide benefits; yet excessive concentrations are dangerous and can block mitochondrial respiration, thereby damaging and killing cells [20].

3 Mechanism of RONSS

RONSS contribute to redox signaling and help regulate homeostasis at cellular and tissue levels. Under stress conditions—environmental (e.g., UV exposure), metabolic (e.g., hyperglycemia), or pathological (e.g., muscular dystrophy, mechanical ventilation)—ROS levels rise, leading to abnormal signaling [21]. The broader role of reactive species beyond antioxidant systems is only recently being explored and remains unclear. Most RSSs are unstable and highly reactive. H_2S reacts with sulfenic acid (RSOH) or RSNO to form persulfides (RSSH), generating polysulfides and reflecting the chemical complexity of RSS and their interaction with RONS [22]. OS, RNS, and RSS exist in both radical and non-radical forms. Notably, ROS regulate antioxidant gene expression and key signaling molecules like MAPKs, PI3K, PTEN, and protein tyrosine phosphatases, influencing processes such as proliferation and survival [23].

3.1 ROS Metabolism

Oxygen is essential for life but can become toxic by forming ROS, which cause oxidative stress [24]. In mitochondria, the electron transport chain (ETC) typically reduces oxygen efficiently, but univalent reduction forms ROS. This starts with the one-electron reduction of O_2 to O_2^-, which generates secondary ROS via enzymes or metals. O_2^- is mostly dismutated to H_2O_2, which can produce reactive HO•. Residual O_2^- reacts with nitric oxide (NO) to form $ONOO^-$ or participates in Fenton and Haber–Weiss reactions, generating harmful radicals like HO• and HO_2• [25]. Fenton reactions, mainly in lysosomes, induce lysosomal stress and impair metabolism and autophagy [26]. In the Haber–Weiss reaction, H_2O_2 and O_2^- yield OH•, OH^-, and O_2. Antioxidants like GPXs and PRDX detoxify H_2O_2 using the GSH/GSSG system [27].

3.2 *RNS Metabolism*

RNS metabolism, tied to iron homeostasis, involves NO production from L-arginine by NOS, using O_2 and NADPH to yield NO, NADP, and L-citrulline in the citrulline-NO cycle. NO, a key signaling molecule, can also form via xanthine oxidase or nonenzymatically. It reacts with O_2 to produce NO_2, or with superoxide (O_2^-) to form peroxynitrite ($ONOO^-$). NO_2 further reacts with NO to form N_2O_3, which decomposes or converts to nitrite (NO_2^-). Nitrite oxidizes to nitrate (NO_3^-), while nitrate and nitrite can be reduced back to NO by nitrate and nitrite reductases [24, 28–30].

3.3 *RSS Metabolism*

ROS are deactivated by cellular thiols, forming disulfides. RSS, including hydrogen sulfide (H_2S), persulfides, and polysulfides, originate from sulfur biomolecules and play crucial roles in cellular signaling and redox homeostasis [31]. During oxidative stress, elevated peroxides and O_2^- polymerize to form $ONOO^-$ and generate RSS. RSS arise from GSH or GSSG oxidation by ROS like peroxides, S-nitrosoglutathione, and $ONOO^-$. RSS, highly reactive with thiols, are unlikely to be sequestered by electron donors like ascorbate. Disulfide-S-oxides impact redox-sensitive proteins and deplete GSH, shifting redox balance toward disulfide stress [24, 32].

3.4 *Molecular Pathways*

3.4.1 Nrf2 (Nuclear Factor (Erythroid-Derived 2)-Like 2), Keap1 (Kelch-Like ECH-Associated Protein 1), and ARE (Antioxidant Response Element) Pathway

The Nrf2/Keap1/ARE pathway is essential for defending against oxidative and chemical stress, maintaining redox balance. Under normal conditions, Keap1 targets Nrf2 for degradation. During stress, Keap1's cysteine residues are modified, releasing Nrf2 to translocate into the nucleus, where it binds AREs with Maf proteins and activates antioxidant gene expression [33–35]. This includes catalase, SOD, and GPx, which reduce ROS-induced damage. RNS like NO and peroxynitrite also contribute to oxidative stress. RSS, particularly H_2S, offer cytoprotective effects. H_2S activates Nrf2 via Keap1 S-sulfhydration, enhancing antioxidant defenses and showing therapeutic promise for diabetes-related vascular damage and sulfur mustard-induced lung injury [36–38].

3.4.2 Nuclear Factor Kappa-Light-Chain-Enhancer of Activated B Cells (NF-KB)/Hypoxia-Inducible Factor-1 (HIF-1)

NF-κB and HIF-1 are key regulators of oxidative stress and inflammation. NF-κB, activated by signals like TNF-α, moves to the nucleus to trigger pro-inflammatory cytokines and ROS-producing enzymes [39, 40]. During hypoxia, excess ROS stabilizes HIF-1α, enhancing NF-κB activity and sustaining inflammation [41]. HIF-1α also promotes mitochondrial ROS regulation and angiogenesis via TNF-α and VEGF, with NF-κB boosting its expression [40, 42]. NF-κB regulates iNOS, increasing NO and contributing to vascular damage in inflammation [43]. In coronary flow syndrome, low NO and high NF-κB activity correlate with elevated IL-1β [44]. Altered RSS levels, particularly reduced H_2S in high-fat diets, activate mTOR/IKK/NF-κB signaling, driving hypothalamic inflammation, neuron loss, and obesity. H_2S donors like sodium hydrosulfide (NaHS) can restore balance by suppressing NF-κB [45].

3.4.3 Mitogen-Activated Protein Kinase/Phosphoinositide 3-Kinase/Protein Kinase B (MAPK/PI3K/AKT) Signaling Pathways

The MAPK and PI3K/AKT pathways are essential for cell survival, proliferation, and apoptosis. MAPKs–ERKs, JNKs, and p38–transmit signals from the membrane to the nucleus, promoting either survival (ERKs) or cell death (JNKs, p38). Overactivation can lead to uncontrolled growth or excessive gene expression. MAPKs and MKPs regulate stress responses, including those triggered by ROS [46–48]. ROS activate MAPKs by modifying proteins or receptors, e.g., activating ERK via EGFR [49]. Intracellular kinases like ASK-1 are also triggered by oxidative stress [50], while MKPs deactivate MAPKs, ROS can inhibit MKPs by oxidizing their catalytic cysteines, sustaining JNK activation [51]. Low ROS promotes MKP expression, whereas high ROS suppresses MKP activity [52]. The PI3K/AKT pathway boosts antioxidant defenses, regulates detoxification genes [53] and controls autophagy—a survival strategy in stressed cancer cells. Targeting this pathway may exploit cancer cells' autophagy dependence and oxidative stress sensitivity [54]. H_2S, a gaseous signaling molecule, also shows therapeutic promise via its effects on the PI3K pathway in chronic cardiomyopathy [55, 56].

3.5 Cellular Responses to Oxidative Stress

3.5.1 Necroptosis

Necroptosis is a regulated cell death process driven by RIPK1, RIPK3, and MLKL, forming the necrosome. Triggered by infection, inflammation, or cellular damage, necroptosis leads to organelle enlargement, loss of cellular integrity, release of cellular contents, and inflammation. It differs from apoptosis and is implicated in conditions like neurodegeneration, cancer, and cardiovascular diseases [57]. Mitochondrial redox homeostasis is crucial in necroptosis by enhancing ROS production. ROS promote necrosome formation by stabilizing RIPK1 and RIPK3 interactions, activating pro-necrotic kinases. Despite mitochondrial ROS involvement, mitochondria are not essential for necroptosis in some cases [58, 59]. RNS, particularly NO, influence cell death by modulating Fas receptor oxidation and JNK activation. RNS-induced cell death bypasses caspase activity but involves mitochondrial depolarization and nuclear condensation [60, 61]. H_2S is a therapeutic agent with anti-apoptotic, anti-inflammatory, and antioxidant effects. It regulates necroptosis by targeting oxidative stress pathways. In diabetic cardiomyopathy, H_2S suppresses necroptosis by inhibiting mitochondrial ROS. It also mitigates cognitive dysfunction in Parkinson's disease by downregulating necroptosis proteins via the cGAS/STING pathway [62, 63] and promotes skin wound healing in diabetes by reducing inflammation and necroptosis [64].

3.5.2 Apoptosis

Apoptosis, a regulated form of cell death, maintains tissue balance by removing damaged cells [65]. It is initiated via intrinsic, extrinsic, and perforin/granzyme pathways, with ROS playing a key role, particularly in the intrinsic pathway [66]. ROS oxidize polyunsaturated fatty acids (PUFAs) in mitochondrial membranes, causing membrane swelling and opening the mitochondrial mega channel. This releases cytochrome c, activating caspase-3 and triggering cell death [67]. ROS also activate p53, promoting pro-apoptotic proteins like Bax and Bak [68]. The extrinsic pathway involves TNF-α binding to receptors, forming a death-inducing signaling complex (DISC), which activates caspase-8 and further caspases, leading to cell death [69]. ROS contribute to cancer progression by promoting tumor invasion and survival via NOX-mediated ROS buildup and ferroptosis-related gene activation [70]. RNS influences apoptosis through complex mechanisms [71], RNS can induce DNA damage and activate p53 to promote cell death, but they can also inhibit caspase activity and upregulate survival genes [72]. In conditions like hyperoxia-induced lung injury, RNS modulate mitochondrial apoptosis pathways [73]. ROS and RNS interact with JNK and p38 MAPK, impacting survival and death processes. NO inhibits nuclear factor kappa B (NF-κB), protecting against apoptosis and necrosis in hepatocytes [74]. H_2S/RSS can either promote or inhibit cell death depending on the concentration and cellular environment [75]. At physiological levels, H_2S boosts antioxidant defenses and activates stress pathways like Nrf [76, 77]. Conversely, higher concentrations of H_2S can induce apoptosis by disrupting mitochondrial function, particularly in cancer cells [78]. H_2S regulates cell death via sulfane sulfur species such as persulfides and polysulfides, which modulate signaling proteins through post-translational modifications like protein persulfidation. These modifications can either promote or inhibit apoptosis depending on the cellular context [79].

4 RONSS in Neurodegenerative Diseases

4.1 RONSS in Alzheimer's Disease

Alzheimer's disease (AD) is a neurodegenerative disorder characterized by memory loss, cognitive decline, and neuronal death. Oxidative stress from reactive oxygen, nitrogen, and sulfur species (RONSS) accelerates amyloid plaque formation and tau hyperphosphorylation. Aβ42 binds metals like copper and iron, increasing ROS and worsening neurotoxicity. Mitochondrial dysfunction—AD's main ROS source—impairs calcium balance, energy production, and damages cells [80]. ROS degrade mitochondrial DNA and trigger apoptosis.

Oxidative stress also promotes neuroinflammation. Activated microglia generate superoxide via NOX and $ONOO^-$; astrocytes release cytokines (TNF-α, IL-1β), further harming neurons and mitochondria [81–83]. Misfolding of Aβ and tau proteins is linked to oxidative damage [84, 85], ROS harm DNA, disrupt repair, and promote neuronal death [86]. Lipid peroxidation damages membranes, producing toxic byproducts like 4-hydroxynonenal (4-HNE) and malondialdehyde (MDA) [87]. Antioxidant enzymes such as SOD, catalase, and glutathione peroxidase are reduced in AD. Nrf2 dysfunction worsens oxidative stress [80–83]. Animal models (e.g., SOD1/SOD2-deficient mice) show how oxidative damage drives neurodegeneration. Therapeutics like melatonin, coenzyme Q10, and N-acetylcysteine may counteract ROS, support mitochondria, and improve cognition [82, 83, 86].

4.2 Role of ROSNSS in Parkinson's Disease

Parkinson's disease (PD) is a chronic neurodegenerative disorder marked by the loss of dopamine-producing neurons in the substantia nigra. RONSS play a key role in its progression [88, 89]. Mitochondrial dysfunction leads to excessive ROS and RNS, creating a damaging feedback loop. RNS, particularly peroxynitrite, modifies proteins like α-synuclein, promoting its toxic aggregation into Lewy bodies—PD's pathological hallmark. Hydrogen sulfide (H_2S) offers cellular protection at low levels but may contribute to disease when imbalanced [90]. Genetic alterations in PARK2, PINK1, and DJ-1 genes impair mitochondrial quality control and antioxidant defense. Oxidative stress also disrupts the PI3K/Akt pathway, reducing cell survival, while nuclear factor erythroid 2-related factor 2/antioxidant response element (Nrf2/ARE) pathway activity declines. Overexpression of NOS2 increases nitric oxide and RNS, exacerbating damage. Mutations or overexpression of SNCA gene, which encodes α-synuclein, further amplify ROS and protein aggregation [92, 93]. Animal models, particularly MPTP-treated mice, mimic PD by selectively destroying dopaminergic neurons and elevating ROS. These models are crucial for studying oxidative stress in PD and developing targeted therapies [94, 95].

4.3 Role of RONSS in Huntington's Disease

Huntington's disease (HD) is a severe neurodegenerative condition caused by a genetic mutation in the huntingtin (HTT) gene, which leads to the production of a defective protein called mutant huntingtin (mHTT). mHTT damages HD's mitochondria, which generates significant amounts of RNS and ROS. This produces oxidative stress and overcomes the mechanism of antioxidant protection damaging the neurons [96, 97]. Mitochondrial dysfunction is the key mechanism by which ROS and RNS induce HD. mHTT alters the electron transport system and causes electron leakage, hence increasing ROS and RNS which also throws off redox signaling pathways, which are essential for regulating mortality, inflammation, and brain plasticity. ROS initiate pro-apoptotic signaling, resulting in neuronal death, while RNS nitrosylate proteins, disrupting their function and contributing to neurodegeneration. In HD, NF-κB and Nrf2 pathways, which are responsible for governing antioxidant and inflammation systems, are impaired worsening oxidative damage and inflammation, accelerating neurodegeneration [98–101]. Several genes involved in oxidative stress pathways are affected in HD. Redox-regulation genes such as SOD1 and SOD2, glutathione peroxidase, and catalase are typically dysregulated in HD, reducing the cell's ability to neutralize ROS and RNS [96, 97]. Animal models, particularly transgenic rodents that express mHTT, have been extensively employed to evaluate the functionalities of ROS and RNS in HD. Genetic and pharmacological approaches that reduce ROS levels have been shown in animal studies to prevent neurodegeneration, highlighting the causal role of oxidative stress in Huntington's disease [98, 99].

4.4 Role of RONSS in Haemorrhagic Stroke

Haemorrhagic stroke, particularly intracerebral haemorrhage (ICH), involves complex mechanisms where RONSS play critical roles. These molecules exacerbate cerebral damage by inducing oxidative stress, disrupting cellular balance, reducing mitochondrial activity, and promoting iron release from haemoglobin, which further generates reactive species [102, 103]. RSS, though less investigated, are likely to interact with ROS and RNS, hence influencing cellular damage and neuronal death [102, 104]. Haemoglobin from damaged red blood cells also has a significant impact by causing ROS, especially by means of iron-driven mechanisms producing quite dangerous hydroxyl radicals [105]. The genes Nrf2, HO-1, and SOD are crucial in controlling oxidative stress; their deficiency could lead to damage [106]. Animal studies have shown that antioxidants like edaravone and natural compounds such as curcumin and resveratrol can reduce oxidative damage and improve outcomes. However, the exact role of reactive species in causing haemorrhagic stroke remains unclear, warranting further investigation [102–104].

5 Role of RONSS in Cardiovascular Diseases

Reactive oxygen and nitrogen species (RONSS) serve dual roles as harmful stressors and essential signaling molecules, influencing gene expression, cell migration, differentiation, and proliferation. When RONSS production surpasses antioxidant defenses, oxidative stress occurs, leading to inflammation, cell death, and biomolecular damage that worsen cardiovascular diseases [107, 108]. Excess ROS reduces NO bioavailability, promoting vasoconstriction, inflammation, and endothelial dysfunction. RNS, like peroxynitrite, modify proteins and lipids, impairing cellular functions and contributing to atherosclerosis. RSS, such as hydrogen sulfide, supports vasodilation and cytoprotection at physiological levels but causes oxidative stress and vascular damage when imbalanced [109]. Although lower H2S pathways have been associated with vascular diseases, RSS and H_2S play less clearly defined roles. Together, ROS, RNS, and RSS induce oxidative stress, inflammation, and cellular damage that fuel cardiovascular diseases [110].

5.1 Role of RONSS in Coronary Heart Disease

Coronary heart disease (CHD) is a multifactorial condition driven by endothelial dysfunction, oxidative stress, and inflammation [111]. Risk factors like diabetes, hypertension, and hypercholesterolemia are closely linked to ROS. Mitochondria, especially via the electron transport chain and Nox4, are key ROS sources. Excess ROS leads to mitochondrial dysfunction, apoptosis, fibrosis, and cardiac damage, promoting disease progression [112]. NO supports vascular health by promoting vasodilation, inhibiting platelets, and reducing inflammation. In CHD, oxidative stress reduces NO levels and impairs its function. ROS react with NO to form peroxynitrite, a harmful RNS that damages cellular components and contributes to endothelial dysfunction and atherosclerosis. Oxidative stress also impairs NO synthesis and activates inflammatory pathways that worsen CHD [113]. Low NO further increases leukocyte adhesion and reduces vasodilation (116). Overactivation of the renin-angiotensin system (RAS) in hypertension and heart failure raises angiotensin II levels, stimulating NADPH oxidase and increasing ROS. This further reduces NO, causing oxidative stress, inflammation, and vascular damage [114, 115]. H_2S, are less studied but protective in CHD. H_2S enhances NO, reduces ROS, and suppresses inflammation. Deficient H_2S impairs vasodilation and mitochondrial function. It also activates Nrf2 and inhibits vascular smooth muscle growth via PI3K/Akt [[114, 116].

5.2 Role of RONSS in Ischemic Heart Disease (IHD)

Ischemic heart disease (IHD), mainly caused by atherosclerotic plaque buildup in coronary arteries, remains a major global health issue. During ischemia–reperfusion, ROS induce oxidative stress, mitochondrial dysfunction, and inflammation, accelerating IHD progression. Oxygen deprivation disrupts mitochondrial electron transport, causing electron buildup in complexes I and III. Excess ROS damage lipids, proteins, and DNA, leading to cardiomyocyte death [117, 118]. ROS also activate redox-sensitive pathways like NF-κB and JNK, increasing pro-inflammatory cytokines (e.g., IL-6) and matrix metalloproteinase (MMPs), which destabilize plaques and promote artery blockage [119, 120]. RNS worsen IHD via nitrosative stress and endothelial dysfunction. Oxidative stress leads to NO and $ONOO^-$ overproduction due to iNOS upregulation and eNOS uncoupling. $ONOO^-$ damages cellular components, impairs mitochondria, and reduces ATP. RNS also lower NO availability, causing vasoconstriction and plaque instability. Chronic nitrosative stress activates NF-κB, driving inflammation and vascular injury [121, 122]. RSS—including H_2S, RSSH, and polysulfides—play protective roles in IHD. H_2S reduces oxidative stress and promotes S-persulfidation, enhancing cell defenses. Impaired RSS pathways lower H_2S levels, increasing vascular damage and inflammation. RSS donors (e.g., diallyl trisulfide, polysulfides) given during ischemia–reperfusion reduce infarct size, preserve mitochondria, and improve heart function [123, 124].

5.3 Role of RONSS in Microvascular Dysfunction

Coronary Microvascular Dysfunction (CMD) involves abnormalities in small coronary vessels that impair blood flow despite clear large arteries. It stems from endothelial and smooth muscle dysfunction, leading to reduced vasodilation, increased vasoconstriction, and inflammation. Mitochondrial dysfunction increases ROS, which lowers NO bioavailability by forming $ONOO^-$, damaging endothelial cells and impairing vasodilation [33]. ROS also disrupt the endothelial barrier and raise vascular permeability through oxidative stress and inflammation. During acute inflammation, activated granulocytes release ROS that damage endothelium, causing leakage. Inhibiting neutrophil infiltration or using superoxide scavengers like SOD reduces this damage, highlighting leukocyte adhesion as a major factor [125]. RNS further contribute to CMD by forming $ONOO^-$, which damages proteins, lipids, and endothelial cells, reducing NO and activating NF-κB. This increases adhesion molecules and cytokines, worsening inflammation and vascular injury [126]. RNS-driven mitochondrial damage also lowers perfusion and induces vascular remodeling, as seen in ischemia–reperfusion injury [127]. RSS, particularly H_2S, affect redox balance, vascular tone, and endothelial function. Reduced H_2S levels in CMD lead to capillary stiffness, impaired regeneration, and vascular dysfunction. H_2S supports mitochondrial function by limiting ROS production, and its signaling pathways are linked to reduced inflammation and fibrosis, suggesting therapeutic potential in systemic microvascular diseases [126, 128].

6 Role of RONSS in Muscular Diseases

6.1 Role of RONSS in Skeletal Muscles and Muscle Myogenesis

ROS play key roles in skeletal muscle physiology and myogenesis. Moderate ROS levels activate signaling pathways like MAPK, NF-κB, and AMPK, supporting muscle contraction, adaptation to exercise, and cell repair [129]. They regulate muscle progenitor cell growth and differentiation via transcription factors like MyoD and myogenin [130, 131]. However, excessive ROS production causes oxidative stress, suppressing myogenesis, inducing muscle injury, and contributing to conditions like muscular dystrophy, sarcopenia, and age-related muscle wasting [130, 132]. NO regulates blood flow, muscle contraction, mitochondrial biogenesis, and myogenesis by regulating satellite cells' activation and differentiation [129, 132]. H_2S and other RSS are currently recognized to play significant roles as muscle physiology regulators. H_2S is responsible for enhancing muscle repair and regeneration, limiting oxidative stress, and regulating redox signaling [131, 133]. H_2S is further acting in synergistically with RNS and ROS in regulating redox balance to support healthy adaptation and muscle function [131, 133].

6.2 Role of RONSS in Muscular Dystrophies

Muscular dystrophies (MDs), mainly Duchenne (DMD), cause progressive muscle loss. Reactive oxygen and nitrogen species (RONSS) contribute to oxidative stress, mitochondrial damage, and muscle dysfunction. Oxidative stress oxidizes protein thiols, harming muscle and cell interactions. Dysferlinopathies show faulty membrane repair and more oxidative damage [134]. In DMD, mitochondrial dysfunction causes excess ROS that overwhelm antioxidants, leading to muscle damage. ROS and RNS aid signaling when acute but cause harm if chronic. NRF2 pathway disruption worsens oxidative stress. Targeting mitochondria and antioxidants may protect muscles [135]. Mitochondrial dysfunction in DMD leads to excessive ROS production. While ROS are essential for normal cellular signaling, high levels become cytotoxic. Targeting mitochondrial function and antioxidant pathways may help regulate ROS and protect muscle. In skeletal muscle, ROS and RNS have a dual role in fatigue and recovery–acute increases support adaptation, but chronic elevation causes oxidative damage and contractile dysfunction, worsening muscle weakness in MDs [136, 137]. NO deficiency contributes to vascular disease and reduced muscle blood flow, accelerating disease progression. Skeletal muscle shows severe effects from ROS and RNS, including chronic inflammation, fibrosis, and impaired regeneration—hallmarks of muscular dystrophies (MDs). Recently, RSS have also been linked to MDs due to their roles in modulating oxidative stress and inflammation. Reactive species play a dual role in MDs: driving oxidative damage and muscle degradation, while also participating in key signaling pathways [134, 140].

6.3 Role of RONSS in Sarcopenia

Sarcopenia, the age-related decline in skeletal muscle mass, strength, and function, is driven by interactions among reactive oxygen, nitrogen, and sulfur species (RONSS). These molecules play dual roles in maintaining muscle health and contributing to its degeneration. ROS, including O_2^- and H_2O_2, are primarily produced in mitochondria during oxidative phosphorylation. At physiological levels, they regulate muscle growth, adaptation, and repair via pathways like Nrf2 and PGC-1α, which enhance antioxidant defenses and mitochondrial biogenesis [141, 142]. Aging disrupts this balance, as increased ROS and mitochondrial dysfunction overwhelm antioxidant systems, inducing oxidative stress. Excess ROS activate pathways like the ubiquitin–proteasome system and NF-κB, promoting muscle atrophy [143, 144]. RNS, especially nitric oxide (NO), support normal muscle function by modulating blood flow, mitochondrial activity, and glucose uptake. Aging increases iNOS expression and NO production, leading to peroxynitrite ($ONOO^-$) formation through reactions with ROS. $ONOO^-$ causes nitrosative stress, impairs calcium handling, and alters protein structure, reducing contractile efficiency. RNS also activate apoptotic pathways and hinder muscle regeneration, contributing to sarcopenia [144, 145]. H_2S, help maintain redox balance and cellular signaling. Produced by CSE and CBS, H_2S supports mitochondrial function, reduces oxidative stress, and has anti-inflammatory effects. With age, H_2S levels decline, worsening mitochondrial dysfunction and inflammation. Excess ROS/RNS drive lipid peroxidation, destabilizing membranes and triggering apoptosis via byproducts like MDA and 4-HNE. Redox imbalance also impairs autophagy, leading to the buildup of damaged organelles. Age-related declines in autophagy further contribute to muscle dysfunction and sarcopenia. Understanding redox regulation is key to developing effective sarcopenia therapies [140, 143, 145–147].

6.4 Role of RONSS in Critical Illness Myopathy

Critical illness myopathy (CIM) commonly affects long-term ventilated or systemically inflamed patients. It is driven by oxidative stress, mitochondrial damage, and inflammation due to excess RONSS. ROS, including O_2^- and H_2O_2, are elevated during critical illness from systemic inflammation, ischemia–reperfusion, and immobility [148–151]. These ROS damage proteins, lipids, and DNA, contributing to muscle atrophy and dysfunction. Superoxide also activates pro-inflammatory pathways and impairs mitochondrial function, disrupting the electron transport chain, reducing ATP, and worsening muscle weakness. Additionally, ROS activate NF-κB, increasing inflammation [148, 150, 151]. RNS, such as NO and $ONOO^-$, are elevated by oxidative stress and inflammation. NO from iNOS reacts with superoxide to form $ONOO^-$, leading to nitrosative stress, protein/lipid nitration, and impaired muscle contraction. RNS also disrupt mitochondrial function by inhibiting enzymes like cytochrome c oxidase, reducing energy production [148, 150]. RSS, particularly H_2S, can be protective or harmful depending on the quantity and environment. At physiological levels, H_2S acts as an antioxidant and supports mitochondrial function. However, its disruption promotes oxidative stress and inflammation. High H_2S levels may inhibit mitochondrial respiration, while lower levels support ATP synthesis. Further research is needed to clarify the roles of ROS, RNS, and RSS in CIM and identify effective treatments [148–152].

7 Role of RONSS in Cancer

Cancer is a complex disease influenced by its microenvironment and regulated by various intracellular and extracellular stimuli. Reactive species (RONSS) play a key role in cancer development and spread. Many cancers accumulate RONSS due to an imbalance between ROS production and antioxidant activity, which alters signal transduction and gene expression, driving cell invasion, migration, proliferation, and apoptosis [153]. ROS act as both essential signaling molecules and harmful agents. Elevated ROS levels in cancer favor pro-tumorigenic signaling and cell proliferation, driving carcinogenesis in cancers such as lung, breast, colorectal, cervical, and hepatocellular carcinomas. Excess ROS can activate oncogenes and inhibit tumor suppressor genes like p53 and KRAS, accelerating cancer development [154, 155]. ROS and RNS contribute to cancer by acting as secondary messengers in signaling pathways, but at high levels, they can also induce apoptosis and cellular senescence [156]. Lower RONSS levels promote cancer cell migration and proliferation, while higher levels can have cytotoxic effects, aiding cancer therapy [157]. Cancer cells produce more ROS than normal cells, which hyperactivate pro-tumorigenic signaling pathways. ROS inhibit pathways like PI3K/Akt and transcription factors like FOXO, which upregulate antioxidants and increase ROS. Elevated MAPK levels also contribute to tumor metastasis [158, 159]. Oxidative stress damages tumor suppressor genes and oncogenes, causing mutations. A key DNA lesion, 8-OHdG, mispairs guanine with adenine instead of cytosine, leading to mutations in oncogenes and tumor suppressor genes, making it a biomarker for oxidative stress [160]. Recent research suggests H_2S can regulate the cell cycle, promoting proliferation in colon cancer and hepatoma cells by activating ERK and AKT pathways [161, 162].

7.1 Role of RONSS in Lung Cancer

Lung cancer is the second most common type of cancer diagnosed worldwide and the primary cause of death due to cancer. Tobacco use, air pollution, and other factors are among the main causes of lung cancer incidence and mortality worldwide. The lungs are especially vulnerable to ROS exposure because they are the body's main organ for gas exchange and due to their continuous exposure to atmospheric oxygen and various environmental pollutants. Exposure to pollution-causing agents and allergens can raise ROS production, which can cause inflammation of the windpipe and asthma. To fight bacterial and viral invasion, the respiratory tract produces excess ROS during infection; however, these must be adequately scavenged to avoid tissue damage [163, 164]. Inhaled toxins generate ROS/RNS, causing oxidative stress and damaging lipids, DNA, and proteins. This triggers inflammation and the release of cytokines. Cigarette smoke contains nitric oxide, which reacts with superoxide to form $ONOO^-$, and tar, which contains free radicals that generate O_2^- and H_2O_2. Monocytes and PMNs flood the lungs in response, but chronic exposure to toxins or pathogens can lead to ROS/RNS overload, causing lung damage and chronic inflammation [163]. RSS also influence lung cancer development by regulating redox balance, mitochondrial function, and chemoresistance. Lung cancer tissues overexpress some enzymes which generate H_2S, such as 3-mercaptopyruvate sulfurtransferase (3-MST), CBS, and CSE. These enzymes play a role in different biochemical processes that produce H_2S as a byproduct. CBS and CSE convert homocysteine into cystathionine, which is then broken down into cysteine, which leads to the formation of H_2S. 3-MST transforms 3-mercaptopyruvate into H_2S. This promotes tumor growth via angiogenesis, proliferation, and apoptosis resistance through ERK and AKT pathways. Elevated H_2S also supports mitochondrial DNA repair and tumor survival. RSS further modulate signaling pathways linked to angiogenesis and tumor progression [165, 166].

7.2 Role of RONSS in Breast Cancer

Breast cancer is the most common cancer in women. Elevated ROS in breast cancer cells and the tumor microenvironment (TME) promote tumor growth, metastasis, genomic instability, inflammation, apoptosis resistance, metabolic reprogramming, and EMT—where epithelial cells lose polarity and adhesion, aiding proliferation [167]. Hypoxic ROS increase tumor-repopulating cell (TRC) proliferation and cancer stemness, suggesting therapeutic potential [159, 164]. ROS also influence metastasis by modulating the antioxidant system and activating MAPK and PI3K/AKT pathways, affecting the cell cycle, adhesion, and metalloproteinase activity. Hypoxia in the TME raises ROS levels and triggers HIF-1α expression, creating a *"ROS-HIF-1α-ROS"* feedback loop that drives cancer progression [167]. $ONOO^-$, formed by reactive species interaction, induces oxidative and nitrosative stress. Chronic low levels activate NF-κB, promote inflammation, and cause DNA mutations, supporting tumor growth. However, $ONOO^-$ can also eliminate cancer cells and inhibit tumor growth, making it a promising therapeutic target [168]. H_2S contributes to breast cancer by disrupting redox balance, promoting cell growth, angiogenesis, and drug resistance. Enzymes like CBS and CSE generate H_2S, activating NF-κB and Nrf2 survival pathways. High H_2S levels support tumor growth and therapy resistance. Inhibiting CBS/CSE can sensitize cancer cells to chemotherapy and oxidative stress [162, 169].

7.3 Role of RONSS in Prostate Cancer

Prostate cancer (PCa) is the second most common cancer in men and a leading cause of cancer-related deaths, mainly affecting those aged 45–60 [155, 170]. ROS contribute significantly to PCa progression by triggering transcription factors like SNAIL, which induces EMT and further elevates ROS, creating a feedback loop [164]. Age-related increases in ROS, especially through PI3K/AKT/mTOR activation, disrupt cellular homeostasis and promote tumor development. ROS can modify cysteine residues, impairing tumor suppressors like PTEN. Elevated ROS in PCa cells also results from mitochondrial dysfunction, glucose metabolism, enzyme activation, and increased p66Shc activity [171]. Low concentrations of the hormone H_2S can activate the growth of prostate cancer cells through pathways such as ERK and AKT, whereas oxidative stress and mitochondrial damage can suppress the tumor growth [162, 172].

8 Role of RONSS in Other Diseases

RONSS and metal ions are vital in processes like signal transduction, inflammation, and neurodegeneration. While essential for normal cellular function, their excessive accumulation contributes to stress-related diseases. Their complex interplay underscores the need for accurate quantification in cells, tissues, and organisms to better understand their roles in health and disease [173]. ROS, RNS, and RSS are reactive molecules that, in excess, cause oxidative, nitrosative, and sulfur-mediated stress, damaging proteins, lipids, and DNA. Moderate levels can disrupt cell function, triggering senescence, abnormal growth, and inflammation, while high levels induce cell death via oncosis, apoptosis, or autophagy [174]. Oxidative stress is central to many diseases, with RONSS serving as key biomarkers. Understanding their fluctuations supports early diagnosis, disease monitoring, and drug development [175]. Fluorescent probes now enable real-time RONSS monitoring *in vitro* and *in vivo* [175]. Though crucial for redox signaling, RONSS overproduction—especially in mitochondria—leads to cellular damage linked to cancer, cardiovascular, neurological, and inflammatory diseases. Their diverse roles can be targeted therapeutically.

Precise, timely detection of RONSS is essential for understanding disease onset and guiding early interventions. Therapies addressing oxidative and nitrosative stress have shown promise in managing various conditions [107].

8.1 Diabetes

Inflammation plays a central role in diabetes progression and its complications. Type 1 diabetes stems from autoimmune or viral inflammation, while type 2 diabetes involves chronic inflammation due to insulin resistance and abnormal glucose metabolism [176]. Oxidative stress, caused by excess ROS and RNS overwhelming antioxidant defenses, is closely linked to diabetes development. Superoxide anion ($O_2^{\cdot}-$), a primary ROS, contributes significantly to diabetic complications [175]. ROS and RNS influence insulin signaling—initially aiding insulin action, but later impairing it and promoting insulin resistance, increasing type 2 diabetes risk [177]. H_2S affects insulin secretion, oxidative stress, and pancreatic function. Diabetes alters H_2S levels depending on tissue type and disease stage. In diabetic models like ZDF and STZ rats, H_2S and its synthesizing enzymes (CBS, CSE) are elevated in the liver and pancreas. In ZDF rats, excessive pancreatic H_2S activates K^+ATP channels, impairing insulin release and causing β-cell dysfunction [178]. Emerging studies also highlight the role of redox imbalance—particularly between ROS and RSS—in regulating ferroptosis and diabetes progression, linking oxidative stress to worsening metabolic dysfunction [179].

8.2 Inflammatory Bowel Disease (IBD)

Inflammatory Bowel Disease (IBD), including ulcerative colitis (UC) and Crohn's disease, involves chronic intestinal inflammation driven by oxidative stress, immune dysregulation, and epithelial barrier damage [180]. Excessive ROS and RNS lead to oxidative and nitrosative stress, damaging cellular components, disrupting tissue homeostasis, and promoting inflammation via pro-inflammatory cytokines and signaling pathways [181]. This imbalance is worsened by immune dysfunction and infiltration of M1 macrophages, which further elevate RONS, degrade gut proteins, and weaken the mucosal barrier [182]. $ONOO^-$, a potent RONS, exacerbates inflammation by damaging epithelial cells and impairing immune regulation in IBD [183]. Gut microbiota alterations—such as increased sulfate-reducing bacteria, fewer butyrate producers, reduced H_2S—detoxifying enzymes, and high dietary sulfate intake—raise intestinal H_2S levels, contributing to inflammation and disease progression [184].

8.3 *Chronic Kidney Disease (CKD)*

Chronic kidney disease (CKD) results from repeated injury to the glomerular or tubular epithelium, triggering cycles of inflammation, repair, and fibrosis that lead to nephron loss and end-stage kidney disease (ESKD). Oxidative stress—caused by excess ROS and RNS—plays a key role in this progression, disrupting cellular signaling and contributing to tissue damage. While elevated ROS levels are well-documented in CKD, their exact origins and mechanisms remain under investigation, offering potential for early intervention strategies [185]. CKD is also marked by persistent inflammation and declining antioxidant defenses, which intensify as kidney function worsens. This imbalance leads to elevated cytokines and stress markers, contributing to uremic complications. Managing oxidative stress and inflammation is essential to slow CKD progression and improve outcomes [186]. Beyond kidney damage, ROS contribute to blood-related issues such as anemia and hypoxia by damaging red blood cells. They also promote abnormal clotting through platelet and immune cell activation, increasing the risk of thromboembolism, stroke, and cardiovascular complications. Controlling oxidative stress is vital for protecting both renal and circulatory health [185, 187]. RSS have emerged as key redox regulators with protective roles in CKD. They support antioxidant defenses and may help reduce oxidative stress, offering therapeutic potential for preserving kidney function [188].

8.4 *Hypertension*

Excessive ROS and RNS production is closely linked to hypertension, stroke, angina, and other cardiovascular diseases. Generated mainly by the mitochondrial respiratory chain, NADPH oxidase, and xanthine oxidase, these species contribute to oxidative stress, a key driver of endothelial dysfunction, vascular inflammation, lipid peroxidation, and tissue damage [189]. RNS, derived from NO via NOS, include peroxynitrite and 3-nitrotyrosine, which disrupt normal cellular functions. While NO helps regulate blood pressure and cell signaling, its interaction with cytochrome c oxidase can impair mitochondrial respiration. Oxidative and nitrosative stress play dual roles—beneficial at physiological levels but harmful when elevated, contributing to hypertension by impairing vascular function [190]. Alongside NO, H_2S regulates blood vessel tone by acting on smooth muscle and endothelial cells. Both molecules modulate enzyme function and protein structure, supporting vascular homeostasis. Imbalances in this system can worsen vascular dysfunction and drive cardiovascular disease, suggesting therapeutic potential in targeting H_2S-NO interactions [191].

8.5 Obesity

Obesity is a chronic, multifactorial disease marked by excessive fat accumulation. White adipose tissue acts as both an energy store and endocrine organ, producing adipokines that regulate inflammation and metabolism. Pro-inflammatory adipokines like IL-6 and appetite-regulating leptin also stimulate ROS production, contributing to oxidative stress (OS) and obesity-related complications [192]. Elevated ROS levels are strongly linked to insulin resistance and type 2 diabetes, yet ROS also play essential roles in cell signaling and may help maintain insulin sensitivity. Researchers assess ROS effects by measuring damage to biomolecules or analyzing antioxidant defenses. Advanced techniques use biosensors to detect ROS forms precisely. Recognizing ROS's dual role is crucial for managing obesity and metabolic health [193]. Over the past 40 years, rising obesity rates have fueled higher risks of diabetes and heart disease. Nitric oxide (NO), a key regulator of metabolism and adiposity, is reduced in obese individuals due to lower NOS activity and ROS interaction. In particular, decreased endothelial NOS (eNOS) activity worsens metabolic dysfunction in obesity and diabetes [194]. H_2S from the gut increases intestinal permeability and inflammation, promoting systemic inflammation and insulin resistance. This may impair adipose function and contribute to obesity and related metabolic disorders [195].

9 Conclusion

Customizing a treatment plan calls for an understanding of the interactions among ROS, RNS, and RSS. Recent studies show how redox-based pharmacological treatments, enzyme inhibitors, and antioxidants modify these species. H_2S-releasing chemicals are being studied for their potential cytoprotective effects in neurodegenerative and metabolic illnesses, while NO• donors are used in cardiovascular treatment to increase vasodilation. The development of fluorescent probes has also made real-time imaging of ROS, RNS, and RSS dynamics in biological systems practicable, therefore increasing our knowledge of their activity in health and disease. This chapter analyzes the contribution of the class of ROS, RNS, and RSS to cellular signaling and redox regulation, as well as their roles in disease, comparing results across neurodegenerative, cardiovascular, and metabolic diseases.

References

1. Dutta, S., Sengupta, P., Samrot, A. V. (2024). Physiological and pathological functions of Reactive Nitrogen Species (RNS) and Reactive Sulphur Species (RSS) on male reproductive functions. *Journal of Integrated Science and Technology, 12*(3), 755–755 [cited 2025 Mar 28]. Available from: https://pubs.thesciencein.org/journal/index.php/jist/article/view/a755

2. Giles, G. I., Nasim, M. J., Ali, W., Jacob, C. (2017). The reactive sulfur species concept: 15 Years on. *Antioxidants, 6*(2), 38 [cited 2025 Mar 29]. Available from: https://www.mdpi.com/2076-3921/6/2/38/htm
3. Song, P., Zou, M. H. (2015). Roles of reactive oxygen species in physiology and pathology. *Atherosclerosis: Risks, Mechanisms, and Therapies*, 379–92 [cited 2025 Mar 29]. Available from: https://doi.org/10.1002/9781118828533.ch30
4. Pourova, J., Kottova, M., Voprsalova, M., Pour, M. (2010). Reactive oxygen and nitrogen species in normal physiological processes. *Acta Physiologica, 198*(1), 15–35 [cited 2025 Mar 29]. Available from: https://doi.org/10.1111/j.1748-1716.2009.02039.x
5. Gruhlke, M. C. H., & Slusarenko, A. J. (2012). The biology of reactive sulfur species (RSS). *Plant Physiology and Biochemistry., 1*(59), 98–107.
6. Hou, J. T., Yu, K. K., Sunwoo, K., Kim, W. Y., Koo, S., Wang, J., et al. (2020). Fluorescent imaging of reactive oxygen and nitrogen species associated with pathophysiological processes. *Chemical, 6*(4), 832–866.
7. Kolluru, G. K., Shen, X., Kevil, C. G. (2020). Reactive sulfur species: A new redox player in cardiovascular pathophysiology. *Arteriosclerosis Thrombosis and Vascular Biology, 40*(4), 874–84 [cited 2025 Mar 29]. Available from: https://pubmed.ncbi.nlm.nih.gov/32131614/
8. Wu, L., Sedgwick, A. C., Sun, X., Bull, S. D., He, X. P., James, T. D. (2019). Reaction-based fluorescent probes for the detection and imaging of reactive oxygen, nitrogen, and sulfur species. *Accounts are Chemical Research, 52*(9), 2582–97 [cited 2025 Mar 29]. Available from: https://doi.org/10.1021/acs.accounts.9b00302
9. Hernansanz-Agustín, P., Enríquez, J. A. (2021). Generation of reactive oxygen species by mitochondria. *Antioxidants, 10*(3), 415 [cited 2025 Mar 29]. Available from: https://www.mdpi.com/2076-3921/10/3/415/htm
10. Canton, M., Sánchez-Rodríguez, R., Spera, I., Venegas, F. C., Favia, M., Viola, A., et al. (2021). Reactive oxygen species in macrophages: Sources and targets. *Front Immunology, 12*, 734229 [cited 2025 Mar 29]. Available from: www.frontiersin.org
11. Tejero, J., Shiva, S., Gladwin, M. T. (2019). Sources of vascular nitric oxide and reactive oxygen species and their regulation. *Physiological Reviews, 99*(1), 311–79 [cited 2025 Mar 29]. Available from: https://pubmed.ncbi.nlm.nih.gov/30379623/
12. Daiber, A., Di Lisa, F., Oelze, M., Kröller-Schön, S., Steven, S., Schulz, E., et al. (2016). Crosstalk of mitochondria with NADPH oxidase via reactive oxygen and nitrogen species signalling and its role for vascular function. *British Journal Pharmacology, 174*(12), 1670 [cited 2025 Mar 29]. Available from: https://pmc.ncbi.nlm.nih.gov/articles/PMC5446573/
13. Ida, T., Sawa, T., Ihara, H., Tsuchiya, Y., Watanabe, Y., Kumagai, Y., et al. (2014). Reactive cysteine persulfides and S-polythiolation regulate oxidative stress and redox signaling. *Proceeding of the National Academy of Science USA, 111*(21), 7606–11 [cited 2025 Mar 29]. Available from: https://doi.org/10.1073/pnas.1321232111
14. Kimura, H. (2014). Hydrogen sulfide and polysulfides as biological mediators. *Molecules, 19*(10), 16146–57 [cited 2025 Mar 29]. Available from: https://pubmed.ncbi.nlm.nih.gov/25302704/
15. Fransen, M., Lismont, C., Walton, P. (2017). The peroxisome-mitochondria connection: How and why? *International Journal Molecule Science, 18*(6) [cited 2025 Mar 20]. Available from: https://pubmed.ncbi.nlm.nih.gov/28538669/
16. Li, X. D., Wang, X. Y., Xu, M. E., Jiang, Y., Yan, T., Wang, X. C. (2020). Progress on the usage of the rotifer Brachionus plicatilis in marine ecotoxicology: A review. *Aquatic Toxicology, 229* [cited 2025 Mar 20]. Available from: https://pubmed.ncbi.nlm.nih.gov/33197688/
17. Lundberg, J. O., Carlström, M., Weitzberg, E. (2018). Metabolic effects of dietary nitrate in health and disease. *Cell Metabolism, 28*(1), 9–22 [cited 2025 Mar 20]. Available from: https://pubmed.ncbi.nlm.nih.gov/29972800/
18. Brook, R. D., Rajagopalan, S., Pope, C. A., Brook, J. R., Bhatnagar, A., Diez-Roux, A. V., et al. (2010). Particulate matter air pollution and cardiovascular disease: An update to the scientific statement from the American Heart Association. *Circulation, 121*(21), 2331–78 [cited 2025 Mar 20]. Available from: https://pubmed.ncbi.nlm.nih.gov/20458016/

19. Paul, B. D., Snyder, S. H. (2018). Gasotransmitter hydrogen sulfide signaling in neuronal health and disease. *Biochemical Pharmacology, 149*, 101–9 [cited 2025 Mar 20]. Available from: https://pubmed.ncbi.nlm.nih.gov/29203369/
20. Szabo, C. (2021). Hydrogen sulfide, an endogenous stimulator of mitochondrial function in cancer cells. *Cells, 10*(2), 1–19 [cited 2025 Mar 20]. Available from: https://pubmed.ncbi.nlm.nih.gov/33499368/
21. Hong, Y., Boiti, A., Vallone, D., Foulkes, N. S. (2024). Reactive oxygen species signaling and oxidative stress: Transcriptional regulation and evolution. *Antioxidants, 13*(3), 312 [cited 2025 Apr 3]. Available from: https://pmc.ncbi.nlm.nih.gov/articles/PMC10967436/
22. Kolluru, G. K., Shen, X., Kevil, C. G. (2020). Reactive sulfur species: A new redox player in cardiovascular pathophysiology. *Arteriosclerosis Thrombosis and Vascular Biology, 40*(4), 874 [cited 2025 Mar 20]. Available from: https://pmc.ncbi.nlm.nih.gov/articles/PMC7098439/
23. Ray, P. D., Huang, B. W., Tsuji, Y. (2012). Reactive oxygen species (ROS) homeostasis and redox regulation in cellular signaling. *Cell Signal, 24*(5), 981–90 [cited 2025 Mar 20]. Available from: https://pubmed.ncbi.nlm.nih.gov/22286106/
24. Chiurchiù, V., MacCarrone, M. (2011). Chronic inflammatory disorders and their redox control: From molecular mechanisms to therapeutic opportunities. *Antioxidants Redox Signaling, 15*(9), 2605–41 [cited 2025 Mar 25]. Available from: https://doi.org/10.1089/ars.2010.3547
25. Kajarabille, N., Latunde-Dada, G. O. (2019). Programmed cell-death by Ferroptosis: antioxidants as mitigators. *International Journal of Molecular Sciences, 20*(19), 4968 [cited 2025 Apr 3]. Available from: https://www.mdpi.com/1422-0067/20/19/4968/htm
26. Domagala, A., Fidyt, K., Bobrowicz, M., Stachura, J., Szczygiel, K., Firczuk, M. (2018). Typical and atypical inducers of lysosomal cell death: A promising anticancer strategy. *International Journal of Molecular Sciences, 19*(8), 2256 [cited 2025 Apr 3]. Available from: https://www.mdpi.com/1422-0067/19/8/2256/htm
27. Malard, E., Valable, S., Bernaudin, M., Pérès, E., Chatre, L. (2021). The reactive species interactome in the brain. *Antioxid Redox Signal, 35*(14), 1176–206 [cited 2025 Apr 3]. Available from: https://doi.org/10.1089/ars.2020.8238
28. Farah, C., Michel, L. Y. M., Balligand, J. L. (2018). Nitric oxide signalling in cardiovascular health and disease. *Nature Reviews Cardiology, 15*(5), 292–316 [cited 2025 Apr 4]. Available from: https://www.nature.com/articles/nrcardio.2017.224
29. Farah, C., Michel, L. Y. M., Balligand, J. L. (2018). Nitric oxide signalling in cardiovascular health and disease. *Nature Reviews Cardiology, 15*(5), 292–316 [cited 2025 Apr 4]. Available from: https://pubmed.ncbi.nlm.nih.gov/29388567/
30. Kapil, V., Khambata, R. S., Jones, D. A., Rathod, K., Primus, C., Massimo, G., et al. (2020). The noncanonical pathway for in vivo nitric oxide generation: The nitrate-nitrite-nitric oxide pathway. *Pharmacology Reviews, 72*(3), 692–766 [cited 2025 Apr 4]. Available from: https://pubmed.ncbi.nlm.nih.gov/32576603/
31. Georgiou-Siafis, S. K., Tsiftsoglou, A. S. (2023). The key role of GSH in keeping the redox balance in mammalian cells: Mechanisms and significance of GSH in detoxification via formation of conjugates. *Antioxidants, 12*(11), 1953 [cited 2025 Apr 4]. Available from: https://www.mdpi.com/2076-3921/12/11/1953/htm
32. Baba, S. P., Bhatnagar, A. (2018). Role of thiols in oxidative stress. *Current Opinion Toxicology, 7*, 133 [cited 2025 Apr 4]. Available from: https://pmc.ncbi.nlm.nih.gov/articles/PMC6188637/
33. Baird, L., Yamamoto, M. (2020). The molecular mechanisms regulating the KEAP1-NRF2 pathway. *Molecular Cell Biology, 40*(13), e00099–20 [cited 2025 Apr 7]. Available from: https://pmc.ncbi.nlm.nih.gov/articles/PMC7296212/
34. Bellezza, I., Giambanco, I., Minelli, A., Donato, R. (2018). Nrf2-Keap1 signaling in oxidative and reductive stress. *Biochimica et Biophysica Acta (BBA)–Molecular Cell Research, 1865*(5), 721–33

35. Liu, S., Pi, J., & Zhang, Q. (2022). Signal amplification in the KEAP1-NRF2-ARE antioxidant response pathway. *Redox Biology, 1*(54), Article 102389.
36. Calvert, J. W., Jha, S., Gundewar, S., Elrod, J. W., Ramachandran, A., Pattillo, C. B., et al. (2009). Hydrogen sulfide mediates cardioprotection through Nrf2 signaling. *Circulation Research, 105*(4), 365 [cited 2025 Apr 7]. Available from: https://pmc.ncbi.nlm.nih.gov/articles/PMC2735849/
37. Xie, L., Gu, Y., Wen, M., Zhao, S., Wang, W., Ma, Y., et al. (2016). Hydrogen sulfide induces Keap1 S-sulfhydration and suppresses diabetes-accelerated atherosclerosis via Nrf2 activation. *Diabetes, 65*(10), 3171–84 [cited 2025 Apr 7]. Available from: https://pmc.ncbi.nlm.nih.gov/articles/PMC8928786/
38. Meng, W., Pei, Z., Feng, Y., Zhao, J., Chen, Y., Shi, W., et al. (2017). Neglected role of hydrogen sulfide in sulfur mustard poisoning: Keap1 S-sulfhydration and subsequent Nrf2 pathway activation. *Scientific Reports 2017, 7*(1), 1–17 [cited 2025 Apr 7]. Available from: https://www.nature.com/articles/s41598-017-09648-6
39. D'Ignazio, L., Bandarra, D., Rocha, S. (2015). NF-κB and HIF crosstalk in immune responses. *FEBS Journal, 283*(3), 413 [cited 2025 Apr 7]. Available from: https://pmc.ncbi.nlm.nih.gov/articles/PMC4864946/
40. Cai, F., Chen, W., Zhao, R., Liu, Y. (2023). Mechanisms of Nrf2 and NF-κB pathways in diabetic wound and potential treatment strategies. *Molecular Biology Reports 2023, 50*(6), 5355–67 [cited 2025 Apr 7]. Available from: https://link.springer.com/article/https://doi.org/10.1007/s11033-023-08392-7
41. Tang, Y. Y., Wang, D. C., Wang, Y. Q., Huang, A. F., & Xu, W. D. (2023). Emerging role of hypoxia-inducible factor-1α in inflammatory autoimmune diseases: A comprehensive review. *Frontiers in Immunology, 25*(13), 1073971.
42. Semenza, G. L. (2010). Hypoxia-Inducible factor 1: Regulator of mitochondrial metabolism and mediator of ischemic preconditioning. *Biochimica et Biophysica Acta, 1813*(7), 1263 [cited 2025 Apr 7]. Available from: https://pmc.ncbi.nlm.nih.gov/articles/PMC3010308/
43. Tejero, J., Shiva, S., Gladwin, M. T. (2019). Sources of vascular nitric oxide and reactive oxygen species and their regulation. *Physiological Reviews, 99*(1), 311–79 [cited 2025 Apr 7]. Available from: https://doi.org/10.1152/physrev.00036.2017
44. Roshanravan, N., Shabestari, A. N., Alamdari, N. M., Ostadrahimi, A., Separham, A., Parvizi, R., et al. (2021). A novel inflammatory signaling pathway in patients with slow coronary flow: NF-κB/IL-1β/nitric oxide. *Cytokine, 1*(143), Article 155511.
45. Zhao, M., Cheng, Y., Wang, X., Cui, X., Cheng, X., Fu, Q., et al. (2022). Hydrogen sulfide attenuates high-fat diet-induced obesity: Involvement of mTOR/IKK/NF-κB signaling pathway. *Molecular Neurobiology, 59*(11), 6903–17 [cited 2025 Apr 7]. Available from: https://doi.org/10.1007/s12035-022-03004-0
46. Son, Y., Cheong, Y. K., Kim, N. H., Chung, H. T., Kang, D. G., Pae, H. O. (2011). Mitogen-activated protein kinases and reactive oxygen species: How can ROS activate MAPK pathways? *Journal Signal Transduction, 2011*(1), 792639 [cited 2025 Apr 7]. Available from: https://doi.org/10.1155/2011/792639
47. Yue, J., López, J. M. (2020). Understanding MAPK signaling pathways in apoptosis. *International Journal of Molecular Sciences, 21*(7), 2346 [cited 2025 Apr 7]. Available from: https://www.mdpi.com/1422-0067/21/7/2346/htm
48. Son, Y., Kim, S., Chung, H. T., & Pae, H. O. (2013). Reactive oxygen species in the activation of MAP kinases. *Methods in Enzymology, 1*(528), 27–48.
49. Winter-Vann, A. M., Johnson, G. L. (2007). Integrated activation of MAP3Ks balances cell fate in response to stress. *Journal Cell Biochemical, 102*(4), 848–58 [cited 2025 Apr 7]. Available from: https://doi.org/10.1002/jcb.21522
50. Nagai, H., Noguchi, T., Takeda, K., & Ichijo, H. (2007). Pathophysiological roles of ASK1-MAP Kinase signaling pathways. *BMB Reports, 40*(1), 1–6.
51. Kamata, H., Honda, S. I., Maeda, S., Chang, L., Hirata, H., Karin, M. (2005). Reactive oxygen species promote TNFα-induced death and sustained JNK activation by inhibiting MAP kinase phosphatases. *Cell, 120*(5), 649–61 [cited 2025 Apr 7]. Available from: https://www.cell.com/action/showFullText?pii=S009286740500190X

52. Hong, Y., Boiti, A., Vallone, D., Foulkes, N. S. (2024). Reactive oxygen species signaling and oxidative stress: Transcriptional regulation and evolution. *Antioxidants, 13*(3), 312 [cited 2025 Apr 7]. Available from: https://www.mdpi.com/2076-3921/13/3/312/htm
53. Kma, L., Baruah, T. J. (2022). The interplay of ROS and the PI3K/Akt pathway in autophagy regulation. Biotechnology *Applied Biochemistry, 69*(1), 248–64 [cited 2025 Apr 7]. Available from: https://doi.org/10.1002/bab.2104
54. He, Y., Sun, M. M., Zhang, G. G., Yang, J., Chen, K. S., Xu, W. W., et al. (2021). Targeting PI3K/Akt signal transduction for cancer therapy. *Signal Transduction and Targeted Therapy, 6*(1), 1–17 [cited 2025 Apr 7]. Available from: https://www.nature.com/articles/s41392-021-00828-5
55. Andrés, C. M. C, Pérez de la Lastra, J. M., Andrés Juan, C., Plou, F. J., Pérez-Lebeña, E. (2023). Chemistry of hydrogen sulfide—Pathological and physiological functions in mammalian cells. *Cells, 12*(23), 2684 [cited 2025 Apr 7]. Available from: https://pmc.ncbi.nlm.nih.gov/articles/PMC10705518/
56. Liang, Y. F., You, Q. X., Chen, S. Y., Ni, L., Meng, X. L., Gao, J. X., et al. (2024). The impact of hydrogen sulfide in the paraventricular nucleus on the MAPK pathway in high salt–induced hypertension. *Journal Cardiovascular Pharmacology, 84*(4), 468 [cited 2025 Apr 7]. Available from: https://pmc.ncbi.nlm.nih.gov/articles/PMC11446517/
57. He, R., Wang, Z., Dong, S., Chen, Z., Zhou, W. (2022). Understanding necroptosis in pancreatic diseases. *Biomolecules, 12*(6), 828 [cited 2025 Apr 6]. Available from: https://www.mdpi.com/2218-273X/12/6/828/htm
58. Rius-Pérez, S. (2023). P53 at the crossroad between mitochondrial reactive oxygen species and necroptosis. *Free Radical Biology and Medicine, 1*(207), 183–193.
59. Zhang, Y., Su, S. S., Zhao, S., Yang, Z., Zhong, C. Q., Chen, X., et al. (2017). RIP1 autophosphorylation is promoted by mitochondrial ROS and is essential for RIP3 recruitment into necrosome. *Nature Communication, 8* [cited 2025 Apr 6]. Available from: https://pubmed.ncbi.nlm.nih.gov/28176780/
60. Dutta, S., Sengupta, P., Das, S., Slama, P., Roychoudhury, S. (2022). Reactive nitrogen species and male reproduction: Physiological and pathological aspects. *International Journal of Molecular Sciences, 23*(18), 10574 [cited 2025 Apr 6]. Available from: https://www.mdpi.com/1422-0067/23/18/10574/htm
61. Shrivastava, P., Pantano, C., Watkin, R., McElhinney, B., Guala, A., Poynter, M. L., et al. (2004). Reactive nitrogen species-induced cell death requires Fas-dependent activation of c-Jun N-Terminal Kinase. *Molecular Cell Biology, 24*(15), 6763 [cited 2025 Apr 6]. Available from: https://pmc.ncbi.nlm.nih.gov/articles/PMC444859/
62. Guo, S., Zhang, Y., Lian, J., Su, C., Wang, H. (2024). The role of hydrogen sulfide in the regulation of necroptosis across various pathological processes. *Molecular and Cellular Biochemistry, 480*(4), 1999–2013 [cited 2025 Apr 6]. Available from: https://doi.org/10.1007/s11010-024-05090-1
63. Le, H. X., Hu, Y., Jiang, W., Jiang, J. M., Zou, W., Zhang, P., et al. (2025). Suppression of cGAS/STING pathway-triggered necroptosis in the hippocampus relates H2S to attenuate cognitive dysfunction of Parkinson's disease. *Experimental Neurology, 1*(385), Article 115093.
64. He, Z., Zhu, Y., Ma, H., Shen, Q., Chen, X., Wang, X., et al. (2024). Hydrogen sulfide regulates macrophage polarization and necroptosis to accelerate diabetic skin wound healing. *International Immunopharmacology, 10*(132), Article 111990.
65. Redza-Dutordoir, M., Averill-Bates, D. A. (2016). Activation of apoptosis signalling pathways by reactive oxygen species. *Biochimica et Biophysica Acta, 1863*(12), 2977–92 [cited 2025 Apr 4]. Available from: https://pubmed.ncbi.nlm.nih.gov/27646922/
66. Chu, Q., Gu, X., Zheng, Q., Wang, J., Zhu, H. (2021). Mitochondrial mechanisms of apoptosis and necroptosis in liver diseases. *Analytical Cellular Pathology, 2021*(1), 8900122 [cited 2025 Apr 4]. Available from: https://doi.org/10.1155/2021/8900122
67. Wang, B., Wang, Y., Zhang, J., Hu, C., Jiang, J., Li, Y., et al. (2023). ROS-induced lipid peroxidation modulates cell death outcome: Mechanisms behind apoptosis, autophagy, and

ferroptosis. *Archives of Toxicology, 97*(6):1439–51 [cited 2025 Apr 4]. Available from: https://doi.org/10.1007/s00204-023-03476-6

68. Dwivedi, P. (2021). ROS mediated apoptotic pathways in primary effusion lymphoma: Comment on induction of apoptosis by Shikonin through ROS-mediated intrinsic and extrinsic pathways in primary effusion lymphoma. *Translational Oncology, 14*(7), Article 101061.
69. Dwivedi, P. (2021). ROS mediated apoptotic pathways in primary effusion lymphoma: Comment on induction of apoptosis by Shikonin through ROS-mediated intrinsic and extrinsic pathways in primary effusion lymphoma. *Translational Oncology, 14*(7), 101061 [cited 2025 Apr 4]. Available from: https://pmc.ncbi.nlm.nih.gov/articles/PMC8050935/
70. Baldi, S., He, Y., Ivanov, I., Sun, Y., Feng, W., Refat, M., et al. (2022). Novel characterization discoveries of ferroptosis-associated molecules in COAD microenvironment based TCGA data. *Frontiers Molecular Bioscience, 9*, 1102735 [cited 2025 Apr 4]. Available from: https://portal.gdc.com
71. Roberts, R. A., Smithm R. A., Safe, S., Szabo, C., Tjalkens, R. B., Robertson, F. M. (2010). Toxicological and pathophysiological roles of reactive oxygen and nitrogen species. *Toxicology, 276*(2), 85–94 [cited 2025 Apr 6]. Available from: https://pubmed.ncbi.nlm.nih.gov/20643181/
72. Pourova, J., Kottova, M., Voprsalova, M., Pour, M. (2010). Reactive oxygen and nitrogen species in normal physiological processes. *Acta Physiologica, 198*(1), 15–35 [cited 2025 Apr 6]. Available from: https://doi.org/10.1111/j.1748-1716.2009.02039.x
73. Zou, D., Li, J., Fan, Q., Zheng, X., Deng, J., Wang, S. (2019). Reactive oxygen and nitrogen species induce cell apoptosis via a mitochondria-dependent pathway in hyperoxia lung injury. *Journal Cell Biochemical, 120*(4), 4837–50 [cited 2025 Apr 6]. Available from: https://pubmed.ncbi.nlm.nih.gov/30592322/
74. Roberts, R. A., Smith, R. A., Safe, S., Szabo, C., Tjalkens, R. B., Robertson, F. M. (2010). Toxicological and pathophysiological roles of reactive oxygen and nitrogen species. *Toxicology, 276*(2), 85–94 [cited 2025 Apr 6]. Available from: https://researchexperts.utmb.edu/en/publications/toxicological-and-pathophysiological-roles-of-reactive-oxygen-and
75. Khattak, S., Rauf, M. A., Khan, N. H., Zhang, Q. Q., Chen, H. J., Muhammad, P., et al. (2022). Hydrogen sulfide biology and its role in cancer. *Molecules, 27*(11), 3389 [cited 2025 Apr 6]. Available from: https://www.mdpi.com/1420-3049/27/11/3389/htm
76. Braunstein, I., Engelman, R., Yitzhaki, O., Ziv, T., Galardon, E., & Benhar, M. (2020). Opposing effects of polysulfides and thioredoxin on apoptosis through caspase persulfidation. *Journal of Biological Chemistry, 295*(11), 3590–3600.
77. Munteanu, C., Turnea, M. A., Rotariu, M. (2023). Hydrogen sulfide: An emerging regulator of oxidative stress and cellular homeostasis—A comprehensive one-year review. *Antioxidants, 12*(9), 1737 [cited 2025 Apr 6]. Available from: https://pmc.ncbi.nlm.nih.gov/articles/PMC10526107/
78. Szabo, C. (2021). Hydrogen sulfide, an endogenous stimulator of mitochondrial function in cancer cells. *Cells, 10*(2), 220 [cited 2025 Apr 6]. Available from: https://www.mdpi.com/2073-4409/10/2/220/htm
79. Filipovic, M. R., Zivanovic, J., Alvarez, B., Banerjee, R. (2017). Chemical biology of H2S signaling through Persulfidation. *Chemical Reviews, 118*(3), 1253 [cited 2025 Apr 6]. Available from: https://pmc.ncbi.nlm.nih.gov/articles/PMC6029264/
80. Sharma, C., Kim, S. R. (2021). Linking oxidative stress and Proteinopathy in Alzheimer's Disease. *Antioxidants, 10*(8), 1231 [cited 2025 Mar 21]. Available from: https://www.mdpi.com/2076-3921/10/8/1231/htm
81. Cenini, G., Lloret, A., Cascella, R. (2019). Oxidative stress in neurodegenerative diseases: From a mitochondrial point of view. *Oxidative Medicine Cell Longevity, 2019* [cited 2025 Mar 21]. Available from: https://pubmed.ncbi.nlm.nih.gov/31210837/
82. Liu, J., McDaid, L., Araque, A., Wade, J., Harkin, J., Karim, S., et al. (2019). GABA regulation of burst firing in hippocampal astrocyte neural circuit: A biophysical model. *Frontiers in Cellular Neuroscience, 23*(13), Article 460520.

83. Lewuillon, C., Guillemette, A., Titah, S., Shaik, F. A., Jouy, N., Labiad, O., et al. (2022). Involvement of ORAI1/SOCE in human AML cell lines and primary cells according to ABCB1 activity, LSC compartment and potential resistance to Ara-C exposure. *International Journal Molecular Science, 23*(10), 5555 [cited 2025 Mar 21]. Available from: https://www.mdpi.com/1422-0067/23/10/5555/htm
84. Teleanu, D. M., Niculescu, A. G., Lungu, I. I., Radu, C. I., Vladâcenco, O., Roza, E., et al. (2022). An overview of oxidative stress, neuroinflammation, and neurodegenerative diseases. *International Journal Molecular Science, 23*(11), 5938 [cited 2025 Mar 21]. Available from: https://pmc.ncbi.nlm.nih.gov/articles/PMC9180653/
85. Mao, Q., Wang, X., Chen, B., Fan, L., Wang, S., Zhang, Y., et al. (2020). KTN1 variants underlying putamen gray matter volumes and Parkinson's Disease. *Frontiers Neuroscience, 14*, 497728 [cited 2025 Mar 21]. Available from: www.frontiersin.org
86. Foret, M. K., Orciani, C., Welikovitch, L. A., Huang, C., Cuello, A. C., Do Carmo, S. (2024). Early oxidative stress and DNA damage in Aβ-burdened hippocampal neurons in an Alzheimer's-like transgenic rat model. *Communications Biology, 7*(1), 1–17 [cited 2025 Mar 21]. Available from: https://www.nature.com/articles/s42003-024-06552-4
87. Zhang, Z., Xue, P., Bendlin, B. B., Zetterberg, H., De Felice, F., Tan, X., et al. (2024). Melatonin: A potential nighttime guardian against Alzheimer's. *Molecular Psychiatry, 30*(1), 237–50 [cited 2025 Mar 21]. Available from: https://www.nature.com/articles/s41380-024-02691-6
88. Stykel, M. G., Ryan, S. D. (2022). Nitrosative stress in Parkinson's disease. *NPJ Parkinsons Disease, 8*(1) [cited 2025 Mar 21]. Available from: https://pubmed.ncbi.nlm.nih.gov/35953517/
89. De Lazzari, F., Bubacco, L., Whitworth, A. J., Bisaglia, M. (2025). Superoxide radical dismutation as new therapeutic strategy in Parkinson's Disease. *Aging Disease, 9*(4), 716 [cited 2025 Mar 21]. Available from: https://pmc.ncbi.nlm.nih.gov/articles/PMC6065289/
90. Kim, T. Y., Leem, E., Lee, J. M., Kim, S. R. (2025). Control of reactive oxygen species for the prevention of Parkinson's Disease: The possible application of flavonoids. *Antioxidants, 9*(7), 583 [cited 2025 Mar 21]. Available from: https://pmc.ncbi.nlm.nih.gov/articles/PMC7402123/
91. Rizor, A., Pajarillo, E., Johnson, J., Aschner, M., Lee, E. (2019). Astrocytic oxidative/nitrosative stress contributes to Parkinson's Disease pathogenesis: The dual role of reactive astrocytes. *Antioxidants, 8*(8), 265 [cited 2025 Mar 21]. Available from: https://pmc.ncbi.nlm.nih.gov/articles/PMC6719180/
92. Chang, K. H., Chen, C. M. (2020). The role of oxidative stress in Parkinson's Disease. *Antioxidants, 9*(7), 597 [cited 2025 Mar 21]. Available from: https://www.mdpi.com/2076-3921/9/7/597/htm
93. Ye, Y., Xu, J., Shen, H., Yu, Z., Chen, G. (2024). Neuroprotective effects of hydrogen sulfide in Parkinson's disease. *Medical Gas Research, 14*(3), 145–8 [cited 2025 Mar 21]. Available from: https://journals.lww.com/mgar/fulltext/2024/14030/neuroprotective_effects_of_hydrogen_sulfide_in.13.aspx
94. Coleman, C., & Martin, I. (2022). Unraveling Parkinson's Disease neurodegeneration: Does aging hold the clues? *Journal of Parkinson's Disease, 12*(8), 2321–2338.
95. Iciek, M., Bilska-Wilkosz, A., Kozdrowicki, M., Górny, M. (2023). Reactive sulfur species in human diseases. https://home.liebertpub.com/ars, 39(13–15):1000–23 [cited 2025 Mar 21]. Available from: https://doi.org/10.1089/ars.2023.0261
96. Zheng, J., Winderickx, J., Franssens, V., Liu, B. (2018). A mitochondria-associated oxidative stress perspective on Huntington's Disease. *Frontiers in Molecular Neuroscience, 11*, 384654 [cited 2025 Mar 21]. Available from: www.frontiersin.org
97. Kumar, A., Ratan, R. R. (2016). Oxidative stress and Huntington's Disease: The good, the bad, and the ugly. *Journal Huntingtons Disease, 5*(3), 217 [cited 2025 Mar 21]. Available from: https://pmc.ncbi.nlm.nih.gov/articles/PMC5310831/
98. Paul, B. D., Snyder, S. H. (2019). Impaired redox signaling in Huntington's disease: Therapeutic implications. *Frontiers in Molecular Neuroscience, 12*, 445778 [cited 2025 Mar 21]. Available from: www.frontiersin.org

99. Bono-Yagüe, J., Gómez-Escribano, A. P., Millán, J. M., Vázquez-Manrique, R. P. (2020). Reactive species in Huntington Disease: Are they really the radicals you want to catch? *Antioxidants, 9*(7), 577 [cited 2025 Mar 21]. Available from: https://pmc.ncbi.nlm.nih.gov/articles/PMC7401865/
100. Olufunmilayo, E. O., Gerke-Duncan, M. B., Holsinger, R. M. D. (2023). Oxidative stress and antioxidants in neurodegenerative disorders. *Antioxidants, 12*(2), 517 [cited 2025 Mar 21]. Available from: https://www.mdpi.com/2076-3921/12/2/517/htm
101. Singh, A., Kukreti, R., Saso, L., Kukreti, S. (2019). Oxidative stress: A key modulator in neurodegenerative diseases. *Molecules, 24*(8), 1583 [cited 2025 Mar 21]. Available from: https://www.mdpi.com/1420-3049/24/8/1583/htm
102. Zhang, Y., Khan, S., Liu, Y., Wu, G., Yong, V. W., Xue, M. (2022). Oxidative stress following intracerebral hemorrhage: From molecular mechanisms to therapeutic targets. *Frontiers Immunology, 13*, 847246 [cited 2025 Mar 21]. Available from: https://pmc.ncbi.nlm.nih.gov/articles/PMC8959663/
103. Zheng, Y., Li, R., Fan, X. (2022). Targeting oxidative stress in intracerebral hemorrhage: Prospects of the natural products approach. *Antioxidants, 11*(9), 1811 [cited 2025 Mar 21]. Available from: https://www.mdpi.com/2076-3921/11/9/1811/htm
104. Maida, C. D., Norrito, R. L., Rizzica, S., Mazzola, M., Scarantino, E. R., Tuttolomondo. A. (2024). Molecular pathogenesis of ischemic and hemorrhagic strokes: Background and therapeutic approaches. *International Journal of Molecular Sciences 2024, 25*(12), 6297 [cited 2025 Mar 21]. Available from: https://www.mdpi.com/1422-0067/25/12/6297/htm
105. Rajan, R., Divya, K. P., Kandadai, R. M., Yadav, R., Satagopam, V. P., Madhusoodanan, U. K., et al. (2020). Genetic architecture of Parkinson's disease in the Indian population: Harnessing genetic diversity to address critical gaps in Parkinson's disease research. *Frontiers in Neurology, 11*, 1–11[cited 2025 Mar 21]. Available from: www.frontiersin.org
106. Slemmer JE, Shacka JJ, Sweeney MI, Weber JT (2025). Antioxidants and free radical scavengers for the treatment of stroke, traumatic brain injury and aging. *Current Medicinal Chemistry, 15*(4), 404–14 [cited 2025 Mar 21]. Available from: https://pubmed.ncbi.nlm.nih.gov/18288995/
107. Baruah, P., Padhi, D., Moorthy, H., Ramesh, M., Govindaraju, T. (2025). Navigating the dichotomy of reactive oxygen, nitrogen, and sulfur species: Detection strategies and therapeutic interventions. *RSC Chemical Biology, 6*(3), 338–57 [cited 2025 Mar 17]. Available from: https://pubs.rsc.org/en/content/articlehtml/2025/cb/d5cb00006h
108. Ushio-Fukai, M., Ash, D., Nagarkoti, S., Belin De Chantemèle, E. J., Fulton, D. J. R., Fukai, T. (2021). Interplay between reactive oxygen/reactive nitrogen species and metabolism in vascular biology and disease, *34*(16):1319–54. https://home.liebertpub.com/ars [cited 2025 Mar 16]. Available from: https://doi.org/10.1089/ars.2020.8161
109. Solanki, K., Bezsonov, E., Orekhov, A., Parihar, S. P., Vaja, S., White, F. A., et al. (2024). Effect of reactive oxygen, nitrogen, and sulfur species on signaling pathways in atherosclerosis. *Vascular Pharmacology, 1*(154), Article 107282.
110. Baruah, P., Padhi, D., Moorthy, H., Ramesh, M., Govindaraju, T. (2025). Navigating the dichotomy of reactive oxygen, nitrogen, and sulfur species: Detection strategies and therapeutic interventions. *RSC Chemical Biology*
111. Zhou, S., Zhou, X., Zhou., Siwei, Li, Y., Qian, Z., Zeng, C., et al. (2022). Clinical significance of HSCARG for atherosclerotic coronary heart disease and reduced ROS-oxidative stress in in vivo and in vitro models via p47phox by NF-κB activity. *Brazilian Journal of Cardiovascular Surgery, 37*(5), 727–36 [cited 2025 Mar 16]. Available from: https://orcid.org/0000-0002-4257-0584
112. Afanas'Ev, I. (2011). ROS and RNS signaling in heart disorders: Could antioxidant treatment be successful? *Oxidative Medicine and Cellular Longevity, 2011*(1), 293769 [cited 2025 Mar 28]. Available from: https://doi.org/10.1155/2011/293769
113. Fu, X., Lu, H., Gao, M., Li, P., He, Y., He, Y., et al. (2024). Nitric oxide in the cardio-cerebrovascular system: Source, regulation and application. *Nitric Oxide, 1*(152), 48–57.

114. Mehta, J. K., Kaur, G., Buttar, H. S., Bagabir, H. A., Bagabir, R. A., Bagabir, S. A., et al. (2023). Role of the renin–angiotensin system in the pathophysiology of coronary heart disease and heart failure: Diagnostic biomarkers and therapy with drugs and natural products. *Frontiers in Physiology, 23*(14), 1034170.
115. Ekholm, M., Kahan, T. (2021). The impact of the Renin-Angiotensin-Aldosterone system on inflammation, Coagulation, and Atherothrombotic complications, and to aggravated COVID-19. *Frontiers in Pharmacology, 12*, 640185 [cited 2025 Mar 28]. Available from: www.frontiersin.org
116. Młynarska, E., Hajdys, J., Czarnik, W., Fularski, P., Leszto, K., Majchrowicz, G., et al. (2024). The role of antioxidants in the therapy of cardiovascular diseases—A literature review. *Nutrients, 16*(16), 2587 [cited 2025 Mar 16]. Available from: https://www.mdpi.com/2072-6643/16/16/2587/htm
117. Severino, P., D'amato, A., Pucci, M., Infusino, F., Adamo, F., Birtolo, L. I., et al. (2020). Ischemic heart disease pathophysiology paradigms overview: From plaque activation to microvascular dysfunction. *International Journal of Molecular Sciences, 21*(21), 8118 [cited 2025 Mar 28]. Available from: https://www.mdpi.com/1422-0067/21/21/8118/htm
118. Zhang, Y., Jiang, M., & Wang, T. (2024). Reactive oxygen species (ROS)-responsive biomaterials for treating myocardial ischemia-reperfusion injury. *Front Bioeng Biotechnol., 2*(12), 1469393.
119. Chen, W., Li, D. (2020). Reactive oxygen species (ROS)-responsive nanomedicine for solving ischemia-reperfusion injury. *Frontiers in Chemistry, 8*, 576088 [cited 2025 Mar 16]. Available from: www.frontiersin.org
120. Moris, D., Spartalis, M., Spartalis, E., Karachaliou, G. S., Karaolanis, G. I., Tsourouflis, G., et al. (2017). The role of reactive oxygen species in the pathophysiology of cardiovascular diseases and the clinical significance of myocardial redox. *Annals of Translation Medicine, 5*(16), 326–326 [cited 2025 Mar 16]. Available from: https://atm.amegroups.org/article/view/16040/html
121. Karoui, A., Crochemore, C., Harouki, N., Corbière, C., Preterre, D., Vendeville, C., et al. (2020). Nitrogen dioxide inhalation exposures induce cardiac mitochondrial reactive oxygen species production, impair mitochondrial function and promote coronary endothelial dysfunction. *International Journal of Environmental Research and Public Health, 17*(15), 5526 [cited 2025 Mar 16]. Available from: https://www.mdpi.com/1660-4601/17/15/5526/htm
122. Pérez-Torres, I., Manzano-Pech, L., Rubio-Ruíz, M. E., Soto, M. E., Guarner-Lans, V. (2020). *Nitrosative stress and its association with Cardiometabolic disorders. Molecules, 25*(11), 2555 [cited 2025 Mar 16]. Available from: https://www.mdpi.com/1420-3049/25/11/2555/htm
123. Pharoah, B. M., Khodade, V. S., Eremiev, A., Bao, E., Liu, T., O'rourke, B., et al. (2022). Hydropersulfides (RSSH) outperform post-conditioning and other reactive sulfur species in limiting Ischemia–Reperfusion injury in the isolated mouse heart. *Antioxidants, 11*(5), 1010 [cited 2025 Mar 16]. Available from: https://www.mdpi.com/2076-3921/11/5/1010/htm
124. Kolluru, G. K., Shen, X., Kevil, C. G. (2020). Reactive sulfur species: A new redox player in cardiovascular pathophysiology. *Arteriosclerosis Thrombosis and Vascular Biology, 40*(4), 874–84 [cited 2025 Mar 16]. Available from: https://doi.org/10.1161/ATVBAHA.120.314084
125. He, P., Talukder, M. A. H, Gao, F. (2020). Oxidative stress and microvessel barrier dysfunction. *Frontiers in Physiology, 11*, 504567 [cited 2025 Mar 16]. Available from: www.frontiersin.org
126. Feuer, D. S., Handberg, E. M., Mehrad, B., Wei, J., Bairey Merz, C. N., Pepine, C. J., et al. (2022). Microvascular dysfunction as a systemic disease: A review of the evidence. *American Journal of Medicine, 135*(9), 1059–68 [cited 2025 Mar 16]. Available from: https://www.amjmed.com/action/showFullText?pii=S0002934322003412
127. Rehan, R., Yong, A., Ng, M., Weaver, J., & Puranik, R. (2023). Coronary microvascular dysfunction: A review of recent progress and clinical implications. *Frontiers Cardiovascular Medcine, 26*(10), 1111721.
128. Masi, S., Rizzoni, D., Taddei, S., Widmer, R. J., Montezano, A. C., Lüscher, T. F., et al. (2021). Assessment and pathophysiology of microvascular disease: Recent progress and clinical implications. *European Heart Journal, 42*(26), 2590–604 [cited 2025 Mar 16]. Available from: https://doi.org/10.1093/eurheartj/ehaa857

129. Ulla, A., Nikawa, T. (2023). Regulatory and pathophysiological roles of reactive oxygen species in skeletal muscle. Biochemistry and Molecular Biology Journal, 9(4):38–53 [cited 2025 Mar 15]. Available from: https://www.primescholars.com/articles/regulatory-and-pathophysiological-roles-of-reactive-oxygen-species-in-skeletal-muscle-122932.html
130. Lian, D., Chen, M. M., Wu, H., Deng, S., Hu, X. (2022). The role of oxidative stress in skeletal Muscle Myogenesis and Muscle Disease. *Antioxidants, 11*(4), 755 [cited 2025 Mar 15]. Available from: https://www.mdpi.com/2076-3921/11/4/755/htm
131. Barbieri, E., Sestili, P. (2012). Reactive oxygen species in skeletal muscle signaling. *Journal Signal Transduction, 2012*(1), 982794 [cited 2025 Mar 15]. Available from: https://doi.org/10.1155/2012/982794
132. Zulfahmidah, Z., Safei, I. (2022). The role of reactive oxygen species in muscle: Beneficial/harmful. *Green Medical Journal, 4*(3), 84–92 [cited 2025 Mar 15]. Available from: https://greenmedicaljournal.umi.ac.id/index.php/gmj/article/view/129
133. Lee, S. C., Lee, N. H., Patel, K. D., Jun, S. K., Park, J. H., Knowles, J. C., et al. (2021). A study on myogenesis by regulation of reactive oxygen species and cytotoxic activity by selenium nanoparticles. *Antioxidants (Basel), 10*(11) [cited 2025 Mar 15]. Available from: https://pubmed.ncbi.nlm.nih.gov/34829599/
134. Terrill, J. R., Radley-Crabb, H. G., Iwasaki, T., Lemckert, F. A., Arthur, P. G., Grounds, M. D. (2013). Oxidative stress and pathology in muscular dystrophies: Focus on protein thiol oxidation and Dysferlinopathies. FEBS Journal, 280(17), 4149–64 [cited 2025 Mar 15]. Available from: https://doi.org/10.1111/febs.12142
135. Petrillo, S., Pelosi, L., Piemonte, F., Travaglini, L., Forcina, L., Catteruccia, M., et al. (2017). Oxidative stress in Duchenne muscular dystrophy: Focus on the NRF2 redox pathway. *Human Molecular Genetics, 26*(14), 2781–90 [cited 2025 Mar 15]. Available from: https://doi.org/10.1093/hmg/ddx173
136. Casati, S. R., Cervia, D., Roux-Biejat, P., Moscheni, C., Perrotta, C., De Palma, C. (2024). Mitochondria and Reactive Oxygen Species: The therapeutic balance of powers for Duchenne Muscular Dystrophy. *Cells, 13*(7), 574 [cited 2025 Mar 15]. Available from: https://www.mdpi.com/2073-4409/13/7/574/htm
137. Cheng, A. J., Yamada, T., Rassier, D. E., Andersson, D. C., Westerblad, H., Lanner, J. T. (2016). Reactive oxygen/nitrogen species and contractile function in skeletal muscle during fatigue and recovery. *Journal Physiology, 594*(18), 5149 [cited 2025 Mar 15]. Available from: https://pmc.ncbi.nlm.nih.gov/articles/PMC5023717/
138. Bottinelli, R., Westerblad, H. (2011). Reactive oxygen and nitrogen species in skeletal muscle: acute and long-term effects. *Journal Physiology, 589*(Pt 9), 2117 [cited 2025 Mar 15]. Available from: https://pmc.ncbi.nlm.nih.gov/articles/PMC3098690/
139. Ellwood, R. A., Slade, L., Lewis, J., Torregrossa, R., Sudevan, S., Piasecki, M., et al. (2022). Sulfur amino acid supplementation displays therapeutic potential in a C. elegans model of Duchenne muscular dystrophy. *Communications Biology, 5*(1), 1–12 [cited 2025 Mar 15]. Available from: https://www.nature.com/articles/s42003-022-04212-z
140. Iciek, M., Bilska-Wilkosz, A., Kozdrowicki, M., Górny, M. (2022). Reactive sulfur species and their significance in health and disease. Bioscience Reports, 42(9), BSR20221006 [cited 2025 Mar 15]. Available from: https://pmc.ncbi.nlm.nih.gov/articles/PMC9484011/
141. Foreman, N. A., Hesse, A. S., Ji, L. L. (2025). Redox signaling and sarcopenia: Searching for the primary suspect. *International Journal of Molecular Sciences, 22*(16), 9045 [cited 2025 Mar 15]. Available from: https://www.mdpi.com/1422-0067/22/16/9045/htm
142. Damiano, S., Muscariello, E., La Rosa, G., Di Maro, M., Mondola, P., Santillo, M. (2019). Dual role of reactive oxygen species in muscle function: Can Antioxidant Dietary supplements counteract age-related sarcopenia? International Journal Molecule Science, 20(15) [cited 2025 Mar 15]. Available from: https://pubmed.ncbi.nlm.nih.gov/31387214/
143. Chen, M., Wang, Y., Deng, S., Lian, Z., & Yu, K. (2022). Skeletal muscle oxidative stress and inflammation in aging: Focus on antioxidant and anti-inflammatory therapy. *Front Cell Dev Biol., 30*(10), Article 964130.

144. Fulle, S., Protasi, F., Tano, G. D., Pietrangelo, T., Beltramin, A., Boncompagni, S., et al. (2004). The contribution of reactive oxygen species to sarcopenia and muscle ageing. *Experimental Gerontology, 39*(1), 17–24[cited 2025 Mar 15]. Available from: https://pubmed.ncbi.nlm.nih.gov/14724060/
145. Sung, H. K., Murugathasan, M., Abdul-Sater, A. A., Sweeney, G. (2023). Autophagy deficiency exacerbates iron overload induced reactive oxygen species production and apoptotic cell death in skeletal muscle cells. *Cell Death and Disease 2023, 14*(4), 1–13 [cited 2025 Mar 15]. Available from: https://www.nature.com/articles/s41419-022-05484-3
146. Nemes, R., Koltai, E., Taylor, A. W., Suzuki, K., Gyori, F., Radak, Z. (2018). Reactive oxygen and nitrogen species regulate key metabolic, anabolic, and catabolic pathways in skeletal muscle. *Antioxidants, 7*(7), 85 [cited 2025 Mar 15]. Available from: https://www.mdpi.com/2076-3921/7/7/85/htm
147. Lu, Y., Li, T., Shu, Y., Lu, C., Luo, Z., Wang, J., et al. (2025). Lipid peroxidation and sarcopenia: Molecular mechanisms and potential therapeutic approaches. *Frontiers Medicine (Lausanne), 3*(12), 1525205.
148. Cheung, K., Rathbone, A., Melanson, M., Trier, J., Ritsma, B. R., Allen, M. D. (2021). Pathophysiology and management of critical illness polyneuropathy and myopathy. Journal Applied Physiology, 130(5):1479–89 [cited 2025 Mar 15]. Available from: https://pubmed.ncbi.nlm.nih.gov/33734888/
149. Shepherd, S., Batra, A., Lerner, D. P. (2017). Review of critical illness myopathy and neuropathy. *Neurohospitalist, 7*(1), 41–8 [cited 2025 Mar 15]. Available from: https://doi.org/10.1177/1941874416663279
150. Zhou, C., Wu, L., Ni, F., Ji, W., Wu, J., Zhang, H. (2014) Critical illness polyneuropathy and myopathy: A systematic review. *Neural Regeneration Research, 9*(1), 101–10 [cited 2025 Mar 15]. Available from: https://journals.lww.com/nrronline/fulltext/2014/09010/critical_illness_polyneuropathy_and_myopathy__a.15.aspx
151. Andrés, C. M. C., Pérez de la Lastra, J. M., Andrés Juan, C., Plou, F. J., Pérez-Lebeña, E. (2023). Superoxide anion chemistry—Its role at the core of the innate immunity. *International Journal of Molecular Sciences, 24*(3), 1841 [cited 2025 Mar 15]. Available from: https://www.mdpi.com/1422-0067/24/3/1841/htm
152. Cirino, G., Szabo, C., Papapetropoulos, A. (2023). Physiological roles of hydrogen sulfide in mammalian cells, tissues, and organs. *Physiology Review, 103*(1), 31–276 [cited 2025 Mar 15]. Available from: https://journals.physiology.org/doi/https://doi.org/10.1152/physrev.00028.2021
153. Moldogazieva, N. T., Lutsenko, S. V., Terentiev, A. A. (2018). Reactive oxygen and nitrogen species-induced protein modifications: Implication in carcinogenesis and anticancer therapy. *Cancer Resource, 78*(21), 6040–7 [cited 2025 Mar 17]. Available from: https://pubmed.ncbi.nlm.nih.gov/30327380/
154. Nakamura, H., Takada, K. (2021). Reactive oxygen species in cancer: Current findings and future directions. *Cancer Science, 112*
155. Verma, P., Rishi, B., George, N. G., Kushwaha, N., Dhandha, H., Kaur, M., et al. (2023). Recent advances and future directions in Etiopatho genesis and mechanisms of reactive oxygen species in cancer treatment. *Pathology and Oncology Research, 18*(29), 1611415.
156. Valko, M., Rhodes, C. J., Moncol, J., Izakovic, M., & Mazur, M. (2006). Free radicals, metals and antioxidants in oxidative stress-induced cancer. *Chemico-Biological Interactions, 160*(1), 1–40.
157. Long, Y., Shi, H., Ye, J., Qi, X. (2025). Exploring strategies to prevent and treat ovarian cancer in terms of oxidative stress and antioxidants. *Antioxidants, 14*(1), 114 [cited 2025 Mar 19]. Available from: https://www.mdpi.com/2076-3921/14/1/114/htm
158. Singh, R., Manna, P. P. (2022). Reactive oxygen species in cancer progression and its role in therapeutics. *Exploration of Medicine, 3*
159. Khan, A. Q., Rashid, K., AlAmodi, A. A., Agha, M. V., Akhtar, S., Hakeem, I., et al. (2021). Reactive oxygen species (ROS) in cancer pathogenesis and therapy: An update on the role of ROS in anticancer action of benzophenanthridine alkaloids. *Biomedicine and Pharmacotherapy, 1*(143), Article 112142.

160. Jelic, M. D., Mandic, A. D., Maricic, S. M., Srdjenovic, B. U. (2021). Oxidative stress and its role in cancer. *Journal Cancer Research Therapeutics, 17*(1), 22–8 [cited 2025 Mar 19]. Available from: https://journals.lww.com/cancerjournal/fulltext/2021/17010/oxidative_stress_and_its_role_in_cancer.4.aspx
161. Iciek, M., Bilska-Wilkosz, A., Kozdrowicki, M., Górny, M. (2022). Reactive sulfur species and their significance in health and disease. Bioscience Reports, 42(9):BSR20221006 [cited 2025 Mar 20]. Available from: https://pmc.ncbi.nlm.nih.gov/articles/PMC9484011/
162. Khattak, S., Rauf, M. A., Khan, N. H., Zhang, Q. Q., Chen, H. J., Muhammad, P., et al. (2022). Hydrogen sulfide biology and its role in cancer. *Molecules, 27*(11), 3389 [cited 2025 Mar 20]. Available from: https://www.mdpi.com/1420-3049/27/11/3389/htm
163. AruIJothi, K. N., Kumaran, K., Senthil, S., Nidhu, A. B., Munaff, N., Janitri, V. B., et al. (2022). Implications of reactive oxygen species in lung cancer and exploiting it for therapeutic interventions. *Medical Oncology, 40*(1), 1–19 [cited 2025 Mar 19]. Available from: https://doi.org/10.1007/s12032-022-01900-y
164. AruIJothi, K. N., Kumaran, K., Senthil, S., Nidhu, A. B., Munaff, N., Janitri, V. B., et al. (2022). Implications of reactive oxygen species in lung cancer and exploiting it for therapeutic interventions. *Medical Oncology, 40*(1), 1–19 [cited 2025 Mar 16]. Available from: https://doi.org/10.1007/s12032-022-01900-y
165. Yang, Y. L., Zhang, K., Zhou, Z. T., Jiang, Z. L., Liu, Y., Zhang, Y. X., et al. (2022). The role of hydrogen sulfide in the development and progression of lung cancer. Molecules, 27(24):9005 [cited 2025 Mar 20]. Available from: https://pmc.ncbi.nlm.nih.gov/articles/PMC9787608/
166. Bilska-Wilkosz, A., Iciek, M. (2022). Reactive sulfur species (RSS) in physiological and pathological conditions and in therapy. Antioxidants, 11(8), 1576 [cited 2025 Mar 20]. Available from: https://pmc.ncbi.nlm.nih.gov/articles/PMC9405001/
167. Malla, R. R., Surepalli, N., Farran, B., Malhotra, S. V., & Nagaraju, G. P. (2021). Reactive oxygen species (ROS): Critical roles in breast tumor microenvironment. *Critical Reviews in Oncology Hematology, 1*(160), Article 103285.
168. Wang, F., Li, C., Xu, X., Zhang, W., & He, Z. (2024). Enhanced breast cancer treatment using phototherapy and RNS therapy with macrophage membrane-coated liposomes. *Colloids and Surfaces. B, Biointerfaces, 1*(239), Article 113961.
169. Li, M., Liu, Y., Deng, Y., Pan, L., Fu, H., Han, X., et al. (2021). Therapeutic potential of endogenous hydrogen sulfide inhibition in breast cancer. *Oncology Reports, 45*(5), 68 [cited 2025 Mar 20]. Available from: https://pmc.ncbi.nlm.nih.gov/articles/PMC8020202/
170. Nguyen-Nielsen, M., Borre, M. (2016). Diagnostic and therapeutic strategies for prostate cancer. Seminars Nuclear Medicine, 46(6):484–90 [cited 2025 Mar 20]. Available from: https://pubmed.ncbi.nlm.nih.gov/27825428/
171. Han, C., Wang, Z., Xu, Y., Chen, S., Han, Y., Li, L., et al. (2020). Roles of reactive oxygen species in biological behaviors of prostate cancer. *Biomedicine Resource International, 2020*, 1269624 [cited 2025 Mar 16]. Available from: https://pmc.ncbi.nlm.nih.gov/articles/PMC7538255/
172. Machado-Neto, J. A., Azevedo Cerqueira, A. R., Veríssimo-Filho, S., Muscará, M. N., Pereira Costa, S. K., Lopes, L. R. (2024). Hydrogen sulfide signaling in the tumor microenvironment: Implications in cancer progression and therapy. https://home.liebertpub.com/ars, *40*(4–6), 250–71 [cited 2025 Mar 20]. Available from: https://doi.org/10.1089/ars.2021.0256
173. Zareen, W., Ahmed, N., Raza, S., Ali Khan, M., & Shafiq, Z. (2024). Recent development in dual function fluorescence probes for HOCl and interaction with different bioactive molecules. *Talanta, 1*(277), Article 126374.
174. Li, R., Jia, Z., Trush, M. A. (2016). Defining ROS in biology and medicine. *Reactive Oxygen Species (Apex), 1*(1), 9 [cited 2025 Mar 17]. Available from: https://pmc.ncbi.nlm.nih.gov/articles/PMC5921829/
175. Geng, Y., Wang, Z., Zhou, J., Zhu, M., Liu, J., James, T. D. (2023). Recent progress in the development of fluorescent probes for imaging pathological oxidative stress. *Chemical Society Reviews, 52*(11), 3873–926 [cited 2025 Mar 17]. Available from: https://pubs.rsc.org/en/content/articlehtml/2023/cs/d2cs00172a

176. Forrester, J. V., Kuffova, L., Delibegovic, M. (2020). The role of inflammation in diabetic retinopathy. *Frontiers Immunology, 11,* 583687 [cited 2025 Mar 17]. Available from: www.frontiersin.org
177. Arman, M. D. S. I. (2019). Free radical, oxidative stress and diabetes mellitus: A mini review. *Discovery Phytomedicine, 6*(3)
178. Zhu, L., Yang, B., Ma, D., Wang, L., Duan, W. (2020). Hydrogen sulfide, adipose tissue and diabetes mellitus. *Diabetes, Metabolic Syndrome and Obesity, 13,* 1873–86 [cited 2025 Mar 20]. Available from: https://www.tandfonline.com/action/journalInformation?journalCode=dmso20
179. He, L., Wei, X., Zhang, W., Xu, N., Wu, J., Yu, F., et al. (2025). Fabrication of a redox-reversible near-infrared fluorogenic probe for ferroptosis process monitoring and the early diagnosis of diabetes. Analytical Chemistry [cited 2025 Mar 20], Available from: https://doi.org/10.1021/acs.analchem.4c05927
180. Yang, Y., Zhang, C., Lin, L., Li, Q., Wang, M., Zhang, Y., et al. (2024). Multifunctional MnGA nanozymes for the treatment of ulcerative colitis by reducing intestinal inflammation and regulating the intestinal flora. ACS Applied Material Interfaces [cited 2025 Mar 20], Available from: https://doi.org/10.1021/acsami.4c14291
181. Muro, P., Zhang, L., Li, S., Zhao, Z., Jin, T., Mao, F., et al. (2024). The emerging role of oxidative stress in inflammatory bowel disease. *Frontiers in Endocrinology (Lausanne), 15*(15), 1390351.
182. Li, Q., Lin, L., Zhang, C., Zhang, H., Ma, Y., Qian, H., et al. (2024). The progression of inorganic nanoparticles and natural products for inflammatory bowel disease. *Journal of Nanobiotechnology, 22*(1), 1–28 [cited 2025 Mar 20]. Available from: https://link.springer.com/articles/https://doi.org/10.1186/s12951-023-02246-x
183. Deng, Y., Shi, X., Hu, X., Xu, L., Liu, X., Gao, G., et al. (2023). A chemiluminescent probe for imaging peroxynitrite in inflammatory cells and tissues. *Analytical Chemistry, 95*(16), 6496–500 [cited 2025 Mar 20]. Available from: https://doi.org/10.1021/acs.analchem.3c01124
184. Stummer, N., Feichtinger, R. G., Weghuber, D., Kofler, B., Schneider, A. M. (2023). Role of hydrogen sulfide in inflammatory bowel disease. *Antioxidants, 12*(8), 1570 [cited 2025 Mar 20]. Available from: https://www.mdpi.com/2076-3921/12/8/1570/htm
185. Irazabal, M. V., Torres, V. E. (2020). Reactive oxygen species and redox signaling in chronic kidney disease. *Cells, 9*(6), 1342 [cited 2025 Mar 20]. Available from: https://www.mdpi.com/2073-4409/9/6/1342/htm
186. Rapa, S. F., Di Iorio, B. R., Campiglia, P., Heidland, A., Marzocco, S. (2019). Inflammation and oxidative stress in chronic kidney disease—Potential therapeutic role of minerals, vitamins and plant-derived metabolites. *International Journal of Molecular Sciences, 21*(1), 263 [cited 2025 Mar 20]. Available from: https://www.mdpi.com/1422-0067/21/1/263/htm
187. Daenen, K., Andries, A., Mekahli, D., Van Schepdael, A., Jouret, F., Bammens, B. (2018). Oxidative stress in chronic kidney disease. *Pediatric Nephrology, 34*(6), 975–91 [cited 2025 Mar 20]. Available from: https://doi.org/10.1007/s00467-018-4005-4
188. Iciek, M., Bilska-Wilkosz, A., Kozdrowicki, M., Górny, M. (2023). Reactive sulfur species in human diseases. https://home.liebertpub.com/ars, 39(13–15), 1000–23 [cited 2025 Mar 20]. Available from: https://doi.org/10.1089/ars.2023.0261
189. Sárközy, M., Kovács, Z. Z. A., Kovács, M. G., Gáspár, R., Szűcs, G., Dux, L. (2018). Mechanisms and modulation of oxidative/nitrative stress in type 4 cardio-renal syndrome and renal sarcopenia. *Frontiers Physiology, 9,* 412088 [cited 2025 Mar 20]. Available from: www.frontiersin.org
190. Sorriento, D., De Luca, N., Trimarco, B., Iaccarino, G. (2018). The antioxidant therapy: New insights in the treatment of hypertension. *Frontiers Physiology, 9*(MAR), 348746 [cited 2025 Mar 20]. Available from: www.frontiersin.org
191. Cacanyiova S, Berenyiova A, Kristek F. The Role of Hydrogen Sulphide in Blood Pressure Regulation. Physiol Res [Internet]. 2016 [cited 2025 Mar 20];65:273–89. Available from: www.biomed.cas.cz/physiolres

192. Fernández-Sánchez A, Madrigal-Santillán E, Bautista M, Esquivel-Soto J, Morales-González Á, Esquivel-Chirino C, et al. Inflammation, Oxidative Stress, and Obesity. International Journal of Molecular Sciences 2011, Vol 12, Pages 3117–3132 [Internet]. 2011 May 13 [cited 2025 Mar 20];12(5):3117–32. Available from: https://www.mdpi.com/1422-0067/12/5/3117/htm
193. McMurray F, Patten DA, Harper ME. Reactive Oxygen Species and Oxidative Stress in Obesity—Recent Findings and Empirical Approaches. Obesity [Internet]. 2016 Nov 1 [cited 2025 Mar 20];24(11):2301–10. Available from: https://onlinelibrary.wiley.com/doi/full/https://doi.org/10.1002/oby.21654
194. Sansbury BE, Hill BG. Anti-obesogenic role of endothelial nitric oxide synthase. Vitam Horm [Internet]. 2014 [cited 2025 Mar 20];96:323. Available from: https://pmc.ncbi.nlm.nih.gov/articles/PMC4410351/
195. Liu, X., Wan, X., Zhang, L., Li, Y., Ao, Y., Zhuang, P., et al. (2023). The sulfur microbial diet and increased risk of obesity: Findings from a population-based prospective cohort study. *Clinical Nutrition., 42*(5), 764–772.

A Better Understanding of Antioxidant Mechanisms in Living Systems and Lifestyle

A. Ayeshamariam, S. Irshad Ahamed, J. C. Balachandar, M. Ismail Fathima, and N. M. I. Alhaji

Abstract It has been proposed that a growing number of diseases are caused by oxidants or radicals. Uncertainty often surrounds the relationship between certain factors and illness, making it hard to discern cause from effect. It is still helpful to learn about the chemicals or enzymes that may block, impede, or prevent radical-initiated processes. Therefore, it is becoming increasingly important to understand which substances can function as antioxidants, where they are in the body, and how they might work. As we age, our bodies become less effective at handling stress, which can lead to serious health issues due to the buildup of damaged proteins and disrupted metal balances that fuel oxidative damage. It has been observed that this stress response is compromised with age. There is compelling evidence that humans' ability to withstand oxidative stress generally decreases with age. Research has shown that as organisms age, their genetic response to stress changes, often involving increased activity of stress-related genes. As individuals age, their stress response mechanisms seem to intensify in terms of quantity but weaken in efficacy. This disruption throws off the delicate balance between oxidation and reduction in the body's tissues and blood, resulting in damage to lipids, proteins, and nucleic acids. Such oxidative harm is associated with alterations in the structure and function of vital molecules and contributes to the onset of various health issues, including cancer and

---

A. Ayeshamariam (✉)
PG & Research Department of Physics, Khadir Mohideen College (Affiliated to Bharathidasan University, Tiruchirappalli), Adirampattinam, Tamil Nadu, India
e-mail: ayeshamariamkmc@gmail.com

S. Irshad Ahamed
Department of General Surgery, Pondicherry Institute of Medical Sciences, Puducherry, Tamil Nadu, India

J. C. Balachandar
Dr. Bala's School of Fitness & Preventive Medicine, Faridabad, Delhi-NCR, India

M. Ismail Fathima
Department of Physics, Velammal College of Engineering and Technology, Virahanoor, Madurai, Tamil Nadu, India

N. M. I. Alhaji
PG & Research Department of Chemistry, Khadir Mohideen College (Affiliated to Bharathidasan University, Tiruchirappalli), Adirampattinam, Tamil Nadu, India

K. Kasinathan and A. Muthukumaran (eds.), *Recent Advances in Oxidative Stress Associated Chronic Diseases Volume 2*, https://doi.org/10.1007/978-981-95-3758-7_8

cardiovascular disease. By accelerating the recovery of motor function, reducing neuronal impairment, and triggering vascular mechanisms in the central nervous system, regular exercise appears to prevent brain damage in a number of anatomical areas, including the cerebellum, brain stems, motor cortex, and hippocampus regions. There is increasing evidence that *"unhealthy"* lifestyle choices raise the risk of disease by contributing to an unbalanced antioxidative equilibrium that results in oxidative stress. The specific mechanism by which the most important endogenous and exogenous antioxidants in each class work to prevent or block particular factors that lead to oxidative damage in cells is then covered.

Keywords Antioxidative equilibrium · Reactive oxygen and nitrogen species · Alzheimer's and Parkinson's diseases · Mycoplasma pneumoniae infections · Impaired mitochondrial function · Inflammatory bowel disease

1 Introduction

Oxygen free radicals are constantly produced by cells when they are in an oxygen-rich environment. ROS, which include reactive species and oxygen-related free radicals, is produced by aerobic metabolism. ROS can form through both enzymatic and non-enzymatic reactions. Various cellular processes give rise to free radicals. The production of energy in mitochondria, the process of phagocytosis, and the synthesis of prostaglandins are all sources. Moreover, the cytochrome P450 system and specific enzymes like xanthine oxidase and certain peroxidases contribute to the generation of reactive species. The interaction of oxygen with organic compounds and exposure to ionizing radiation also play a role in producing these highly reactive molecules. The buildup of reactive molecules in cells can trigger or worsen several diseases, including diabetes, obesity, and associated conditions like vision loss, kidney failure, and cardiovascular disease [1]. While reactive oxygen and nitrogen species play roles in normal cell function and defense, their overproduction can be detrimental. Maintaining a balance between the generation and elimination of these reactive species is crucial for preventing or mitigating various health issues [2]. While certain molecules play a role in normal cell signaling, excessive levels can harm cellular components, including proteins, lipids, and DNA, under conditions of oxidative stress. This damage contributes to the development of various diseases, such as diabetes, obesity, and neurodegenerative disorders, including Alzheimer's and Parkinson's diseases [3]. The accumulation of advanced glycation end-products (AGEs) and their interaction with receptor for advanced glycation end-products (RAGEs) can create a self-reinforcing cycle of oxidative stress. In people with diabetes and obesity, this AGE-RAGE interaction may also play a role in the development of pancreatic cancer. Moreover, high levels of AGEs are correlated with oxidative stress and an increased risk of various diseases, including diabetes, obesity, and Alzheimer's [4]. Lipid peroxidation products, such as Malondialdehyde assay (MDA) and Hydromechanical neonatal encephalopathy (HNE), resulting from oxidative stress, can alter

DNA and promote cancer. This has sparked interest in developing medications that neutralize these products and antioxidants that mitigate lipid peroxidation, potentially preventing cancer-causing changes [5]. RAGE inhibitors show potential for addressing various health issues, including cancer and neurodegenerative diseases. However, their development is complicated by the receptor's intricate binding mechanism. Ongoing clinical trials are assessing the efficacy and safety of these inhibitors in treating specific conditions. There is increasing interest in this area as a way to develop effective treatments for a variety of neurological disorders since nanozymes that are made to specifically enter neuronal cells are effective at lowering oxidative stress. One recent advancement in this area is the imaging of neuroinflammation sites on the surface of near-infrared probes using single-atom-based nanozymes. Nanozyme materials made from redox-active metals such as Mn, Co, Zn, and Pt reduce neuroinflammation, kill tumor cells, and aid in diabetic wound healing by absorbing reactive oxygen species (ROS) or acting as antimicrobials. These nanozymes can also be engineered to cross the blood–brain barrier (BBB) and show excellent selectivity for entry into the neuronal cells, which is useful when treating neurological conditions like traumatic brain injury (TBI). In addition to being generated nonenzymatically by cellular metabolism and redox processes catalyzed by transition metal ions (Fenton reaction), reactive oxygen species (ROS) are physiologically produced during the respiratory cycle of the mitochondria. The antioxidant enzymes ascorbate oxidase, catalase, and Mn superoxide dismutase (Mn SOD) control the levels of ROS in the mitochondria, which are primarily where ROS is generated [6]. A single-atom Mn catalyst that sequesters ROS restores neurological function, lowers neuroinflammation, and promotes BBB reconstruction. This single-atom Mn catalyst was placed in a near-infrared-II (1500 nm–1700 nm) silver telluride (Ag_2Te) quantum dot as an imaging probe to monitor the neuroinflammation caused by ROS. In this nanozyme-type catalytic system, the redox-active Mn transforms the superoxide radical anions ($O_2\cdot-$) and hydroxyl radicals ($\cdot OH$) into hydroxyl anions ($-OH$) and dioxygen (O_2), which are subsequently transformed into H_2O by removing a proton from the adjacent acidic sites. The BBB is seriously damaged in TBI, and excessive amounts of ROS such as hydrogen peroxide (H_2O_2), hydroxyl radicals ($\cdot OH$), and superoxide radical anions ($O_2\cdot -$) are generated. ROS-induced destruction of neuronal cells leads to long-term health effects and, in some cases, the onset of AD [7]. An analogous technique that uses Mn doped onto the near IR-II (1500–1700 nm) active PbS/CdS quantum dot (QD) imaging agent has neuroprotective effects by limiting the excess ROS in TBI brains and preventing the release of pro-inflammatory factors. Currently, therapeutic candidates employing single atom based catalytic nanozymes are being developed for the treatment of neurological disorders such as Alzheimer's disease (AD) and TBI [8]. Furthermore, nanozymes that can cross the blood–brain barrier and eliminate misfolded proteins may be developed; these could be effective treatments, particularly for neurological disorders. A better understanding of the mechanisms by which antioxidants function and the sites and times at which they are most effective may lead to a more rational approach that yields greater pharmacological success.

Here, we discuss the relationship between oxidative stress, redox signaling, and disease; how oxidative stress can cause pathology; how antioxidant defenses work; what limits their effectiveness; and how dietary factors, physiological signaling, and potential pharmaceutical intervention can strengthen antioxidant defenses [9]. The precise biochemical mechanisms underlying oxidative stress-induced aging remain unclear to us. Cellular senescence, a defense mechanism that halts cell division when replication-related damage occurs, is thought to be brought on by elevated Recepteur d'Origine Nantais (RONs) levels. When senescent cells develop an irreversible senescence-associated secretory phenotype (SASP), they release soluble substances like growth factors, chemokines, and interleukins [10]. Table 1, in particular, shows an interesting trend, Liang et al., (2021) claim that oxidative stress plays a role in the onset and spread of cancers like colorectal, lung, and breast cancer. The importance of antioxidant defenses in preventing cellular damage and improving overall health was highlighted by their findings [11]. Reactive oxygen species (ROS) and angiogenesis were examined by Ushio-Fukai and Nakamura (2008), who also suggested that nicotinamide adenine dinucleotide phosphate (NADPH) oxidase might be a target for cancer treatment. Their results demonstrated how important it is to comprehend ROS pathways in order to develop effective cancer treatments. [12]. Ushio-Fukai and Nakamura (2008) looked into how reactive oxygen species (ROS) affect angiogenesis and found that NADPH oxidase could be a target for cancer treatment. Their results demonstrated how important it is to understand ROS pathways in order to develop effective cancer treatments [13]. The distinct mitochondrial metabolism and dynamics in prostate cancer were examined by Fontana et al., (2023). Their results demonstrated how important it is to understand how mitochondria function in the development and spread of cancer [14]. Recent studies have explored the impact of oxidative stress on various biological processes. Research by Liu et al., (2023) highlighted the role of oxidative stress in skin cell damage, inflammation, and cancer, suggesting plant polyphenols as potential therapeutic agents. Another study by Wang (2020) examined the metabolic role of Yes-associated protein (YAP) in liver regeneration, shedding light on the complex mechanisms involved. Additionally, Alves-Silva et al., (2023) found that plant polyphenols may help reduce inflammation and oxidative stress, indicating their potential therapeutic benefits [15–17]. Arora et al., (2024) investigated how pesticides affect the development and spread of cancer. Their results highlighted the need for more research into the mechanisms underlying this association and the potential influence of ambient oxidative stressors on the development of cancer [18]. Since there is a significant difference in the behavior of cells in normal tissue and affected tissue, two stages of cell targeting techniques, such as passive targeting and active targeting, have historically been employed. When compared to conventional methods, the use of nanoformulation for cancer treatment is unquestionably a significant advancement. Undoubtedly, nanoformulation is a significant advancement. Cellular aging and oxidative stress are complex processes involving various molecular mechanisms. Reduced activity of certain enzymes can contribute to cardiac aging, while sirtuin dysregulation can lead to increased inflammation and cancer promotion. Cellular senescence is tightly regulated by specific signaling pathways. Reactive oxygen species (ROS) are naturally produced in cells and play a role in physiological control but can also cause oxidative damage. Mitochondrial electron transport is a significant source of ROS, which can harm biological macromolecules and interact with antioxidant vitamins and enzymes [19].

The role of reactive oxygen species (ROS) in the development of rheumatoid arthritis (RA) was examined by Gellerman et al., (2007). Their study highlighted how ROS contribute to the development of RA and offered potential treatments that focus on oxidative stress as shown in Table 2. The significance of additional research into the intricate relationships among ROS, inflammation, and joint destruction in RA was emphasized by the authors [20]. Poznyak et al., (2024) examined Systemic Lupus Erythematosus and mitochondrial dysfunction. With a focus on oxidative stress and mitochondrial dysfunction, their findings provide insight into potential treatment options for Systemic Lupus Erythematosus (SLE). The authors emphasized the importance of further research into effective treatments as well as the need to comprehend the intricate mechanisms underlying the pathogenesis of SLE [21].

A study by Noordzij et al., (2018) explored the use of PET/CT scans in diagnosing and treating liver cancer. The research demonstrated the importance of precise imaging in guiding treatment decisions for patients with hepatocellular carcinoma [22]. Fibrocystin's function in the biliary epithelium and its connection to the production of connective tissue growth factor (CTGF) and IL-8 were investigated by Tsunoda et al., (2019). According to their research, fibrocystin deficiency promotes CTGF synthesis and IL-8-dependent proliferation, both of which advance liver disease as shown in Table 3. Although ROS was not specifically addressed in the study, it did shed light on the intricate mechanisms underlying liver disease [23].

Table 1 Various diseases and its ROS system

Diseases	ROS source	Findings from the researchers	Refs.
Breast cancer	Mitochondrial respiratory chain	Increased ROS production	[11]
Lung cancer	NADPH oxidase	Elevated ROS levels	[12]
Colorectal cancer	Cytochrome P450 system	Impaired antioxidant defenses	[13]
Prostate cancer	Mitochondrial respiratory chain	Increased ROS production	[14]
Skin cancer	UV radiation	Increased ROS production	[15]
Pancreatic cancer	NADPH oxidase	NADPH oxidase	[16]
Ovarian cancer	Mitochondrial respiratory chain	Increased ROS production	[17]
Brain cancer	Cytochrome P450 system	Impaired antioxidant defenses	[18]

Table 2 List of autoimmune diseases

Disease	ROS source	Finding	Refs.
Rheumatoid arthritis (RA)	NADPH oxidase	Increased ROS production	[20]
Lupus	Mitochondrial respiratory chain	Impaired mitochondrial function	[21]

Table 3 Overview of liver diseases

Disease	ROS Source	Finding	Refs.
Hepatitis	NADPH oxidase	Increased ROS production	[22]
Liver cancer	Mitochondrial respiratory chain I	Impaired mitochondrial function	[23]

In order to support the polio endgame, Zipursky et al., (2017) discussed the lessons they learned from overseeing the planning and execution of the introduction of the inactivated polio vaccine. In order to meet global health goals, their research emphasized the importance of efficient planning, coordination, and implementation [24]. Brealey et al., (2002) investigated the connection between the severity and outcome of septic shock and mitochondrial dysfunction as shown in Table 4. Their study highlighted the potential involvement of mitochondrial function in sepsis by finding that it is linked to worse outcomes and increased severity in septic shock patients [25].

The effects of diet, sex, and mating status on longevity in Mediterranean fruit flies (Ceratitis capitata) were investigated by Davies et al., (2005). Their study found that these variables affect longevity, providing insights into the biology of aging in this species [26]. Wang et al., (2017) looked into how oxidative stress affected cell division, particularly how it suppressed the formation of mitotic spindles mediated by Aurora A. Their research revealed the likely mechanisms causing cell cycle disruption by demonstrating that oxidative stress results in mitotic arrest as shown in Table 5 [27].

Anderson et al., (2009) investigated redox balance and mitochondrial metabolism in the atriums of human hearts with Type 2 diabetes. Their research provided insight into the likely causes of diabetes cardiomyopathy by identifying substrate-specific abnormalities in mitochondrial metabolism and redox balance as shown in Table 6 [28]. Cosentino et al., (2003) looked into how high glucose affected human endothelial cells and found that it changed the prostanoid profile and increased cyclooxygenase-2 activity. Their research showed that reactive oxygen species and protein kinase C are involved in this mechanism [29]. Endothelial dysfunction in the aging kidney was examined by Borri et al., (2025), who provided insights into

Table 4 Factors that contribute to infectious diseases

Disease	ROS Source	Finding	Ref
Tuberculosis	NADPH oxidase	Increased ROS production	[24]
Sepsis	Mitochondrial respiratory chain	Impaired mitochondrial function	[25]

Table 5 Aging factor-related disease caused by oxidative stress

Disease	ROS Source	Finding	Refs.
Cellular senescence	Mitochondrial respiratory chain	Impaired mitochondrial function	[26]
Telomere shortening	NADPH oxidase	Increased ROS production	[27]

the basic mechanisms underlying this process. Their research clarifies the vascular changes that come with aging, which can guide strategies for preventing or managing kidney disease associated with aging [30]. Tangvarasittichai et al., (2018) investigated the relationship among diabetic retinopathy, oxidative stress, and ocular disease, with an emphasis on the potential mechanisms underlying these conditions [31]. In their 2019 evaluation, Eldman et al., concentrated on the etiology, risk factors, clinical symptoms, and available treatments for diabetic neuropathy. The study emphasized the complex interplay between genetic, vascular, and metabolic factors that lead to this disease. In a study by Wang et al., it was found that the antioxidant activity of 1, 25-dihydroxyvitamin D3 prevents the oxidation of the kidneys, which may lower inflammation [32, 33].

The function of reactive oxygen species produced by NADPH oxidase in kidney fibrosis was investigated by Bondi et al., (2018). Their results highlight the possible significance of focusing on this pathway in the treatment of kidney disease since they demonstrate that reactive oxygen species produced by NADPH oxidase contribute to kidney fibrosis [34]. The importance of mitochondrial dysfunction in diabetic renal disease was examined by Sharma et al., (2013) as shown in Table 7. Their results highlighted the need for more research into the underlying mechanism and the significance of mitochondrial dysfunction in the onset and progression of diabetic kidney disease [35]. The role of NADPH oxidase in hypertensive nephrosclerosis was examined by Wei et al., (2016). According to their research, NADPH oxidase plays a role in hypertensive nephrosclerosis, suggesting that blocking this enzyme could be a viable treatment strategy for kidney disease linked to hypertension [36]. In their 2019 study, Wang et al., examined mitochondrial dysfunction in glomerulonephritis. Their results underscored the significance of mitochondrial dysfunction in the onset and advancement of glomerulonephritis, underscoring the need for further research into the underlying mechanisms [37]. Li et al., (2018) looked into how NADPH oxidase contributes to the development of renal cell carcinoma. According to their research, NADPH oxidase accelerates the development of renal cell carcinoma, suggesting that blocking this enzyme could be a useful treatment strategy [38].

Table 6 Diabetes categories are listed in various ways

Disease	ROS Source	Finding	Refs.
Type 2 diabetes	Mitochondrial respiratory chain	Impaired insulin signaling	[28]
Type 1 diabetes	NADPH oxidase	Increased ROS production	[29]
Diabetic retinopathy	NADPH oxidase	Elevated ROS levels	[30]
Diabetic neuropathy	Mitochondrial respiratory chain	Impaired mitochondrial function	[31]
Diabetic cardiomyopathy	NADPH oxidase	Increased ROS production	[32]
Diabetic nephropathy	Cytochrome P450 system	Impaired antioxidant defenses	[33]

Table 7 Renal diseases play a key role

Disease	ROS Source	Finding	Refs.
Kidney fibrosis	NADPH oxidase	Increased ROS production	[34]
Diabetic nephropathy	Mitochondrial respiratory chain	Impaired mitochondrial function	[35]
Hypertensive nephrosclerosis	NADPH oxidase	Elevated ROS levels	[36]
Glomerulonephritis	Mitochondrial respiratory chain	Impaired mitochondrial function	[37]
Renal cell carcinoma	NADPH oxidase	Increased ROS production	[38]

The role of reactive oxygen species produced by NADPH oxidase in age-related macular degeneration was investigated by Kaneko et al., (2017). According to their research, age-related macular degeneration is exacerbated by reactive oxygen species produced by NADPH oxidase, which highlights the potential benefit of treating this condition by focusing on this pathway [39].

Tawfik et al., (2018) investigated how mitochondrial dysfunction contributes to the development of diabetic retinopathy as shown in Table 8. Their research showed that mitochondrial dysfunction plays a role in the development of diabetic retinopathy, highlighting the necessity of more investigation into the underlying mechanisms and possible treatment approaches [40].

Research by Sivamani et al., (2019) explored the role of reactive oxygen species (ROS) produced by NADPH oxidase in psoriasis development. The study found that ROS generated by NADPH oxidase contribute to the development of psoriasis, suggesting that targeting this pathway may be a potential therapeutic approach for treating the condition as shown in Table 9 [41]. The role of mitochondrial dysfunction in the etiology of vitiligo was examined by Dell'Anna et al., (2018). Their research showed that vitiligo may be caused by mitochondrial dysfunction, underscoring the need for more research into the underlying mechanisms and possible treatment approaches [42].

Table 8 Eye diseases grouped by category

Disease	ROS source	Finding	Refs.
Age-related macular	Degeneration NADPH oxidase	Increased ROS production	[39]
Diabetic retinopathy	Mitochondrial respiratory chain	Impaired mitochondrial function	[40]

Table 9 Factors contributing to skin diseases

Disease	ROS source	Finding	Refs.
Psoriasis	NADPH oxidase	Increased ROS production	[41]
Vitiligo	Mitochondrial respiratory chain	Impaired mitochondrial function	[42]

Table 10 Respiratory diseases play an important role in the development of these diseases

Disease	ROS source	Finding	Refs.
COPD	NADPH oxidase	Increased ROS production	[43]
Asthma	Mitochondrial respiratory chain	Impaired mitochondrial function	[44]
Pulmonary fibrosis	NADPH oxidase	Elevated ROS levels	[45]
Lung cancer	NADPH oxidase	Increased ROS production	[46]
Cystic fibrosis	Mitochondrial respiratory chain	Increased ROS production	[47]
Pneumonia	NADPH oxidase	Increased ROS production	[48]
Bronchiectasis	Mitochondrial respiratory chain	Impaired mitochondrial function	[49]

The role of reactive oxygen species (ROS) in promoting NETosis was investigated by Azzouz and Palaniyar (2024). Their research indicates that oxidative DNA damage, DNA repair, and chromatin decondensation are the mechanisms by which ROS induce NETosis [43]. In patients with salt-sensitive hypertension, Mutchler et al., (2025) looked into the epigenetic regulation of innate and adaptive immune cells. Their research revealed that the development of salt-sensitive hypertension is significantly influenced by epigenetic regulation [44]. A long-term follow-up study on CD19 CAR treatment for acute lymphoblastic leukemia was conducted by Park et al., (2018). According to their findings, acute lymphoblastic leukemia can be effectively treated with CD19 CAR therapy [45]. The 2018 study by Divangahi highlights the significance of training and tolerance in the fight against tuberculosis (TB), highlighting the fact that these factors are necessary to eradicate TB. The study offers fresh perspectives on managing and preventing tuberculosis while highlighting the importance of education and resilience in the fight against the illness as shown in Table 10 [46]. The role of protein S-glutathionylation and glutaredoxin-1 in redox regulation was investigated by Matsui et al., (2020). Their research showed that protein S-glutathionylation and glutaredoxin-1 are essential for redox regulation [47]. Guo et al., (2024) looked into the connection between Mycoplasma pneumoniae infections and long-term lung conditions like bronchiectasis, COPD, and asthma. The researchers found a connection between chronic lung diseases and Mycoplasma pneumoniae infections. [48]. According to Li et al., (2018), who investigated the role of sirtuins in sepsis, these proteins are crucial to the onset of sepsis, a potentially fatal condition brought on by an overreaction of the immune system to an infection. According to their research, sirtuins may be a promising therapeutic target for sepsis [49].

2 Defense Mechanisms Against Oxidative Stress

Heavy metals and environmental factors can generate free radicals, leading to oxidative stress and damage to biological components. Cellular sources of reactive species include the mitochondrial electron transport chain, endoplasmic reticulum, and specific enzyme systems. ROS and RNS play roles in cell signaling, regulating processes like cell survival and proliferation. However, excessive levels can cause DNA damage and contribute to diseases like cancer. Signaling pathways, such as NF-κB, are also impacted by reactive species [50].

3 Neurodegenerative Disorders Due to Oxidative Stress

Dopaminergic neurons are particularly vulnerable to oxidative stress due to dopamine metabolism. Neurodegenerative diseases, such as Alzheimer's disease, Parkinson's disease, and Huntington's disease, are associated with mitochondrial dysfunction, calcium dysregulation, and oxidative damage. The accumulation of oxidized proteins, lipids, and DNA in brain tissue highlights the role of oxidative stress in these diseases. Specific neuronal populations are susceptible to oxidative stress, leading to selective neuronal loss and protein misfolding. Lipid peroxidation is a sensitive indicator of oxidative stress, particularly affecting unsaturated lipids. The process involves the formation of highly reactive lipid peroxyl radicals, leading to a chain reaction of oxidative damage. Breakdown products, such as F2-isoprostanes and 4-hydroxy-2-nonenal (HNE), are formed as a result. Oxidative stress can also lead to protein modification and aggregation, contributing to neurodegenerative diseases as shown in Table 11. Highly oxidized and cross-linked proteins can inhibit proteasomal activity, further exacerbating protein accumulation. There is evidence that oxidative stress and protein aggregation trigger a cascade of events leading to cell death in neurodegenerative diseases [51].

Ansari et al., (2017) found that reactive oxygen species (ROS) derived from cytochrome P450 aid in the progression of Alzheimer's disease, highlighting the possible importance of oxidative stress in neurodegeneration [52]. The significance of mitochondrial health in neurodegenerative diseases was highlighted by Lin et al.,

Table 11 Diseases are categorized based on their neurodegenerative characteristics

Disease	ROS source	Finding	Refs.
Alzheimer's	CytochromeP450 system	Elevated ROS levels	[52]
Parkinson's	Mitochondrial respiratory chain	Impaired mitochondrial function	[53]
Huntington's	NADPH oxidase	Increased ROS production	[54]
Amyotrophic lateral sclerosis	Mitochondrial respiratory chain	Impaired mitochondrial function	[55]
Multiple sclerosis	NADPH oxidase	Elevated ROS levels	[56]
Friedreich's ataxia	Mitochondrial respiratory chain	Impaired mitochondrial function	[57]
Spinocerebellar Ataxia	Cytochrome P450 system	Impaired antioxidant defenses	[58]

(2017), who found that mitochondrial dysfunction speeds up the progression of Parkinson's disease. [53]. According to Klepac et al., (2017), ROS produced by NADPH oxidase accelerates the course of Huntington's disease, suggesting that oxidative stress plays a major part in the etiology of the disease [54]. Cozzolino et al., (2018) found that the development of ALS is influenced by mitochondrial dysfunction. This highlights the importance of mitochondrial function for brain health and provides insight into potential future treatments for ALS [55]. According to Zhang et al., (2019), ROS produced by NADPH oxidase hastens the onset of multiple sclerosis. This raises the possibility that oxidative stress contributes to the genesis of disease and identifies potential targets for treatment. [56]. Li et al., (2019) found that Friedreich's ataxia progression is associated with mitochondrial dysfunction. This highlights the significance of mitochondrial health in neurodegenerative diseases and provides insight into potential treatment approaches [57]. According to Lin et al., (2020), spinocerebellar ataxia develops when antioxidant defenses are compromised. This highlights the importance of antioxidant defenses for brain health and provides information on potential therapeutic approaches [58].

4 Oxidative Stress and Lifestyle of Human

Exposure to toxins and pollutants can lead to the formation of pro-oxidants, causing DNA damage and disrupting enzymatic activity, which in turn affects gene expression. Modern lifestyles, including consumption of processed foods and exposure to chemicals, contribute to oxidative stress. This stress is linked to various chronic diseases, reactive oxygen species play a role in disease development. Free radicals are essential for certain biological processes, such as the destruction of microorganisms by immune cells. They also participate in cellular signaling pathways, helping to maintain homeostasis. At moderate levels, ROS regulate various cellular functions, research has shown that factors like age and smoking significantly impact oxidative stress markers, such as urine 8-isoprostane. Age also affects urine 8-OHdG levels [59]. Research has shown that factors influencing oxidative stress markers vary between males and females. In males, body weight significantly affects urine 8-OHdG levels, whereas in females, age, smoking, body weight, and levels of Hs-CRP and LDL-cholesterol have a substantial impact. Physical inactivity has a profound metabolic impact on the human body, affecting multiple organ systems and increasing the risk of chronic diseases like Type II diabetes, cancer, and cardiovascular disease. Conversely, regular exercise can improve metabolic status, insulin sensitivity, and weight management. However, oxidative stress during exercise may increase with age due to various cellular and biochemical changes. These changes can limit the body's ability to defend against oxidative damage. Certain dietary components, such as antioxidants found in red wine and the Mediterranean diet, have been shown to help mitigate oxidative stress. The combination of these dietary elements can increase levels of beneficial compounds like beta-carotene, vitamin C, and polyphenols in the body [60].

5 Antioxidant Capacity of Healthy Subjects

Antioxidants can be categorized based on several key characteristics. These include their origin (endogenous or exogenous), solubility (hydrosoluble or liposoluble), and function (primary or secondary). Primary antioxidants typically work by neutralizing free radicals, while secondary antioxidants employ various mechanisms, such as absorbing UV radiation, chelating metals, or inhibiting oxidative enzymes. Understanding these classifications can provide insight into how antioxidants function and their potential applications. Researchers used a biological test to assess the impact of lifestyle factors on overall antioxidant status. The test measured the resistance of whole blood to free-radical damage, aiming to identify factors that contribute to variations in antioxidant levels. However, a lower antioxidant potential was linked to moderate consumption of vegetables, low consumption of fruit, low consumption of fish, alcohol consumption, psychological stress, tobacco smoking, and, to a lesser extent, high exposure to natural UV light. Thus, it was determined which *"unhealthy"* and *"healthy"* lifestyle choices were responsible for the variation in each person's antioxidant capacity. The study found that certain lifestyle factors linked to cancer and cardiovascular disease risk were associated with lower antioxidant levels. Assessing an individual's resistance to oxidative stress could help identify those at risk for diseases related to free radical damage. Oxidative stress is also thought to play a role in aging, contributing to cellular damage and decline. Some research suggests that oxidative stress can impair cellular function, disrupt gene regulation, and reduce energy production. Additionally, artificial antioxidants are commonly used in food preservation. These molecules are added to food to stop lipid and protein oxidation, and their antioxidant effects may spread throughout the body. We only found a weak link between increased antioxidant potential and high fruit consumption; a lower individual antioxidant capacity is probably linearly related to moderate vegetable intake. According to this research, lowering dietary antioxidants is a more straightforward method of impairing the regulated antioxidant defense system than raising them [61].

6 Factors Affecting Antioxidants of Human Health

UV radiation is a key factor in generating reactive oxygen species (ROS) in the skin and hair. This type of radiation is divided into Ultraviolet A radiation (UVA), Ultraviolet B radiation (UVB), and Ultraviolet C radiation (UVC), with the ozone layer absorbing most UVC. UVA and UVB have different effects on the skin, with UVB's shorter wavelengths causing damage to the skin's surface and UVA's longer wavelengths penetrating deeper into skin tissues. Like skin, UVA rays can cause up to five times as much damage by penetrating deeper into the cortex of hair shafts. While UVB-induced protein loss results in more superficial morphological damage to the cuticle of hair shafts, UVA-induced biochemical damage alters the color of hair. Pollution is a major environmental factor that also contributes to the generation of ROS. One of the main sources of ultrafine particles—defined as particles with a diameter of less than 0.1 μm—is vehicle exhaust emissions. In particular, ultrafine particles have been shown to settle in the mitochondria of macrophages and epithelial cells and have a greater ability to enter tissues. Although the majority of environmental pollution agents can produce ROS, the type of particulate matter is influenced by factors such as geographic location, mild climate, and chemical composition. The impact of lifestyle factors, such as diet and exercise, on oxidative stress has been a subject of research. Physical activity, for instance, can stimulate the production of ROS, although the underlying mechanisms are complex and multifaceted. The relationship between exercise and oxidative stress may depend on various factors, including the type and intensity of exercise, and further research is needed to fully understand this dynamic as shown in Table 12 [62]. A growing body of evidence suggests that diet plays a crucial role in mitigating oxidative stress. Certain nutrients, including vitamins A, C, and E, as well as carotenoids found in fruits and vegetables, have been shown to have antioxidant properties. These compounds can help protect cells from damage by neutralizing free radicals and reducing oxidative reactions. Conversely, exposure to environmental stressors like smoking and UV radiation can exacerbate oxidative stress [63].

In fact, Leslie et al., (2017) examined the effects of antiresorptive therapy on women's fracture risk in relation to changes in trabecular bone score (TBS). Their results suggest that TBS may not be a reliable indicator of fracture risk in this population since changes in TBS do not predict fracture risk in women [64]. The importance of galectin 9 in experimental lupus models was discussed by Wada and Zeggar (2018), who emphasized the disease's complexity. The necessity of further research into the mechanisms underlying the significance of galectin 9 in lupus was emphasized in their discussion [65]. Additionally, unhealthy lifestyle choices like drinking and smoking have been connected to oxidative stress levels 2. A byproduct of the metabolism of ethanol, acetaldehyde, adds to the increase in ROS brought on by both acute and long-term alcohol consumption. Cigarette smoking, which is also known to accelerate the aging of the skin and hair, is one particular health risk factor. H_2O_2 and $\cdot OH$, two free radical compounds found in cigarette smoke, can create intracellular ROS and are inhaled directly into the respiratory system. Similar to alcohol, cigarette smoke reduces antioxidant availability, which increases oxidative stress levels. Parkes et al., (2023) investigated the association between ulcerative colitis long-term clinical and patient-reported outcomes and mucosal healing endpoints. Their results showed a strong correlation, highlighting how crucial mucosal healing is to ulcerative colitis patients' successful outcomes [66]. Chang and colleagues

Table 12 Diseases of the bones and their behavioral effects

Disease	ROS source	Finding	Refs.
Osteoporosis	NADPH oxidase	Increased ROS production	[64]
Rheumatoid arthritis	Mitochondrial respiratory chain	Impaired mitochondrial function	[65]

(2022) examined metabolic reprogramming in gastric cancer in response to mitochondrial DNA alterations and mitochondrial dysfunction. Their research revealed new treatment targets by demonstrating the critical role metabolic reprogramming plays in the development and spread of gastric adenocarcinoma as shown in Table 13 [67]. From conception to death, Metcalfe and Alonso-Alvarez (2010) examined the effects of oxidative stress on phenotypes. Research highlights the importance of the interplay between antioxidant defenses and ROS in shaping an organism's life trajectory. Oxidative stress can have devastating effects, and studying its causes and consequences can provide valuable insights into the aging process and disease development [68]. Cells produce an enzyme called superoxide dismutase (SOD) that helps neutralize harmful substances. SOD converts superoxides into hydrogen peroxide, which can still harm cells if not further neutralized. Fortunately, other antioxidants like glutathione and catalase can render hydrogen peroxide harmless by converting it into water. However, the presence of certain metal ions can interfere with this process, leading to the formation of more toxic substances [68].

Lipopolysaccharide stimulates the development of mouse endometriosis-like lesions through the nuclear factor-kappa B pathway, according to research by Azuma et al., (2017). This study highlights the potential role that immunological responses and inflammation may play in the development of endometriosis [69]. The relationship between hypoxia, epithelial-mesenchymal transition, and TET-mediated epigenetic modifications was examined by Kao et al., (2016). According to their research, hypoxia can result in epithelial-mesenchymal transition and epigenetic changes, both of which may hasten the course of disease [70]. The Nrf2-Keap1 pathway is a key regulator of antioxidant expression as shown in Table 14. When not under stress, Nrf2 is degraded due to its association with Keap1. During oxidative stress, Nrf2 dissociates from Keap1, translocate to the nucleus, and triggers the production of antioxidants and mitochondrial defense genes.

In their study of the functional vascular contributions to cognitive impairment and dementia (VCID), Toth et al., (2016) emphasized the importance of endothelial damage, neurovascular uncoupling in aging, and cerebral autoregulatory dysfunction [71]. With an emphasis on how aging contributes to the development of the

Table 13 Guide to ordering gastroenterological diseases

Disease	ROS source	Finding	Refs.
Inflammatory bowel disease	NADPH oxidase	Increased ROS production	[66]
Gastric cancer impaired	Mitochondrial respiratory chain	Mitochondrial function	[67]

Table 14 Reproductive diseases and their functioning

Disease	ROS source	Finding	Refs.
Endometriosis	NADPH oxidase	Increased ROS production	[69]
Male infertility	Mitochondrial respiratory chain	Impaired mitochondrial function	[70]

Table 15 Order for cardiovascular diseases

Disease	ROS source	Finding	Refs.
Atherosclerosis	NADPH oxidase	Increased ROS production	[71]
Hypertension	Mitochondrial respiratory chain	Impaired mitochondrial function	[72]
Heart failure	NADPH oxidase	Elevated ROS levels	[73]
Cardiac arrhythmias	Mitochondrial respiratory chain	Impaired mitochondrial function	[74]
Stroke	NADPH oxidase	Increased ROS production	[75]
Cardiomyopathy	Mitochondrial respiratory chain I	Impaired mitochondrial function	[76]
Atrial fibrillation	NADPH oxidase	Elevated ROS levels	[77]

disease, Fang et al., (2025) investigated the molecular mechanisms and available treatments for systemic aging-related heart failure [72]. Zimmerman et al., (2004) highlighted the importance of this pathway in blood pressure regulation by finding that Rac1-dependent NADPH oxidase is necessary for angiotensin II's cardiovascular and dipsogenic effects in the brain as shown in Table 15 [73]. The importance of mitochondrial dysfunction in raising arrhythmic triggers and substrates, as well as potential anti-arrhythmic drug targets, was investigated by Saadeh and Fazmin (2021) [74]. According to Heymes et al., (2003), myocardial NADPH oxidase activity rises in heart failure in humans, suggesting that this enzyme may contribute to the onset of the condition [75]. The importance of AMPK in controlling mitochondrial activity and cardiovascular health was highlighted by Wu and Zou's (2020) investigation of the relationship between AMPK, mitochondrial function, and cardiovascular disease [76].

7 Conclusion

Throughout evolution, one of the major mechanisms relating to human lifespan and aging Zou et al., (2020). Heart disease, mitochondrial activity, and AMPK, Molecular Sciences International Journal. The stress response needs to be carefully controlled for healthy aging because it can directly impair cells and tissues and could be the first stage of molecular pathogenesis. In various diseases, an excessive oxidative stress response can trigger or worsen the condition, much like its impact on neurological disorders where it contributes to cell death. This highlights the importance of balancing oxidative stress responses to prevent disease progression. Because oxidative stress is made worse by environmental stressors like pollution, UVR, and lifestyle choices, people have begun including exogenous antioxidants in their diets.

References

1. Sies, H., Berndt, C., & Jones, D. P. (2017). Oxidative stress. *Annual Review of Biochemistry, 86*(1), 715–748.
2. Preiser, J. C. (2012). Oxidative stress. *Journal of Parenteral and Enteral Nutrition, 36*(2), 147–154.
3. Di Meo, S., Reed, T. T., Venditti, P., & Victor, V. M. (2016). Role of ROS and RNS sources in physiological and pathological conditions. *Oxidative Medicine and Cellular Longevity, 2016*(1), 1245049.
4. Rojas-Gutierrez, E., Muñoz-Arenas, G., Treviño, S., Espinosa, B., Chavez, R., Rojas, K., Flores, G., Díaz, A., & Guevara, J. (2017). Alzheimer's disease and metabolic syndrome: A link from oxidative stress and inflammation to neurodegeneration. *Synapse (New York, N. Y.), 71*(10), Article e21990.
5. Forman, H. J., & Zhang, H. (2021). Targeting oxidative stress in disease: Promise and limitations of antioxidant therapy. *Nature Reviews Drug Discovery, 20*(9), 689–709.
6. Wei, L., Li, Z., Guo, S., Ma, H., Shi, Y., An, X., Huang, K., Xiong, L., Xue, T., Zhang, Z., & Yao, K. (2024). Human papillomavirus infection affects treatment outcomes and the immune microenvironment in patients with advanced penile squamous cell carcinoma receiving programmed cell death protein 1 inhibitor–based combination therapy. *Cancer, 130*(9), 1650–1662.
7. Arumugasamy, S. K., Chellasamy, G., Murugan, N., Govindaraju, S., Yun, K., & Choi, M. J. (2024). Synthesis and surface engineering of Ag chalcogenide quantum dots for near-infrared biophotonic applications. In *Advances in colloid and interface science* (p. 103245).
8. Manjceevan, A., & Bandara, J. (2016). Robust surface passivation of trap sites in PbS q-dots by controlling the thickness of CdS layers in PbS/CdS quantum dot solar cells. *Solar Energy Materials and Solar Cells, 147*, 157–163.
9. Georgiou, N. A., Garssen, J., & Witkamp, R. F. (2011). Pharma–nutrition interface: The gap is narrowing. *European Journal of Pharmacology, 651*(1–3), 1–8.
10. Lopez-Crisosto, C., Pennanen, C., Vasquez-Trincado, C., Morales, P. E., Bravo-Sagua, R., Quest, A. F., Chiong, M., & Lavandero, S. (2017). Sarcoplasmic reticulum–mitochondria communication in cardiovascular pathophysiology. *Nature Reviews Cardiology, 14*(6), 342–360.
11. Liang, W., He, X., Bi, J., Hu, T., & Sun, Y. (2021). Role of reactive oxygen species in tumors based on the 'seed and soil'theory: A complex interaction. *Oncology Reports, 46*(3), 208.
12. Ushio-Fukai, M., & Nakamura, Y. (2008). Reactive oxygen species and angiogenesis: NADPH oxidase as target for cancer therapy. *Cancer Letters, 266*(1), 37–52.
13. Neganova, M., Liu, J., Aleksandrova, Y., Klochkov, S., & Fan, R. (2021). Therapeutic influence on important targets associated with chronic inflammation and oxidative stress in cancer treatment. *Cancers, 13*(23), 6062.
14. Fontana, F., Anselmi, M., & Limonta, P. (2023). Unraveling the peculiar features of mitochondrial metabolism and dynamics in prostate cancer. *Cancers, 15*(4), 1192.
15. Liu, H. M., Cheng, M. Y., Xun, M. H., Zhao, Z. W., Zhang, Y., Tang, W., Cheng, J., Ni, J., & Wang, W. (2023). Possible mechanisms of oxidative stress-induced skin cellular senescence, inflammation, and cancer and the therapeutic potential of plant polyphenols. *International Journal of Molecular Sciences, 24*(4), 3755.
16. Wang, Y. (2020). *The metabolic role of YAP in liver regeneration* (Doctoral dissertation, The Chinese University of Hong Kong (Hong Kong)).
17. Alves-Silva, J. M., Pedreiro, S., Cavaleiro, C., Cruz, M. T., Figueirinha, A., & Salgueiro, L. (2023). Effect of *Thymbra capitata* (L.) Cav. on inflammation, senescence and cell migration. *Nutrients, 15*(8), 1930.
18. Arora, Y., Kumar, M., & Kumar, D. (2024). The role of pesticides in cancer initiation and progression: A comprehensive review. *Chemical Biology Letters, 11*(3), 669–669 (Oxidative stressors and their impact on living organisms)

19. Valavanidis, A., Vlahogianni, T., Dassenakis, M., & Scoullos, M. (2006). Molecular biomarkers of oxidative stress in aquatic organisms in relation to toxic environmental pollutants. *Ecotoxicology and Environmental Safety, 64*(2), 178–189.
20. Gelderman, K. A., Hultqvist, M., Olsson, L. M., Bauer, K., Pizzolla, A., Olofsson, P., & Holmdahl, R. (2007). Rheumatoid arthritis: The role of reactive oxygen species in disease development and therapeutic strategies. *Antioxidants & Redox Signaling, 9*(10), 1541–1568.
21. Poznyak, A. V., Orekhov, N. A., Churov, A. V., Starodubtseva, I. A., Beloyartsev, D. F., Kovyanova, T. I., Sukhorukov, V. N., & Orekhov, A. N. (2024). Mitochondrial dysfunction in systemic lupus erythematosus: Insights and therapeutic potential. *Diseases, 12*(9), 226.
22. Noordzij, W., de Jong, K. P., & de Meijer, V. E. (2018). Response to: Positron emission tomography/computed tomography with 18F-fluorocholine improve tumor staging and treatment allocation in patients with hepatocellular carcinoma. *Journal of Hepatology, 69*(2), 554–555.
23. Tsunoda, T., Kakinuma, S., Miyoshi, M., Kamiya, A., Kaneko, S., Sato, A., Tsuchiya, J., Nitta, S., Kawai-Kitahata, F., Murakawa, M., & Itsui, Y. (2019). Loss of fibrocystin promotes interleukin-8-dependent proliferation and CTGF production of biliary epithelium. *Journal of Hepatology, 71*(1), 143–152.
24. Zipursky, S., Patel, M., Farrell, M., Gonzalez, A. R., Kachra, T., Folly, Y., Kurji, F., Veira, C. L., Wootton, E., & Hampton, L. M. (2017). Lessons learned from managing the planning and implementation of inactivated polio vaccine introduction in support of the polio endgame. *The Journal of Infectious Diseases, 216*(suppl_1), S15–S23.
25. Brealey, D., Brand, M., Hargreaves, I., Heales, S., Land, J., Smolenski, R., Davies, N. A., Cooper, C. E., & Singer, M. (2002). Association between mitochondrial dysfunction and severity and outcome of septic shock. *The Lancet, 360*(9328), 219–223.
26. Davies, S., Kattel, R., Bhatia, B., Petherwick, A., & Chapman, T. (2005). The effect of diet, sex and mating status on longevity in Mediterranean fruit flies (Ceratitis capitata), Diptera: Tephritidae. *Experimental Gerontology, 40*(10), 784–792.
27. Wang, G. F., Dong, Q., Bai, Y., Yuan, J., Xu, Q., Cao, C., & Liu, X. (2017). Oxidative stress induces mitotic arrest by inhibiting Aurora A-involved mitotic spindle formation. *Free Radical Biology and Medicine, 103*, 177–187.
28. Anderson, E. J., Kypson, A. P., Rodriguez, E., Anderson, C. A., Lehr, E. J., & Neufer, P. D. (2009). Substrate-specific derangements in mitochondrial metabolism and redox balance in the atrium of the type 2 diabetic human heart. *Journal of the American College of Cardiology, 54*(20), 1891–1898.
29. Cosentino, F., Eto, M., De Paolis, P., van der Loo, B., Bachschmid, M., Ullrich, V., Kouroedov, A., Delli Gatti, C., Joch, H., Volpe, M., & Lüscher, T. F. (2003). High glucose causes upregulation of cyclooxygenase-2 and alters prostanoid profile in human endothelial cells: Role of protein kinase C and reactive oxygen species. *Circulation, 107*(7), 1017–1023.
30. Borri, M., Jacobs, M. E., Carmeliet, P., Rabelink, T. J., & Dumas, S. J. (2025). Endothelial dysfunction in the aging kidney. *American Journal of Physiology-Renal Physiology.*
31. Tangvarasittichai, O., & Tangvarasittichai, S. (2018). Oxidative stress, ocular disease and diabetes retinopathy. *Current Pharmaceutical Design, 24*(40), 4726–4741.
32. Feldman, E. L., Callaghan, B. C., Pop-Busui, R., Zochodne, D. W., Wright, D. E., Bennett, D. L., Bril, V., Russell, J. W., & Viswanathan, V. (2019). Diabetic neuropathy. *Nature Reviews Disease Primers, 5*(1), 41.
33. Wang, D., Li, Y., Wang, N., Luo, G., Wang, J., Luo, C., Yu, W., & Hao, L. (2020). *1α, 25-dihydroxyvitamin D3 prevents renal oxidative damage via the PARP1/SIRT1/NOX4 pathway in Zucker diabetic fatty rats.*
34. Bondi, C. D., et al. (2018). NADPH oxidase-derived reactive oxygen species contribute to kidney fibrosis. *American Journal of Physiology-Renal Physiology, 315*(4), F734–F743. https://doi.org/10.1152/ajprenal.00228.2018
35. Sharma, K., et al. (2013). Mitochondrial dysfunction in diabetic kidney disease. *Seminars in Nephrology, 33*(4), 393–404. https://doi.org/10.1016/j.semnephrol.2013.05.007
36. Wei, Q., et al. (2016). NADPH oxidase contributes to hypertensive nephrosclerosis. *American Journal of Hypertension, 29*(5), 621–628. https://doi.org/10.1093/ajh/hpv187

37. Wang, X., et al. (2019). Mitochondrial dysfunction in glomerulonephritis. *Journal of Nephrology, 32*(3), 347–355. https://doi.org/10.1007/s40620-019-00603-9
38. Li, M., et al. (2018). NADPH oxidase promotes renal cell carcinoma progression. *Oncotarget, 9*(38), 25171–25184. https://doi.org/10.18632/oncotarget.25341
39. Kaneko, H., et al. (2017). NADPH oxidase-derived reactive oxygen species contribute to age-related macular degeneration. *Investigative Ophthalmology & Visual Science, 58*(10), 6389–6398. https://doi.org/10.1167/iovs.17-21942
40. Tawfik, A., et al. (2018). Mitochondrial dysfunction contributes to diabetic retinopathy progression. *American Journal of Pathology, 188*(11), 2599–2613. https://doi.org/10.1016/j.ajpath.2018.07.014
41. Sivamani, R. K., et al. (2019). NADPH oxidase-derived reactive oxygen species contribute to psoriasis progression. *Journal of Investigative Dermatology, 139*(1), 211-219.e3. https://doi.org/10.1016/j.jid.2018.05.035
42. Dell'Anna, M. L., et al. (2018). Mitochondrial dysfunction in vitiligo: A potential role in disease pathogenesis. *Journal of Investigative Dermatology, 138*(1), 141-148.e2. https://doi.org/10.1016/j.jid.2017.03.037
43. Azzouz, D., & Palaniyar, N. (2024). How do ROS induce NETosis? Oxidative DNA damage, DNA repair, and chromatin decondensation. *Biomolecules, 14*(10), 1307.
44. Mutchler, A. L., Haynes, A. P., Saleem, M., Jamison, S., Khan, M. M., Ertuglu, L., & Kirabo, A. (2025). Epigenetic regulation of innate and adaptive immune cells in salt-sensitive hypertension. *Circulation Research, 136*(2), 232–254.
45. Park, J. H., Rivière, I., Gonen, M., Wang, X., Sénéchal, B., Curran, K. J., Sauter, C., Wang, Y., Santomasso, B., Mead, E., & Roshal, M. (2018). Long-term follow-up of CD19 CAR therapy in acute lymphoblastic leukemia. *New England Journal of Medicine, 378*(5), 449–459.
46. Divangahi, M. (2018). Are tolerance and training required to end TB? *Nature Reviews Immunology, 18*(11), 661–663.
47. Matsui, R., Ferran, B., Oh, A., Croteau, D., Shao, D., Han, J., Pimentel, D. R., & Bachschmid, M. M. (2020). Redox regulation via glutaredoxin-1 and protein S-glutathionylation. *Antioxidants & Redox Signaling, 32*(10), 677–700.
48. Guo, Z. Q., Gu, S. Y., Tian, Z. H., & Du, B. Y. (2024). A comprehensive review of Mycoplasma pneumoniae infection in chronic lung diseases: Recent advances in understanding asthma, COPD, and bronchiectasis. *Frontiers in Medicine, 11*, 1437731.
49. Li, L., Chen, Z., Fu, W., Cai, S., & Zeng, Z. (2018). Emerging evidence concerning the role of sirtuins in sepsis. *Critical Care Research and Practice, 2018*(1), 5489571.
50. Wink, D. A., Hines, H. B., Cheng, R. Y., Switzer, C. H., Flores-Santana, W., Vitek, M. P., Ridnour, L. A., & Colton, C. A. (2011). Nitric oxide and redox mechanisms in the immune response. *Journal of Leukocyte Biology, 89*(6), 873–891.
51. Chen, X., Guo, C., & Kong, J. (2012). Oxidative stress in neurodegenerative diseases. *Neural Regeneration Research, 7*(5), 376–385.
52. Ansari, et al. (2017). Cytochrome P450-derived ROS contributes to Alzheimer's disease progression. *Neurobiology of Aging, 53*, 111–122.
53. Lin, et al. (2017). Mitochondrial dysfunction contributes to Parkinson's disease progression. *Journal of Neurochemistry, 140*(3), 435–446.
54. Klepac, et al. (2017). NADPH oxidase-derived ROS promotes Huntington's disease progression. *Human Molecular Genetics, 26*(10), 1841–1853.
55. Cozzolino, et al. (2018). Mitochondrial dysfunction contributes to amyotrophic lateral sclerosis progression. *Neurology, 90*(11), e957–e966.
56. Zhang, et al. (2019). NADPH oxidase-derived ROS promotes multiple sclerosis progression. *Journal of Neuroimmunology, 337*, 577–585.
57. Li, et al. (2019). Mitochondrial dysfunction contributes to Friedreich's ataxia progression. *Neurobiology of Disease, 130*, Article 104483.
58. Lin, et al. (2020). Impaired antioxidant defenses contribute to spinocerebellar ataxia progression. *Neurochemistry International, 137*, Article 104718.

59. Dato, S., Crocco, P., D'Aquila, P., De Rango, F., Bellizzi, D., Rose, G., & Passarino, G. (2013). Exploring the role of genetic variability and lifestyle in oxidative stress response for healthy aging and longevity. *International Journal of Molecular Sciences, 14*(8), 16443–16472.
60. Sharifi-Rad, M., Anil Kumar, N. V., Zucca, P., Varoni, E. M., Dini, L., Panzarini, E., Rajkovic, J., Tsouh Fokou, P. V., Azzini, E., Peluso, I., & Prakash Mishra, A. (2020). Lifestyle, oxidative stress, and antioxidants: Back and forth in the pathophysiology of chronic diseases. *Frontiers in Physiology, 11*, Article 552535.
61. Protano, C., Andreoli, R., Mutti, A., Petti, S., & Vitali, M. (2014). Biomarkers of oxidative stress to nucleic acids: Background levels and effects of body mass index and life-style factors in an urban paediatric population. *Science of the Total Environment, 500*, 44–51.
62. Carraro, E., Schilirò, T., Biorci, F., Romanazzi, V., Degan, R., Buonocore, D., Verri, M., Dossena, M., Bonetta, S., & Gilli, G. (2018). Physical activity, lifestyle factors and oxidative stress in middle age healthy subjects. *International Journal of Environmental Research and Public Health, 15*(6), 1152.
63. Pisoschi, A. M., & Pop, A. (2015). The role of antioxidants in the chemistry of oxidative stress: A review. *European Journal of Medicinal Chemistry, 97*, 55–74.
64. Leslie, W. D., Majumdar, S. R., Morin, S. N., Hans, D., & Lix, L. M. (2017). Change in trabecular bone score (TBS) with antiresorptive therapy does not predict fracture in women: The Manitoba BMD cohort. *Journal of Bone and Mineral Research, 32*(3), 618–623.
65. Wada, J., & Zeggar, S. (2018). Perplexing role of galectin 9 in experimental lupus models: Comment on the article by Zeggar et al Reply. *Arthiritis & Rheumatology, 70*(9), 1531–1532.
66. Parkes, G., Ungaro, R. C., Danese, S., Abreu, M. T., Arenson, E., Zhou, W., Ilo, D., Laroux, F. S., Deng, H., Sanchez Gonzalez, Y., & Peyrin-Biroulet, L. (2023). Correlation of mucosal healing endpoints with long-term clinical and patient-reported outcomes in ulcerative colitis. *Journal of Gastroenterology, 58*(10), 990–1002.
67. Chang, T. C., Lee, H. T., Pan, S. C., Cho, S. H., Cheng, C., Ou, L. H., Lin, C. I., Lin, C. S., & Wei, Y. H. (2022). Metabolic reprogramming in response to alterations of mitochondrial DNA and mitochondrial dysfunction in gastric adenocarcinoma. *International Journal of Molecular Sciences, 23*(3), 1857.
68. Metcalfe, N. B., & Alonso-Alvarez, C. (2010). Oxidative stress as a life-history constraint: The role of reactive oxygen species in shaping phenotypes from conception to death. *Functional Ecology, 24*(5), 984–996.
69. Azuma, Y., Taniguchi, F., Nakamura, K., Nagira, K., Khine, Y. M., Kiyama, T., Uegaki, T., Izawa, M., & Harada, T. (2017). Lipopolysaccharide promotes the development of murine endometriosis-like lesions via the nuclear factor-kappa B pathway. *American Journal of Reproductive Immunology, 77*(4), Article e12631.
70. Kao, S. H., Wu, K. J., & Lee, W. H. (2016). Hypoxia, epithelial-mesenchymal transition, and TET-mediated epigenetic changes. *Journal of clinical medicine, 5*(2), 24.
71. Toth, P., Tarantini, S., Csiszar, A., Ungvari, Z., & Zoltan Ungvari, M. D. (2016). *Functional vascular contributions to cognitive impairment and dementia (VCID): Mechanisms and consequences of cerebral autoregulatory dysfunction, endothelial impairment and 2 neurovascular uncoupling in aging 3.*
72. Fang, Z., Raza, U., Song, J., Lu, J., Yao, S., Liu, X., Zhang, W., & Li, S. (2025). Systemic aging fuels heart failure: Molecular mechanisms and therapeutic avenues. *ESC Heart Failure, 12*(2), 1059–1080.
73. Zimmerman, M. C., Dunlay, R. P., Lazartigues, E., Zhang, Y., Sharma, R. V., Engelhardt, J. F., & Davisson, R. L. (2004). Requirement for Rac1-dependent NADPH oxidase in the cardiovascular and dipsogenic actions of angiotensin II in the brain. *Circulation Research, 95*(5), 532–539.
74. Saadeh, K., & Fazmin, I. T. (2021). Mitochondrial dysfunction increases arrhythmic triggers and substrates; potential anti-arrhythmic pharmacological targets. *Frontiers in Cardiovascular Medicine, 8*, Article 646932.
75. Heymes, C., Bendall, J. K., Ratajczak, P., Cave, A. C., Samuel, J. L., Hasenfuss, G., & Shah, A. M. (2003). Increased myocardial NADPH oxidase activity in human heart failure. *Journal of the American College of Cardiology, 41*(12), 2164–2171.

76. Wu, S., & Zou, M. H. (2020). AMPK, mitochondrial function, and cardiovascular disease. *International Journal of Molecular Sciences, 21*(14), 4987.
77. Liu, X., Fu, Y., Wang, J., Wu, D., Li, S., Wang, C., Yang, Z., & Zhou, E. (2021). β-Conglycinin induces the formation of neutrophil extracellular traps dependent on NADPH oxidase-derived ROS, PAD4, ERK1/2 and p38 signaling pathways in mice. *Food & Function, 12*(1), 154–161.

Recent Advances in the Use of Antioxidants as Preservative Agents in Food, Fodder, and Cosmetics

Pandian Chithamparam, Harsath Ram Govindan Radhakrishnan, and Azhaguchamy Muthukumaran

Abstract Antioxidants are the agents that prevent the oxidation of molecules in the metabolism of living things. Generally, antioxidants play an essential role in inhibiting oxidation by the release mechanism of a singlet oxygen atom, and that oxygen atom can neutralize another nascent oxygen atom because the nascent oxygen can form free radicals. These free radicals are a chain-like reaction; they can collapse the proper metabolic activity. Hence, the antioxidants are essential to regulate the metabolism. These antioxidants are broadly divided into two groups based on their sources. One is natural and another one is synthetic. The natural antioxidants fall into specific subdivisions based on their nature of origin. Some vitamins act as antioxidants, like vitamins A, B-complex, C, D, and E. Minerals act as antioxidants, such as selenium, copper, iron, and zinc. Flavonoids, phenols, and tannin-like phytochemicals are also examples of antioxidants. This chapter reviewed how antioxidants act as a food, fodder, and cosmetics preserving agent. Antioxidants restrict oxidation and eradicate the free radical formation in living things; hence, they can preserve biologically based things and oxidize foods and oxidizable things. In the food industries, some vitamins and selenium-like minerals are used as preserving agents to resist free radical formation and maintain their physiological characteristics. In animal food, antioxidants, including vitamin E and polyphenols, are used as preservatives for fodder. In cosmetics, polyphenols, vitamins, and selenium play an antioxidative role. The antioxidants are nontoxic and essential to living things to maintain metabolism and the quality of the products. In some cases, these antioxidants are also used as packing material for food products. Hence, antioxidants are irreplaceable for preserving food, fodder, and cosmetics.

Keywords Antioxidants are nontoxic and essential to life · Metabolic function · Product quality and reliability · Indispensable for preserving food · Cosmetics · Fodder · Livestock feed

---

P. Chithamparam · H. R. G. Radhakrishnan · A. Muthukumaran (✉)
Department of Biotechnology, Kalasalingam Academy of Research and Education, Krishnankoil, Tamil Nadu, India
e-mail: a.muthukumaran@klu.ac.in

K. Kasinathan and A. Muthukumaran (eds.), *Recent Advances in Oxidative Stress Associated Chronic Diseases Volume 2*, https://doi.org/10.1007/978-981-95-3758-7_9

1 Introduction

Antioxidants can oxidize themselves in place of or in addition to other molecules. These substances or frameworks can engage with free radicals and discontinue the chain reaction before they damage essential components [11]. It is widely acknowledged in biology that an antioxidant is any molecule that can slow down or stop the action of oxidants [4]. Oxidation is transferring electrons from a molecule to an oxidizing substance. Free radicals are highly reactive, with one or more unpaired electrons in their outermost shell. Once produced, the chain reaction begins [26]. Antioxidants interact with these free radicals, oxidizing themselves to stop further oxidation reactions and stopping the chain reaction by eliminating free radical intermediates [11]. Free radicals are chemical species with one or more unpaired electrons in their last electron layer. These include reactive oxygen species (ROS), superoxide radicals (O_2), hydroxyl radical (HO), and peroxyl radical (ROO), hydrogen peroxide (H_2O_2), and singlet oxygen ($^1O_2$), which are frequently produced as byproducts of biological reactions, primarily due to exposure to pollution and ultraviolet radiation [12]. The process consists of three steps: initiation, propagation, and termination. The first step involves the formation of radicals, while the second involves their reaction with other molecules. Finally, the third step involves their transformation into other products [6]. Antioxidants prevent free radical damage and promote overall health [46]. These antioxidants offer defense against harm from free radicals, which have been linked to the emergence of numerous chronic illnesses, such as cancer, inflammation, heart disease, aging, cardiovascular diseases, and anemia [49]. The perfect antioxidant should bind redox metals at physiologically appropriate amounts, remove free radicals, and be easily absorbed. It should work in both aqueous and membrane domains and positively impact gene expression [23]. When an antioxidant encounters a free radical in the bloodstream, it naturally mixes and converts it to harmless water and oxygen [16].

2 Antioxidants Can Function on Three Levels in the Body

- Prevention: minimizing the generation of reactive species, such as desferrioxamine,
- Interception—Scavenging reactive species with catalytic and non-catalytic molecules such as ascorbic acid and alpha-tocopherol,
- Repairing damaged target molecules like glutathione [22].

Antioxidants are also commonly used in dietary supplements to uphold health and avoid diseases like cancer and coronary heart disease [16]. Antioxidants have numerous industrial applications, including food and cosmetic preservatives, and they prevent rubber and fuel degradation [16]. Leukocytes and other phagocytes use potent oxidants such as NO, O_2, H_2O_2, and OCl to kill bacteria, parasites, and viruses. However, if left unchecked, these phagocytes can cause oxidative damage, DNA mutation, and carcinogenesis [4].

3 Types of Antioxidants

The antioxidants are classified into two types based on their origins they are shown in Fig. 1, respectively, (i) natural antioxidants (ii) synthetic antioxidants.

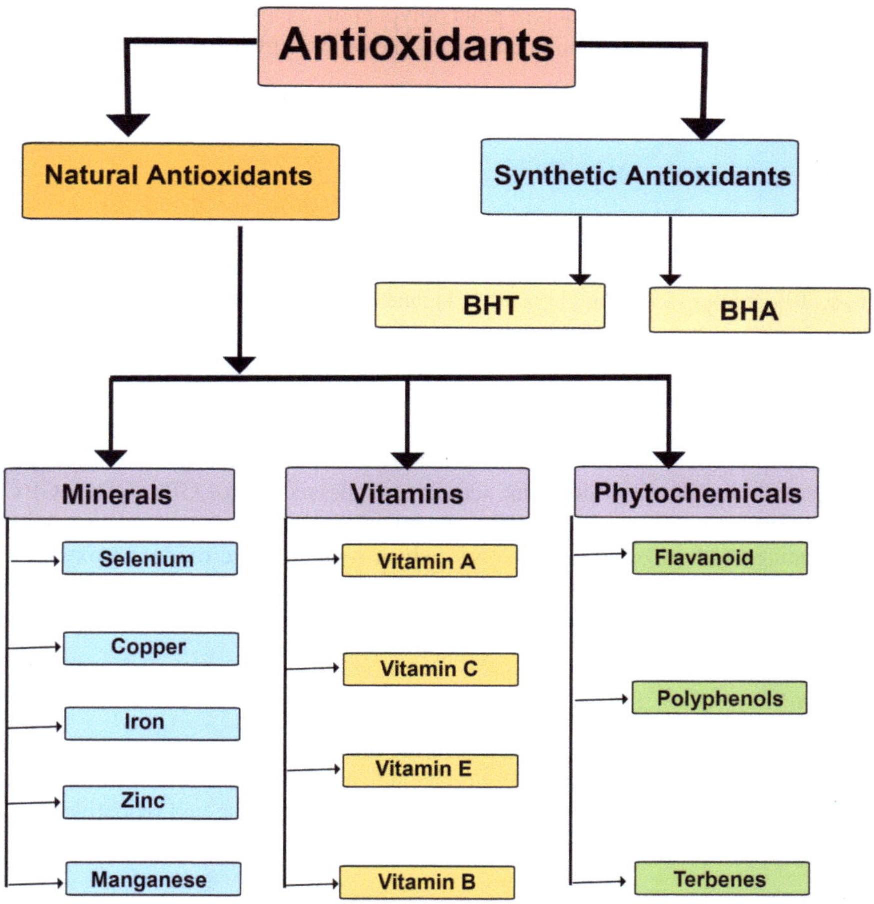

Fig. 1 Types and subtypes of natural and synthetic antioxidants

3.1 Natural Antioxidants

Natural antioxidants are compounds in food that stop reactions like sourness, color change, and disruption. Plant species, diversity, extraction and/or processing techniques, and growth circumstances all affect the activity of natural antioxidants primarily sourced from plant sources. They can be found in all plants, animal tissues, and microbes. High antioxidant activity is found in natural antioxidants, particularly in red, orange, and purple fruits and vegetables [49]. Antioxidants can be found in various foods, such as grains, legumes, fruits, vegetables, tea, and wine. The antioxidants and polyphenols found in dietary plants have been linked to their preventative qualities. Secondary metabolites of plant polyphenols have over 8000 structural variations and comprise many compounds with one or more hydroxyl moieties attached to their aromatic ring(s) [22].

3.1.1 Antioxidants Minerals

Antioxidant minerals are a cofactor of antioxidant enzymes. Their absence will undoubtedly affect the metabolism of carbohydrates [16]. Examples include selenium [8, 9, 30], copper [45], iron [1], Zinc [21], and manganese [35].

- *Selenium*

  Selenium is mainly known for its antioxidant capabilities. This minimal mineral substance is necessary for producing and operating around 20–40 enzymes. Most enzymes prevent cellular damage caused by ROS/free radicals [46]. Selenoenzymes with significant antioxidant activity include six sets of GPx: GPx1, GPx3, GPx4, GPx5, and GPx6. These GPx are crucial in defending cells against oxidative damage caused by ROS and RNS, which include nitric oxide, peroxynitrite, hydrogen peroxide, superoxide, and hydroxyl radicals [4]. Selenium functions similarly to vitamin E, can be used in its stead, and lowers the risk of cardiovascular disease and cancer [26].

- *Iron*

  In the biological system, iron is one of the prevalent trace metals shown to bind with proteins. Free iron levels are typically quite low, and low levels of iron-binding proteins encourage the generation of ROS, lipid peroxidation, and oxidative stress. Therefore, taking iron supplements lowers oxidative stress [26].

3.1.2 Antioxidant Vitamins

It is needed for most of the body's metabolic functions[16]. They incorporate Vitamins A, C, E, and B. (i) *Vitamin A:* and its derivatives are potent anti-aging treatments.

Retinoids govern cell death, differentiation, and proliferation. Retinoids have anti-wrinkle effects that increase keratinocyte proliferation, protect the epidermis, restrict water loss, and prevent collagen deprivation. Vitamin A is present mainly in egg yolk and liver. Beta-carotene, an antecedent to vitamin A, is a lipid-soluble antioxidant that effectively neutralizes singlet oxygen, a greatly reactive free radical [17]. (ii) *Vitamin C:* Since vitamin C is an electron donor (reducing agent or antioxidant), this property likely accounts for all its molecular and biochemical activities. Eleven enzymes receive electrons from vitamin C. Vitamin C from gastric juice may stop the production of nitroso compounds, which can potentially cause mutation. Although there is no proof of a cause-and-effect link, high vitamin C ingestion is correlated with a lower hazard of stomach cancer. Ex vivo, vitamin C shields low-density lipoproteins from oxidation; it may have a comparable effect in circulation [4]. (iii) *Vitamin E:* The most potent chain-breaking antioxidant inside the cell membrane, shielding membrane fatty acids from lipid peroxidation, is vitamin E, a significant lipid-soluble antioxidant. One of the most significant lipid-soluble prime defensive antioxidants is vitamin E. It quickly transforms a lipid peroxyl radical into a lipid hydroperoxide and a vitamin E radical by transferring its phenolic H-atom. Potent antioxidants called tocopherols (vitamin E) and tocotrienols (provitamin E) give red palm olein (RPO) oxidative stability and aid in maintaining the stability of the oil's carotenoids and other quality indicators [46]. (iv) *Vitamin B:* Vitamin B is richest sources in whole grains and green leafy vegetables. Panthenol is the alcohol form of pantothenic acid, also known as vitamin B5. [17].

3.1.3 Phytochemicals

These phenolic substances are neither vitamins nor minerals [16]. Although phenolic chemicals, particularly flavonoids, have a variety of biological activities, the three most significant ones are their antioxidant, capillary-protective, and tumor-stage-specific inhibitory effects. Because of their ability to donate electrons, phenolics can scavenge reactive oxygen species. Their stability in various settings and the quantity and arrangement of hydroxyl groups determine how effective they are as antioxidants. Phenolic compounds showed more antioxidant activity than vitamins in numerous in vitro investigations [4]. (i) *Flavonoids:* support healthy cells, antioxidant activity, and the body's natural tissue growth and rejuvenation. They may help minimize some of the symptoms of aging by reducing oxidative stress for the water-based component of the cell in conjunction with vitamin C. Over 4000 distinct flavonoids exist, and combining several kinds is the most efficient way to use them. Red and black grapes, oranges, lemons, grapefruits, cranberries, kale, beets, berries, and green tea are examples of food sources [46]. (ii) *Polyphenols and terpenes:* are the most often found phyto-antioxidants, distinguished by their molecular weight, polarity, and solubility. The quantity and position of -OH groups on the benzene ring determine their antioxidant properties. Phenolic groups reduce lipid peroxidation, which affects protein phosphorylation. The majority prevalent polyphenols are flavonoids and stilbenes, while carotenoids are the most abundant terpenes and operate as single oxygen quenchers [17].

4 Synthetic Antioxidants

Synthetic antioxidants are chemically manufactured and supplementary to food as preservatives to avoid lipid oxidation. They do not exist naturally [40]. Preserving food, pharmaceutical, and cosmetic products using artificial substances with antioxidant properties has long been standard practice. In the food sector, butylated 4-hydroxytoluene (BHT) and butylated 4-hydroxyanisole (BHA) are the most widely utilized synthetic antioxidants [10]. BHT and BHA share similar names, structures, and antioxidant actions and are frequently employed in fats and oils. Despite this, BHT and BHA are on the chemicals considered "*generally accepted as safe*" list. Studies on long-term toxicity suggest that high amounts of BHT may encourage tumor growth [4]. BHA antioxidants are mono-phenolic and lipid-soluble, making them more effective for animal fat lipid oxidation than vegetable oil. BHT, a mono-phenolic fat-soluble antioxidant, is more constant than BHA at elevated temperatures and works synergistically. Many commercially available antioxidant formulations include both antioxidants [26].

5 Preservative

A preservative must work well with some ingredients, including solvents, surfactants, fragrances, active substances, and other food additives. Certain preservatives may become inactive due to another compound's antagonistic activity. The addition of non-ionic surfactants can change the antibacterial activity of parabens. When preservative concentrations are high, solid materials, or organic solids like cellulose and starch can absorb them. However, some sunscreen filters and polyols can increase the effectiveness of preservatives. Chelating compounds that operate as preservative boosters, including ethylenediaminetetraacetic acid (EDTA), are well known. Compatibility on a physical level is also crucial. The preservative must not change the final formula's organoleptic properties and be colorless and odorless [3]. Preservatives should generally be sufficiently soluble in water, have a wide range of activity independent of pH, and not react with other substances [43]. Preservatives protect cosmetic items from microbes but can still degrade due to oxidation. Natural emollients primarily cause oxidative alterations in the lipid phase through autoxidation, as in most vegetable oils, these reactions are responsible as shown in Fig. 2. The development of hydroperoxides and carbonyl molecules can result in a rotten flavor [43]. Antioxidants, such as natural polyphenols, can compensate for the effect. However, further research is needed to optimize their dosage and investigate potential synergistic or antagonistic effects with preservatives [43].

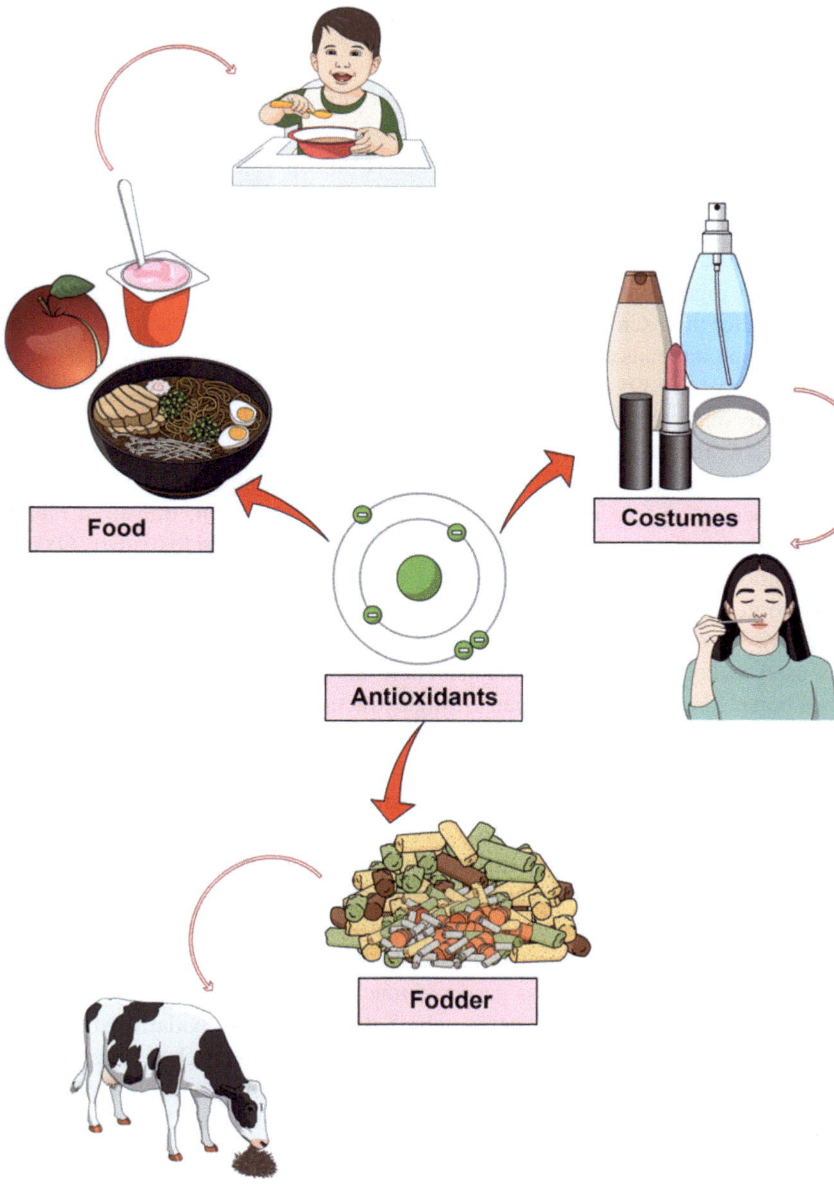

Fig. 2 Antioxidants as preservatives in food, fodder and costumes

5.1 Antioxidants as Preservatives in Food

Most foods are composed of several chemical components that are readily oxidized. Lipids, including fats, oils, and waxes, have the highest propensity to lose electrons. Auto-oxidation of lipids in food. When exposed to light, heat, ionizing radiation, metal ions, or metal-protein catalysts, food can lose its color, flavor, texture, quality, and safety. Triglycerides make up the chemical makeup of fats found in food, and oxidation at the unsaturated sites of the triglycerides causes food to become rancid [4]. Every dietary ingredient undergoes metabolic reactions and is vulnerable to bacteria, which can alter their sensory qualities and create harmful compounds. Antioxidants prevent the spontaneous oxidation of fats, oils, flavorings, vitamins, and colors when exposed to oxygen [39]. Proteins are free amino acids. Their antioxidant potential depends only on the kind of reactive groups, such as hydroxyl, carboxyl, and aromatic units, that are present. For instance, when these groups are in nonionized form, they can offer hydrogen atoms that can quench free radicals, and after ionization, extra electrons neutralize free radicals [2]. The presence of oxygen, transition metals, moisture, heat, and light are some of the elements that cause food to oxidize. Oxygen and transition metals should be eliminated or sequestered to stop, reduce, or slow down the rate at which food oxidizes [39]. Shea nuts, the seeds of shea trees (*Vitellaria paradoxa*), were used to make shea butter. When making chocolate, the extracted fat is used as cocoa butter equivalents. There are a lot of unsaturated fatty acids in shea butter. Oxidation can happen during extraction, post-harvest processing, and storage. The study examined the antioxidant properties of shea butter using synthetic butylated hydroxytoluene (BHT) and a few natural antioxidants, including gallic acid and rosmarinic acid. Shea butter was mixed with 0.02% of each antioxidant, and the mixtures were then heated to 90 °C for 0, 72, and 144 h to maximize oxidation. Oxidation measurements showed that adding antioxidants prevented the shea butter from oxidizing. Antioxidants added to shea butter did not significantly alter the levels of primary fatty acids or the criteria used to detect oxidation. Ultimately, the author concluded that natural antioxidants were nearly as effective as synthetic ones [29]. Selenium research has garnered much attention lately due to its critical function in antioxidant selenoproteins, which defend against oxidative stress from excess reactive oxygen species (ROS) [44]. The selenium-enriched green tea and regular green tea were contrasted in an earlier investigation. The author conducted this investigation in the fall, dividing a 1000 m^2 area into two 500 m^2 sections: one for the control and one for the selenium spraying region. The author sprayed the fertilizer form of selenium on the green tea leaves for up to seven weeks in the selenium spraying area. Both sections of green tea leaves were then gathered and roasted independently once the green tea plants produced new buddings. Vitamin C, polyphenols, color, odor, and taste were tested on the selenium-enriched and control tea. The selenium-enriched tea outperformed the control tea by a large margin. These two distinct leaves have the same amount of vitamin C, polyphenols, color, taste, and aroma over the first sixty days. The selenium-enriched tea's vitamin C and polyphenol content is higher than the control tea after the storage period. Due to its higher chlorophyll concentration, the selenium-enriched tea maintained its color, but the control tea did not. Due to their unstable nature during storage, control tea has less aroma and a bitter taste than selenium-enriched tea, which has a higher aroma and a sweeter taste after 90 days as shown in Fig. 3 [19].

In the previous study, chitosan and propolis, which are industrially useful, were extracted from the bees. To create an edible coating for catfish fillets that can be stored for an extended period, chitosan and propolis were mixed with selenium

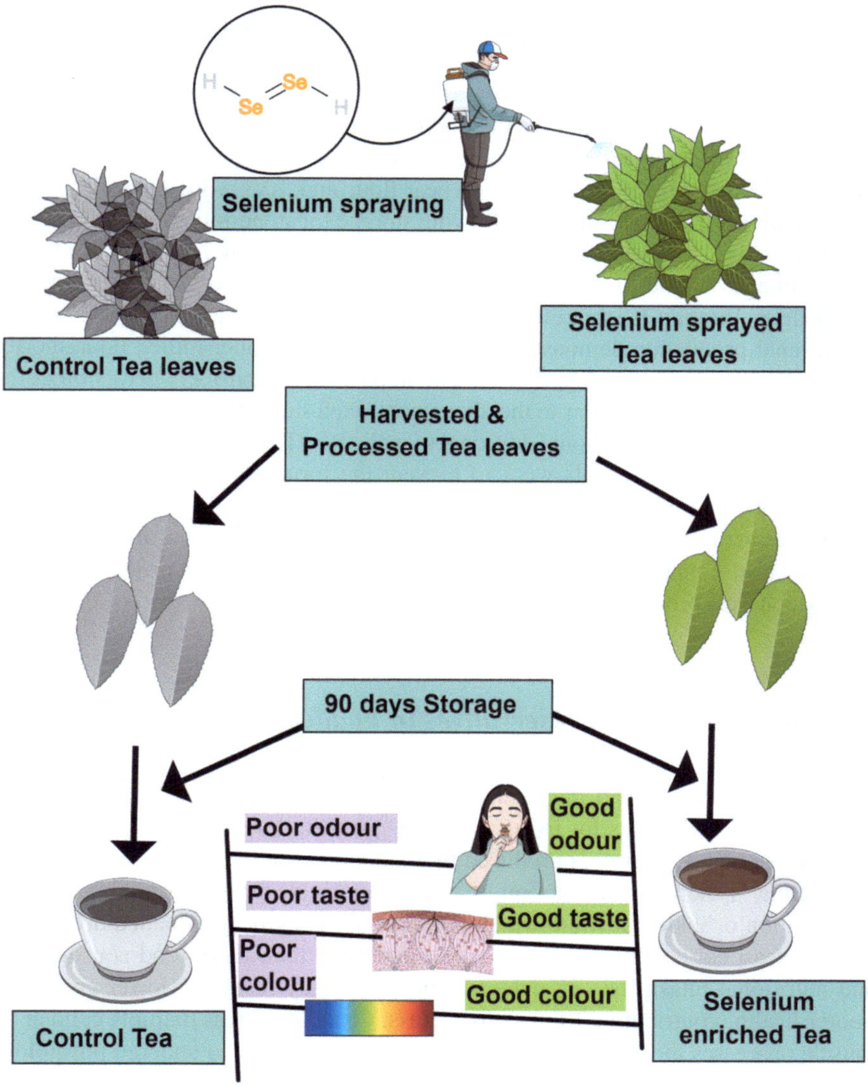

Fig. 3 Comparison between selenium-enriched tea and control tea

nanoparticles. Propolis and chitosan were collected and purified from the bees for this investigation. After being created via the biosynthetic approach, the propolis-mediated selenium nanoparticles were mixed with chitosan collected by bees and processed to create an edible component or composite. The control samples are dipped in distilled water, while the catfish fillets are submerged and drained in the solution for three hours at 2–8 °C. Then, on the 0th and 7th days of cold storage, specific microbiological, chemical, and sensory analyses were performed on the edible-coated fish fillets. In addition to being consistent in terms of shape, color, and odor across sensory categories, the edible-coated fish significantly inhibit the growth of germs, yeasts, and molds after storage as shown in Fig. 4. Ultimately, the author concluded that this bee chitosan propolis-mediated SeNPs composite solution might be an environmentally and financially favorable catfish fillet preserving agent [48].

The earlier study examined the effects of vitamin E on roller-milled and ground wheat flour. The total vitamin E content, measured in vitamin E equivalents (α-TE), was 18.7 α-TE for stone-milled wheat flour and 10.8 α-TE for roller-milled wheat flour one day after milling. Following storage for stone-milled wheat flour, the overall loss was 24%, whereas for roller-milled wheat flour, it was 50%. According to these findings, vitamin E, abundant in wheat germ, is an antioxidant in stone-milled wheat flour. Vitamin E is an efficient antioxidant in wheat germ, as evidenced by the hexanal production demonstrating lipid oxidation in roller-milled flour shortly after milling as shown in Fig. 5 [31]. The impact of natural tocopherols on corn-flake lipid oxidation is similar to the abovementioned study. Corn flakes are made in industry, both with and without additional tocopherols. The sensory analysis demonstrated how well tocopherols work to prevent the formation of an off-flavor over storage. Tocopherols' antioxidant properties prevented the headspace from evolving into more intense oxidation, particularly the subsequent oxidation of hexanal into hexanoic acid. Finally, the study of triacylglycerol polymerization products verified that tocopherols prevent lipid oxidative degradation [34].

5.2 Antioxidants as Preservatives in Fodder

Food provides macro and micronutrients and non-nutrients such as water, cutin, lignin, colors, and waxes for living creatures' growth, development, and proliferation. Most notably, energy is produced by oxidation events in tissue compartments and cellular systems' consumption of oxygen molecules (O_2). Farm animals face malnutrition or undernutrition, which is frequently caused by extended drought or adverse seasonal conditions (hot or cold) that affect their daily food consumption and nutritional intake, i.e., an unbiased diet. This may result in slower growth or poorer health, contributing to oxidative stress. Oxidative stress disrupts cellular or muscle processes through various events, allowing the body to undergo physiological or pathological alterations that influence energy metabolism and nutrition partitioning. Antioxidants can function as prooxidants based on the animal's productive status, metabolic requirement, and oxidative stress conditions. Antioxidants from

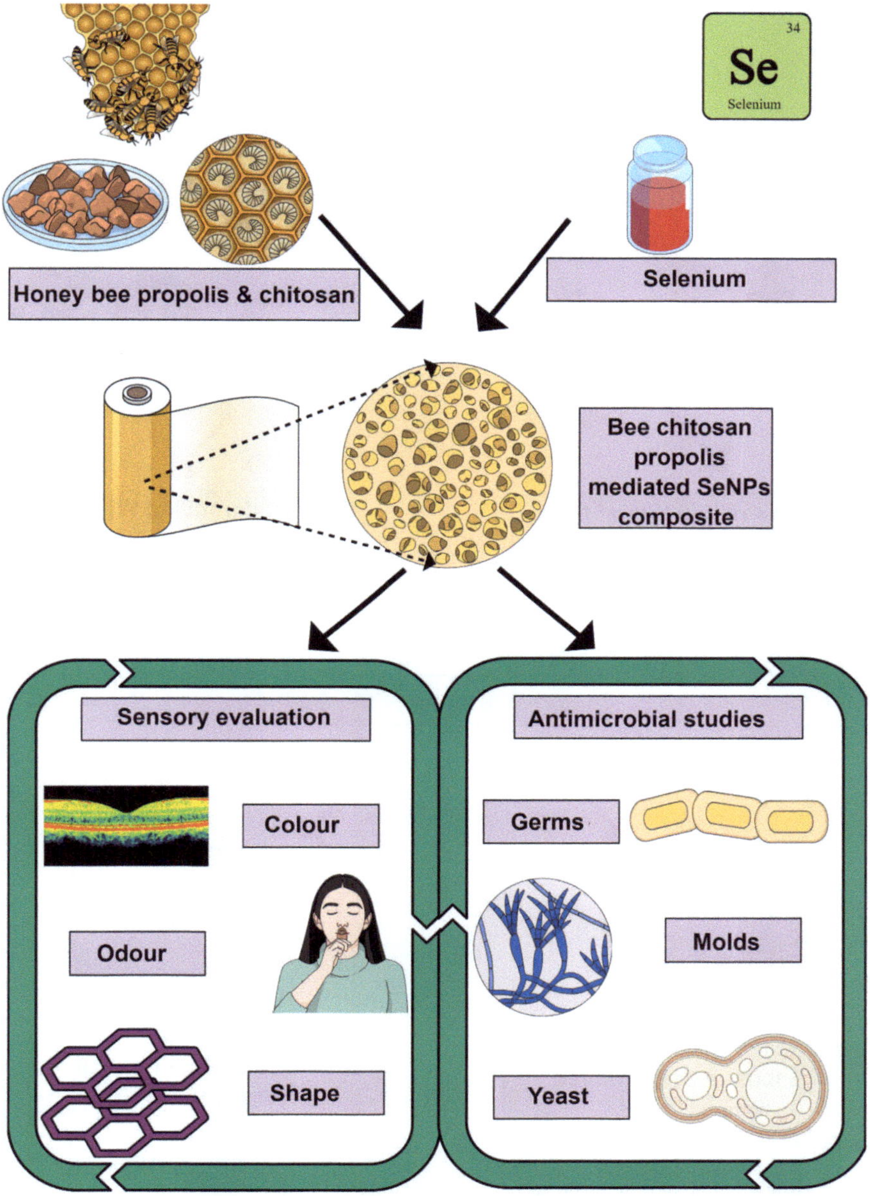

Fig. 4 Synthesis and evaluation of bee propolis and chitosan-mediated Se biocomposite

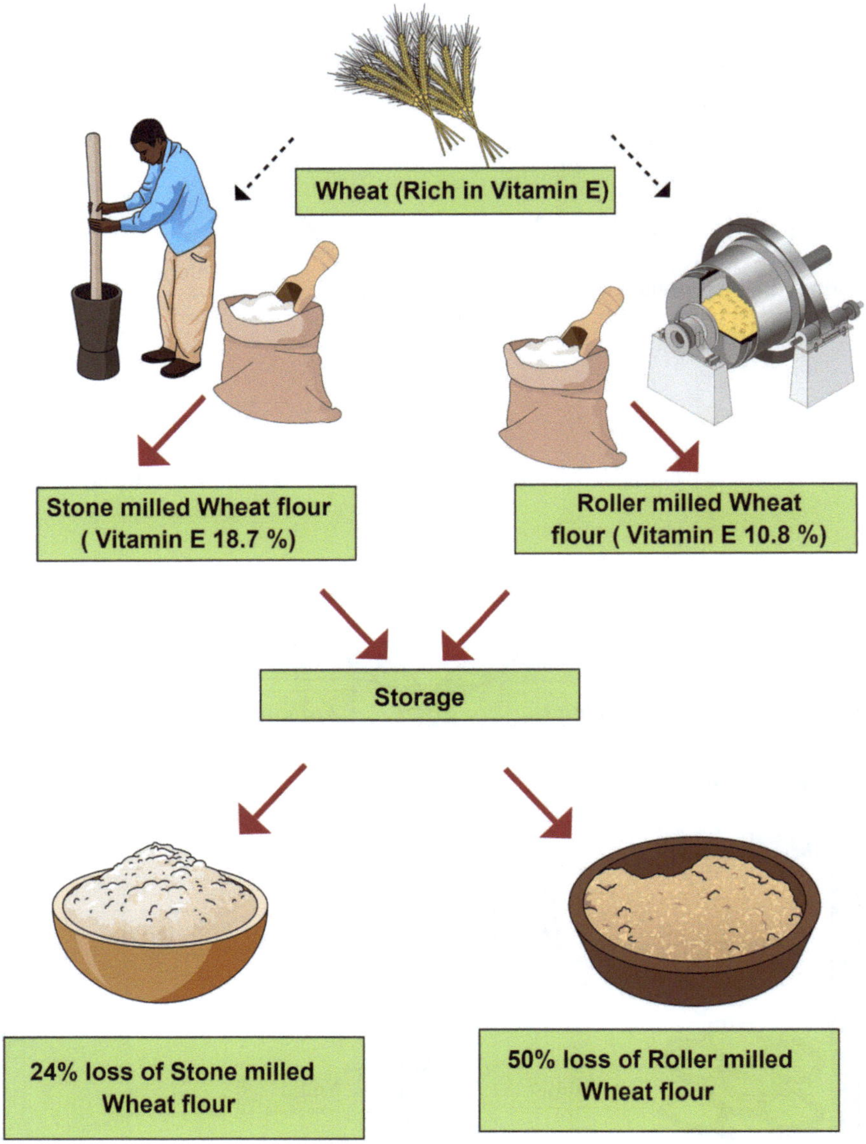

Fig. 5 Role of vitamin E in the preservation of two different milled wheat flour

feed (meal) play a significant role in animal health and milk productivity as shown in Fig. 6 [38].

Recent research aims to enhance antioxidant defense during the transition to combat organic diseases such as ketosis, hypocalcaemia, and mastitis. Supplementing

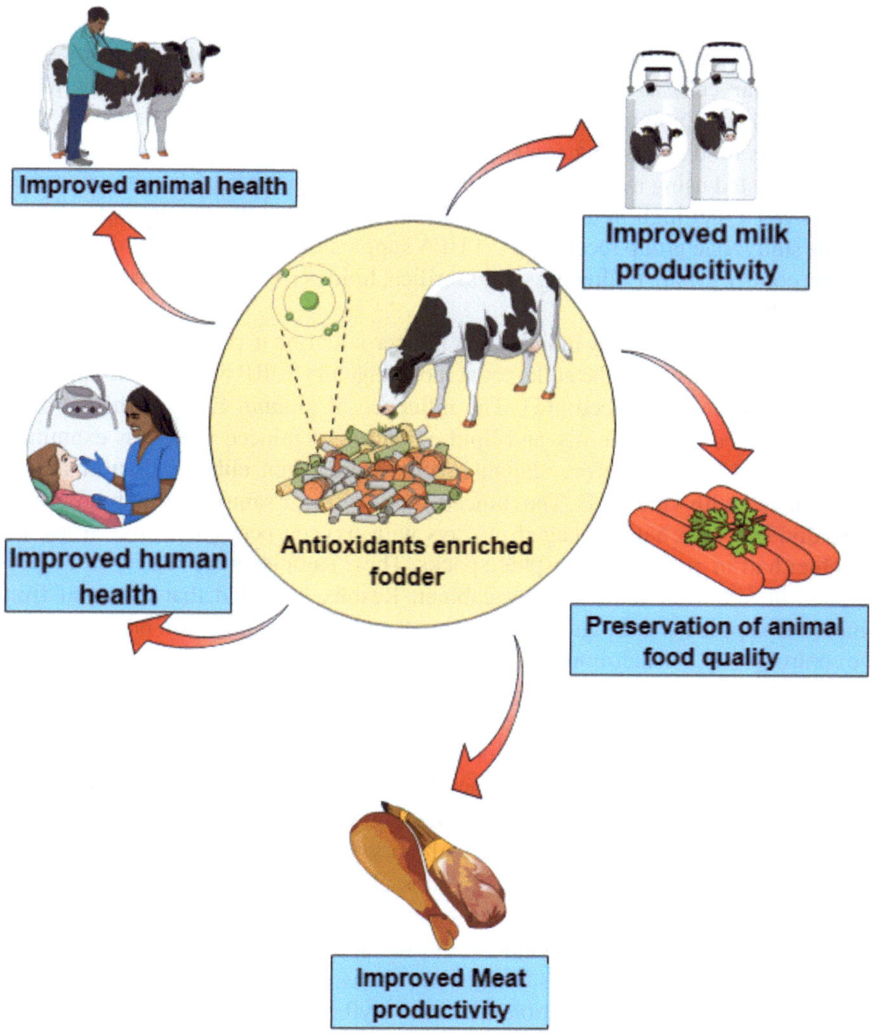

Fig. 6 Various benefits of antioxidants in animal fodder

with antioxidants can improve the cow's health during a delicate time, like the transition period, but it can also give value to the finished product (milk or meat) that benefits the customer's health [7]. The author described skeletal myopathy in adult horses, cattle, and chickens as a pathology linked to selenium and vitamin E deficiencies. There is currently conflicting data regarding the effectiveness of vitamin E and selenium supplementation in reducing placental retention in dairy cattle. Selenium has recently been shown to play a biochemical role in the GSH-Px enzyme, which works in cells with vitamin E to regulate peroxidation [33]. According to one study, n-3 PUFA is beneficial for domestic animals, but it can be oxidized by lipid peroxidation, leading to profound structural and functional changes and harmful metabolites in mammalian cells as shown in Fig. 7. Therefore, the author chose a few plants and herbs rich in polyphenols (*PERP polyphenol-enriched plants* that have been employed as food antioxidants to stop lipid oxidation. Therefore, the author of this study suggested using new natural antioxidants to stop lipoperoxidation in domestic animals fed a diet high in n-3 PUFAs. The author divided the rats into two groups for this study: one group received n-3 PUFA supplements, and the other did not; this was the control group. Following their sacrifice, lipid peroxidation assessments were performed on their plasma.

The findings of this study indicate that lipoperoxidation in plasma can be inhibited by up to 36%. As a result, the author concluded that this PERP would make appealing antioxidants for animal feed [15]. The influence of vitamin E supplementation in pig diets on the color stability and lipid oxidation of minced pork was examined. For the last 12 weeks before slaughter, pigs were given either a control diet or one enriched with vitamin E. The minced shoulder meat samples from both groups were divided, with half packaged in trays with oxygen-permeable overwraps and the other half in modified atmosphere packs. These samples were stored for 10 days at 7 °C in a well-lit retail display cabinet. Results indicated that the meat from vitamin E-supplemented pigs showed more excellent lipid oxidation resistance than the control group. Furthermore, gas packaging increased lipid oxidation in the control meat, while the meat from the supplemented pigs remained stable. Color stability in gas-packed meat was similar for both dietary groups, but the oxygen-permeable overwraps adversely affected the color stability of the vitamin E-enriched meat. The shelf life of both types of meat was found to be similar. Thus, adding vitamin E to pig diets can enhance pork's stability against lipid oxidation [18]. The study related to chickens were fed a diet enriched with antioxidant-rich foods like oregano, α-tocopherol, and ascorbic acid. After that, the chickens were either transported and electrically stunned at the slaughterhouse (*stress*) or killed by cervical dislocation in the stable (*no stress*). Supplementation did not affect the activities of antioxidative enzymes in the liver, iliotibialis (IL), or pectoralis major (PM). However, ascorbic acid α-tocopherol supplementation increased erythrocyte stability, a more complex model system for assessing oxidative status, and it also tended to rise following oregano treatment. Reduced TBA-reactive chemicals (TBARS) in the liver and PM of ascorbic acid-α-tocopherol-supplemented chickens, as well as in the liver of oregano-supplemented chickens, were indicative of this improved antioxidative status in non-stressed birds when compared to non-stressed control birds. In PM and IL, the control animals' TBARS levels rose in response to stress. Supplementing with ascorbic acid-α tocopherol shielded IL while supplementing with oregano shielded PM from stress-induced rises in TBARS. This variation in muscle action could be a sign of various defense mechanisms. Therefore, the author concluded that giving hens supplements of oregano, α-tocopherol, and ascorbic acid protects them from stress-induced increases in TBARS in various muscles [47].

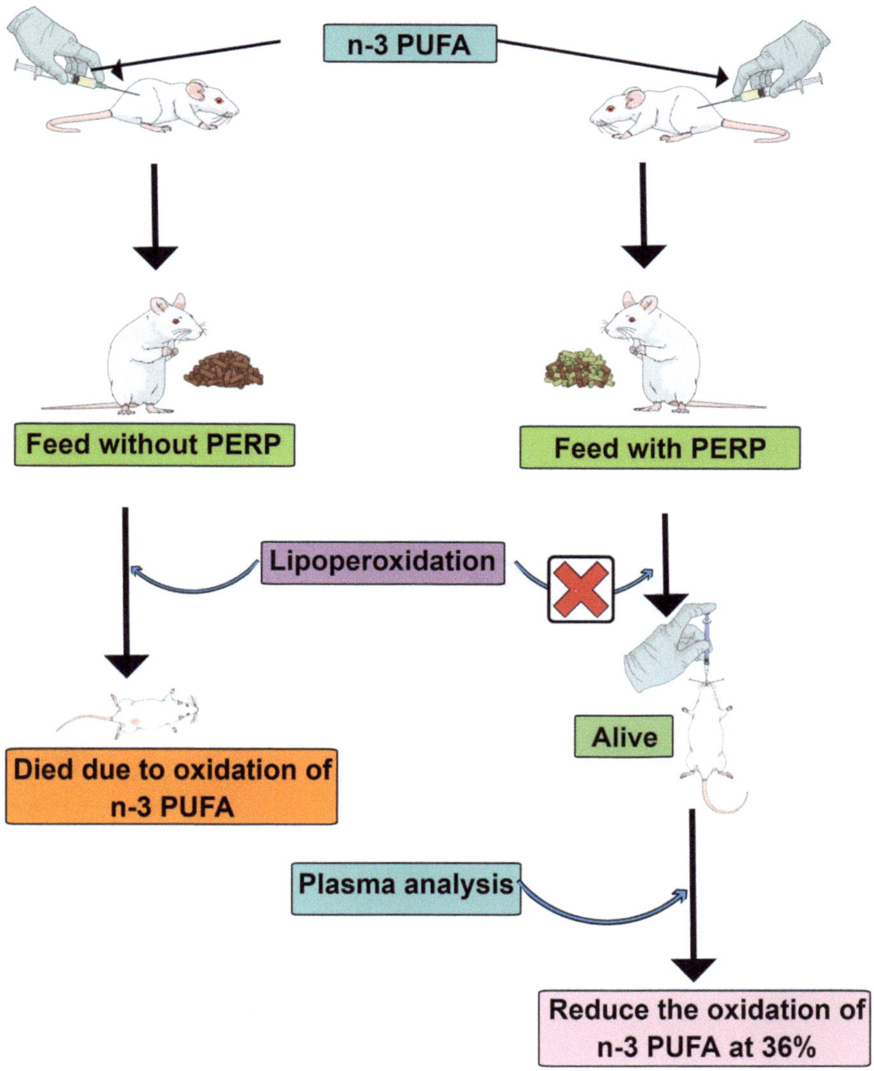

Fig. 7 Antioxidant activity of PERP in animal feed

5.3 Antioxidants as Preservatives in Cosmetics

The skin is the body's primary protective system, as it is constantly exposed to the outside world. Skin thickness varies based on various factors, including location, exposure to UV rays, injuries, and age [28]. Skin protects the body against radiation, heat, cold, overheating, reactive oxygen species (ROS), and chemicals as shown in Fig. 8. Its passive (protection) and active (secretion and sensing) functions are separated [25].

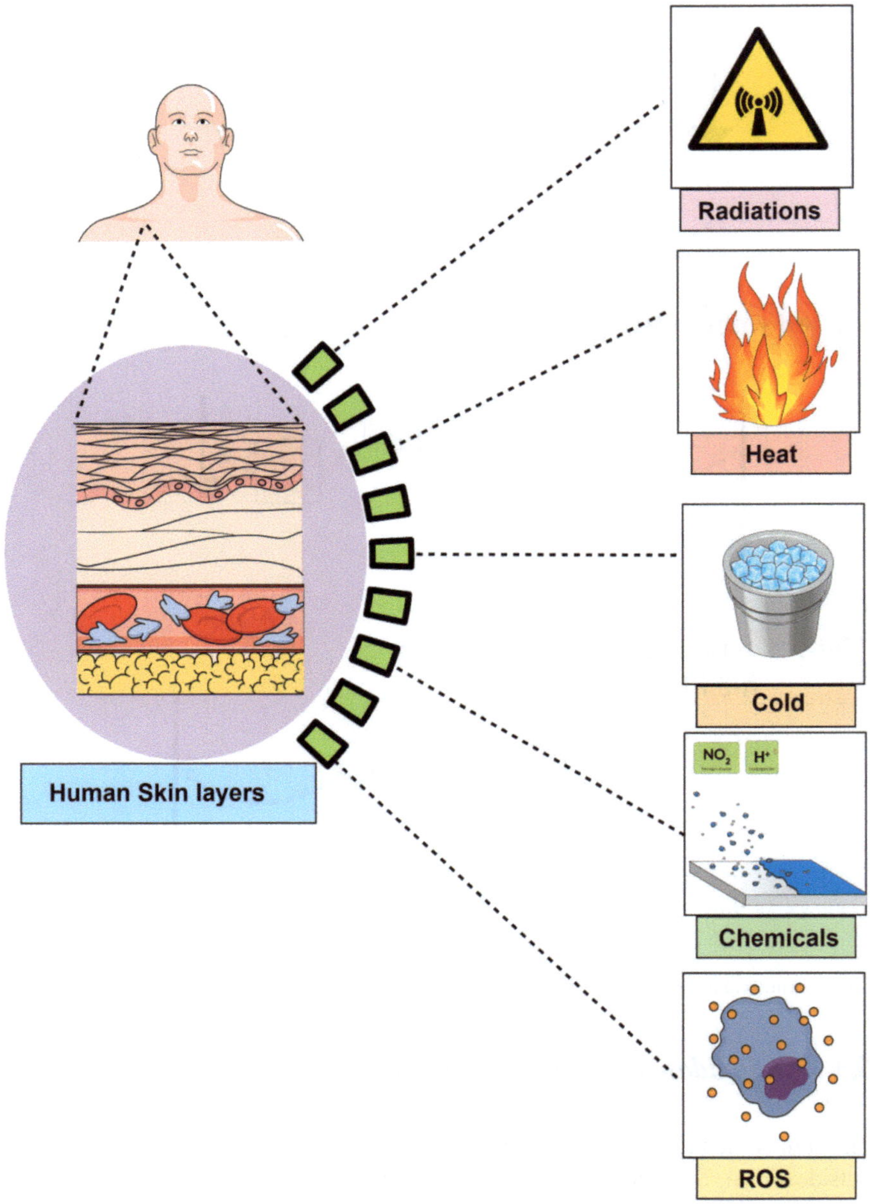

Fig. 8 Protection of skin from various external sources

In contrast to externally applied medications, cosmetic preparations shouldn't alter the composition and functionality of the tissues to which they are applied. Utilized technologies, production methods, and intricate cosmetics recipes all affect the quality of end products. It should be clear to manufacturers that the products they put on the market must be safe. A cosmetic product must be safe if it has sufficient microbiological purity and stability and has no banned chemicals or permitted ingredients in excessive amounts [28]. Although parabens were previously considered harmless at permissible amounts, recent scientific research suggests that their use in cosmetics can negatively affect the human body [28]. Additionally, the possibility of contamination with certain toxic materials, such as heavy metals, toxic microorganisms, microbial toxins, pesticides, and residual solvents, must be assessed in cosmetics that contain plant extracts as one of their active ingredients. Examples of these include a-tocopherol from wheat germ and sunflower, Aloesin from Aloe vera, Anthocyanin, Proanthocyanin from Berry species, Arbutin from *Vaccinium vitis-idaea*, and ascorbic acid from rosehip oil. The content of all these toxic components must be below their maximum allowed levels [24]. Antioxidants should be applied to the skin as a pretreatment before exposure to the sun. When applied topically before UV rays (UVR) exposure, synthetic and *"natural"* antioxidants have been shown to have substantial photoprotective benefits in human trials. Antioxidants only successfully stop UVR-induced skin damage when present in appropriate quantities, at the site of damage, and during oxidative stress. UVR-induced skin damage begins quickly [37]. Antioxidant preservatives are used in fat-based cosmetics like lipstick and moisturizer to avoid rancidity [4]. Since antioxidants are typically unstable, intensely pigmented, and susceptible to hydrolysis and photodegradation when exposed to oxygen, creating compelling cosmetic compositions and preserving their acceptability and aesthetic validity is exceedingly challenging. To achieve the intended activity, the antioxidant content in a cosmetic composition needs to be stable. Their hue shouldn't alter during manufacturing and storage operations to maintain their antioxidant properties and the product's visual appeal [24]. Oral administration of EGCG significantly increases the minimal erythema dose to UV and prevents disruption of the epidermal barrier function, suggesting that EGCG strengthens the skin's tolerance to UV-initiating stress as shown in Fig. 9. Polyphenols produced by plants are distinguished based on the presence of phenol units in their molecular structure [27].

One of the biological activities of antioxidants is the prevention or reduction of photoaging. Because it can stop continuous lipid peroxidation by reducing the lipid peroxyl radical to hydroperoxide, this antioxidant shields skin from the damaging effects of oxidative stress. Topical tocopherol supplementation can also aid in restoring the endogenous antioxidants lost due to UV exposure. Its free aromatic hydroxyl group is thought to be responsible for its antioxidant properties. Following topical treatment, reports of a high concentration in the stratum corneum have been reported. It has been discovered that tocopherol applied topically reduces lipid peroxidation and photocarcinogenesis [41]. The skin contains natural melanins. Melanins are pigments that consume oxygen and comprise heterogeneous polymers that interact with free radicals and reactive oxygen species to undergo redox

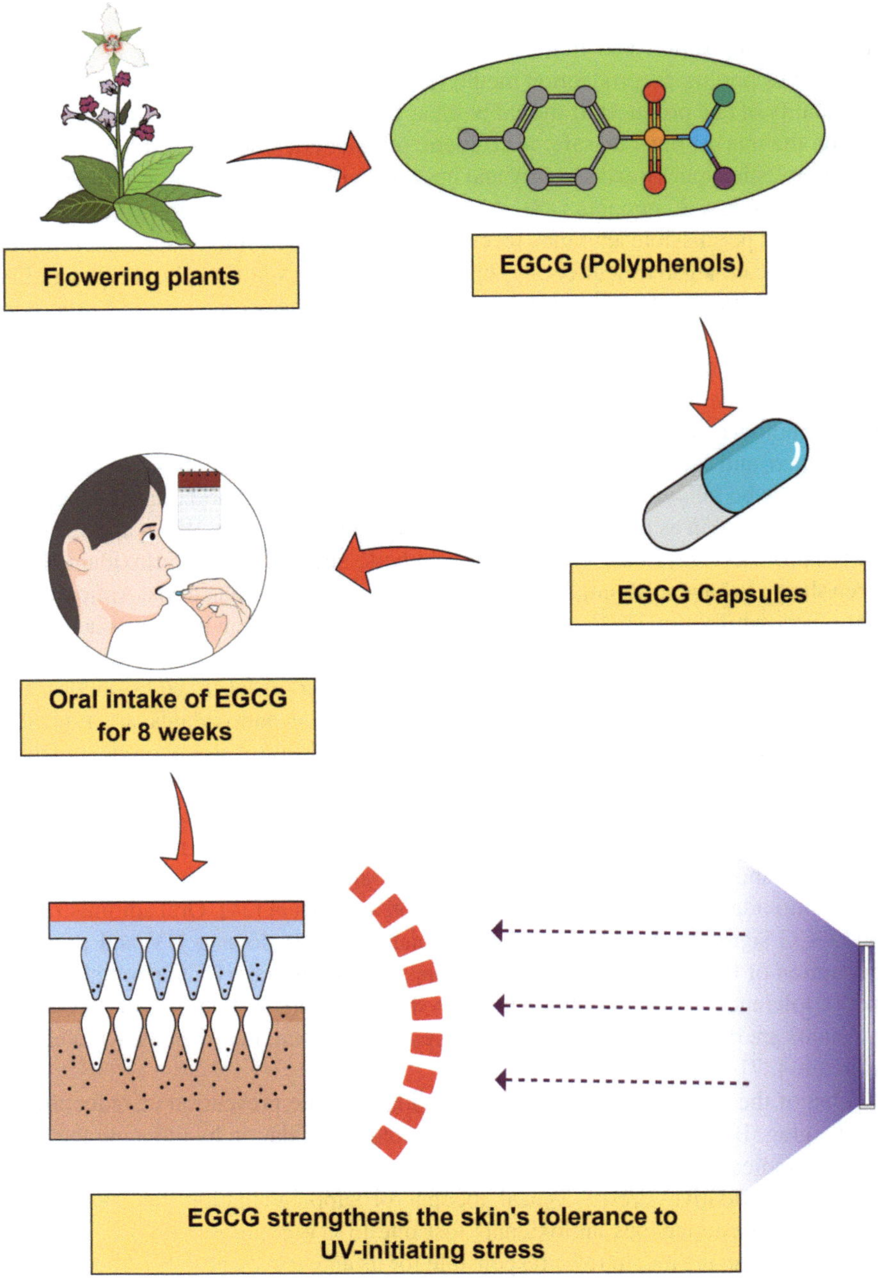

Fig. 9 Polyphenol EGCG mediated strengthens the skin's tolerance against UV radiation

reactions and consume oxygen [5, 13]. The previous study compared biotechnologically generated melanin (P1, P2, P3, P4, and L1, L2, L3, L4) with synthetic melanin. They employ male and female mice's liver microsomal suspensions. The lipid peroxidation (LPO) assay employs malano-di-aldehyde as a marker. The 1 ml sample was then mixed with melanin and/or vitamin E. Then, 1 mM ferric chloride and 5 mM adenosine-5-diphosphate were added to initiate the peroxidative damage, and the mixture was incubated at 37 °C for an hour. After the incubation, the reactions are terminated by the addition of 20% (w/v) trichloroacetic acid and 1% (w/v) 2-thiobarbituric acid, then they are heated up to 90 °C, after which all samples are centrifuged finally, MDA levels were determined by measuring the development of a pink shade at 535 nm. The MDA concentration was determined using a 1.5 $\times 10^5$ M^{-1} molecular extinction value. According to the findings of this investigation, biomelanin P4 (58%) was nearly as effective as Vitamin E (57%) in inhibiting LPO production. The assay is then repeated with a combination of vitamin E and melanin P4, vitamin C with melanin P4, and vitamins E, C, and melanin P4. Among these, Vitamin E and Melanin P4 block LPO by up to 78% [20]. The major selenium poisoning includes face flushing, tremors, muscle soreness, severe weariness, irritability, foul breath, sensitivity to inflammation, discolored fingernails, mottled teeth, nerve lesions, and diarrhea. Acute Se intoxication can cause severe intestinal and neurological symptoms, heart attack, kidney failure, and even death [36]. The free radicals cause more complications of diabetes mellitus. However, the mixture of vitamin E, C, and selenium reduces the adverse effects of free radicals, thus increasing non-enzymatic glycosylation and preventing skin disorders in diabetes mice. Therefore, [42] conducted a study with diabetic albino mice compared with healthy mice (control). Vitamin E, vitamin C and selenium combinations were injected into healthy and diabetic mice. At the end of the experimental periods, the blood samples were collected from mice and they underwent GSH (glutathione), LPO (lipid peroxidation) and NEG (non-enzymatic glycosylation) estimations. The results are mentioned in Table 1.

Table 1 Quantifications of GSH, LPO, and NEG of three different groups of mice

Mice	Glutathione—GSH (%)	Lipid peroxidation—LPO (%)	Non-enzymatic glycosylation—NEG (%)
Healthy albino (positive control)	25.14	1.56	7.40
Diabetic (negative control)	18.71	2.38	16.58
Diabetic + (vitamin E, vitamin C, and selenium)	24.89	1.98	8.71

According to these results, the GSH and NEG values show that the antioxidants of vitamin E, Vitamin C, and selenium injected mice effectively control the lipid peroxidation and induce non-enzymatic glycosylation and also increase the glutathione level compared with control and negative control mice. The author finally concluded that vitamin E, vitamin C and selenium play an excellent role in preventing the cells from oxidative threats and these antioxidants increase skin tissue NEG levels provoked by diabetes as shown in Fig. 10 [42].

In a previous study, the author evaluated the effects of SkinAx2TM, an oral supplement rich in specific antioxidants, on improving skin radiance in women. Environmental factors have an impact on skin aging, resulting in a decrease in skin radiance. Nutrition, particularly antioxidants, may help fight against skin degradation. For eight weeks, the subjects received a daily oral supplement of 150 mg of an antioxidant-rich formulation that included zinc, vitamin C, grape seed extract rich in flavanol monomers, and melon concentrate rich in superoxide dismutase. Every subject acted as her own master. The test has been used to assess the skin radiance components of transparency (T), brightness (B), luminosity (L), and facial skin coloring (C). Clinical evaluation has been used to evaluate facial skin flaws. Clinical evaluation and cutometer measurements have been used to gauge firmness. Lastly, an auto-questionnaire has been used to gauge individuals' satisfaction with various aspects of skin brightness and the supplement's overall effectiveness. The results showed that supplementation significantly improved the *"olive"* and *"red pink"* tones of the skin. A ~ 25.9% rise in luminosity was observed. However, the supplementing did not affect transparency or brightness. Following consumption of the antioxidant-rich mixture, facial skin flaws were considerably lessened. In fact, following oral medication, dark circles, redness, and spots considerably decreased. It has been demonstrated that firmness and elasticity are enhanced. Subjects reported improvements in the skin on their faces and overall satisfaction with the product (~82.4%). Additionally, ~ 64.7% said they looked better after taking the supplements. Lastly, the author concluded that the antioxidant-rich oral supplement improved skin radiance in women with dull complexions by decreasing skin coloration, boosting face luminosity, decreasing flaws, and improving skin firmness [14]. One of the most significant risks to skin health is exposure to air pollution, which speeds up skin mainly aging through oxidative stress mechanisms. Skin protection techniques are required since reducing skin exposure to air contaminants is challenging, particularly in metropolitan areas. It has been discovered that plant phenolic compounds help reduce inflammation and cellular oxidative stress brought on by various air pollutants, and a dietary strategy centered around these compounds may offer an effective defense. The author studied the efficacy of a commercially available polyphenol-rich dietary supplement (Zeropollution®) in lowering pollution-induced oxidative stress and improving various skin metrics associated with skin aging in Caucasian and Asian volunteers exposed to air pollution. Zero pollution consists of four standardized herbal extracts: Olea europaea leaf, *Lippia citriodora*, Rosmarinus officinalis, and Sophora japonica. To evaluate the dietary supplement's effectiveness, 100 outdoor workers in Milan, a polluted urban area of Europe, participated in a double-blind, randomized, parallel-group trial. Measurements were made of the skin's radiance

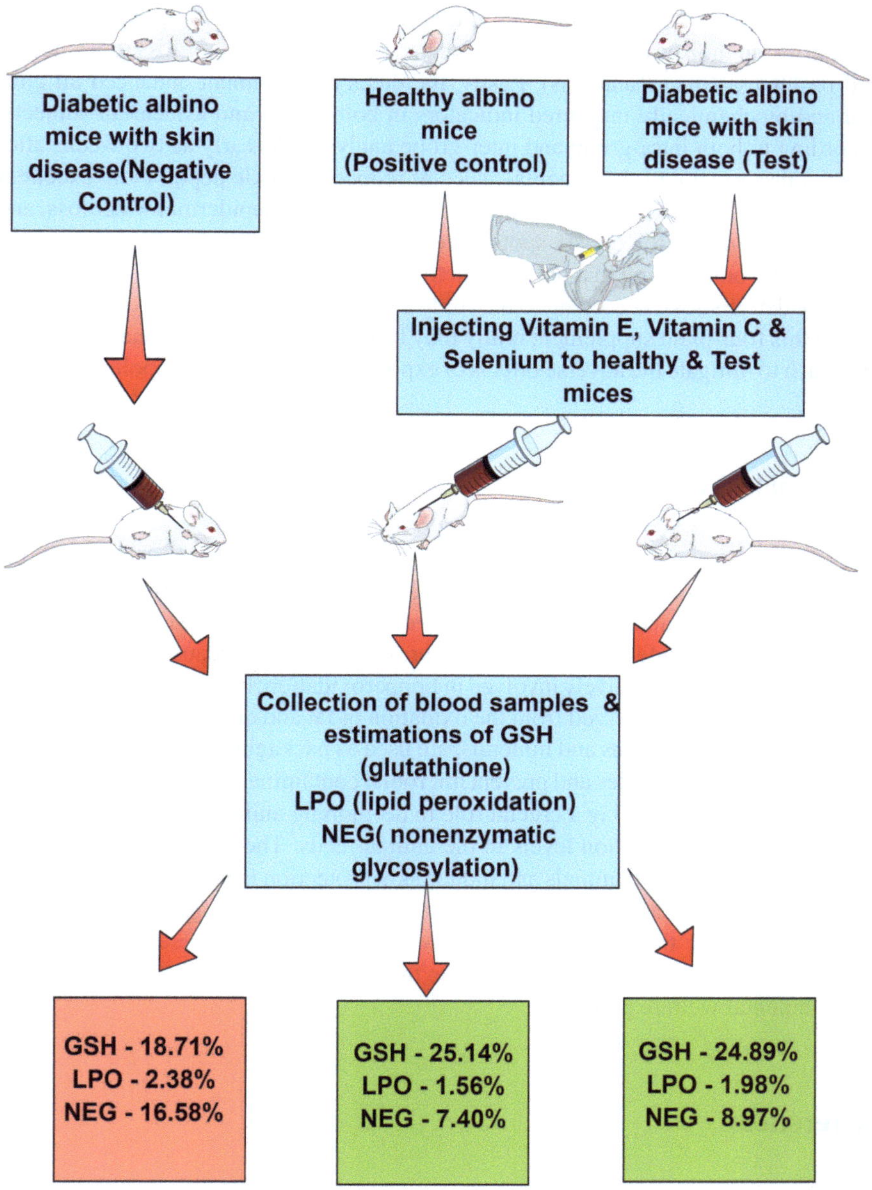

Fig. 10 Actions of antioxidants such as vitamin C, vitamin E, and selenium against diabetic skin diseases caused by free radicals

and color, elasticity, sebum content, oxidative damage on the skin (lipoperoxides content), skin moisturization (chronometer), transepidermal water loss (tewameter), and roughness (image analysis). Lastly, the dietary supplement enhanced all clinical and biochemically measured indicators in both Asian and Caucasian subjects, according to both intra-group and inter-group analyses. As early as two weeks after product use, several of the benefits—such as reduced wrinkle depth, more excellent elasticity and firmness, better skin moisturization and transepidermal water loss, and decreased pigmentation of dark spots—were statistically significant. Both Asian and Caucasian women living in a polluted metropolitan region showed less skin damage from oxidative stress, according to the study's findings, the author stated. As a result, taking this four-plant supplement orally may be regarded as an additional nutritional approach to mitigate the adverse effects of exposure to environmental pollutants [32].

6 Conclusion

The well-known fact is that antioxidants are irreplaceable molecules in the metabolism of living things. It can liberate the single oxygen atom to neutralize the nascent oxygen and eradicate the reactive oxygen species. However, this chapter discussed the other side of antioxidants as a preserving agent. Because of their antioxidant property, they are actively involved in lipoperoxidation; hence, the oil-containing food sample can be preserved from the oxidation of fat and oils. In some cases, these antioxidants, like vitamins and minerals, are used as packaging materials to maintain the food's sensory qualities and prevent microbial contamination. In the fodder fields, vitamins and selenium play a crucial role in developing animal nutrition values and reducing the lipid oxidation levels in the animal feeds. The selenium and vitamins improve the skincare of animals and induce skin protection from external factors like pollutants and UV radiation. Hence, antioxidants actively engage in food, fodder, and cosmetics for the above reasons. This chapter uses much research to investigate the other dimensions of antioxidants that can be used to develop human and environmental welfare products.

References

1. Adjimani, J. P., & Asare, P. (2015). Antioxidant and free radical scavenging activity of iron chelators. *Toxicology Reports, 2*, 721–728. https://doi.org/10.1016/j.toxrep.2015.04.005
2. Aluko, R. E. (2015). *Amino acids, peptides, and proteins as antioxidants for food preservation*. Elsevier Ltd. https://doi.org/10.1016/B978-1-78242-089-7.00005-1
3. Angelopoulou, P., Giaouris, E., & Gardikis, K. (2022). Applications and prospects of nanotechnology in food and cosmetics preservation. *Nanomaterials, 12*(7), 1–28. https://doi.org/10.3390/nano12071196

4. Atta, E. M., Mohamed, N. H., & Abdelgawad, A. A. M. (2017). Antioxidants: An overview on the natural and synthetic types. *European Chemical Bulletin, 6*(8), 365. https://doi.org/10.17628/ecb.2017.6.365-375
5. Bustamante, J., Bredeston, L., Malanga, G., & Mordoh, J. (1993). Role of Melanin as a scavenger of active oxygen species. *Pigment Cell Research, 6*(5), 348–353. https://doi.org/10.1111/j.1600-0749.1993.tb00612.x
6. Carocho, M., Morales, P., & Ferreira, I. C. F. R. (2018). Antioxidants: Reviewing the chemistry, food applications, legislation and role as preservatives. *Trends in Food Science and Technology, 71*, 107–120. https://doi.org/10.1016/j.tifs.2017.11.008
7. Castillo, C., Pereira, V., Abuelo, Á., & Hernández, J. (2013). Effect of supplementation with antioxidants on the quality of bovine milk and meat production. *The Scientific World Journal*. https://doi.org/10.1155/2013/616098
8. Chandramohan, S., Sundar, K., & Muthukumaran, A. (2019). Hollow selenium nanoparticles from potato extract and investigation of its biological properties and developmental toxicity in zebrafish embryos. *IET Nanobiotechnology, 13*(3), 275–281. https://doi.org/10.1049/iet-nbt.2018.5228
9. Chandramohan, S., Krishnan, S., & Muthukumaran, A. (2019). Reducing agents influence the shapes of selenium nanoparticles (SeNPs) and subsequently their antibacterial and antioxidant activity. *Materials Research Express, 27*, 0–31. https://doi.org/10.1088/2053-1591/ab29d6
10. Chung, J. G. (1999). Effects of butylated hydroxyanisole (BHA) and butylated hydroxytoluene (BHT) on the acetylation of 2-aminofluorene and DNA-2-aminofluorene adducts in the rat. *Toxicological Sciences, 51*(2), 202–210. https://doi.org/10.1093/toxsci/51.2.202
11. Costa, R., & Santos, L. (2017). Delivery systems for cosmetics—From manufacturing to the skin of natural antioxidants. *Powder Technology, 322*, 402–416. https://doi.org/10.1016/j.powtec.2017.07.086
12. de Lima Cherubim, D. J., Buzanello Martins, C. V., Oliveira Fariña, L., & da Silva de Lucca, R. A. (2020). Polyphenols as natural antioxidants in cosmetics applications. *Journal of Cosmetic Dermatology, 19*(1), 33–37. https://doi.org/10.1111/jocd.13093
13. Dossgg, L. L., & Memula, N. (1982). The radio responsiveness of melanoma. *International Journal of Radiation Oncology, Biology, Physics, 8*(7), 1131–1134. https://doi.org/10.1016/0360-3016(82)90060-8
14. Dumoulin, M., Gaudout, D., & Lemaire, B. (2016). Clinical effects of an oral supplement rich in antioxidants on skin radiance in women. *Clinical, Cosmetic and Investigational Dermatology, 9*, 315–324. https://doi.org/10.2147/CCID.S118920
15. Gladine, C., Morand, C., Rock, E., Bauchart, D., & Durand, D. (2007). Plant extracts rich in polyphenols (PERP) are efficient antioxidants to prevent lipoperoxidation in plasma lipids from animals fed n—3 PUFA supplemented diets. *Animal Feed Science and Technology, 136*(3–4), 281–296. https://doi.org/10.1016/j.anifeedsci.2006.09.003
16. Hamid, A. A., Aiyelaagbe, O. O., Usman, L. A., Ameen, O. M., & Lawal, A. (2010). Antioxidants: Its medicinal and pharmacological applications. *African Journal of Pure and Applied Chemistry, 4*(8), 142–151. http://www.academicjournals.org/AJPAC
17. Hoang, H. T., Moon, J. Y., & Lee, Y. C. (2021). Natural antioxidants from plant extracts in skincare cosmetics: Recent applications, challenges and perspectives. *Cosmetics, 8*(4), 1–24. https://doi.org/10.3390/cosmetics8040106
18. Houben, J. H., & Gerris, C. V. M. (1998). Effect of the dietary supplementation with Vitamin E on colour stability of packaged, sliced pasteurized ham. *Meat Science, 50*(4), 421–428. https://doi.org/10.1016/S0309-1740(98)00054-0
19. Huang, Y., Xu, J., & Hu, Q. (2005). Effect of Selenium on preservation quality of green tea during autumn tea-processing season. *Journal of Agricultural and Food Chemistry, 53*(19), 7444–7447. https://doi.org/10.1021/jf048314j
20. Kalka, K., Mukhtar, H., Turowski-Wanke, A., & Merk, H. (2000). Biomelanin antioxidants in cosmetics: Assessment based on inhibition of lipid peroxidation. *Skin Pharmacology and Applied Skin Physiology, 13*(3–4), 143–149. https://doi.org/10.1159/000029919

21. Khan, A. R., Huang, K., Jinzhong, Z., Zhu, T., Morsi, Y., Aldalbahi, A., El-Newehy, M., Yan, X., & Mo, X. (2021). Exploration of the antibacterial and wound healing potential of a PLGA/silk fibroin based electrospun membrane loaded with zinc oxide nanoparticles. *Journal of Materials Chemistry B, 9*(5), 1452–1465. https://doi.org/10.1039/d0tb02822c
22. Kunwar, A., & Priyadarsini, K. (2011). Free radicals, oxidative stress and importance of antioxidants in human health. *Journal of Medical and Allied Sciences, 1*(2), 53–60.
23. Kurutas, E. B. (2016). The importance of antioxidants which play the role in cellular response against oxidative/nitrosative stress: Current state. *Nutrition Journal, 15*(1), 1–22. https://doi.org/10.1186/s12937-016-0186-5
24. Kusumawati, I., & Indrayanto, G. (2013). Natural antioxidants in cosmetics. In *Studies in Natural Products Chemistry* (1st ed., Vol. 40). Copyright © 2013 Elsevier B.V. All rights reserved. https://doi.org/10.1016/B978-0-444-59603-1.00015-1
25. Lee, S. H., Jeong, S. K., & Ahn, S. K. (2006). An update of the defensive barrier function of skin. *Yonsei Medical Journal, 47*(3), 293–306. https://doi.org/10.3349/ymj.2006.47.3.293
26. Mamta, K. M., Dhillon, G. S., & Brar, S. K. (2014). Antioxidants. In *Biotransformation of waste biomass into high value biochemicals* (Vol. 9781461480, pp. 118–138). https://doi.org/10.1007/978-1-4614-8005-1
27. Masaki, H. (2010). Role of antioxidants in the skin: Anti-aging effects. *Journal of Dermatological Science, 58*(2), 85–90. https://doi.org/10.1016/j.jdermsci.2010.03.003
28. Matwiejczuk, N., Galicka, A., & Brzóska, M. M. (2020). Review of the safety of application of cosmetic products containing parabens. *Journal of Applied Toxicology, 40*(1), 176–210. https://doi.org/10.1002/jat.3917
29. Nahm, H., Juliani, H., & Simon, J. (2012). Effects of selected synthetic and natural antioxidants on the oxidative stability of shea butter (*Vitellaria paradoxa* subsp. paradoxa). *Journal of Medicinally, 1*(2), 5–10. http://scholarworks.umass.edu/cgi/viewcontent.cgi?article=1003&context=jmap
30. Naveenkumar, S., Alagumanikumaran, N., Kaviyarasu, K., & Muthukumaran, A. (2024). Influence of encapsulated sodium alginates and pectin on selenium nanoparticles and efficient cardioprotective effect in C_2C_{12} cell line. *Journal of Nanoparticle Research, 26*(3), 1–16. https://doi.org/10.1007/s11051-024-05956-x
31. Nielsen, M. M., & Hansen, Å. (2008). Stability of vitamin E in wheat flour and whole wheat flour during storage. *Cereal Chemistry, 85*(6), 716–720. https://doi.org/10.1094/CCHEM-85-6-0716
32. Nobile, V., Schiano, I., Peral, A., Giardina, S., Spartà, E., & Caturla, N. (2021). Antioxidant and reduced skin-ageing effects of a polyphenol-enriched dietary supplement in response to air pollution: A randomized, double-blind, placebo-controlled study. *Food and Nutrition Research, 65*, 1–19. https://doi.org/10.29219/fnr.v65.5619
33. ohn F. Van Vleet, D. (1980). Current knowledge of selenium—Vitamin E deficiency in domestic animals. *Journal of the American Veterinary Medical Association, 176*(4), 321–325.
34. Paradiso, V. M., Summo, C., Trani, A., & Caponio, F. (2008). An effort to improve the shelf life of breakfast cereals using natural mixed tocopherols. *Journal of Cereal Science, 47*(2), 322–330. https://doi.org/10.1016/j.jcs.2007.04.009
35. Pardhiya, S., Priyadarshini, E., & Rajamani, P. (2020). In vitro antioxidant activity of synthesized BSA conjugated manganese dioxide nanoparticles. *SN Applied Sciences, 2*(9), 1–12. https://doi.org/10.1007/s42452-020-03407-5
36. Pincemail, J., & Meziane, S. (2022). On the potential role of the antioxidant couple vitamin E/selenium taken by the oral route in skin and hair health. *Antioxidants, 11*(11), 1–26. https://doi.org/10.3390/antiox11112270
37. Poljsak, B., Dahmane, R., & Godic, A. (2013). Skin and antioxidants. *Journal of Cosmetic and Laser Therapy, 15*(2), 107–113. https://doi.org/10.3109/14764172.2012.758380
38. Ponnampalam, E. N., Kiani, A., Santhiravel, S., Holman, B. W. B., Lauridsen, C., & Dunshea, F. R. (2022). The importance of dietary antioxidants on oxidative stress, meat and milk production, and their preservative aspects in farm animals: Antioxidant action, animal health, and product quality—Invited review. *Animals, 12*(23). https://doi.org/10.3390/ani12233279

39. Santos-Sánchez, N. F., Salas-Coronado, R., Valadez-Blanco, R., Hernández-Carlos, B., & Guadarrama-Mendoza, P. C. (2017). Natural antioxidant extracts as food preservatives. *Acta Scientiarum Polonorum, Technologia Alimentaria, 16*(4), 361–370. https://doi.org/10.17306/J.AFS.0530
40. Shahidi, F., Janitha, P. K., & Wanasundara, P. D. (1992). Phenolic antioxidants. *Critical Reviews in Food Science and Nutrition, 32*(1), 67–103. https://doi.org/10.1080/10408399209527581
41. Silva, S., Ferreira, M., Oliveira, A. S., Magalhães, C., Sousa, M. E., Pinto, M., Sousa Lobo, J. M., & Almeida, I. F. (2019). Evolution of the use of antioxidants in anti-ageing cosmetics. *International Journal of Cosmetic Science, 41*(4), 378–386. https://doi.org/10.1111/ics.12551
42. Sokmen, B. B., Basaraner, H., & Yanardag, R. (2013). Combined effects of treatment with vitamin C, vitamin e and Selenium on the skin of diabetic rats. *Human and Experimental Toxicology, 32*(4), 379–384. https://doi.org/10.1177/0960327112454897
43. Springer, A., & Ziegler, H. (2022). The role of preservatives and multifunctionals on the oxidation of cosmetic O/W emulsions. *Cosmetics, 9*(3). https://doi.org/10.3390/cosmetics9030059
44. Tinggi, U. (2008). *Selenium: Its role as antioxidant in human health* (pp. 102–108). https://doi.org/10.1007/s12199-007-0019-4
45. Wu, S., Rajeshkumar, S., Madasamy, M., & Mahendran, V. (2020). Green synthesis of copper nanoparticles using *Cissus vitiginea* and its antioxidant and antibacterial activity against urinary tract infection pathogens. *Artificial Cells, Nanomedicine and Biotechnology, 48*(1), 1153–1158. https://doi.org/10.1080/21691401.2020.1817053
46. Yadav, A., Kumari, R., Yadav, A., Mishra, J. P., Srivatva, S., & Prabha, S. (2016). Antioxidants and its functions in human body—A Review. *Research Environment Life Science, 9*(11), 1328–1331.
47. Young, J. F., Stagsted, J., Jensen, S. K., Karlsson, A. H., & Henckel, P. (2003). Ascorbic acid, α-tocopherol, and oregano supplements reduce stress-induced deterioration of chicken meat quality. *Poultry Science, 82*(8), 1343–1351. https://doi.org/10.1093/ps/82.8.1343
48. Youssef, D. M., Alshubaily, F. A., Tayel, A. A., Alghuthaymi, M. A., & Al-Saman, M. A. (2022). Application of nanocomposites from bees products and nano-selenium in edible coating for catfish fillets biopreservation. *Polymers, 14*(12). https://doi.org/10.3390/polym14122378
49. Zehiroglu, C., & Ozturk Sarikaya, S. B. (2019). The importance of antioxidants and place in today's scientific and technological studies. *Journal of Food Science and Technology, 56*(11), 4757–4774. https://doi.org/10.1007/s13197-019-03952-x

Exploring Microbial Antioxidant Metabolism: Mechanisms, Recent Advances, and Biotechnological Applications

Dharanidharan Murugan, S. Victor Vasanth Singh, M. Lavanya, Karpaga Swathi Marieswaran, Pandian Chithamparam, and Azhaguchamy Muthukumaran

Abstract Microbial antioxidant metabolism is a vital area of research due to its profound implications in essential biology and applied sciences. This chapter explores how microorganisms counteract oxidative stress, encompassing enzymatic systems, including superoxide dismutase and catalase, and non-enzymatic antioxidants, including glutathione and secondary metabolites. The evolutionary diversifications and regulatory networks that decorate microbial survival below oxidative pressure were also examined. A substantial cognizance has been located on microbial metabolites, detailing their antioxidant residences, various types, and mechanisms of motion, along with biotechnological strategies to optimize their production for industrial and therapeutic functions. Recent advances within the area are highlighted, such as metagenomics and multiomics approaches that resolve microbial diversity and characteristics, synthetic biology for engineering antioxidant pathways, and the position of microbiome studies in know-how host–microbe interactions. Innovations are also mentioned, including nanotechnology and exploring extremophilic microbes for novel antioxidants. This chapter discusses the industrial and biomedical packages of microbial antioxidants, from prescription drugs and healthcare to bioelectronics and biosensing. Despite those advancements, demanding situations exist, including scalability, metabolic engineering hurdles, and ecological issues. The chapter outlines future study opportunities to explore and emphasizes the need for interdisciplinary collaboration to harness microbial antioxidants for sustainable and modern answers.

Keywords Microbial antioxidants · Oxidative stress · Biotechnological applications · Multi-omics approaches

---

D. Murugan · S. V. V. Singh · M. Lavanya · K. S. Marieswaran · P. Chithamparam · A. Muthukumaran (✉)
Department of Biotechnology, Kalasalingam Academy of Research and Education, Krishnankoil, Tamil Nadu, India
e-mail: a.muthukumaran@klu.ac.in

K. Kasinathan and A. Muthukumaran (eds.), *Recent Advances in Oxidative Stress Associated Chronic Diseases Volume 2*, https://doi.org/10.1007/978-981-95-3758-7_10

1 Introduction

All living organisms undergo cellular metabolism, which produces byproducts such as reactive oxygen species (ROS). The *"free radical theory of aging"* states that the leading causes of aging are ROS derived from cellular metabolism dysfunction [1]. ROS originate from oxygen and include free radicals, which are highly reactive due to their unpaired electrons, and non-free radical species that possess oxidizing properties but lack free radicals. Free radicals include superoxide anion radicals ($O2^{\bullet -}$), singlet oxygen, hydroxyl radicals (•OH), nitric acid (•NO), and non-free radical species, such as hydrogen peroxide (H_2O_2), hypochlorous acid (HOCl), and peroxynitrite ($ONOO^-$) [2]. ROS are byproducts of cellular metabolism, particularly mitochondrial oxidative phosphorylation. Normal ROS levels regulate cell proliferation and differentiation. Under stressful conditions, the concentration of ROS is higher, leading to oxidative stress. Oxidative stress occurs when there is an imbalance between the production of ROS and the body's antioxidant defenses, leading to impaired cellular structure and damage to DNA and proteins [3, 4]. Mitochondria, NADPH oxidases, cytochrome p450, uncoupled NO synthase, xanthine oxidase (XO), cyclooxygenases, the endoplasmic reticulum, and peroxidases are all intracellular sources of ROS. Among these, mitochondria and NADPH oxidases are the primary contributors, interacting with each other and playing a significant role in the development of oxidative stress [5]. Antioxidants neutralize ROS by donating electrons to and stabilizing free radicals, thereby preventing cellular oxidative stress. Based on their origin, antioxidants are classified as either natural or synthetic. Natural antioxidants are produced either inside the human body (endogenous) or are supplemented with human diet (exogenous) antioxidants. Superoxide dismutase (SOD), glutathione peroxidase (GPx), and catalase are examples of enzymatic antioxidants produced within the body, while non-enzymatic antioxidants include substances like glutathione, lipoic acid, and specific metal-binding proteins such as ferritin, albumin, and ceruloplasmin. In contrast, exogenous antioxidants, including vitamin C, polyphenols, Vitamin E, vitamin A, carotenoids, and flavonoids, are found in fruits, vegetables, microbial metabolites, and tea [6, 7]. β-Caratenoids are phytoconstitutions found in yellow, orange, and red-colored vegetables and fruits that reduces heart disease, enhance the immune system, prevent certain types of cancer, and protect against age-dependent macular degeneration as shown in Fig. 1.

Phenolic compounds present in green tea are known to lower the risk of cancer and neurodegenerative and cardiovascular diseases. They are also found in marine algae, positively affecting hypertension, dyslipidemia, obesity, and diabetes. Pectin extracted from citrus peel prevents cardiovascular and oxidative damage, coronary heart disease, and diabetes mellitus. Anthocyanins are pigments in fruits and vegetables that prevent cancer, diabetes, neuronal disease, and inflammation [8]. To prevent lipid peroxidation, synthetic antioxidants like butylated hydrotoluene, butylated hydroanisole, and tert-butylated hydroquinone are used as food additives, but these are not widely used because they form some toxic and carcinogenic substances during degradation [9]. The mechanism of antioxidants includes scavenging free

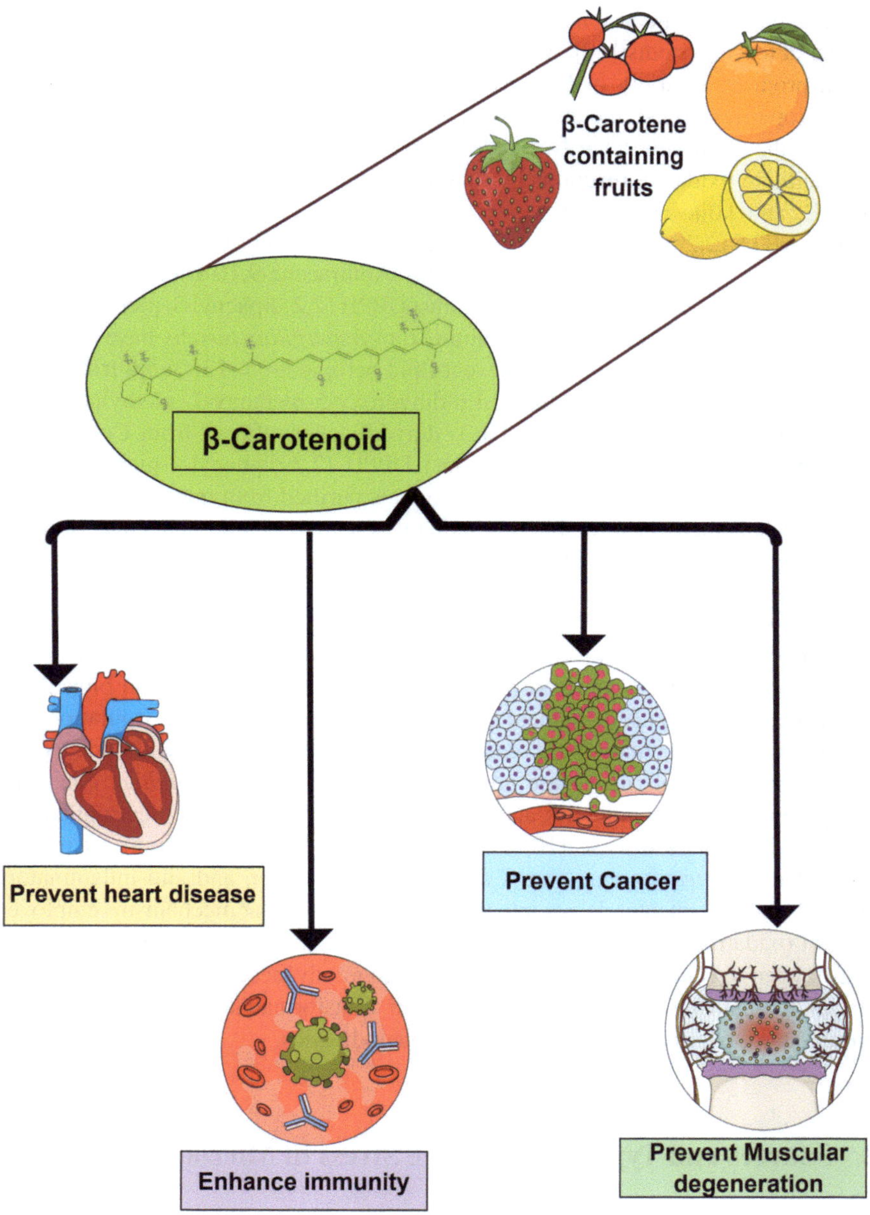

Fig. 1 β-Caratenoids sources and their applications

radicals, chelating metal ions that promote oxidative reactions, and activating the body's defense systems [10]. Microorganisms, such as bacteria, algae, yeast, and fungi, produce a variety of bioactive compounds (metabolites) that have antioxidant properties. These metabolites are widely used in food and pharmaceuticals because they can neutralize free radicals, boost the body's antioxidant defenses, and facilitate oxidative reactions by binding metal ions. Versicolone A, phomapyrone C, and kojic acid from *Aspergillus Versicolor A* have scavenging activity against ABTS (2,2′-azino-bis (3-ethylbenzothiazoline-6-sulfonic acid)) radical [11]. (Z)-1-((1-Hydroxypenta-2,4-Dien-1-Yl) Oxy)Anthracene-9,10-Dione from *Nocardiopsis alba* have scavenging activity against DPPH (2,2-diphenyl-1-picrylhydrazyl) radicals [12]. Terreic acid and terremutin from *Aspergillus terreus* have hydrogen peroxide scavenging activity and prevent lipid peroxidation [13]. 4,5,6-trihydroxy-7-methyl-1,3-dihydroisobenzofuran, 4,6-dihydroxy-5-methoxy-7- methylphthalide, 4,6-dihydroxy-5-methoxy-7-methyl-1,3- dihydroisobenzofuran, from *Cephalosporium* sp. have scavenging activity against DPPH (2,2-diphenyl-1-picrylhydrazyl) radical [14]. These are examples of antioxidants purified from microbial sources.

Recent advances in microbial antioxidant research include nanoencapsulation, metabolic engineering using CRISPR, metagenomics, and multiomics approaches. Phycocyanin (PC) from *Spirulina platensis* is a soluble pigment protein with antioxidant properties. However, it is sensitive to pH, temperature, and light, leading to color loss and reduced bioactivity. To overcome this, PC has been encapsulated in liposomes to reduce their solubility and improve their stability and bioactivity. Encapsulation of PC using alginate and chitosan by the sharp-hole coagulation bath method reduces the temperature sensitivity of PC [15]. Microbial antioxidants also play a significant role in disease prevention. *Akkermansia muciniphila,* a bacteria that produces polyphenols, has antioxidant properties that reduce neuroinflammation in Alzheimer's disease [16, 17]. Antioxidant fucoxanthine from microalgae *Phaeodactylum tricornutum* shows antiobesity, anticancer, and anti-inflammatory effects [18]. This chapter delves into the microbial molecular mechanisms employed to fight oxidative stress, highlights key microbial metabolites that neutralize reactive oxygen species (ROS), reviews cutting-edge advanced technology in microbial antioxidant research, and discusses their transformative applications in food preservation, pharmaceuticals and sustainable industries.

2 Sources and Types of Oxidative Stress in Microorganisms

Oxidative stress is a type of physical challenge faced by microorganisms due to an imbalance in the release of ROS and the ability of the organism to detox those reactive intermediates. This has become a significant interest to the research community in studying the occurrence of damage to a cell. ROS includes Superoxide ions ($O_2{\bullet}^-$), hydroxyl radicals (OH•), alcoxyle (•OR), hydrogen peroxide (H_2O_2), nitric oxide (NO•), and peroxyl radical (•OOR), nitrogen dioxide (NO_2), oxidative stress and damage occurs when their concentrations are higher than the antioxidant defense system of the cell. In microorganisms, oxidative stress is caused by both endogenous metabolic processes and exogenous environmental factors as shown in Fig. 2, each of which results in cellular damage that impairs growth and viability [18]. Oxidative stress in microorganisms occurs in various forms, each with distinct biochemical consequences. Though less reactive than other ROS, superoxide radicals can inactivate enzymes with iron-sulfur (Fe-S) cluster, disrupting significant metabolic pathways such as the amino acid biosynthesis and the tricarboxylic acid (TCA) cycle. Hydrogen peroxide, being more stable and membrane-permeable, can oxidize thiol groups in proteins, leading to enzyme inactivation and protein aggregation. The OH• radical is the most reactive ROS, indiscriminately attacking DNA, lipids, and proteins, causing DNA strand breaks, protein carbonylation, and lipid peroxidation [20]. Lipid peroxidation, a significant consequence of oxidative stress, compromises membrane integrity and generates reactive aldehydes such as malondialdehyde (MDA), further damaging cellular components. Protein carbonylation, an irreversible oxidative modification, leads to loss of enzymatic function and promotes protein aggregation, which is particularly detrimental under prolonged stress conditions. Oxidative DNA damage, including strand breaks and base modifications, can result in mutagenesis or cell death if not repaired by nucleotide excision repair (NER) pathways or base excision repair (BER) [21] as shown in Fig. 3.

2.1 Endogenous Sources of Oxidative Stress

In microorganisms, the primary source of oxidative stress is the endogenous source of ROS generation by aerobic respiration, where the electron transport chain (ETC) leaks electrons that react with oxygen molecules to make superoxide radicals. These fundamentalists are later converted into hydrogen peroxide by SOD and further into hydroxyl radicals through Fenton reactions in the availability of metals such as copper and iron. Recent studies have reported metabolic enzymes' extensive role in generating ROS as sub-products of their catalytic activities, including NADPH oxidases and flavoproteins [22]. Additionally, the incomplete oxidation of the substrate during metabolic flow can cause the accumulation of ROS in the cell, especially in nutrient-limited conditions where metabolic pathways are disabled. Another significant endogenous contributor to oxidative stress is the activity of redox-active enzymes involved in detoxification and biosynthesis. For example, cytochrome P450 enzymes, which are important for the oxidation of organic compounds, inadvertently produce ROS as side products. Similarly, oxidation of low flavin and quinone can generate ROS during electron shuttling, especially when electron acceptors are rare. These endogenous sources underline the imperative of oxidative stress in microbial physiology, requiring a strong antioxidant system to reduce damage [23].

- *NADPH Oxidases*

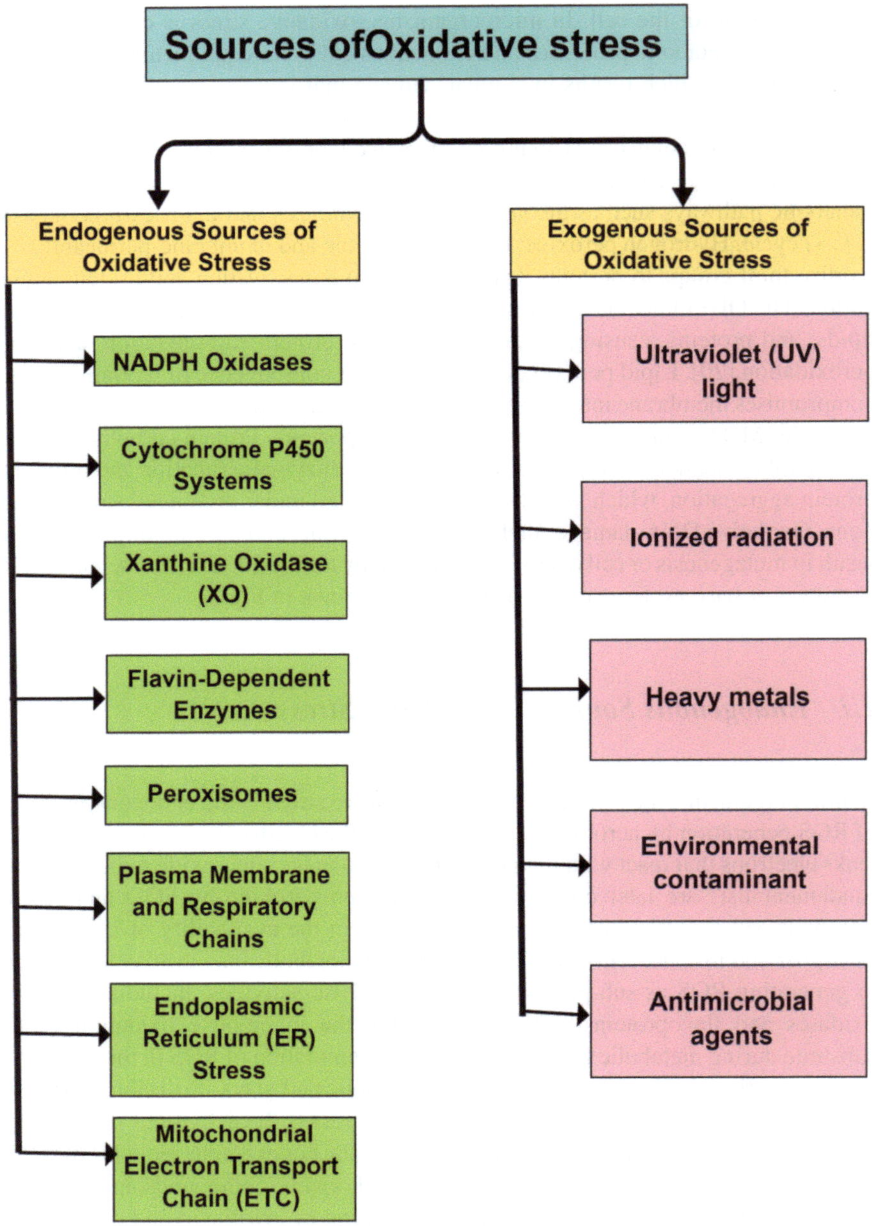

Fig. 2 Various sources of endogenous and exogenous oxidative stress

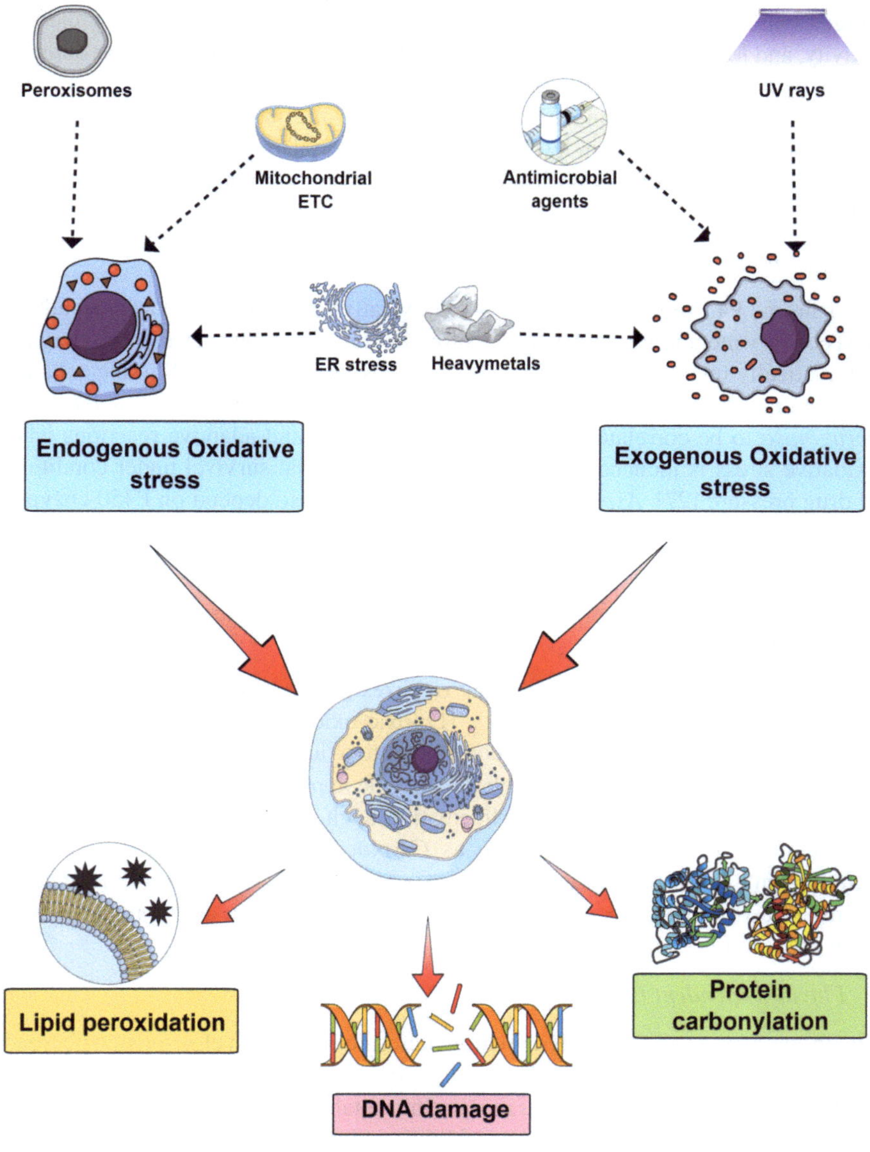

Fig. 3 Consequences of endogenous and exogenous oxidative stress in cells

NADPH oxidase is a crucial enzyme that contribute to electron transfer to molecular oxygen and superoxide radical (O_2^-) generation from NADPH. In *Streptococcus pneumoniae*, enzymes similar to NOX (NOX-like) regulate oxidative stress and wiring through host-pathogen interaction post-infections. In addition, ROS generated by NOX increases the pathogenicity of microorganisms. It is needed in the pathogenic fungus *Candida albicans* to develop hyphae and biofilm central to the fungus's pathogenicity [24, 25]. Redox homeostasis in cells is associated with NOX activity, and loss of it will produce excess ROS, initiating oxidative cellular damage. NOXs consist of seven members: NOX1, NOX2, NOX3, NOX4, NOX5, DUOX1, and DUOX2, which are found in cellular membranes that produce superoxide radicals (O_2^-). NOX2 generates high levels of ROS for bacterial eradication. NOX regulates gene expression, promotes proliferation, synthesizes hormones, and differentiates cells [26].

- *Cytochrome P450 Systems*

Cytochrome P450 enzymes catalyze many oxidative processes, such as biosynthesis of secondary metabolites and detoxification. In these catalytic cycles, these enzymes automatically release electrons that react with oxygen to form ROS in the cell. The generation of ROS through the P450 pathway has been established in *Pseudomonas aeruginosa* to be correlated with antibiotic tolerance and adaptive reactions when oxidative stress is induced, leading to elevated bacterial survival under conditions of drug pressure [27]. *Aspergillus fumigatus* species also depend on P450 enzymes for the biosynthesis of ergosterol, which generates ROS as a byproduct at one point. This ROS generation has an impact on the membrane integrity and can even be a cause of stress for fungi [28].

- *Xanthine Oxidase (XO)*

Xanthine oxidase is a flavoprotein that oxidizes purines and produces superoxide and hydrogen peroxide as byproducts. Hypoxic oxidative activity was reported in *Escherichia coli*, where the enzyme contributes to oxidative DNA damage and mutagenesis. The ROS formation through the enzyme applies most usefully to bacterial populations in conditions of oxygen fluctuation, such as those encountered in biofilm communities. Pathogenic fungi such as *Cryptococcus neoformans* use xanthine dehydrogenase-like Aox1 alternate oxidase (a homologous enzyme) to generate ROS during infection, which can contribute to host tissue invasion and immune evasion from the host [29, 30].

- *Flavin-Dependent Enzymes*

Flavin enzymes, such as NADH dehydrogenase, are engaged in electron transfer reactions and can catalyze ROS development under low electron transport conditions. In antibiotic stress in *Staphylococcus aureus*, the metabolism resolving flavin-medieval leads to an increase in the synthesis of ROS, which either strengthens bactericidal action or may be responsible for mutation and drug resistance. Maintaining a balance between the production and neutralization of ROS is crucial for microbial survival under stress [31].

- *Mitochondrial Electron Transport Chain (ETC)*

The electron transport chain (ETC) of mitochondria is a primary source of ROS in eukaryotic microbes such as yeasts. Electron leakage from complexes I and III of the ETC is an oxygen reaction that produces superoxide radicals. ROS of *Saccharomyces cerevisiae* is implicated in aging control and apoptosis and controls cell lifespan. Bacteria such as *Mycobacterium tuberculosis* possess ETC-like machinery for producing ROS under hypoxia and are shown to be involved in bacterial multiplication and drug resistance [32].

- *Peroxisomes*

Peroxisomes are organelles involved in fatty acid oxidation and detoxification, with hydrogen peroxide as a byproduct. ROS produced in peroxisomes control antifungal resistance in pathogenic fungi like *Candida glabrata* by controlling cellular redox levels. Peroxisomal metabolism cross-talks with oxidative stress responses, reflecting the importance of the organelle for stress adaptation in microbes [33].

- *Endoplasmic Reticulum (ER) Stress*

The endoplasmic reticulum is also where proteins are folded, and any defect in the process will produce ROS by oxidases such as Ero1p and protein disulfide isomerases (PDIs). ROS produced inside *Cryptococcus neoformans* from the ER has been found to regulate the pathogen's virulence determinants for host cell infection. The sequestration of misfolded proteins in the endoplasmic reticulum (ER) triggers oxidative stress responses, which can lead to either cell survival or death, depending on the stress level [34].

- *Plasma Membrane and Respiratory Chains*

Plasma membrane-dwelling bacterial respiratory chain enzymes can generate ROS under specific conditions. In *Pseudomonas aeruginosa*, cytochrome oxidase enzymes generate ROS during stress resistance, biofilm development, and bacterial signaling. ROS generation on the membrane is dynamically controlled for microbial survival under dynamic conditions [35].

2.2 Exogenous Sources of Oxidative Stress

Beyond internal metabolic processes, microorganisms face oxidative stress from various environmental factors. One of the primary external sources is exposure to ultraviolet (UV) radiation, which induces the formation of ROS through the stimulus of water photolysis and endogenous photosensitizers such as flavins and porphyrins. UV-induced oxidative damage is especially harmful to photosynthetic microorganisms, including cyanobacteria, where it disrupts the photosynthesis device and produces singlet oxygen ($^1O_2$), a very catastrophic ROS [19]. Chemical stress, including heavy metals, antibiotics, and pollutants, also increases oxidative stress in the microbial population. Heavy metals, such as cadmium, mercury, and arsenic, disrupt cellular redox, leading to unchecked ROS accumulation by reducing glutathione (GSH) and preventing antioxidant enzymes. Antibiotics, especially those that target bacterial cell walls or protein synthesis, have been shown to induce oxidative stress as a secondary effect. For example, β-lactam antibiotics trigger hydroxyl radical formation by iron metabolism and redox balance. Similarly, environmental pollutants like polycyclic aromatic hydrocarbons (PAHs) and pesticides generate ROS through redox cycling, adding oxidative stress to microbial communities [19, 36]. Intracellularly, ROS is produced as a byproduct of metabolism, while externally, they originate from external factors such as ultraviolet (UV) radiation, ionization radiation, heavy metals, environmental toxins, and antimicrobial agents from various sources. These external pools are essential to pay attention to in clinical microbiology, industrial microbiology, and microbial ecology because oxidative stress is at the core of microbial survival, antibiotic resistance, and pathological effects' origin [37].

- **Ultraviolet (UV) Light**

UV radiation is an external oxidative stress compound mainly found in aquatic and terrestrial microorganisms, which is mainly highlighted by sunlight. UV-A wavelength of 315 nm–400 nm and UV-B wavelength of 280 nm–315 nm are of biological importance, with the ability to enter the microbial cells and generate ROS through photochemistry. Oxidative stress is also induced by endogenous photosensitizers such as ultraviolet radiation, flavin, porphyrin, and NADH, which excite molecular oxygen (O_2) to form singlet oxygen and superoxide ions ($O_2{\bullet}^-$) [38]. DNA damage from ultraviolet radiation causes indirect ROS generation by repairing cyclobutane pyrimidine dimers (CPD) and 6–4 photoproducts. UV-induced oxidative stress was first observed to stimulate antioxidant enzymes like SOD and CAT in *Escherichia coli* and *Pseudomonas aeruginosa* as a protective mechanism. Phototrophic bacteria and cyanobacterial electron transport chains (ETC's) are also interrupted by UV, resulting in ROS leakage by interfering with redox reactions. UV light radiation, for prolonged risk, induces lipid peroxidation enzymes, protein carbonylation, and cell death and, therefore, is an environmental stress factor in microbial existence [39].

- **Ionized Radiation**

Ionization radiation induces oxidative stress in microorganisms through direct and indirect mechanisms facilitated by X-rays and gamma rays. Direct ionization damage to cellular materials by high-energy photons or particles, and indirect damage via water radiolysis and hydroxyl fundamentalist (•OH) generation, hydrogen peroxide (H_2O_2), and hydrated electron (E^-aq). Hydroxyl radical (•OH) is highly cytotoxic as it has DNA strands and proteins, and it peroxidizes the lipid membrane's capacity as it is very reactive. These microorganisms, such as *Deinococcus radiodurans,* are ionizing radiation-resistant and are toxic to ROS due to the efficient repair of bacterial DNA and high intracellular concentrations of manganese ions [40]. Ionization radiation induces widespread oxidative damage in most bacteria, such as human pathological bacteria *Staphylococcus aureus,* and such bacteria undergo mutations or are fatal. Ionizing radiation is used in the food sector for food sterilization and radiotherapy. The health sector exploits its murder ability by inducing oxidative stress in microorganisms [41].

- **Heavy Metals**

Heavy metals such as cadmium (Cd), mercury (Hg), arsenic (As), and lead (Pb) are strong indicators of oxidative stress in microbial communities. Cellular redox reactions involving redox-active metals like iron (Fe) and copper (Cu) disrupt homeostasis by catalyzing the formation of highly reactive •OH radicals through Fenton reactions. Non-redox-active metal cadmium indirectly promotes ROS formation by inhibiting scavenging and antioxidant enzymes through glutathione (GSH). Cadmium causes mitochondrial injury by generating superoxide in *Saccharomyces cerevisiae* [42]. Arsenite causes oxidative stress due to arsenic exposure to the thiol protein disintegration by reacting with arsenite oxidase. Out of metal flow, metal ions for metal effects for metal ion complexions (e.g., *Cupriavidus metallidurans* CZC Operon) to metal effects, ranging from metal effects to metal efflux system for metal ion complexions to metal effects. The mechanism becomes oxidized when there is an increase in metal concentration above normal concentration, and thus causes oxidative damage in addition to development suppression [43, 44].

- **Environmental Contaminant**

Industrial chemical contaminants such as polycyclic aromatic hydrocarbons (PAHS), pesticides, and toxic compound nanoparticles caused oxidative stress in the microbial population. Bacterial enzymes, such as cytochrome P450 monooxygenases and dioxygenases, metabolize PAHs like Benzo[a]pyrene (BaP), producing ROS as byproducts. Separately, TiO_2 and ZnO nanoparticles promote oxidative stress and ROS generation via surface-mediated mechanisms (e.g., photocatalysis or metal ion release). UV light exposure to the release of TiO_2 nanoparticles increases the lysis of OH and microbial membranes. Pesticides such as glyphosates stopped the disintegration of microbial redox homeostasis by stopping the shikimate pathway and inspired ROS generation on a wholesale scale a second time. Bioremediation microbe *Pseudomonas putida* enhanced the antioxidant defense mechanism to detoxify ROS. Chronic exposure enriches oxidative damage with microbial community structure and disintegration of function [45, 46].

- **Antimicrobial Agents**

Antibiotics and disinfectants kill microbes by partly inducing oxidative stress. Beta-lactams disrupt cell wall biosynthesis, while fluoroquinolones inhibit DNA gyrase. Both classes can lead to metabolic dysregulation and excessive ROS production, contributing to cell death. Experimental evidence in the *Escherichia coli* model suggests that bactericidal antibiotic drugs trigger the production of hydroxyl radicals by a puff reaction, causing cell death. Silver nanoparticles (AgNPs) also release silver ions, which reduce thiol groups and glutathione (GSH) levels, thus increasing ROS production [47]. The action of antimicrobial resistance (AMR) is defined by the upgradation of the antioxidant defense mechanisms, which makes superoxide dismutase production less effective in methicillin-resistant *Staphylococcus aureus* (MRSA) [48].

3 Microbial Defense Mechanisms Against Oxidative Stress

Microorganisms employ a sophisticated array of antioxidant defenses to counteract oxidative damage. Enzymatic antioxidants such as SOD, CAT, and peroxidases (e.g., glutathione peroxidase) are crucial in detoxifying ROS. Non-enzymatic antioxidants, including glutathione (GSH), thioredoxins, and mycothiol, maintain cellular redox balance by scavenging free radicals and repairing oxidized biomolecules. Additionally, some bacteria produce specialized molecules like carotenoids and melanin, which act as ROS quenchers and protect against UV-induced oxidative stress [49]. Transcriptional regulators such as OxyR, SoxR, and PerR sense oxidative stress and activate the expression of antioxidant genes, ensuring a rapid and coordinated defense response. In extremophiles, unique adaptations such as increased membrane rigidity and the accumulation of compatible solutes further enhance oxidative stress resistance. Understanding these defense mechanisms provides insights into microbial resilience and has implications for developing antimicrobial strategies and engineering stress-tolerant strains for industrial applications [50]. Oxidative stress in microorganisms arises from endogenous metabolic processes and exogenous environmental insults, contributing to cellular damage through ROS-mediated pathways. While endogenous sources like aerobic respiration and enzyme activities are unavoidable, exogenous factors such as UV radiation, heavy metals, and antibiotics exacerbate oxidative burdens. The resulting damage to DNA, lipids, and proteins. DNA underscores the necessity of robust antioxidant defenses, which vary across microbial species and environmental niches [51].

3.1 Enzymatic Antioxidants

Microorganisms use enzymes to bloom in a severe environment, such as electromagnetic radiation, high salinity, or a pollutant-rich environment. These microbial enzymes, especially antioxidative types, contain some surprising biotechnological applications, including food preservation, pharmaceuticals, and environmental cleansing. Researchers constantly reveal new microbial antioxidants through metagenomics and enzyme engineering, helping improve their effectiveness and professional use [52] (Fig. 4). These enzymatic antioxidants are a unique protein that microorganisms produce to resist ROS-induced oxidative stress. These enzymes protect microbial cells from DNA, protein, and lipid damage. Some key microbial antioxidant enzymes include SOD, which converts harmful superoxide radicals into H_2O_2; CAT and POD, which breaks hydrogen peroxide into oxygen and water; and GPX, which helps to reduce the use of glutathione as a cofactor [53].

- Superoxide Dismutase (SOD)

SOD is a vital metalloenzyme that catalyzes the conversion of superoxide radicals ($O_2{\bullet}^-$) into molecular oxygen (O_2) and hydrogen peroxide (H_2O_2), thereby reducing

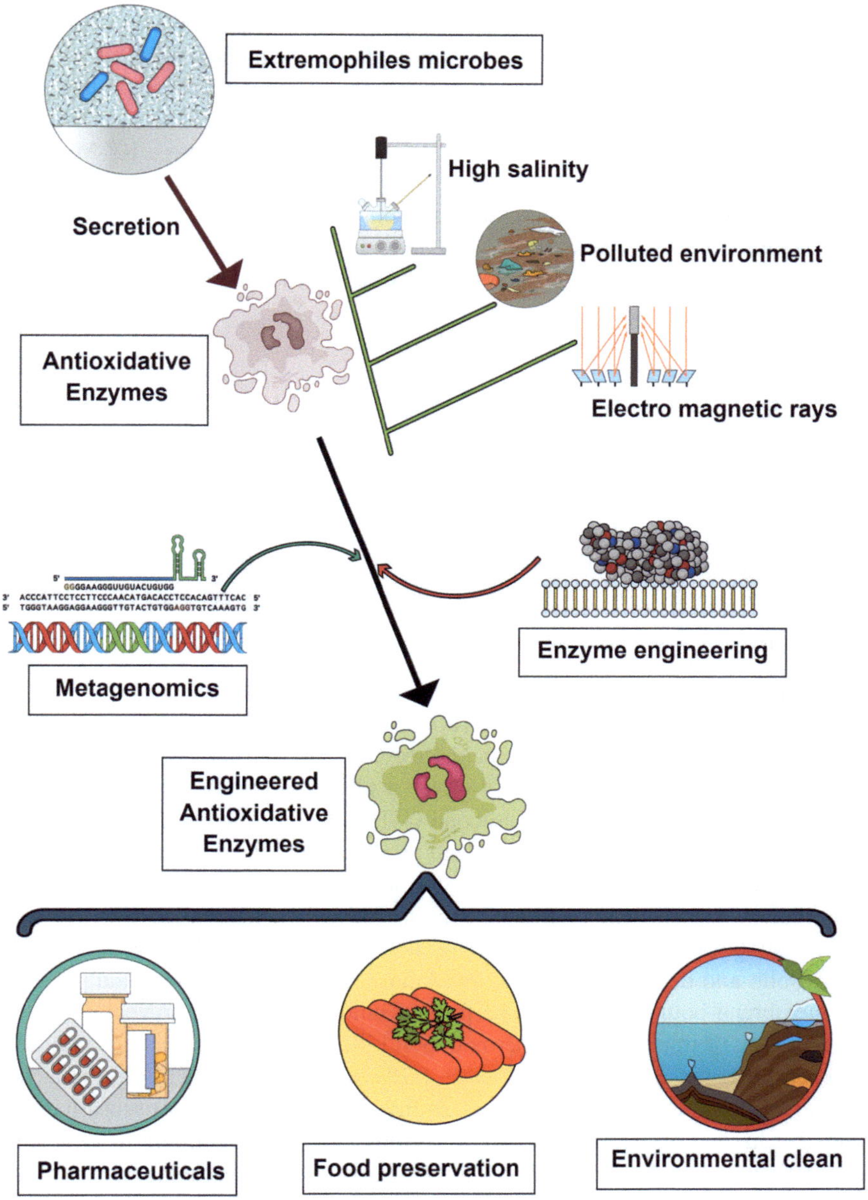

Fig. 4 Biotechnological application in microbial enzymes and its effects

oxidative stress. The response takes place in two steps: first, the enzyme reduces $O_2\bullet^-$ to H_2O_2 with the valuable resource of oxidizing the metal cofactor (Cu^{2+}/Zn^{2+} in SOD1, Mn^{2+} in SOD2, or Fe^{2+} in SOD3). In the second step, any other $O_2\bullet^-$ oxidizes the metal back to its original state while producing O_2. SOD's performance is due to its capability to perform at near diffusion-restrained fees (~ 10^9 M^{-1} s^{-1}), making it one of the simplest antioxidant enzymes. The generated H_2O_2 is, in addition, detoxified by catalase or peroxidases. SOD isoforms are localized in one-of-a-kind cellular cubicles: cytosolic (SOD1), mitochondrial (SOD2), and extracellular (SOD3). Dysregulation of SOD interest is linked to neurodegenerative illnesses, most cancers, and aging [54, 55].

- Catalase

Catalase is a heme-containing enzyme that breaks down H_2O_2 into water and oxygen, preventing the formation of highly reactive •OH radicals via the Fenton reaction. The catalytic mechanism includes steps: first, H_2O_2 oxidizes the heme Fe (III) to shape Compound I (Fe (IV) = O and a porphyrin cation radical). A second H_2O_2 molecule then reduces Compound I back to Fe (III), liberating H_2O and O_2. Catalase is, in particular, green, with a turnover wide variety of ~ 10^6 reactions according to second, basically localized in peroxisomes, wherein oxidases amply generate H_2O_2 [56].

- Glutathione Peroxidase (GPx)

GPx is a selenocysteine (Sec)-containing enzyme that catalyzes the breakdown of H_2O_2 using reduced GSH as an electron donor. The catalytic cycle begins with selenolate (Sec^-) oxidation to selenenic acid (Sec-OH) through the peroxide substrate. This intermediate reacts with GSH to form a selenosulfide adduct (GPx-Sec-SG), releasing water or alcohol [57]. A second GSH molecule assaults the selenosulfide bond, regenerating the active enzyme and producing oxidized glutathione (GSSG). The high nucleophilicity of Sec, better by its low pKa (~ 5), permits GPx to detoxify peroxides at near-diffusion-confined prices efficiently. Mammalian GPx isoforms (GPx1–8) showcase excellent substrate choices and cellular localizations; GPx1 is cytosolic and decreases H_2O_2, at the same time as GPx4 preferentially detoxifies membrane-bound LOOH and performs a unique function in preventing ferroptosis by way of inhibiting lipid peroxidation [58]. The enzyme's reliance on GSH connects it to the glutathione reductase (GR)-mediated NADPH-based recycling device, ensuring sustained antioxidant ability. GPx is essential in tissues liable to oxidative harm, consisting of the liver, erythrocytes, and brain, and its deficiency or genetic polymorphisms are related to neurodegenerative diseases, cardiovascular problems, and most cancers. The selenoprotein nature of GPx makes it sensitive to selenium deficiency, impairs its synthesis, and exacerbates oxidative stress. Recent structural research shows how GPx's thioredoxin-like fold and catalytic triad (Sec, Gln, Trp) optimize peroxide cut price while publishing-translational adjustments can modulate its activity under pathological conditions [59].

- Heme Oxygenase-1 (HO-1)

HO-1 breaks down heme into biliverdin, carbon monoxide (CO), and unbound iron, exerting antioxidant and anti-inflammatory effects. The reaction requires NADPH-cytochrome P450 reductase and O_2. Biliverdin is ultimately reduced to bilirubin, a fantastic antioxidant, while loose iron induces ferritin synthesis, sequestering superoxide-oxidant Fe^{2+}. HO-1 is upregulated under oxidative stress through the Nrf2/ARE pathway and is protective against ischemia-reperfusion injury and neurodegenerative disorders [60].

- Glutathione Reductase (GR)

GR keeps cell GSH tiers by reducing GSSG to GSH using NADPH as an electron donor. The enzyme's flavin adenine dinucleotide (FAD) cofactor accepts NADPH electrons transferred to a disulfide bridge (Cys58-Cys63 in humans), reducing it to thiols. These thiols then assault GSSG, liberating GSH molecules. GR is essential for sustaining redox homeostasis, and its inhibition results in oxidative stress and cell death by apoptosis [61].

- Glutathione S-Transferase (GST)

GSTs constitute a family of phase II detoxification enzymes that facilitate the conjugation of GSH to electrophilic substrates, which incorporates reactive oxygen species (ROS)-derived lipid peroxidation merchandise, xenobiotics, and carcinogens. This conjugation will increase the solubility of poisonous compounds, facilitating their excretion. The catalytic mechanism entails the deprotonation of GSH via a conserved tyrosine or serine residue within the lively internet website, producing a noticeably reactive thiolate anion (GS^-) [62]. This nucleophile then attacks electrophilic sites on substrates, forming a GSH adduct. Some GST isoforms display off glutathione peroxidase function, lowering lipid hydroperoxides (LOOH) to a bargain much less harmful alcohols (LOH) the usage of GSH. GSTs are classified into cytosolic, mitochondrial, and microsomal subtypes, with cytosolic GSTs divided into alpha, mu, pi, theta, and excellent classes based on sequence homology. Polymorphisms in GST genes (e.g., GSTM1 null genotype) are associated with expanded susceptibility to oxidative stress-associated diseases, which include cancer, neurodegenerative diseases, and cardiovascular illnesses. The expression of GSTs is regulated with the valuable resource of the Nrf2/ARE pathway; it's activated under oxidative stress [61, 63].

- Thioredoxin (Trx)

Thioredoxin (Trx) is a small redox-energetic protein ($\sim$ 12 kDa) vital in retaining cell redox homeostasis. It includes a conserved Cys-Gly-Pro-Cys active web website that undergoes reversible oxidation-reduction. In its reduced (oxidized state), Trx donates electrons to target proteins, decreasing disulfide bonds and regulating their function. Key substrates encompass ribonucleotide reductase (essential for DNA synthesis), peroxiredoxins (Prxs, which detoxify peroxides), transcription elements, and NF-κB and p53. The redox function of thioredoxin (Trx) is interconnected with thioredoxin reductase (TrxR), which restores reduced Trx by utilizing NADPH as an electron donor [64]. Trx, moreover, interacts with apoptosis signal-regulating kinase 1

(ASK1), inhibiting its pro-apoptotic activity under oxidative stress. Mammalian cells have primary Trx structures: Trx1 (cytosolic and nuclear) and Trx2 (mitochondrial). Dysregulation of Trx is implicated in most cancers, neurodegenerative diseases, and inflammatory problems [65].

- Thioredoxin Reductase (TrxR)

Thioredoxin reductase (TrxR) is a selenocysteine (Sec)—containing flavoprotein that reduces oxidized thioredoxin (Trx) with the use of NADPH. It belongs to the pyridine nucleotide-disulfide oxidoreductase family and exists in three isoforms: TrxR1 (cytosolic), TrxR2 (mitochondrial), and TrxR3 (testis-unique). The enzyme includes FAD and a redox-active disulfide (Cys-Sec within the C-terminal motif), which transfers electrons from NADPH to Trx. The catalytic mechanism involves NADPH reducing FAD to $FADH_2$. $FADH_2$ transfers electrons to the N-terminal disulfide (Cys59-Cys64 in human TrxR1), reducing it to a dithiol. The C-terminal selenylsulfide (Cys497-Sec498) is decreased, freeing Sec as a nucleophile that attacks oxidized Trx. The resulting selenenyl sulfide intermediate is resolved through other thiol-disulfide exchanges, regenerating energetic TrxR. TrxR is vital for DNA synthesis (via ribonucleotide reductase), antioxidant defense (via peroxiredoxins), and redox signaling. Its inhibition by gold compounds (e.g., auranofin) or selenium deficiency disrupts redox balance, contributing to sicknesses, together with cancer and neurodegeneration [66, 67].

3.2 Non-enzymatic Antioxidants

Microbial non-enzymic antioxidants are low-molecular-weight materials from microbes that fight in the direction of oxidative stress through ROS and RNS. Glutathione, mycothiol, ergothioneine, carotenoids, phenolics, melanin, and nutrients (inclusive of vitamins C and E) are a few antioxidants that possess a herbal feature in protecting microbial cells in competition to oxidative destruction via real-time scavenging of free radicals, metal ion chelation, or repairing oxidized biomolecules [68]. Compared to enzymic antioxidants (including superoxide dismutase and catalase), non-enzymic antioxidants no longer catalyze biochemical reactions but act as redox buffers or sacrificial markers. Glutathione, as an instance, a thiol-containing tripeptide, governs cellular redox homeostasis by donating electrons to detoxify ROS, while melanin schematically represents a broad-spectrum antioxidant interest through its effective free radical-quenching properties. Microbial biosynthesis of such compounds is perpetually controlled through environmental stressors, such as UV radiation, hypoxia, or toxicity [69].

- Glutathione (GSH)

GSH is a tripeptide and a complex intracellular antioxidant. In most cases, its antioxidant function consists of the system's thiol (-SH) organization, which acts as an electron donor to neutralize ROS, including hydrogen peroxide, hydroxyl radicals, and lipid peroxides [70]. The mechanism consists of two main steps: reduction of glutathione (GSH), reduced ROS, and the following regeneration of GSH by glutathione reduces (GR). In one step, glutathione peroxidase (GPX) catalyzes the reduction of water (H_2O) from GSH to oxidizing GSH to glutathione disulfide (GSSG). Similarly, GSH, with electrophilic pollution, and GSH, with glutathione S-Transferase (GST) electrophilic pollution. Oxidized GSSG then reduces the lower back of GSH via GR, using NDPH as an electron donor. This cycle guarantees glutathione-reduced non-stop delivery. In addition, GSH directly scavenges radicals through defensive mobile components such as non-enzymatic reactions, proteins, lipids, and oxidative DNA damage. Glutathione also contributes to revitalizing other antioxidants, including vitamins C and E, as well as their oxidized forms. Its depletion is linked to oxidative stress-related illnesses, including cancer [71, 72].

- Albumin

Human serum albumin (HSA) is a multifunctional plasma protein that displays antioxidant properties primarily by its free thiol groups (CYS 34), which scavenges ROS and RNS, including hydroxyl radicals (•OH), hypochlorous acid (HOCl), and superoxide (O_2^-). HSA binds to transitions such as copper and iron, closing Fenton reactions that produce significant reactive radicals. Protein acts as a nitric oxide (NO) reservoir, controls its bioavailability, and closes peroxynitrite formation [73]. In addition, HSA binds and transports bilirubin and fatty acids, which contributes more to its antioxidative capacity. Under oxidative pressure, CS 34 passes through sulfinylation (-SOH) or forms disulfides, which can be reduced again using thioredoxins or glutathione systems. The ability to distinguish HSA's season-oxidant metals and neutralize undetected radicals is essential in reducing systemic oxidative damage in conditions such as sepsis, liver disorders, and diabetes. Recent research highlights its function in protection against endothelial dysfunction and atherosclerosis [74].

- Bilirubin

Bilirubin, a heme degradation product, is a powerful endogenous antioxidant that scavenges peroxyl radicals (ROO•) and inhibits lipid peroxidation. Its mechanism includes cyclic redox feedback in which bilirubin is oxidized to biliverdin, after which it is reduced by biliverdin [75]. The lipophilic nature of the bilirubin allows it to integrate into the cell membrane, protecting lipids from oxidative damage. It also challenges the transition metals, closing the catalyst radical pair generation. Bilirubin NADPH oxidase suppresses humor in high concentrations, reducing superoxide production [76].

- Melatonin

Melatonin (N-acetyl-5-methoxytryptamine) is a potent antioxidant primarily synthesized in the pineal gland, with direct and indirect action methods against oxidative stress. Its radical-scavenging capacity is produced by the loss of an electron-rich indole ring, which donates the electrons to neutralize ROS, including hydroxyl radicals (OH), singlet oxygen (^{1}), and pelvis [77]. (AFMK) and cyclic 3-hydroxylatonin.

Unlike the classical antioxidative, melatonin does not engage in mitochondrial Q-cycle electron transfer. Instead, it provides antioxidant protection through direct free radical neutralization (e.g., •OH, $ONOO^-$) and indirect upregulation of antioxidant enzymes. Melatonin upregulates key antioxidant enzymes such as SOD, GPx, and catalase through the activation of nuclear factor erythroid 2-related factor 2 (Nrf2). It also modulates nitric oxide synthase (NOS) and nitric oxide (NO) levels, further reducing oxidative stress [78, 79]. Melatonin reduces electron leakage from the electron transport chain in mitochondria, decreasing superoxide (O_2^-) generation. Its amphiphilic nature protects cellular membranes and cytosolic structures (e.g., proteins/DNA), making it highly effective against oxidative DNA damage and lipid peroxidation.

- Mycothiol

Mycothiol (MSH) is a thiol molecule with a low molecular weight produced by actinomycetes, which functions analogously to glutathione in microorganisms. Its antioxidative mechanism consists of a thiol organization of cysteine, which immediately reduces ROS and detoxifies electrophilic compounds by mycothiol S-Transferase. MSH mixed with oxidized protein (mycothyolation) protects them from irreversible oxidation, and the use of NADPH is revived with the help of mycothyol reduces [80]. It also contributes to the reduction of peroxides by mycothiol peroxidase. Unlike glutathione, MSH contains an acetylated glucosamine-inositol backbone, which improves its balance in a harsh environment. Mycothiol-poor bacteria showcase their essential role in oxidative stress, antibiotics, and heavy metals, and bacterial redoxes in homeostasis. In mycobacteria, mycothiol biosynthesis is aimed at developing novel antibiotics for pathogenic mycobacteria such as tuberculosis [81].

- Coenzyme Q10 or Ubiquinone

The coenzyme Q10, a lipid-soluble benzoquinone, also shows a critical electron carrier as an attractive antioxidant within the mitochondrial respiration chain. In its reduced form (Ubuquinol, $COQ10h_2$), it prevents lipid peroxidation and low-density lipoprotein (LDL) in the cell membrane [82]. Ubiquinol reproduces tocopherol (Vitamin E) from its radical shape, expanding the antioxidant defense network. Unlike other antioxidants, COQ10 is a continuous cytosolic enzyme, such as oxidized (Ubiquinon) and NQO1 upregulation shows therapeutic potential in statin-induced myopathy (via CoQ10 regeneration), cardiovascular diseases, and neurodegenerative disorders. It supports mitochondrial ATP synthesis and antioxidant defenses, which are critical for muscle and neuronal health [83].

3.3 *Evolutionary Adaptations of Microbial Antioxidant Defenses*

Microbial antioxidant protection mechanisms have advanced to counteract numerous stages of oxidative stress confronted by microbes in various milieus. In aerobic environments where oxygen is abundant, microbes are subjected to ROS in the form of superoxide anions (O_2^-), hydroxyl radicals (•OH), and hydrogen peroxide (H_2O_2) [84]. Aerobic microorganisms and archaea, therefore, have potent enzymatic antioxidants such as SOD and catalase that rapidly neutralize ROS and inhibit oxidative damage caused by them to the cell [85]. These enzymes are frequently constitutively expressed at basal levels; however, they are upregulated under oxidative stress, ensuring a robust defense mechanism [86, 87]. Anaerobic microorganisms cultivated below oxygen-constrained conditions, including the gut or deep subsurface sediment, utilize an evolutionary strategy. As they lack conventional aerobic antioxidant enzymes, alternative mechanisms include maximum anaerobic use of thioredoxin and glutathione structures for redox homeostasis. Some of these obligate anaerobes, like the Clostridium species, also have rubrerythrin and flavodilol proteins that can function as scavengers of peroxide to make it through its temporary exposure to oxygen. Somewhat amazingly, anaerobes have been reported to be acquiring antioxidant genes through horizontal gene transfer (HGT) from oxygen-tolerating relatives and found that there is evidence for an evolutionary dynamic adaptation to variable oxygen levels. For example, the gut symbiont *Bacteroides thetaiotaomicron* has acquired a hybrid SOD enzyme that helps it survive against oxidative bursts against host immunity [88]. Extremophiles living in intense situations with high ROS manufacturing have specific antioxidant diversifications. Microbes, such as *Thermus thermophilus* and *Sulfolobus solfataricus*, contain thermostable antioxidant enzymes active under acidic pH conditions and in surfeit temperatures. Microbial metalloenzymes generally comprise metal cofactors (such as iron, manganese, or copper) that impart structural stability under extreme temperatures. Novel ROS-scavenging molecules, including polyamines, which guard cellular systems against oxidative harm, are also utilized by hyperthermophilic archaea. The work with *Pyrococcus furiosus* has located a brand new, surprisingly efficient peroxiredoxin device, operative at temperatures above 100 °C, evidencing the evolutionary sophistication of extremophilic antioxidant protection [86, 87].

3.4 *Regulatory Networks Controlling Microbial Antioxidant Function*

Microbial antioxidant systems' expression and function are precisely regulated through highly evolved genetic and post-translational systems, detecting oxidative stress and engaging proportionate defense. Chief among these transcriptional regulators are OxyR, SoxR, and PerR, which control these responses with significant roles.

OxyR is a well-studied redox-regulated protein from Gram-negative bacteria and has been demonstrated to regulate catalase (katG), alkyl hydroperoxide reductase, and glutathione biosynthesis gene expression upon detection of hydrogen peroxide. Structural studies have indicated that OxyR changes conformation upon oxidation, so the capability to bind to promoter regions and activate transcription exists. Studies in Pseudomonas aeruginosa have recently illustrated that OxyR also controls biofilm formation and virulence, connecting antioxidant defense with pathogenic fitness [50]. In Gram-positive bacteria, the PerR regulon regulates the peroxide stress response. Metalloregulator PerR represses antioxidant genes under non-stress but leaves DNA upon metal-catalyzed oxidation, derepressing targets such as katA and mrgA in *Bacillus subtilis.* A 2023 Nature Microbiology study found that *Staphylococcus aureus* PerR orthologs show strain-specific variability of metal selectivity, which limits peroxide resistance and competence for colonization of the human skin microbiota. This illustrates evolutionary flexibility in regulatory networks for adaptation to host-associated environments [89]. In addition to canonical regulators, redox-sensitive transcription factors and sRNAs fine-tune the antioxidant response. The sRNA OxyS is oxidative-stressed and induced to regulate fhlA and rpoS expression, combining oxidative defense and metabolic and stress-response mechanisms. Likewise, the MarA/SoxS/Rob regulon of Enterobacteriaceae activates SODs and multidrug efflux pumps, demonstrating cross-talk between antibiotic resistance and oxidative stress [90].

4 Microbial Metabolites with Antioxidant Properties

4.1 Phenolic Compounds

The microbial metabolites produced by *Streptomyces cellulosae* strain TES17 have shown significant antioxidant activity. The phenolic and flavonoid content of the strain's extract was measured, and it was found that it demonstrated decent antioxidant activity. The metabolites produced by the *S. cellulosae* strain TES17, which were identified using UPLC analysis, include other phenolic compounds such as gallic acid, catechin, epicatechin, caffeic acid, umbelliferone, coumaric acid, ellagic acid, quercetin, and kaempferol. These compounds exhibit antioxidant activity through various mechanisms such as free radical scavenging, metal chelation, and inhibition of lipid peroxidation. In particular, catechins are identified as the primary phenolic compound with antioxidant activity [91]. The study on fungal endophytes revealed several phenolic compounds that exhibit strong antioxidant properties. Those compounds include quercetin, a flavonoid found in *Nectria haematococca* and *Fusarium chlamydosporum*, which plays an essential role in maintaining oxidative stress by free radical-scavenging activity, caffeic acid, p-hydroxybenzoic (pHBA), and p-coumaric (pCA) acid found in *Alternaria tenuissima,* which have shown DPPH radical-scavenging capacity and were found in *Bipolaris specifera.* Other phenolic acids, such as vanillic, ferulic, and chlorogenic from *Aspergillus terreus,* have potential antioxidant capacity [92]. Cyanobacteria like *Anabaena constricta* extract is rich in polyphenols (gallic acid, chlorogenic acid, caffeic acid, vanillic acid, ferulic acid) and flavonoids (rutin, quercetin, kaempferol), showing high antioxidant activity [93]. Fungi such as *Aspergillus tamarii, Aspergillus oryzae*, and *Aspergillus candidus* also produce phenolics such as ellagic acid (EA), gallic acid (GA), and ferulic acid (FA), which show more excellent antioxidant activity [94] as show in Table 1.

4.2 Exopolysaccharides

Exopolysaccharides produced by lactic acid bacteria (LAB) strains identified from yogurt, such as *Lactobacillus plantarum, Lactobacillus delbrueckii*, and *Lactobacillus fermentum,* have shown greater antioxidant potential through high DPPH radical-scavenging activity and also shown optimal ferric oxide reducing power [95]. Exopolysaccharides obtained from *Lactiplanti bacillus plantarum* VAL6 demonstrate antioxidant activity by showing a significant radical-scavenging activity. It has also shown a total antioxidant capacity of about 8.00 mg AAE/g. These exopolysaccharides possess unique structural characteristics, i.e., they have alpha sugar and beta sugar configurations and a huge amount of mannose. This type of configuration ensures them as a potential alternative to synthetic antioxidants [96]. The Exopolysaccharides Produced by *Lysobacter soyae* sp. nov exhibited radical-scavenging activity of ~ 73.47%. This indicates that EPS produced by *Lysobacter soyae* has notable antioxidant potential. These exopolysaccharides have sugars such as mannose, glucose, galactose, and trace amounts of arabinose [97]. Various strains of *Bacillus* are major producers of exopolysaccharides, which can act as antioxidants. These EPS can neutralize free radicals and reduce ROS overload. The antioxidant activity of the exopolysaccharides produced by different microbial strains varies due to their unique structural arrangements. *Bacillus subtilis* and the marine bacterial strain *Bacillus altitudini* have shown potential antioxidant properties [98]. A novel strain of *Aspergillus* sp. DHE6 was found to be the potential producer of exopolysaccharides under optimized conditions. Exopolysaccharide extract from this strain was purified, and its radical neutralization capacity was assessed using DPPH radical—scavenging activity. It is found that these EPS can potentially neutralize free radicals, making them potent antioxidants [99]. The exopolysaccharides produced by Bacillus albus DM-15 have a non-crystalline structure and are made up of glucose, galactose, xylose, and rhamnose. The EPS demonstrates potent nitric oxide radical scavenging at ~ 58.92%. DPPH radical scavenging assay at ~ 58.17% and ABTS radical scavenging at ~ 70.47%. This property helps in reducing oxidative stress [100]. Levan, a polysaccharide produced by certain microorganisms, can exhibit antioxidant properties. Due to its polysaccharide nature, microbial levan can potentially scavenge free radicals and enhance cellular mechanisms [101].

Table 1 Microbial metabolites and their antioxidant activities

Metabolites	Source: microorganisms	Antioxidant activity	Reference
Phenolic compounds			
Gallic acid	*Streptomyces cellulosae TES17, aspergillus* spp., *Anabaena constricta*	Strong radical scavenger, metal chelation	[91, 93, 94]
Catechin	*Streptomyces cellulosae TES17*	A primary antioxidant prevents lipid oxidation	[91]
Caffeic acid	*Streptomyces cellulosae TES17, Alternaria tenuissima, Anabaena constricta*	DPPH radical scavenging, anti-inflammatory	[91–93]
Quercetin	*Streptomyces cellulosae TES17, Nectriahaematococca, fusarium chlamydosporum, Anabaena constricta*	Potent free radical scavenger	[92, 93]
Kaempferol	*Streptomyces cellulosae TES17, Anabaena constricta*	Enhances cellular antioxidant defense	[91, 93]
Ferulic acid	*Aspergillus terreus, Anabaena constricta, aspergillus* spp.	Prevents oxidative stress-related diseases	[92–94]
Chlorogenic acid	*Aspergillus terreus, anabaena constricta*	Reduces oxidative stress, metal chelator	[92, 93]
p-Hydroxybenzoic acid	*Alternaria tenuissima, Bipolarisspecifera*	DPPH radical scavenging, anti-aging properties	[92]
Exopolysaccharides (EPS)	*Lactobacillus* spp., *bacillus* spp., *Lysobactersoyae, aspergillus* sp.	DPPH radical scavenging, nitric oxide reduction, cellular protection	[95, 97, 99]
Carotenoids			
β-Carotene, lycopene, lutein	*Cyanobacteria (blue-green algae), chlorella, Dunaliella, Haematococcus, Corynebacterium glutamicum, Blakesleatrispora*	Quenches reactive oxygen species (ROS)	[102, 103]
C-Phycocyanin	*Cyanobacteria (blue-green algae)*	Free radical scavenger, anti-inflammatory	[102]
Astaxanthin	*Haematococcuspluvialis*	Superior singlet oxygen quencher	[102]
Torularhodin	*Various oxidative-stressed species*	Increases antioxidant capacity under stress	[104]

(continued)

Table 1 (continued)

Metabolites	Source: microorganisms	Antioxidant activity	Reference
Zeaxanthin	*Flavobacterium* sp., *pseudomonas* sp., *Paracoccuscarotinifaciens*	A potent antioxidant, it prevents cellular damage	[104]
Canthaxanthin	*Dietzianatronolimnaea, Brevibacterium* sp., *Rhodotorulamucilaginosa*	Free radical scavenger protects against oxidative damage	[105]
Short chain fatty acids (SCFAs)			
Acetic acid	*Escherichia coli, Bifidobacterium* spp., *lactobacillus* spp.	Neutralizes ROS supports cellular defense	[106]
Propionic acid	*Propionibacterium* spp., *Veillonella* spp., *Bacteroides* spp.	Enhances antioxidant enzymes	[106]
Butyric acid	*Clostridium butyricum, Faecalibacteriumprausnitzii, Roseburia* spp.	Reduces oxidative stress, promotes gut health	[106]

4.3 Carotenoids

Carotenoid substances like β-Carotene, lycopene, and lutein found in blue-green algae (cyanobacteria) can effectively quench reactive oxygen species (ROS), which can act as efficient antioxidants. C-phycocyanin, a pigment found in blue-green algae, can act as a free radical scavenger and inhibit inflammation. Astaxanthin, a carotenoid found in *Haematococcus pluvialis*, can efficiently quench singlet oxygen and scavenge free radicals. This property has higher antioxidant activity when compared to other carotenoids. Various fungal species, including zygomycetes, ascomycetes, and basidiomycetes, are known for their carotenoid-producing ability. For example, *Rhodotorula* sp. accumulates carotenoids, which have potential antioxidant applications [49]. Microbial sources such as *Chlorella, Dunaliella, Haematococcus, Corynebacterium glutamicum,* and *Blakeslea trispora* are known for producing essential carotenoids like astaxanthin, lutein, and beta-carotene. They show antioxidant activity on ROS. These free radicals can cause cellular damage and several diseases. The antioxidant activity of these carotenoids can reduce the risk by neutralizing the harmful free radicals [102]. Apart from this, Torularhodin is also known for its antioxidant capabilities. Under oxidative stress, species increase their carotenoid content to neutralize the free radicals [103]. Microorganisms like *Flavobacterium* sp., *Pseudomonas* sp., and *Paracoccus carotinifaciens* are major producers of the carotenoid Zeaxanthin. Likewise, another carotenoid, canthaxanthin, is found in organisms like *Dietzia natronolimnaea, Brevibacterium* sp., and *Rhodotorula mucilaginosa.* These carotenoids function as antioxidants by suppressing free radicals and reducing oxidative stress [104].

4.4 Short-Chain Fatty Acids

Short-chain fatty acids like acetic acid produced by *Escherichia coli, Bifidobacterium* spp., and *Lactobacillus spp,* propionic acid produced by *Propionibacterium* spp., *Veillonella* spp., and *Bacteroides spp, and* butyric acid produced by *Clostridium butyricum, Faecalibacterium prausnitzii,* and *Roseburia* spp. have tremendous antioxidant potential. These SCFAs neutralize ROS and increase the antioxidant enzymes, thereby improving cellular defense against oxidative stress [105] as shown in Fig. 5, respectively.

5 Mechanisms of Microbial Antioxidant Metabolites

Antioxidants utilize various mechanisms to prevent cellular damage caused by ROS/free radicals. Microbial antioxidant metabolites are essential for cellular defense against ROS. These metabolites, produced by microorganisms, exhibit diverse mechanisms of action. They are as follows: (i) *Direct Scavenging of Free Radicals:* Antioxidants can directly react with free radicals and effectively neutralize them by donating electrons or protons, stabilizing the radicals and preventing further damage. For example, vitamins C and E donate electrons to free radicals like superoxide ($O_2{\bullet}^-$) and hydroxyl radicals (OH•). (ii) *Enzymatic Reactions:* Enzymatic antioxidants, such as superoxide dismutase (SOD), catalase (CAT), and glutathione peroxidase, catalyze specific reactions that convert free radicals and ROS into less harmful substances. SOD converts superoxide anions to hydrogen peroxide. CAT and glutathione peroxidase then convert H_2O_2 into water and oxygen, reducing the harmful effects. (iii) *Metal Chelation:* Some antioxidants can bind to metal ions (like Fe and Cu ions) that enhance the production of free radicals through Fenton reactions. By chelating these metals, antioxidants reduce the availability of these ions for radical generation, thus preventing oxidative stress [106]. (iv) *Regeneration of Other Antioxidants:* Certain antioxidants can regenerate or reactivate other antioxidants. Ascorbic acid (vitamin C) can regenerate tocopherol (vitamin E) after neutralizing a free radical, enhancing the antioxidant activity. (v) *Preventing Lipid Peroxidation:* Antioxidants can inhibit lipid peroxidation by preventing the free radicals from attacking lipids in cell membranes, reducing cellular damage. Compounds like flavonoids and carotenoids can break the chain reaction of lipid oxidation processes [107]. (vi) *Gene Expression Modulation:* Some antioxidants can modulate gene expression related to oxidative stress responses, enhancing the production and

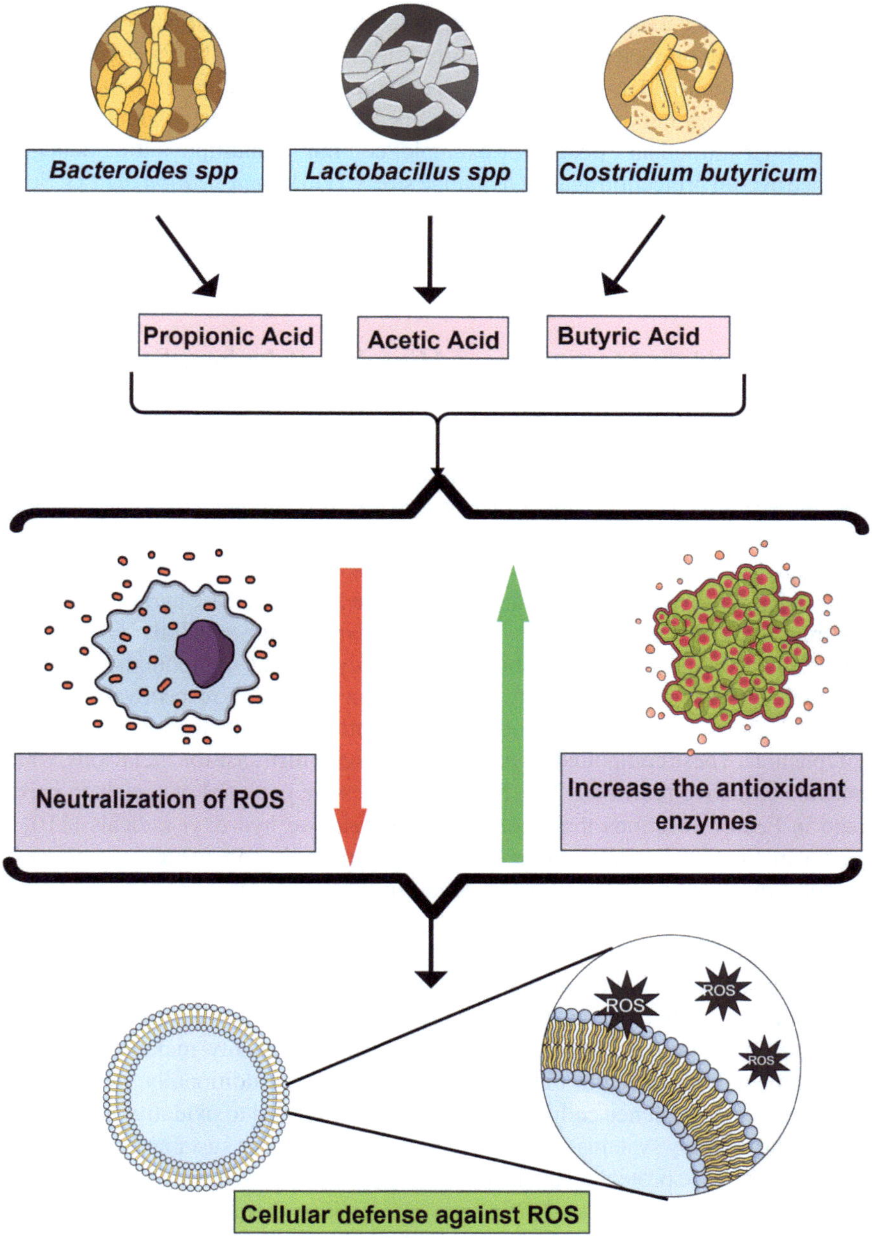

Fig. 5 Short-chain fatty acid-mediated cellular defense against oxidative stress

activity of endogenous antioxidant enzymes. This includes pathways that lead to the upregulation of nuclear factor (erythroid-derived 2)-like 2 (Nrf2), regulating the expression of antioxidants and protective enzymes. (vii) *Synergistic Actions:* Many antioxidants work in combination because their combined effect is greater than the sum of their individual effects. For example, combining multiple antioxidants can enhance overall protection against oxidative stress. (viii) *Sequestration of ROS*: Some antioxidants may sequester ROS by encapsulating them or forming stable complexes that mitigate their reactivity and biological impact [108].

5.1 Antioxidant Mechanism of Microbial Polyphenolic Compounds

Microbial polyphenolic compounds represent a diverse group of secondary metabolites with significant antioxidant properties. These compounds demonstrate their antioxidant activity via multiple mechanisms that protect against oxidative damage and cellular stress. One primary mechanism involves directly scavenging ROS. The hydroxyl groups in polyphenolic structures enable them to release hydrogen atoms to suppress free radicals, thereby preventing oxidative damage to cellular components. This electron delocalization capability allows polyphenolic compounds to stabilize free radicals through aromatic rings [109] as shown in Fig. 6, respectively.

Metal chelation represents another crucial antioxidant mechanism of microbial polyphenols. These compounds can form complexes with transition metal ions, which enhance oxidative reactions. This binding reduces the potential of metals to participate in Fenton reactions that generate highly reactive hydroxyl radicals [110] as shown in Fig. 7. The interaction between polyphenols and metal ions creates a microenvironment with restricted reagents for biological reactions, where trivalent iron or other minerals act as essential cofactors for many common enzymes.

Microbial polyphenols also exert their antioxidant effects through membrane interactions. They can interact with lipid bilayers, altering membrane permeability and disrupting protein chains, which can both protect against oxidative damage and release ROS in targeted applications. This dual functionality makes them versatile compounds in different biological contexts [111]. Additionally, polyphenolic compounds can influence cellular enzyme activities related to oxidative stress. They modify the delivery systems of bioactive molecules, creating sustained release properties that enhance protection against microorganisms. The solubility of polyphenols can be increased by converting them into salts, making them more useful in various food and biomedical applications. The oxidation state of polyphenols affects their activity with oxygen, creating a critical interplay that can reduce the capability of specific bacterial strains relying on the polyphenols. Studies have shown that certain polyphenols, like myricetin, quercetin, rutin, and epicatechin, exhibited more excellent oxidative stability than synthetic antioxidants, highlighting their potential as natural alternatives [112].

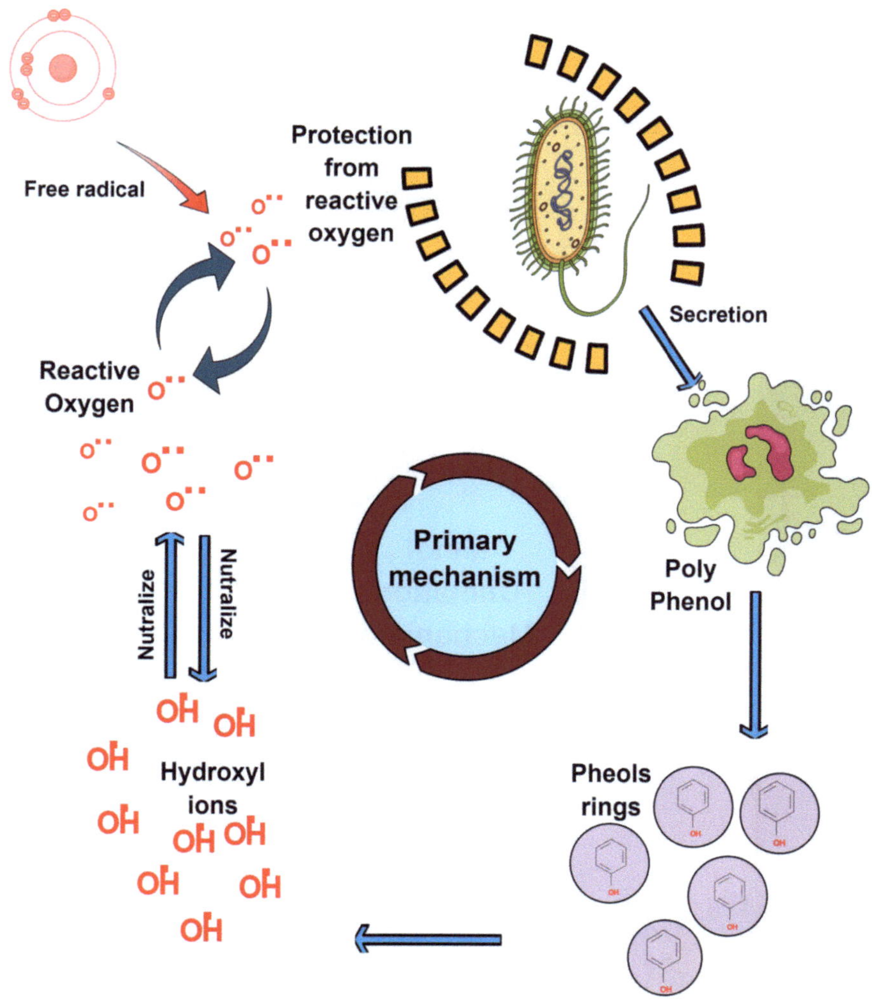

Fig. 6. The primary mechanism of bacterial cell protection from reactive oxygen

5.2 *Antioxidant Mechanism of Flavonoids*

Flavonoids are a subclass of polyphenolic compounds with potent antioxidant activities that operate through multiple complementary mechanisms. Their antioxidant capabilities are primarily attributed to their chemical structure, particularly the configuration of hydroxyl groups and conjugated double bonds. The most common antioxidant mechanism of flavonoids is their ability to neutralize free radicals. Flavonoids neutralize ROS by supplying hydrogen atoms or electrons, effectively reducing the damage to cellular components. This process is facilitated by the presence of phenolic hydroxyl groups that can stabilize free radicals by delocalizing the

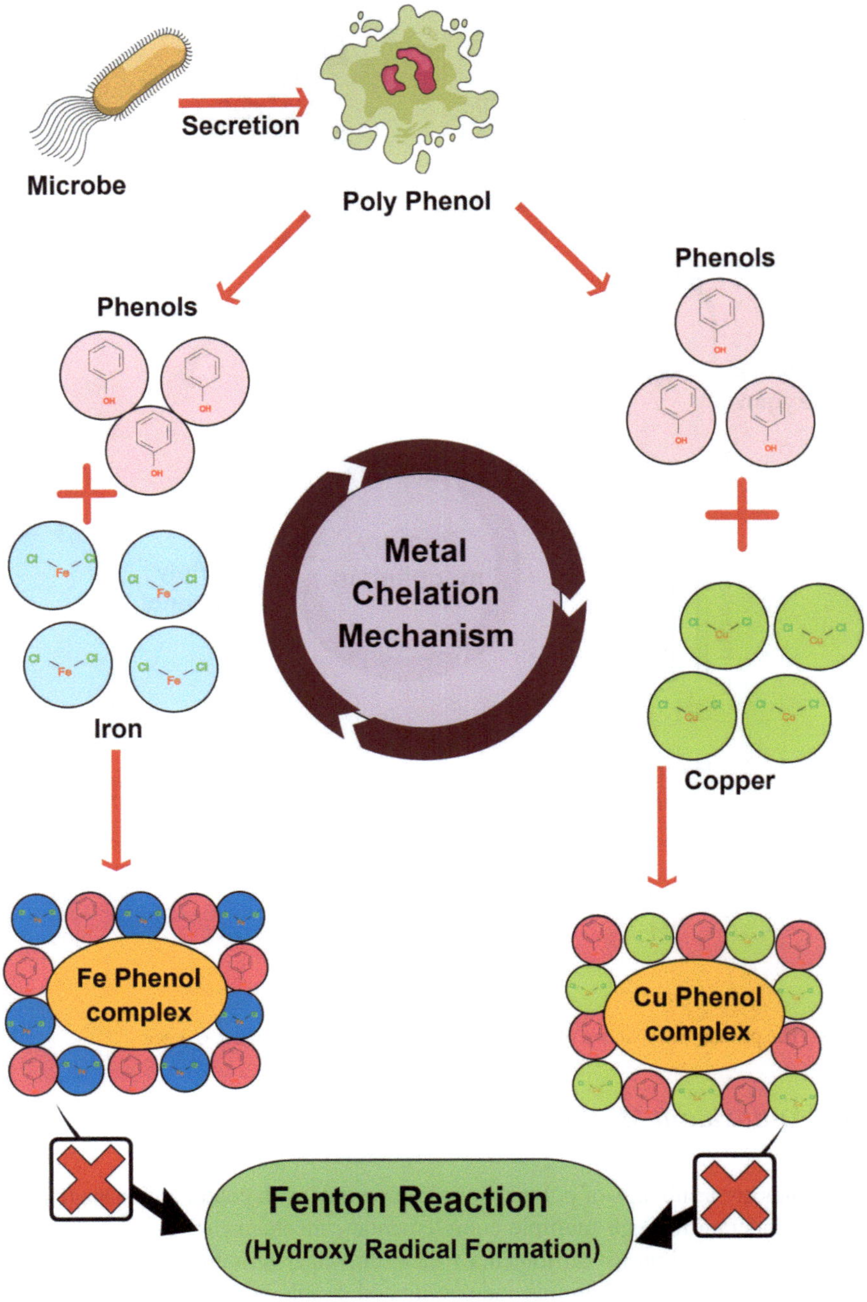

Fig. 7 Metal chelation mechanism for control of the Fenton free radical formation reaction

unpaired electron across the aromatic ring structure [113]. Metal chelation represents another significant antioxidant mechanism of flavonoids. These compounds can bind to transition metals such as iron and copper. This chelation activity inhibits oxidative stress induced by metal ions and prevents the formation of ROS. Flavonoids also inhibit enzymes responsible for generating reactive species, such as xanthine oxidase and NADPH oxidase, thereby reducing the overall generation of ROS in cellular environments. This enzyme inhibition occurs through competitive interaction with ATP at the catalytic sites, disrupting their function and limiting oxidative stress. Another mechanism involves the regeneration of antioxidants, like vitamins C and E, strengthening the overall antioxidant defense mechanism. This recycling capability creates a synergistic antioxidant network that provides comprehensive protection against oxidative damage [114]. Flavonoids play a specific role in ROS homeostasis: acting as antioxidants in normal physiological conditions while promoting pro-oxidants in cancer cells and initiating apoptotic signaling pathways [113]. The relationship between the structure and activity of flavonoids is well understood, with certain structural elements playing an essential role in enhancing antioxidant capacity. A catechol group on the B-ring, a double bond between the C2 and C3 positions, and hydroxyl groups at positions 3 and 5 are fundamental in enhancing their antioxidant potency [115].

5.3 Antioxidant Mechanism of Carotenoids

Carotenoids are fat-soluble pigments produced by various microorganisms known for their potent antioxidant properties. With their unique molecular structure, particularly their extensive conjugated double-bond system, carotenoids convert singlet oxygen to its less reactive triplet ground state ($3O_2$). In this process, carotenoids absorb the energy and release it as heat. Then, they return to their original state to repeat the allowing cycle. Carotenoids can act as antioxidants through chemical quenching, directly reacting with ROS. Carotenoids donate electrons, become oxidized, and are gradually used up [116]. Carotenoids also use hydrogen atom transfer (HAT), transferring both a proton and an electron to stabilize ROS. Their structural features strongly influence the antioxidant capacity of carotenoids. They are broadly categorized into carotenes (*which consist solely of hydrocarbons*) and xanthophylls (*which contain functional groups like hydroxyl, keto, or methoxy*). The pattern of conjugated double bonds is the primary factor affecting antioxidant activity [106]. Stereochemistry plays a crucial role in affecting carotenoid functionality influencing solubility and absorbance. Trans isomers are more rigid, which can lead to greater aggregation and crystallization. It can display more potent antioxidant activity than cis forms. Carotenoids are produced more often as defense mechanism against oxidative damage in microorganisms. Different bacterial species follow specific pathways for carotenoid synthesis, with products having different structures based on the enzymes involved and environmental conditions. The microbial carotenoids prevent cells from damage by neutralizing the ROS [117].

5.4 Antioxidant Mechanism of Exopolysaccharides

Microbial exopolysaccharides (EPS) are secondary metabolites secreted by various microorganisms, including bacteria and fungi, with significant antioxidant properties. These complex carbohydrate polymers exhibit antioxidant activity through several distinctive mechanisms. One primary antioxidant mechanism of EPS involves free radical-scavenging activity. Bacterial EPS can neutralize ROS and reactive nitrogen species (RNS), including hydroxyl radicals, superoxide anions, and peroxynitrite. This scavenging capability is attributed to specific functional groups, including hydroxyl, carboxyl, and sulfate groups, which can donate hydrogen atoms or electrons to stabilize free radicals. Metal ion chelation represents another significant antioxidant mechanism of bacterial EPS. These compounds can form complexes with transition metal ions like iron (Fe^{2+}/Fe^{3+}) and copper (Cu^{2+}), otherwise catalyzing oxidative reactions through Fenton or Haber-Weiss reactions. By sequestering these metal ions, EPS prevents the production of highly reactive hydroxyl radicals and subsequent oxidative damage to biological molecules [28]. The structure–activity relationship is essential in determining microbial EPS's antioxidant potential. Several structural features influence their antioxidant capacity, including monosaccharide composition, molecular weight, and the presence of specific functional groups. Higher molecular weight EPS increases the viscosity by changing its double-helical forms to macroscopic gels, which can enhance its antioxidant activities. Uronic and sulfate groups in EPS structures significantly contribute to their antioxidant capabilities. The negative charges associated with these groups facilitate electron donation and hydrogen atom transfer, improving the ability to neutralize free radicals. EPS with lower molecular weight and extra negative charge can also improve immunological activity [29, 117]. Bacterial EPS serves multiple protective functions beyond direct antioxidant activity. They shield microbial cells from environmental stresses, including oxidative damage, and can reduce the effect of toxic compounds such as bile salts, metal ions, hydrolyzing enzymes and antibiotics. This protective function extends to their ability to withstand temperature, pH, and osmolarity changes, all of which can induce oxidative stress. Bacterial EPS offers significant production costs and large-scale manufacturing advantages due to their abundant metabolic pathways and efficient production capacity compared to polysaccharides from animal and plant sources [117].

5.5 Antioxidant Mechanism of Short-Chain Fatty Acids

Short-chain fatty acids (SCFAs) are organic molecules of fewer than six carbon atoms. The most abundant SCFAs include acetate, propionate, and butyrate, constituting approximately 60, 20, and 20% of the total SCFAs. These microbial metabolites exhibit significant antioxidant properties through various mechanisms. SCFAs involve the activation of the Nrf2 pathway, which is the master regulator of cellular antioxidant defenses. SCFAs, particularly butyrate, can activate this pathway by inhibiting histone deacetylases (HDACs), resulting in the transcription of genes involved in oxidative stress response, detoxification, and cellular protection [118]. Studies indicate sodium butyrate stimulates Nrf2 to mitigate diabetic nephropathy by inhibiting HDAC. This activation enhances the expression of antioxidant enzymes and reduces oxidative damage in kidney tissues. Likely, butyrate lessens diabetes-induced aortic endothelial dysfunction through P300-mediated Nrf2 transcription. The G protein-coupled receptors FFAR2 (also known as GPR43) and FFAR3 (GPR41) mediate many of the biological effects of SCFAs, including their antioxidant activities. Acetate primarily activates FFAR2, while propionate has a higher affinity for FFAR3. These receptors are found in different cells, allowing SCFAs to act throughout the entire body [105]. SCFAs also demonstrate anti-inflammatory properties, contributing indirectly to their antioxidant effects since inflammation and oxidative stress are closely linked. The anti-inflammatory effect helps maintain redox homeostasis and prevents oxidative damage in cardiovascular tissues. In the brain, SCFAs can cross the blood–brain barrier and activate Nrf2, promoting antioxidant defenses in neuronal tissues. The activation of Nrf2 in the brain positively correlates with the presence of SCFA-producing gut bacteria in mice treated with Nrf2 activators, supporting the role of microbial metabolites in gut-brain communication [118].

6 Biotechnological Approaches for Enhancing Microbial Antioxidant Metabolite Production

Several biotechnological strategies are used to boost microbial metabolite production, primarily focusing on metabolic engineering. This involves manipulating metabolic pathways through gene disruption and overexpression to increase the yield of primary metabolites. Direct modification of specific microorganisms is challenging. Therefore, heterologous expression can be used to transfer genes to more easily manageable microbial species, facilitating large-scale production. Random mutagenesis can help find better strains with better production, boosting essential metabolic genes. Enhancing essential coenzymes, such as NADH and ATP, through genetic modifications is crucial for effective synthesis. The discharge of final metabolites from cells reduces stress and encourages continued production. "Omics" technologies help identify clusters involved in metabolite biosynthesis, enabling targeted genetic approaches to improve yields [119]. Biotechnological approaches for producing microbial antioxidants utilize several strategies to enhance compound yield and efficacy. Strain improvement through mutagenesis or genome editing is a significant method, allowing for enhanced yield of desired metabolites. Fermentation optimization is another crucial aspect, where conditions like temperature, pH, and nutrient levels are fine-tuned to maximize microbial secondary metabolite production. Advanced strategies such as fed-batch or continuous fermentation can further boost productivity [118]. Metabolic engineering plays a significant role in redirecting microbial pathways to promote the biosynthesis of specific antioxidants, enabling higher concentrations of these beneficial compounds. Strict quality control measures are vital to maintain the consistency and safety of produced antioxidants, contributing to their reliable application in food and dietary supplements. These strategies improve the efficiency and sustainability of microbial antioxidant production [118]. The co-cultivation of *Aspergillus* sp. CO_2 and *Bacillus* sp. COBZ21 reveals the potential of microbial interactions to stimulate the production of bioactive compounds with significant antimicrobial properties as shown in Fig. 8. Co-culturing can produce unique metabolites that enhance bioactivity by facilitating interactions between microbial species. Furthermore, this approach opens the way for innovative biotechnological applications, such as developing natural preservatives and therapeutics, ultimately contributing to the fight against antibiotic resistance [120].

Antioxidant peptides can be produced through microbial fermentation, which involves inoculating microbial strains into a reactor containing a mixture of concentrated protein, water, and microbial foods like simple sugars. Microbes break down the protein, and peptidases secreted by microbes during fermentation cleave peptide bonds. Lactic acid bacteria, *Bacillus subtilis* A26, *Bacillus amyloliquefaciens* An6, *Streptococcus* spp., *Lactobacillus brevis, Lactobacillus plantarum, Pediococcusacidilactici,* and *Enterococcus gallinarum* are frequently used to produce antioxidant peptides from different protein sources. Some specific proteolytic fungi strains, such as *Aspergillus oryzae*, can produce bioactive peptides with better antioxidant activities. Microbial fermentation is considered economically feasible and more eco-friendly than enzymatic methods for producing bioactive peptides at an industrial level [121].

7 Recent Advances in Microbial Antioxidant Research

7.1 Omics Approaches to Determine Microbial Antioxidants

Omics technologies furnish potent methodologies for identifying and characterizing microbial antioxidants by investigating genetic, transcriptomic, proteomic, and metabolomic datasets. Genomics elucidates genes associated with antioxidant functions, whereas transcriptomics examines the degree of expression of these genes activated under oxidative conditions. Proteomics identifies free radical-neutralizing enzymes like SOD and catalase, while metabolomics detects antioxidant metabolites, such as exopolysaccharides and phenolic compounds. Integrating multiomics strategies facilitates comprehension of microbial antioxidant pathways, thereby contributing to identifying novel bioactive compounds. Such insights help biotechnological usefulness in pharmaceuticals, food preservation, and cosmetics, underscoring the role of microbes as sustainable reservoirs of natural antioxidants as shown in Fig. 9, respectively.

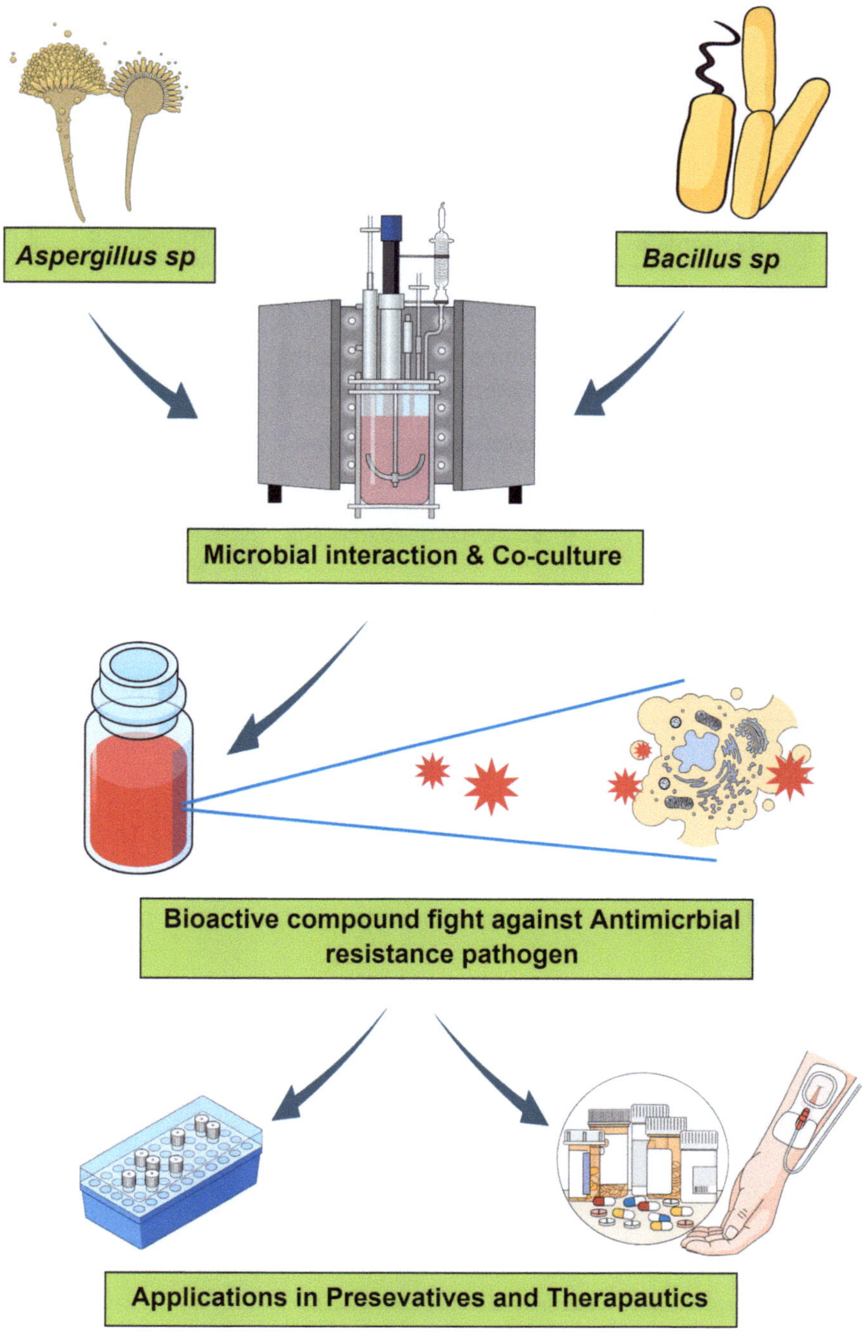

Fig. 8 Biotechnological applications in developing natural preservatives and therapeutics

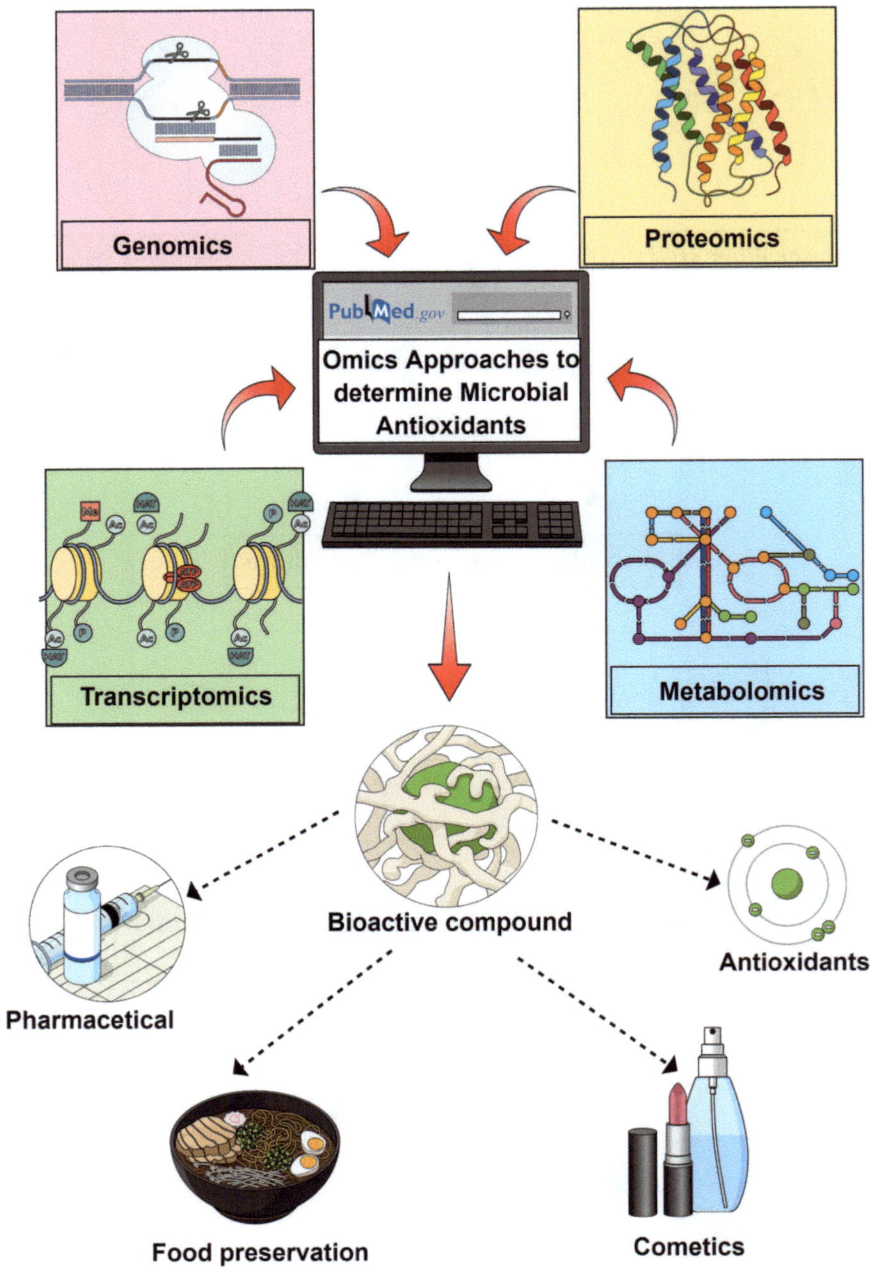

Fig. 9 Omics approaches to determine microbial antioxidants

7.2 Metagenomics

Metagenomics is a method for detecting entire microbial communities found within a given habitat by sequencing the genomes without needing to cultivate individual ones. Genomics analyzes genetic content from individual species, whereas metagenomics investigates the collective genetic material from whole microbial communities. Two methodologies are used to identify microbial populations: marker-sed analysis and shotgun sequencing, which explore the antioxidant-producing microbes from the environment [122]. In the marker gene analysis technique, the primer was designed to attach to conserved genetic regions of interest, such as the 16S rRNA gene in bacteria, which helps identify their phylogenies [123]. A major revolution in the classification of prokaryotic organisms occurred with the realization that evolutionary relationships could be deduced based on differences in the gene sequence. The rRNAs from the small ribosomal subunits (*such as 16S in bacteria*) are now widely used as preferred markers for determining evolutionary relationships among prokaryotes. The small subunit rRNAs (SSU rRNAs) are integral to the ribosome. The genes encoding these rRNAs change gradually over evolutionary periods and do not appear subject to lateral gene transfer. The small subunit rRNAs (SSU rRNAs) contain specific sequences, with certain regions being diverse across species, while others remain conserved. These variations help distinguish closely related taxa, while conserved segments assist in identifying more distantly related microbes. An extensive array of databases is accessible for amplicons focused on this particular region, which facilitates the classification of reads and the construction of taxonomic profiles of a microbiome [124]. Advancements in technology have shifted the studies toward shotgun approaches such as whole metagenome sequencing (WMS) [125]. Shotgun sequencing profiles have the potential to elucidate information regarding the functional capacities of all microbial populations at the DNA level, drawing upon established repositories containing data from earlier genome sequencing efforts. The whole genome was shredded into random fragments using enzymatic treatment, and the ligands were ligated into each fragment, followed by the high-throughput DNA sequencing method. The sequences were reassembled using de novo assembly or mapping of a reference genome to reconstruct the original genome. Computational tools are used to identify genes, functional elements, and taxonomic classifications [126]. Kandasamy et al. investigated *Lacticaseibacillus casei* KACC92338, isolated from unprocessed Korean milk, to evaluate its antioxidant and probiotic characteristics. The strain demonstrated significant oxidative stress resilience, surviving concentrations of H_2O_2 up to 2 mM, while its cell-free supernatant exhibited considerable antioxidant activity, encompassing DPPH scavenging (95.61%), hydroxyl radical scavenging (34.10%), reducing power (2.220), and iron-chelating ability (81.06%). Furthermore, the CFS promoted yeast cell viability in conditions of oxidative stress (76.05% survival at 10 mM hydrogen peroxide). Whole-genome sequencing analysis revealed a genome size of 3,050,901 bp comprising 3048 genes, which included 2981 coding sequences and 67 RNA molecules, along with an average nucleotide identity (ANI) of 96%, thereby confirming its phylogenetic proximity to *L. casei* strains. Functional annotation highlighted the presence of genes associated with the metabolism of carbohydrates and amino acids, genetic regulation, and mechanisms of stress resistance, which encompass tolerance to acidic, bile, thermal, cold, and oxidative stress, in addition to immunomodulatory functions and intestinal adhesion properties. Kandasamy et al. further identified six prophage regions, three CRISPR arrays, and clusters of antimicrobial bacteriocins, suggesting a potential for antimicrobial action. Notably, no detected genes were associated with acquired antibiotic resistance, virulence, or pathogenicity, substantiating its safety profile. Collectively, these findings propose that KACC92338 serves as a robust probiotic candidate for applications in functional foods, healthcare, and skin care, although additional research is warranted further to validate its safety and efficacy [127].

7.3 Metatranscriptomics

Metatranscriptomics is used to identify the active gene expression profiles of microbial communities that play a crucial role in detecting the microbial antioxidant mechanism, such as the detection of transcripts related to antioxidant enzymes and stress response proteins and elucidating information about how the microbes use specific pathways to combat oxidative stress [128]. Meta-transcriptomics employs high-throughput RNA sequencing methodologies to measure the amount of microbial RNA found within the sample. Simultaneously, the mRNA constituent is less than 5% of the entire RNA in the cell. Prior investigations have indicated that specific microbial mRNAs may possess suboptimal ribosome binding sites, leading to inefficient translation, whereas only mRNAs with robust ribosome binding sites are translated with greater frequency. This suggests that a significant portion of the microbial mRNA identified through meta-transcriptomics may not be directly implicated in a specific host phenotype. For instance, research on ammonia-oxidizing organisms has identified a substantial abundance of superoxide dismutase and peroxiredoxin-like transcripts, suggesting an augmented response to oxidative stress within the microbial consortium [126].

7.4 Metaproteomics

Metaproteomics employs advanced mass spectrometry techniques to detect and measure the abundance of proteins produced. The collected data is analyzed using computational workflows that link peptides to metagenomic references to ascertain the most likely microbial taxa responsible for the expression of the proteins [129]. Metaproteomics serves as an indispensable methodology for elucidating microbial responses to oxidative stress. Research conducted by Long et al. investigated the protein expression profiles of the intestinal microbial community in individuals with colorectal cancer versus healthy controls [130]. The researchers analyzed 91,902 peptides and 30,062 protein clusters derived from 195 microbial genera, identifying 341 proteins exhibiting significant abundance differences among individuals with CRC and those without the disease. Among these, proteins associated with oxidative stress responses were highly upregulated in CRC patients, indicating an adaptive microbial reaction to upregulate oxidative conditions in the gut environment. A highly notable antioxidative protein identified was rubrerythrin, which was found in elevated levels in CRC patients. Rubrerythrin, commonly produced by anaerobic bacteria such as *Clostridium* and *Desulfovibrio*, is a nonheme iron protein that reduces hydrogen peroxide to water, thereby protecting microbial cells from oxidative damage. This suggests that gut bacteria may upregulate antioxidant defenses as a reaction to oxidative imbalance, a hallmark of CRC pathophysiology. Additionally, superoxide dismutase (SOD), an enzyme produced by gut microbes such as *Lactobacillus* and *Bifidobacterium*, was also identified. SOD catalyzes the transformation of superoxide molecules into oxygen and H_2O_2, mitigating oxidative stress and preventing injury triggered by reactive oxygen compounds to both microbial and host cells. Another crucial antioxidant enzyme detected in the metaproteomic analysis was catalase, which is expressed by microbes such as *Escherichia coli* and *Staphylococcus species*. Catalase decomposes hydrogen peroxide into H_2O and O_2, an essential defense mechanism against oxidative damage in gut microbiota [131]. The presence of these antioxidant proteins highlights the gut microbiome's ability to regulate oxidative stress and maintain homeostasis. This study underscores the importance of microbial antioxidant mechanisms in shaping gut health and suggests that changes in microbial antioxidant activity may contribute to CRC progression. The identification of antioxidant proteins such as rubrerythrin, SOD, and catalase in CRC patients suggests a potential microbial adaptation to oxidative stress, which could be leveraged for therapeutic interventions targeting the gut microbiome. Metaproteomics provides a comprehensive approach to studying microbial antioxidant metabolism, offering insights into microbial responses to oxidative stress and its connection to the advancement of diseases [132].

7.5 Metabolomics

Metabolomics is more than just a way to comprehend the chemical processes behind producing antioxidants in microbial metabolism. Metabolites are different small biomolecules such as fats, amino compounds, protein fragments, nucleotides, organic compounds, essential nutrients, sulfur-containing compounds, and sugars, making global analysis challenging. This methodology helps evaluate antioxidant production, regulation, and function in microorganisms. The research shows the great importance of microbial antioxidants in the fight against oxidative stress, thereby being used in various industries, medicine, and environmental applications. A variety of analytical techniques are employed in metabolomics research [133]. However, metabolomics data don't tell us which bacteria present produced a metabolite. In addition to that, the fact that several distinct microbes like *Ruminococcaceae* and *Lachnospiraceae* can release identical compounds, such as short-chain fatty acids (SCFAs), also complicates the accurate prediction of the different bacteria that produce specific metabolites, even though the metabolomics data are combined with other *"omic"* data. It is both necessary and, at the same time, not enough to use a single "omic" strategy; instead, it is necessary to combine different "omic" techniques (multiomics) to arrive at correct conclusions [134].

8 Synthetic Biology and Engineered Antioxidant Pathways

8.1 Microbial Production of Potent Antioxidant Carotenoids

Carotenoids are isoprenoid pigments with crucial roles in pigmentation, biocommunication, and provitamin A compounds with strong antioxidative effects. These protective traits with specific roles (e.g., *lutein and zeaxanthin*) help combat age-related decline, malignancies, eye lens opacities, heart-related conditions, and nervous system disorders. Widely used in nutritional additives and edible pigments, nutraceuticals, and cosmetics, carotenoids are in high demand, driving industrial production. Currently, 80–90% are synthetically produced, but concerns over synthetic variants are boosting interest in natural sources. The microbial production of carotenoids focuses on antioxidant potential and industrial-scale biosynthesis using microalgae, bacteria, and fungi, including genetically engineered strains, to enhance yield [135]. Reactive oxygen species (ROS), which include singlet oxygen ($^1\Delta g$), hydroxyl radicals (•OH), and superoxide anions ($\bullet O_2^-$), play a crucial role in the development of long-term degenerative illnesses such as aging, cancer, cataracts, heart disease, and neurological disorders. Incorporating a diet abundant in carotenoids confers considerable protective effects against the onset of such diseases, attributable to the robust antioxidant capabilities inherent in carotenoids. Although present at different levels, all carotenoids are known to suppress processes driven by ROS, including the peroxidation of lipids, superoxide generation triggered by LPS, cell toxicity caused by peroxides, and oxidation of LDL [136]. Carotenoids, with 721 types identified to date [137]. Structurally, they are categorized into two main groups such as hydrocarbon carotenoids, referred to as carotenes, which include compounds like phytoene, lycopene, α-carotene, and β-carotene, and oxygen-containing forms of carotenes, termed xanthophylls, such as lutein, canthaxanthin, and astaxanthin. Functionally, carotenoids are classified as primary or secondary. Main carotenoids like violaxanthin, antheraxanthin, zeaxanthin, lutein, neoxanthin, and β-carotene function as key light-capturing pigments in the photosynthetic process. Non-primary carotenoids such as astaxanthin and canthaxanthin build up under environmental stress and support cellular defense. Additionally, carotenoids are categorized into vitamin A precursors like α,β-carotene, and those that do not serve as vitamin A sources, such as astaxanthin, lutein, and lycopene, with the latter lacking a non-substituted β-ionone ring, preventing conversion to retinal [138]. Beyond these, carotenoids also include apocarotenoids and C50 carotenoids like decaprenoxanthin. Apocarotenoids result from the enzymatic cleavage of carotenoids through the action of cleavage enzymes known as carotenoid dioxygenases, a process initiated by singlet oxygen ($^{1}O_2$) attack. Notably, *Corynebacterium glutamicum* is distinguished for synthesizing the C50 class carotenoids, such as decaprenoxanthin and its sugar-bound forms. *Corynebacterium glutamicum*, a common soil bacterium, is gaining attention for its ability to produce unique C50 carotenoids like 4-decaprenoxanthin and its glycosylated forms. These compounds are widely used in skincare and medical products due to their strong UV protection. *C. glutamicum* is also a major player in the commercial production of amino acids, generating over three million tons annually. Researchers have enhanced its carotenoid production through genetic engineering, resulting in a striking red pigmentation and an impressive 80-fold increase in lycopene content, reaching 2.4 mg/g of dry cell weight (DCW) [102].

8.2 *Biosynthesis of Carotenoids Using Microbial Source*

Carotenoid biosynthesis universally relies on isoprenoid precursors, specifically isopentenyl pyrophosphate (IPP) and its isomer dimethylallyl pyrophosphate (DMAPP). These precursors are synthesized via the 2-C-methyl-D-erythritol 4-phosphate (MEP) pathway, which begins with pyruvate and glyceraldehyde 3-phosphate, as illustrated in Fig. 10 [139]. In microbial systems, the initial step is catalyzed by the enzyme Dxs, which is often targeted to enhance carotenoid production [140]. The subsequent condensation of IPP and DMAPP yields geranyl pyrophosphate (GPP), which is then elongated to form geranylgeranyl diphosphate (GGPP) through the action of GGPP synthase. Following this, phytoene synthase converts GGPP into 15-(Z)-phytoene, which undergoes a series of desaturation and isomerization reactions catalyzed by phytoene desaturase (PDS). These reactions lead to the formation of lycopene via lycopene synthase as shown in Fig. 10, which serves as the first physiologically significant carotenoid and a central intermediate in the biosynthesis of all downstream C_{40} carotenoids [141]. Finally, lycopene undergoes cyclization at both ends by lycopene β-cyclase (LCY-β) to form β-carotene.

8.3 *Production of Commercially Vital Carotenoids*

Extensive carotenoid biosynthesis has been effectively achieved by utilizing diverse high-yield microalgal strains. For instance, *Dunaliella salina* [142] is cultivated in open pond systems for β-carotene, while *Haematococcus pluvialis* is known for its astaxanthin production. Similarly, *Phaeodactylum tricornutum* is recognized for its production of fucoxanthin, *Botryococcus braunii* for synthesizing echinenone, *Scenedesmus species* for lutein, and *Chlorella zofingiensis* for canthaxanthin.

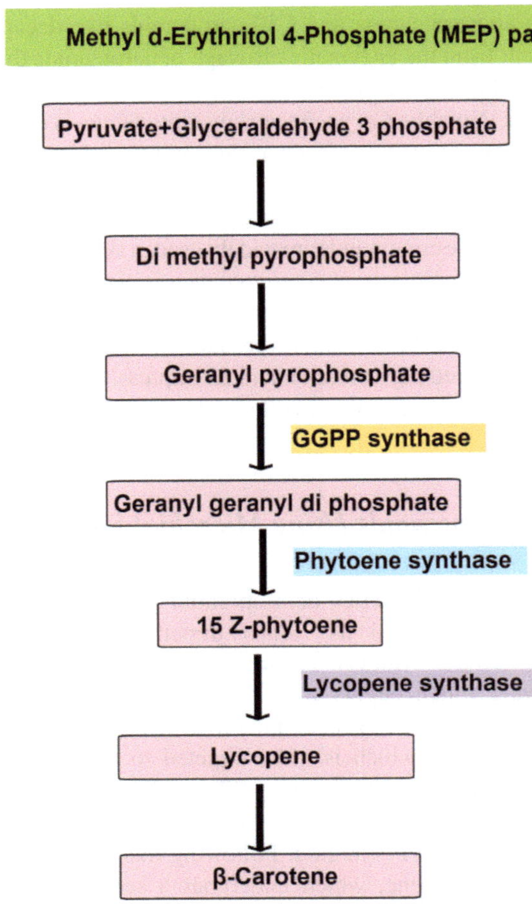

Fig. 10 The major steps in the MEP pathway for the synthesis of carotenoids

Beyond microalgae, certain fungi, bacteria, and yeasts are also utilized for commercial carotenoid production. The fungus *Blakesleatrispora* produces β-carotene and lycopene, while *Mucor circinelloides* specializes in β-carotene synthesis. The yeast *Xanthophyllomyces dendrorhous* (formerly known as *Phaffiarhodozyma*) is exploited for astaxanthin, and bacterial species like *Gordonia jacobea* and *Corynebacterium glutamicum* contribute to the production of canthaxanthin and decaprenoxanthin (C50 carotenoid), respectively. Emerging microbial sources of carotenoids include fungi such as *Fusarium sporotrichioides* and *Neurospora crassa*, as well as bacteria like *Paracoccus carotinifaciens, Agrobacterium aurantiacum,* and *Halobacterium salinarium* [139]. Additionally, several non-carotenoid producing Genetically engineered strains of *Escherichia coli*, *Saccharomyces cerevisiae*, *Xanthophyllomyces dendrorhous*, and *Blakeslea trispora* have been developed to effectively produce economically important carotenoids, including lycopene, canthaxanthin, β-carotene, and astaxanthin [135].

8.4 *Metabolic Engineering Strategies for Heterologous Carotenoid Biosynthesis*

Like the natural production of carotenoids in carotenogenic microbes, various factors and methods have been investigated to achieve efficient and economical synthesis using heterologous systems [143]. Figure 11 shows several strategies that have been utilized to enhance carotenoid production in E. coli. Initially, carotenoid biosynthesis is improved by increasing the availability of essential precursors (pyruvate and glyceraldehyde-3-phosphate) linked to the MEP pathway [140]. Next, the metabolic flow of IPP and DMAPP is boosted by overexpressing key isoprenoid genes of the MEP pathway, such as 1-deoxy-D-xylulose-5-phosphate synthase (Dxs). In the third stage, the main pathways that consume carotenoid precursors are suppressed or inhibited. In the fourth stage, the production of crucial cofactors like adenosine triphosphate (ATP) and nicotinamide adenine dinucleotide phosphate (NADPH) is increased by overexpressing essential genes involved in the tricarboxylic acid (TCA) cycle (e.g., α-ketoglutarate dehydrogenase) and ATP synthesis (e.g., ATP synthase), which meet the energy requirements for biosynthetic activities. The main activities of genetic engineering of microorganisms for the heterologous production of carotenoids are summarized in Table 2.

8.5 *Extremophilic Microbes as a Source of Novel Antioxidants*

Mandelli et al. [149] reported that extremophilic microbes such as *Halococcus morrhuae*, *Halobacterium salinarium*, and *Thermus filiformis* produce carotenoids, which were analyzed using advanced chromatographic and spectrometric methods [150]. These extracts showed antioxidant activity against both radical and non-radical species. Halophiles mainly produced carotenoids like bacterioruberin and its derivatives, while the thermophilic strain yielded zeaxanthin variants. The antioxidant potential of *H. morrhuae* and *H. salinarium* was higher than *T. filiformis*, likely due to structural differences in carotenoids. Boujida et al. [151] isolated ten halophilic strains from hypersaline sites in Morocco that produced exopolymers. Most were affiliated with the *Halomonas* genus, while one (N4) belonged to *Marinobacter* [152]. These strains displayed moderate halophilicity and enzyme activity. The polymers showed both emulsifying and antioxidant properties, with *Marinobacter* sp. N4 yields the most stable and effective formulation. This highlights the potential application of these exopolymers in food, cosmetics, and related industries.

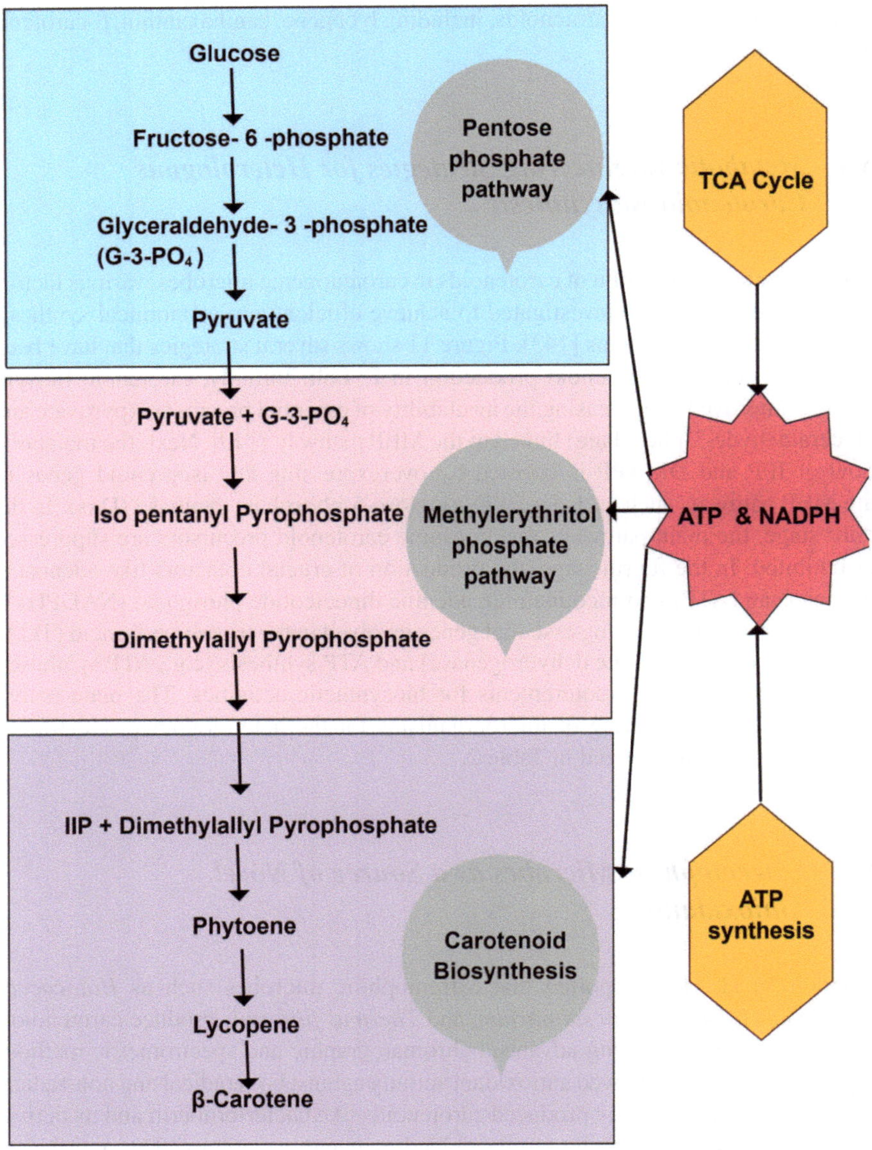

Fig. 11 Schematic illustration of carotene biosynthesis pathway

Table 2 Key developments in the genetic engineering approach in microbes for heterologous carotenoid production

Types of carotenoids	Host organism	Metabolic engineering strategy	Yield	References
Lycopene	*E. coli* CAR001	Enhancement of production was achieved by deactivating the genes responsible for zeaxanthin glucosyltransferase and lycopene β-cyclase, while also modulating the expression of enzymes that contribute to NADPH and ATP generation.	50.6 mg/g DCW	[144]
Canthaxanthin	*Mucor circinelloides*	By using the *CrtW* gene, encoding β-carotene ketolase, was derived from *Paracoccus* sp. strain N81106.	200 μg/g DCW	[145]
Zeaxanthin	*E. Coli* KT2440	Regulated co-expression of *CrtY* and *CrtZ* (β-carotene 3-hydroxylase).	11.95 mg/g DCW	[146]
β-Carotene	*E. Coli* CAR005	The process involved five critical modules: The biosynthesis of β-carotene, the MEP pathway, and three fundamental metabolic pathways, namely ATP production, the TCA cycle, and the pentose phosphate pathway (PPP).	60 mg/g DCW	[147]
Retinol	*E. Coli*	Regulated co-expression of CrtY and CCD1	33 mg/L	[148]

9 Industrial and Biomedical Applications of Microbial Antioxidants

Microbial antioxidants are high-potential biomolecules with enormous potential uses in medical and pharmaceutical applications and biomedical and biosensing instrumentation [49]. They are obtained from microorganisms, such as bacterial, fungal, or microalgal sources, with strong scavenging activity against free radicals, which would greatly help prevent diseases associated with oxidative stress and enhance stability in biomaterials. Given the ever-increasing thrust on oxidative damage to living beings and the inadequacy of synthetic antioxidants, microbial-type antioxidants have made it to the forefront due to their biocompatibility, eco-compatibility, and therapeutic value [153]. Microbial antioxidants have serious applicability at the level of drug industries and in medicine to secure cell components through oxidative damage related pathology of many diseases like neurodegenerative diseases, cardiovascular disease, and cancer-as it generally does with these microcells, including *Lactobacillus*, *Bifidobacterium*, and *Aspergillus*, which produce antioxidants in the forms of carotenoids, flavonoids, and polyphenols, which have proven to show great anti-inflammatory and cytoprotective activities [154]. Such antioxidants are investigated for applicability in drug delivery systems because they can counteract mobile harm caused by oxidative strain and improve treatment efficacies. For instance, the microorganism-derived astaxanthin, one of the most effective antioxidants produced by way of *Phaffiarhodozyma* and *Haematococcuspluvialis*, turned into neuro and cardioprotective through inhibition of lipid peroxidation and enhancement of mitochondrial function [155]. Microbial antioxidants can also be used in regenerative medicinal drugs to perform the design development of antioxidant-loaded biomaterials for stronger wound recuperation, tissue regeneration, and stem mobile therapy. Incorporating microbial antioxidants into nanoparticles, scaffolds, and hydrogels will increase their balance and bioavailability, making viable managed drug shipping and decreasing cytotoxicity in biomedical implants [156]. Outside the pharmaceutical utility, microbial antioxidants have been confirmed to be very promising in bioelectronics and biosensing technology. Antioxidant-primarily based biosensors were intensely studied for markers of oxidative pressure, pollutants from the environment, and metabolic illnesses. Microbial antioxidants, namely phenolic compounds and exopolysaccharides, were used as redox-lively materials in biosensing systems to beautify sign sensitivity and specificity [157]. Examples of such microbial-derived antioxidants consist of superoxide dismutase and catalase fashioned collectively to design enzyme-based biosensors to sense reactive oxygen species (ROS) in medical and environmental diagnostic tracking. This ought to permit the detection of oxidative stress biomarkers, which might be measurable in a roundabout way in chronic diseases, including diabetes, most cancers, and neurodegenerative conditions [158].

9.1 Pharmaceutical and Healthcare Applications

Antioxidants play an essential role in pharmaceutical and healthcare applications because of their potential to neutralize reactive oxygen species (ROS) and mitigate oxidative strain, which is implicated in numerous ailments. Presently, research is increasingly focused on the restoration functionality of antioxidants in treating persistent conditions alongside cardiovascular sicknesses, neurodegenerative problems, and most cancers [107]. For instance, polyphenols like resveratrol and curcumin have been proven to have substantial cardioprotective consequences by lowering oxidative damage to endothelial cells and improving vascular characteristics. These compounds modulate signaling pathways, which include NF-κB and Nrf2, which is probably crucial in infection and antioxidant response. Similarly, flavonoids located in inexperienced tea and berries have been shown to enhance cell protection mechanisms, imparting the capacity to stop atherosclerosis and ischemic heart disorder. Integrating these herbal antioxidants into pharmaceutical formulations has opened new avenues for adjuvant treatments in cardiovascular fitness [159, 160]. In neurodegenerative sicknesses such as Alzheimer's and Parkinson's, oxidative strain is a key contributor to neuronal degeneration. Antioxidants like vitamin E, coenzyme Q10, and alpha-lipoic acid have been investigated for neuroprotective residences. Additionally, nanoparticle-based shipping structures have been superior in decorating the bioavailability of these compounds, ensuring better penetration of the blood–brain barrier [161, 162]. Such improvements underscore the capability of antioxidants in dealing with neurodegenerative situations, although scientific trials are ongoing to set up standardized dosing and long-term efficacy. Antioxidants' anticancer properties have garnered significant attention, especially in lowering chemotherapy-associated side effects and improving remedy effects. For instance, N-acetylcysteine (NAC) has been used to mitigate cisplatin delivered on nephrotoxicity in cancer patients by restoring intracellular glutathione ranges [163]. In dermatology, antioxidants are widely used to deal with skin aging and photodamage due to UV radiation. Topical applications of vitamins C and E with ferulic acid are comprehensible for photoprotection and collagen synthesis [164]. Novel Yogas incorporating selenium and zinc-based antioxidants have effectively reduced inflammatory skin conditions such as psoriasis and eczema. The development of stable, cosmetic-grade antioxidant mixtures has revolutionized skincare, providing preventive and medical benefits against environmental oxidative stresses [165, 166]. The role of antioxidants in metabolic disorders, especially diabetes, has been another area of active research. Oxidative stress is a major driver of insulin resistance and dysfunction of pancreas cells. Studies have shown that antioxidants such as berberine and quercetin improve glucose metabolism by modifying AMPK and PI3K/Act Pathways [167]. In addition, astaxanthin, a powerful carotenoid-related clinical trial, reported significant cuts in HBA1C levels and diabetic patients with systemic inflammation markers [168]. These findings highlight the ability of antioxidants to be helpful therapy in diabetes management, although further research is required to adapt their clinical application.

9.2 *Bioelectronic and Biosensing Applications of Microbial Antioxidants*

Incorporating microbial antioxidants into bioelectronic systems has paved the way for innovative biosensor technologies. Electrochemical biosensors utilizing microbial antioxidants offer several advantages, such as high sensitivity, rapid response times, and ease of miniaturization. These properties make them suitable for various applications, including food quality assessment, drug screening, and toxicity evaluation [169]. One notable bioelectronic application is microbial fuel cell-based biosensors (MFC-Bs). These systems leverage the metabolic activity of bacteria to convert organic substrates into electrical energy, enabling real-time monitoring of biochemical oxygen demand (BOD) and environmental toxicity. Recent advancements in this field have focused on enhancing biosensor performance by incorporating nanomaterials. For instance, nanostructured electrodes significantly improve biosensors' sensitivity, amplifying electrical signals generated during biochemical reactions [170]. Integrating microbial antioxidants into bioelectronic devices has also facilitated the development of sensors capable of detecting oxidative stress markers in biological and environmental samples. Such advancements highlight the potential of microbial antioxidants in creating efficient, cost-effective, and portable biosensors for diverse applications [171]. Microbial antioxidants have emerged as key components in biosensing technologies designed for health diagnostics and environmental monitoring. Whole-cell bacterial biosensors (WCBs) incorporating microbial antioxidant systems have been explored for biomarker detection, offering promising solutions for disease diagnosis and therapeutic monitoring. These biosensors operate within clinically relevant concentration ranges, enabling early detection of pathological conditions associated with oxidative stress [172]. In environmental applications, microbial antioxidant-based biosensors effectively detect pollutants, including heavy metals, organic contaminants, and toxicants. Compared to traditional analytical methods, these biosensors offer several advantages, such as lower costs, reduced technical complexity, and minimal sample preparation. DNA-based electrochemical biosensors represent another significant advancement, utilizing microbial DNA repair and protection mechanisms to assess the genotoxic potential of chemical compounds. Such biosensors hold immense promise for environmental toxicity assessments, offering real-time insights into pollution levels and ecosystem health [157, 173]. Microbial antioxidant-based biosensors have the potential to revolutionize health diagnostics and environmental monitoring [174]. In healthcare, these biosensors enable real-time monitoring of oxidative stress levels, facilitating early disease detection and personalized treatment strategies. Their ability to detect oxidative biomarkers with high specificity makes them valuable tools for managing chronic conditions, such as neurodegenerative disorders and metabolic syndromes [157]. Beyond healthcare and environmental applications, microbial antioxidants are increasingly used in food safety monitoring. Tyrosinase-based biosensors, for instance, can detect catechol and p-hydroquinone levels in food products, offering valuable insights into antioxidant availability. These biosensors provide rapid, on-site testing capabilities, reducing reliance on labor-intensive conventional analytical methods [172].

10 Future Perspectives and Challenges in Microbial Antioxidants

Microbial antioxidants have captured the attention of biotechnology, healthcare, and industry experts, focusing on the potential applications of these compounds in these areas. In the meantime, the technical challenges for large-scale manufacturing, regulatory authorization, and functional characterization have not yet been completely settled. With the help of metabolic engineering and synthetic biology, they bring about the expediency of modifying microbial pathways to a degree that an increase in the production of antioxidants is possible [175]. Modern technologies utilizing omics can characterize the metabolites of antioxidants and their mode of action. The question of safety in the market is still being asked, as the FDA demands safety assessments before selling bio-products. Knowing how these novel antioxidants are structured will give a good understanding of their inherent properties and behavior in various applications. In parallel, utilizing bioreactors and immobilization strategies can escalate the production rate and, thus, the efficiency of the whole process. Only by utilizing research can we discover unrevealed novel antioxidants, microbial diversity in extreme environments being the subject of future bio-prospecting. Employing computing and statistical techniques to analyze large data volumes can significantly speed up the identification process of antioxidants among unknown data clusters and uncover the best combinations for the optimal production of antioxidants [176]. Unfortunately, even though the invention might be the brightest, a lack of public awareness and consumer education about microbial-derived antioxidants are the two factors that could guide them off the right course and could bring up the rejection of these products. The only way to conquer these challenges is through the collaboration between academia and industry, where both parties need to work as a team to make a complex and, in many cases, controversial handshake that will lead to the rollover of the findings by the university to the business side and subsequently to real products. Although there is a demand for the market, future research should focus on understanding the basic course of microbial antioxidants in the human organism so that these can serve the dual purpose of reducing free radicals and healing or preventing diseases. By facing such challenges, the bio sources of these secondary metabolites will align with the digital direction of the industries and will be biobased and much closer to market parameters or even profitable [177].

11 Conclusion

Microbial antioxidant metabolism is a significant defense mechanism that enables microbes to survive under oxidative stress with the help of enzymatic antioxidants, such as SOD, catalase, and peroxidases, as well as non-enzymatic compounds, such as glutathione, flavonoids, and carotenoids. Various microbial metabolites have been reported as antioxidants, which are used to prevent diseases. Recent advances in genomics, proteomics, metagenomics, and metabolomics have elucidated microbes' regulatory pathways and adaptive strategies in diverse environments. Recent advancements in antioxidant research have driven significant biotechnological applications, including developing microbial-based therapeutics, which hold promise in preventing oxidative stress-related diseases like inflammation, neurodegeneration, and cancer. By leveraging microbial antioxidant systems, researchers can innovate sustainable health solutions, underscoring the importance of continued interdisciplinary research to harness microbial potential for disease prevention and to improve human health.

References

1. Harman, D. (1956, July). Aging: A theory based on free radical and radiation chemistry. *Journal of Gerontology, 11*(3), 298–300. https://doi.org/10.1093/geronj/11.3.298
2. Phaniendra, A., Jestadi, D. B., & Periyasamy, L. (2015, January). Free radicals: Properties, sources, targets, and their implication in various diseases. *Indian Journal of Clinical Biochemistry, 30*(1), 11–26. https://doi.org/10.1007/s12291-014-0446-0
3. Hayat, M.A. (2015). Introduction to autophagy: Cancer, other pathologies, inflammation, immunity, infection, and aging, volume 7. In *Autophagy: cancer, other pathologies, inflammation, immunity, infection, and aging, vol. 7* (pp. 1–53). Elsevier Inc. https://doi.org/10.1016/B978-0-12-801043-3.00001-7.
4. Thannickal, V. J., & Fanburg, B. L. (2000). Invited review reactive oxygen species in cell signaling 2000 [Online]. Available http://www.ajplung.orgL1005Downloadedfromjournals.physiology.org/journal/ajplung
5. De Almeida, A. J. P. O., et al. (2022). ROS: Basic concepts, sources, cellular signaling, and its implications in aging pathways. Hindawi Limited. https://doi.org/10.1155/2022/1225578.
6. Panova, I. G., & Tatikolov, A. S. (2023, July). Endogenous and exogenous antioxidants as agents preventing the negative effects of contrast media (contrast-induced nephropathy). *Pharmaceuticals, 16*(8), 1077. https://doi.org/10.3390/ph16081077
7. Atta, E. M., Mohamed, N. H., & Abdelgawad, A. A. M. (2017, September). Antioxidants: An overview on the natural and synthetic types. *European Chemical Bulletin, 6*(8), 365. https://doi.org/10.17628/ecb.2017.6.374-384
8. Zehiroglu, C., & Ozturk Sarikaya, S. B. (2019, November 01). The importance of antioxidants and place in today's scientific and technological studies. Springer https://doi.org/10.1007/s13197-019-03952-x.
9. Nieva-Echevarría, B., Manzanos, M. J., Goicoechea, E., & Guillén, M. D. (2015, January). 2,6-Di-Tert-butyl-Hydroxytoluene and its metabolites in foods. *Comprehensive Reviews in Food Science and Food Safety, 14*(1), 67–80. https://doi.org/10.1111/1541-4337.12121
10. Lü, J. M., Lin, P. H., Yao, Q., & Chen, C. (2010, April). Chemical and molecular mechanisms of antioxidants: Experimental approaches and model systems. *Journal of Cellular and Molecular Medicine, 14*(4), 840–860. https://doi.org/10.1111/j.1582-4934.2009.00897.x

11. Li, T.-X., et al. (2020, May). Antioxidant coumarin and pyrone derivatives from the insect-associated fungus *aspergillus versicolor*. *Natural Product Research, 34*(10), 1360–1365. https://doi.org/10.1080/14786419.2018.1509334
12. Janardhan, A., Kumar, A. P., Viswanath, B., Saigopal, D. V. R., & Narasimha, G. (2014). Production of bioactive compounds by Actinomycetes and their antioxidant properties. *Biotechnology Research International, 2014, March*, 1–8. https://doi.org/10.1155/2014/217030
13. Dewi, R. T., Tachibana, S., & Itoh, K. (2012). Isolation of antioxidant compounds from aspergillus Terreus LS01. *Journal of Microbial and Biochemical Technology Research, 04*(01). https://doi.org/10.4172/1948-5948.1000065
14. Huang, X.-Z., et al. (2012, April). A novel antioxidant Isobenzofuranone derivative from fungus Cephalosporium sp.AL031. *Molecules, 17*(4), 4219–4224. https://doi.org/10.3390/molecules17044219
15. Li, Y., et al. (2022, September 01). Progress of microencapsulated phycocyanin in food and pharma industries: A review. MDPI. https://doi.org/10.3390/molecules27185854.
16. Xu, R., et al. (2023), The role of the probiotic Akkermansiamuciniphila in brain functions: Insights underpinning therapeutic potential, Taylor and Francis Ltd. https://doi.org/10.1080/1040841X.2022.2044286.
17. Kim, Y., Lim, J., & Oh, J. (2024, September 01). Taming neuroinflammation in Alzheimer's disease: The protective role of phytochemicals through the gut−brain axis. *Elsevier Masson s.r.l.* https://doi.org/10.1016/j.biopha.2024.117277.
18. Koo, S. Y., et al. (2019, May). Anti-obesity effect of standardized extract of microalga phaeodactylumtricornutum containing fucoxanthin. *Marine Drugs, 17*(5). https://doi.org/10.3390/md17050311
19. Imlay, J. A. (2019). Where in the world do bacteria experience oxidative stress? https://doi.org/10.1111/1462-2920.14445.
20. Rani, A. et al. (2021) Microorganisms: A potential source of bioactive molecules for antioxidant applications. https://doi.org/10.3390/molecules26041142.
21. Viedma-Poyatos, Á., González-Jiménez, P., Langlois, O., Company-Marín, I., Spickett, C. M. & Pérez-Sala, D. (2021). Protein lipoxidation: Basic concepts and emerging roles. https://doi.org/10.3390/antiox10020295.
22. Srivastava, A. K., Kumar, D., Singh, D., & Singh, R. K. (2022). Mechanism of oxidative stress in carcinogenesis induced by xenobiotics. *Xenobiotics in Chemical Carcinogenesis*. https://doi.org/10.1016/b978-0-323-90560-2.00003-0
23. Fasnacht, M., & Polacek, N. (2021). Oxidative stress in bacteria and the central dogma of molecular biology. https://doi.org/10.3389/fmolb.2021.671037
24. Hernandez-Morfa, M., et al. (2023). The oxidative stress response of Streptococcus pneumoniae: Its contribution to both extracellular and intracellular survival. https://doi.org/10.3389/fmicb.2023.1269843
25. Rossi, D. C. P., et al. (2017). Candida albicans FRE8 encodes a member of the NADPH oxidase family that produces a burst of ROS during fungal morphogenesis. *PLoS Pathogens, 13*(12). https://doi.org/10.1371/journal.ppat.1006763
26. Herb, M. & Schramm, M. (2021) Functions of ros in macrophages and antimicrobial immunity. https://doi.org/10.3390/antiox10020313.
27. Lewis, D. F. V. (2002). Oxidative stress: The role of cytochromes P450 in oxygen activation. https://doi.org/10.1002/jctb.648.
28. Grahl, N., Dinamarco, T. M., Willger, S. D., Goldman, G. H., & Cramer, R. A. (2012). Aspergillus fumigatus mitochondrial electron transport chain mediates oxidative stress homeostasis, hypoxia responses and fungal pathogenesis. *Molecular Microbiology, 84*(2). https://doi.org/10.1111/j.1365-2958.2012.08034.x
29. Davies, K. J. A., & Lin, S. W. (1988). Degradation of oxidatively denatured proteins in Escherichia coli. *Free Radical Biology and Medicine, 5*(4). https://doi.org/10.1016/0891-5849(88)90015-9

30. Akhter, S., McDade, H. C., Gorlach, J. M., Heinrich, G., Cox, G. M., & Perfect, J. R. (2003). Role of alternative oxidase gene in pathogenesis of Cryptococcus neoformans. *Infection and Immunity, 71*(10). https://doi.org/10.1128/IAI.71.10.5794-5802.2003
31. Sinha, T., Naash, M. I., & Al-Ubaidi, M. R. (2020). *Flavins act as a critical liaison between metabolic homeostasis and oxidative stress in the retina.* https://doi.org/10.3389/fcell.2020.00861
32. Zhao, R. Z., Jiang, S., Zhang, L., & Bin Yu, Z. (2019). Mitochondrial electron transport chain, ROS generation and uncoupling (Review). https://doi.org/10.3892/ijmm.2019.4188.
33. Nordgren, M., & Fransen, M. (2014). Peroxisomal metabolism and oxidative stress. https://doi.org/10.1016/j.biochi.2013.07.026.
34. Liu, X., Hussain, R., Mehmood, K., Tang, Z., Zhang, H., & Li, Y. (2022). Mitochondrial-endoplasmic reticulum communication-mediated oxidative stress and autophagy. https://doi.org/10.1155/2022/6459585.
35. Larosa, V., & Remacle, C. (2018). Insights into the respiratory chain and oxidative stress. https://doi.org/10.1042/BSR20171492.
36. Chautrand, T., Souak, D., Chevalier, S., & Duclairoir-Poc, C. (2022). Gram-negative bacterial envelope homeostasis under oxidative and nitrosative stress. https://doi.org/10.3390/microorganisms10050924.
37. Barbosa, K. B. F., Costa, N. M. B., Alfenas, R. D. C. G., De Paula, S. O., Minim, V. P. R., & Bressan, J. (2010). Estresseoxidativo: Conceito, implicações e fatoresmodulatórios. https://doi.org/10.1590/S1415-52732010000400013.
38. Santos, A. L., Gomes, N. C. M., Henriques, I., Almeida, A., Correia, A., & Cunha, Â. (2012). Contribution of reactive oxygen species to UV-B-induced damage in bacteria. *Journal of Photochemistry and Photobiology B, 117*. https://doi.org/10.1016/j.jphotobiol.2012.08.016
39. de Jager, T. L., Cockrell, A. E., & Du Plessis, S. S. (2017). Ultraviolet light induced generation of reactive oxygen species. *Advances in Experimental Medicine and Biology, 996*. https://doi.org/10.1007/978-3-319-56017-5_2
40. Nuszkiewicz, J., Woźniak, A. & Szewczyk-Golec, K. (2020). Ionizing radiation as a source of oxidative stress—The protective role of melatonin and vitamin d. https://doi.org/10.3390/ijms21165804.
41. Azzam, E. I., Jay-Gerin, J. P., & Pain, D. (2012). Ionizing radiation-induced metabolic oxidative stress and prolonged cell injury. https://doi.org/10.1016/j.canlet.2011.12.012
42. Wysocki, R., & Tamás, M. J. (2010). How *Saccharomyces cerevisiae* copes with toxic metals and metalloids. *FEMS Microbiology Reviews.*
43. Khalid, M., Hassani, S., & Abdollahi, M. (2020). Metal-induced oxidative stress: An evidence-based update of advantages and disadvantages. https://doi.org/10.1016/j.cotox.2020.05.006
44. Bhavya, G., Hiremath, K. Y., Jogaiah, S., & Geetha, N. (2022). Heavy metal-induced oxidative stress and alteration in secretory proteins in yeast isolates. *Archives of Microbiology, 204*(3). https://doi.org/10.1007/s00203-022-02756-6
45. Zhang, X., et al. (2020). Oxidative damage mechanism in Saccharomyces cerevisiae cells exposed to tetrachlorobisphenol a. *Environmental Toxicology and Pharmacology, 80*. https://doi.org/10.1016/j.etap.2020.103507
46. Ďurovcová, I., et al. (2021). Acute exposure to bisphenol a causes oxidative stress induction with mitochondrial origin in saccharomyces cerevisiae cells. *Journal of Fungi, 7*(7). https://doi.org/10.3390/jof7070543
47. Vaishampayan, A., & Grohmann, E. (2022). Antimicrobials functioning through ros-mediated mechanisms: Current insights. doi:https://doi.org/10.3390/microorganisms10010061.
48. Mourenza, Á., Gil, J. A., Mateos, L. M., & Letek, M. (2020). Oxidative stress-generating antimicrobials, a novel strategy to overcome antibacterial resistance. https://doi.org/10.3390/antiox9050361.
49. Chandra, P., Sharma, R. K., & Arora, D. S. (2020). Antioxidant compounds from microbial sources: A review. https://doi.org/10.1016/j.foodres.2019.108849
50. Méndez, V., Rodríguez-Castro, L., Durán, R. E., Padrón, G., & Seeger, M. (2022). The OxyR and SoxR transcriptional regulators are involved in a broad oxidative stress response

in Paraburkholderiaxenovorans LB400. *Biological Research, 55*(1). https://doi.org/10.1186/s40659-022-00373-7
51. Sarkar, K., & Sil, P. C. (2022). Antioxidants and immune functions. *Antioxidants Effects in Health: The Bright and the Dark Side*. https://doi.org/10.1016/B978-0-12-819096-8.00057-4
52. Pontonio, E., Verni, M., Dingeo, C., Diaz-De-cerio, E., Pinto, D., & Rizzello, C. G. (2020). Impact of enzymatic and microbial bioprocessing on antioxidant properties of hemp (Cannabis sativa L.). *Antioxidants, 9*(12). https://doi.org/10.3390/antiox9121258
53. Staerck, C. et al. (2017). Microbial antioxidant defense enzymes. https://doi.org/10.1016/j.micpath.2017.06.015.
54. Borgstahl, G. E. O., & Oberley-Deegan, R. E. (2018). Superoxide dismutases (SODs) and SOD mimetics. https://doi.org/10.3390/antiox7110156
55. Horspool, A. M., & Chang, H. C. (2017). Superoxide dismutase SOD-1 modulates C. Elegans pathogen avoidance behavior. *Scientific Reports, 7*. https://doi.org/10.1038/srep45128
56. Anwar, S., et al. (2024). Exploring therapeutic potential of catalase: Strategies in disease prevention and management. *Biomolecules, 14*(6). https://doi.org/10.3390/biom14060697
57. Tabet, F., & Touyz, R. M. (2007). Reactive oxygen species, oxidative stress, and vascular biology in hypertension. *Comprehensive Hypertension*. https://doi.org/10.1016/B978-0-323-03961-1.50033-7
58. Lubos, E., Loscalzo, J., & Handy, D. E. (2011). Glutathione peroxidase-1 in health and disease: from molecular mechanisms to therapeutic opportunities. *Antioxidants & redox signaling*. https://doi.org/10.1089/ars.2010.3586
59. Pei, J., Pan, X., Wei, G., & Hua, Y. (2023). Research progress of glutathione peroxidase family (GPX) in redoxidation. https://doi.org/10.3389/fphar.2023.1147414
60. Consoli, V., Sorrenti, V., Grosso, S., & Vanella, L. (2021). Heme oxygenase-1 signaling and redox homeostasis in physiopathological conditions. https://doi.org/10.3390/biom11040589
61. Couto, N., Wood, J., & Barber, J. (2016). The role of glutathione reductase and related enzymes on cellular redox homoeostasis network. https://doi.org/10.1016/j.freeradbiomed.2016.02.028
62. Sharma, R., Yang, Y., Sharma, A., Awasthi, S., & Awasthi, Y. C. (2004). Antioxidant role of glutathione S-transferases: Protection against oxidant toxicity and regulation of stress-mediated apoptosis. https://doi.org/10.1089/152308604322899350
63. Dasari, S., Ganjayi, M. S., Yellanurkonda, P., Basha, S., & Meriga, B. (2018). Role of glutathione S-transferases in detoxification of a polycyclic aromatic hydrocarbon, methylcholanthrene. https://doi.org/10.1016/j.cbi.2018.08.023
64. Rocca, C., Pasqua, T., Boukhzar, L., Anouar, Y., & Angelone, T. (2019). Progress in the emerging role of selenoproteins in cardiovascular disease: Focus on endoplasmic reticulum-resident selenoproteins. https://doi.org/10.1007/s00018-019-03195-1
65. Lu, J., & Holmgren, A. (2014). The thioredoxin antioxidant system. https://doi.org/10.1016/j.freeradbiomed.2013.07.036
66. Bjørklund, G., Zou, L., Wang, J., Chasapis, C. T., & Peana, M. (2021). Thioredoxin reductase as a pharmacological target. https://doi.org/10.1016/j.phrs.2021.105854
67. Patwardhan, R. S., Sharma, D., & Sandur, S. K. (2022). Thioredoxin reductase: An emerging pharmacologic target for radiosensitization of cancer. https://doi.org/10.1016/j.tranon.2022.101341
68. Rudenko, N. N., Vetoshkina, D. V., Marenkova, T. V., & Borisova-Mubarakshina, M. M. (2023). Antioxidants of non-enzymatic nature: Their function in higher plant cells and the ways of boosting their biosynthesis. https://doi.org/10.3390/antiox12112014
69. Mirończuk-Chodakowska, I., Witkowska, A. M., & Zujko, M. E. (2018). Endogenous non-enzymatic antioxidants in the human body. https://doi.org/10.1016/j.advms.2017.05.005
70. Averill-Bates, D. A. (2023). The antioxidant glutathione. *Vitamins and Hormones, 121*. https://doi.org/10.1016/bs.vh.2022.09.002
71. Poladian, N., Navasardyan, I., Narinyan, W., Orujyan, D., & Venketaraman, V. (2023). Potential role of glutathione antioxidant pathways in the pathophysiology and adjunct treatment of psychiatric disorders. https://doi.org/10.3390/clinpract13040070

72. Labarrere, C. A., & Kassab, G. S. (2022). Glutathione: A Samsonian life-sustaining small molecule that protects against oxidative stress, ageing and damaging inflammation. https://doi.org/10.3389/fnut.2022.1007816
73. Belinskaia, D. A., et al. (2020). The universal soldier: Enzymatic and non-enzymatic antioxidant functions of serum albumin. https://doi.org/10.3390/antiox9100966
74. Belinskaia, D. A., Voronina, P. A., & Goncharov, N. V. (2021). Integrative role of albumin: Evolutionary, biochemical and pathophysiological aspects. *Journal of Evolutionary Biochemistry and Physiology, 57*(6). https://doi.org/10.1134/s002209302106020x
75. Stocker, R., Yamamoto, Y., McDonagh, A. F., Glazer, A. N., & Ames, B. N. (1979). Bilirubin is an antioxidant of possible physiological importance. *Science, 235*(4792), 1987. https://doi.org/10.1126/science.3029864
76. Kim, S. Y., & Park, S. C. (2012, March). Physiological antioxidative network of the bilirubin system in aging and age-related diseases. *Frontiers in Pharmacology, 3*. https://doi.org/10.3389/fphar.2012.00045
77. Reiter, R. J., Tan, D. X., Mayo, J. C., Sainz, R. M., Leon, J., & Czarnocki, Z. (2003). Melatonin as an antioxidant: Biochemical mechanisms and pathophysiological implications in humans. https://doi.org/10.18388/abp.2003_3637
78. Tan, D. X., Manchester, L. C., Esteban-Zubero, E., Zhou, Z., & Reiter, R. J. (2015). Melatonin as a potent and inducible endogenous antioxidant: Synthesis and metabolism. https://doi.org/10.3390/molecules201018886
79. Naghizadeh, M., Reiter, R. J., Kabiri, R., & Moradi, R. (2024). Melatonin improves antioxidant defense mechanism of basil under drought stress. *Horticulture, Environment, and Biotechnology, 65*(1). https://doi.org/10.1007/s13580-022-00507-6
80. Castillo Arteaga, R. D., et al. (2022). Mycothiol peroxidase activity as a part of the self-resistance mechanisms against the antitumor antibiotic Cosmomycin D. *Microbiology Spectrum, 10*(3). https://doi.org/10.1128/spectrum.00493-22
81. Newton, G. L., Buchmeier, N., & Fahey, R. C. (2008). Biosynthesis and functions of Mycothiol, the unique protective thiol of Actinobacteria. *Microbiology and Molecular Biology Reviews, 72*(3). https://doi.org/10.1128/mmbr.00008-08
82. Kędziora-Kornatowska, K., et al. (2010). Effects of coenzyme Q10 supplementation on activities of selected antioxidative enzymes and lipid peroxidation in hypertensive patients treated with indapamide. A pilot study. *Archives of Medical Science, 6*(4). https://doi.org/10.5114/aoms.2010.14461
83. Sifuentes-Franco, S., Sánchez-Macías, D. C., Carrillo-Ibarra, S., Rivera-Valdés, J. J., Zuñiga, L. Y., & Sánchez-López, V. A. (2022). Antioxidant and anti-inflammatory effects of coenzyme Q10 supplementation on infectious diseases. https://doi.org/10.3390/healthcare10030487
84. Jomova, K., et al. (2023). Reactive oxygen species, toxicity, oxidative stress, and antioxidants: chronic diseases and aging. https://doi.org/10.1007/s00204-023-03562-9
85. Kashef, N., & Hamblin, M. R. (2017). Can microbial cells develop resistance to oxidative stress in antimicrobial photodynamic inactivation? *Drug Resistance Updates, 31*. https://doi.org/10.1016/j.drup.2017.07.003
86. Babu, P., Chandel, A. K., & Singh, O. V. (2015). Survival mechanisms of extremophiles. https://doi.org/10.1007/978-3-319-12808-5_2
87. Coker, J. A. (2023). 'All About' Extremophiles. *Faculty Reviews, 12*. https://doi.org/10.12703/r/12-27
88. Ślesak, I., Ślesak, H., Zimak-Piekarczyk, P., & Rozpądek, P. (2016). Enzymatic antioxidant Systems in Early Anaerobes: Theoretical considerations. *Astrobiology, 16*(5). https://doi.org/10.1089/ast.2015.1328
89. Ji, C. J., et al. (2015). Staphylococcus aureus PerR is a hypersensitive hydrogen peroxide sensor using iron-mediated histidine oxidation. *Journal of Biological Chemistry, 290*(33). https://doi.org/10.1074/jbc.M115.664961
90. Fröhlich, K. S., & Gottesman, S. (2018). Small regulatory RNAs in the Enterobacterial response to envelope damage and oxidative stress. *Microbiology Spectrum, 6*(4). https://doi.org/10.1128/microbiolspec.rwr-0022-2018

91. Rani, R., Arora, S., Kaur, J., & Manhas, R. K. (2018). Phenolic compounds as antioxidants and chemopreventive drugs from Streptomyces cellulosae strain TES17 isolated from rhizosphere of Camellia sinensis. *BMC Complementary and Alternative Medicine, 18*(1), 1–15. https://doi.org/10.1186/s12906-018-2154-4
92. Das, M., Prakash, H. S., & Nalini, M. S. (2017). Antioxidative properties of phenolic compounds isolated from the fungal endophytes of Zingiber nimmonii (J. Graham) Dalzell. *Frontiers in Biology, 12*(2), 151–162. https://doi.org/10.1007/s11515-016-1441-z
93. Singh, D. P., Prabha, R., Verma, S., Meena, K. K., & Yandigeri, M. (2017). Antioxidant properties and polyphenolic content in terrestrial cyanobacteria. *3 Biotech, 7*(2), 1–14. https://doi.org/10.1007/s13205-017-0786-6
94. Bhanja Dey, T., Chakraborty, S., Jain, K. K., Sharma, A., & Kuhad, R. C. (2016). Antioxidant phenolics and their microbial production by submerged and solid state fermentation process: A review. *Trends in Food Science and Technology, 53*, 60–74. https://doi.org/10.1016/j.tifs.2016.04.007
95. Adelekan, A. O., Olurin, T. O., & Ezeani, A. O. (2020). Antioxidant activities of exopolysaccharides produced by lactic acid bacteria isolated from commercial yoghurt samples. *Advances in Microbiology, 10*(08), 359–374. https://doi.org/10.4236/aim.2020.108026
96. Nguyen, P. T., & Nguyen, H. T. (2024). Antioxidant property and structural characteristics of exopolysaccharides derived from Lactiplantibacillus plantarum VAL6. *Discover Bacteria.* https://doi.org/10.1007/s44351-024-00004-4
97. Kim, I., et al. (2023). Characterization and antioxidant activity of exopolysaccharides produced by Lysobactersoyae sp. nov isolated from the root of Glycine max L. *Microorganisms, 11*(8). https://doi.org/10.3390/microorganisms11081900
98. Díaz-Cornejo, S., Otero, M. C., Banerjee, A., & Gordillo-Fuenzalida, F. (2023). Biological properties of exopolysaccharides produced by Bacillus spp. *Microbiological Research, 268*(July 2022), 127276. https://doi.org/10.1016/j.micres.2022.127276
99. El-Ghonemy, D. H. (2021). Antioxidant and antimicrobial activities of exopolysaccharides produced by a novel Aspergillus sp. DHE6 under optimized submerged fermentation conditions. *Biocatalysis and agricultural biotechnology, 36, no. September*, 102150. https://doi.org/10.1016/j.bcab.2021.102150
100. Vinothkanna, A., et al. (2022). Exopolysaccharide produced by probiotic Bacillus albus DM-15 isolated from Ayurvedic fermented Dasamoolarishta: Characterization, antioxidant, and anticancer activities. *Frontiers in Microbiology, 13*(March), 1–14. https://doi.org/10.3389/fmicb.2022.832109
101. Domżał-Kędzia, M., Ostrowska, M., Lewińska, A., & Łukaszewicz, M. (2023). Recent developments and applications of microbial Levan, a versatile polysaccharide-based biopolymer. *Molecules, 28*(14). https://doi.org/10.3390/molecules28145407
102. Saini, R. K., & Keum, Y. S. (2017). Progress in microbial carotenoids production. *Indian Journal of Microbiology, 57*(1), 129–130. https://doi.org/10.1007/s12088-016-0637-x
103. Saubenova, M., Rapoport, A., Venkatachalam, M., Dufossé, L., Yermekbay, Z., & Oleinikova, Y. (2024). Production of carotenoids by microorganisms. *Fermentation, 10*(10), 1–16. https://doi.org/10.3390/fermentation10100502
104. Martínez-Cámara, S., Ibañez, A., Rubio, S., Barreiro, C., & Barredo, J.-L. (2021). Main carotenoids produced by microorganisms. *Encyclopedia, 1*(4), 1223–1245. https://doi.org/10.3390/encyclopedia1040093
105. Fusco, W., et al. (2023). Short-chain fatty-acid-producing bacteria: Key components of the human gut microbiota. *Nutrients, 15*(9). https://doi.org/10.3390/nu15092211
106. Azat Aziz, M., Shehab Diab, A., & Abdulrazak Mohammed, A. (2019). Antioxidant categories and mode of action. *Antioxidants.* https://doi.org/10.5772/intechopen.83544
107. Losada-Barreiro, S., Sezgin-Bayindir, Z., Paiva-Martins, F., & Bravo-Díaz, C. (2022). Biochemistry of antioxidants: Mechanisms and pharmaceutical applications. *Biomedicine, 10*(12). https://doi.org/10.3390/biomedicines10123051
108. Hunyadi, A. (2019). The mechanism(s) of action of antioxidants: From scavenging reactive oxygen/nitrogen species to redox signaling and the generation of bioactive secondary

metabolites. *Medicinal Research Reviews, 39*(6), 2505–2533. https://doi.org/10.1002/med.21592
109. De Rossi, L., Rocchetti, G., Lucini, L., & Rebecchi, A. (2025). Antimicrobial potential of polyphenols: Mechanisms of action and microbial responses—A Narrative Review. *Antioxidants, 14*(2). https://doi.org/10.3390/antiox14020200
110. Lakey-Beitia, J., Burillo, A. M., La Penna, G., Hegde, M. L., & Rao, K. S. (2021). Polyphenols as potential metal chelation compounds against Alzheimer's disease. *Journal of Alzheimer's Disease, 82*(s1), S335–S357. https://doi.org/10.3233/JAD-200185
111. Mandal, M. K., & Domb, A. J. (2024). Antimicrobial activities of natural bioactive polyphenols. *Pharmaceutics, 16*(6). https://doi.org/10.3390/pharmaceutics16060718
112. Muscolo, A., Mariateresa, O., Giulio, T., & Mariateresa, R. (2024). Oxidative stress: The role of antioxidant phytochemicals in the prevention and treatment of diseases. *International Journal of Molecular Sciences, 25*(6). https://doi.org/10.3390/ijms25063264
113. Zahra, M., Abrahamse, H., & George, B. P. (2024). Flavonoids: Antioxidant powerhouses and their role in nanomedicine. *Antioxidants, 13*(8). https://doi.org/10.3390/antiox13080922
114. Jomova, K., et al. (2025). Flavonoids and their role in oxidative stress, inflammation, and human diseases. *Chemico-Biological Interactions, 413*(February), 111489. https://doi.org/10.1016/j.cbi.2025.111489
115. Panche, A. N., Diwan, A. D., & Chandra, S. R. (2016). Flavonoids: An overview. *Journal of Nutritional Science, 5*. https://doi.org/10.1017/jns.2016.41
116. Siziya, I. N., Hwang, C. Y., & Seo, M. J. (2022). Antioxidant potential and capacity of microorganism-sourced C30 carotenoids—A review. *Antioxidants, 11*(10). https://doi.org/10.3390/antiox11101963
117. Wang, W., Ju, Y., Liu, N., Shi, S., & Hao, L. (2023). Structural characteristics of microbial exopolysaccharides in association with their biological activities: A review. *Chemical and Biological Technologies in Agriculture, 10*(1), 1–16. https://doi.org/10.1186/s40538-023-00515-3
118. Rusu, A. V., Trif, M., & Rocha, J. M. (2023). Microbial secondary metabolites via fermentation approaches for dietary supplementation formulations. *Molecules, 28*(16). https://doi.org/10.3390/molecules28166020
119. Tamano, K. (2014). Enhancing microbial metabolite and enzyme production: Current strategies and challenges. *Frontiers in Microbiology, 5*(DEC), 1–5. https://doi.org/10.3389/fmicb.2014.00718
120. Hamed, A. A., et al. (2024). Induction of antimicrobial, antioxidant metabolites production by co-cultivation of two red-sea-sponge-associated aspergillus sp. CO2 and bacillus sp. COBZ21. *BMC Biotechnology, 24*(1), 1–13. https://doi.org/10.1186/s12896-024-00830-z
121. Tadesse, S. A., & Emire, S. A. (2020). Production and processing of antioxidant bioactive peptides: A driving force for the functional food market. *Heliyon, 6*(8), e04765. https://doi.org/10.1016/j.heliyon.2020.e04765
122. Daliri, E. B. M., Ofosu, F. K., Chelliah, R., Lee, B. H., & Oh, D. H. (2021, Febraury 01). Challenges and perspective in integrated multi-omics in gut microbiota studies, MDPI AG. https://doi.org/10.3390/biom11020300.
123. Kunin, V., Copeland, A., Lapidus, A., Mavromatis, K., & Hugenholtz, P. (2008, December). A Bioinformatician's guide to metagenomics. *Microbiology and Molecular Biology Reviews, 72*(4), 557–578. https://doi.org/10.1128/mmbr.00009-08
124. Rahbari, M., Rahlfs, S., Jortzik, E., Bogeski, I., & Becker, K. (2017, March). H_2O_2 dynamics in the malaria parasite plasmodium falciparum. *PLoS One, 12*(4). https://doi.org/10.1371/journal
125. Weber, J. L. & Myers, E. W.. Human whole-genome shotgun sequencing
126. Aguiar-Pulido, V., Huang, W., Suarez-Ulloa, V., Cickovski, T., Mathee, K. & Narasimhan, G. (2016, May 12). Metagenomics, metatranscriptomics, and metabolomics approaches for microbiome analysis. Libertas Academica Ltd. https://doi.org/10.4137/EBO.S36436.

127. Kandasamy, S., et al. (2024). Whole genome sequencing of Lacticaseibacillus casei KACC92338 strain with strong antioxidant activity, reveals genes and gene clusters of probiotic and antimicrobial potential. *Frontiers in Microbiology, 15*. https://doi.org/10.3389/fmicb.2024.1458221
128. Bashiardes, S., Zilberman-Schapira, G., & Elinav, E. (2016, April). Use of metatranscriptomics in microbiome research. *Bioinformatics and Biology Insights, 10*, 19–25. https://doi.org/10.4137/BBI.S34610
129. Maron, P. A., Ranjard, L., Mougel, C., & Lemanceau, P. (2007, April). Metaproteomics: A new approach for studying functional microbial ecology. *Microbial Ecology*, 486–493. https://doi.org/10.1007/s00248-006-9196-8
130. Dutilh, B. E., Backus, L., Van Hijum, S. A. F. T., & Tjalsma, H. (2013). Screening metatranscriptomes for toxin genes as functional drivers of human colorectal cancer. Bailliere Tindall Ltd. https://doi.org/10.1016/j.bpg.2013.03.008.
131. Wilmes, P., & Bond, P. L. (2006, Febraury). *Metaproteomics: Studying functional gene expression in microbial ecosystems*. https://doi.org/10.1016/j.tim.2005.12.006
132. Long, S., et al. (2020, December). Metaproteomics characterizes human gut microbiome function in colorectal cancer. *npj Biofilms and Microbiomes, 6*(1). https://doi.org/10.1038/s41522-020-0123-4
133. Zhang, A., Sun, H., Wang, P., Han, Y., & Wang, X. (2012, January 21). Modern analytical techniques in metabolomics analysis. *Royal Society of Chemistry*. https://doi.org/10.1039/c1an15605e
134. Qamar, H., et al. (2024, July). Integrated metabolomics and metagenomics unveiled biomarkers of antioxidant potential in fermented brewer's grains. *Antioxidants, 13*(7). https://doi.org/10.3390/antiox13070872
135. Saini, R. K., & Keum, Y. S. (2019, May 10). Microbial platforms to produce commercially vital carotenoids at industrial scale: an updated review of critical issues. *Springer Verlag*. https://doi.org/10.1007/s10295-018-2104-7
136. Visioli, F., & Artaria, C. (2017, January). Astaxanthin in cardiovascular health and disease: Mechanisms of action, therapeutic merits, and knowledge gaps. In *Food and function* (pp. 39–63). Royal Society of Chemistry. https://doi.org/10.1039/c6fo01721e
137. Nupur, L. N. U., Vats, A., Dhanda, S. K., Raghava, G. P. S., Pinnaka, A. K., & Kumar, A. (2016, May). ProCarDB: A database of bacterial carotenoids. *BMC Microbiology, 16*(1). https://doi.org/10.1186/s12866-016-0715-6
138. Berman, J., et al. (2015, October 26). Nutritionally important carotenoids as consumer products. *Kluwer Academic Publishers*. https://doi.org/10.1007/s11101-014-9373-1
139. Foong, L. C., Loh, C. W. L., Ng, H. S., & Lan, J. C. W. (2021, January 01). Recent development in the production strategies of microbial carotenoids. *Springer Science and Business Media B.V.* https://doi.org/10.1007/s11274-020-02967-3
140. Paniagua-Michel, J., Olmos-Soto, J., & Ruiz, M. A. (2012). Pathways of carotenoid biosynthesis in bacteria and microalgae. *Methods in Molecular Biology, 892*, 1–12. https://doi.org/10.1007/978-1-61779-879-5_1
141. Mussagy, C. U., Winterburn, J., Santos-Ebinuma, V. C., & Pereira, J. F. B. (2019, Febraury 01). Production and extraction of carotenoids produced by microorganisms. *Springer Verlag*. https://doi.org/10.1007/s00253-018-9557-5
142. Rabbani, S., Beyer, P., Lintig, J. V., Hugueney, P., & Kleinig, H. (1998). Induced-Carotene synthesis driven by Triacylglycerol deposition in the unicellular alga Dunaliellabardawil 1.
143. Sandmann, G., & Vioque, A. (1999). Phylogeny of carotenogenic genes and enzymes in cyanobacteria. In *The phototrophic prokaryotes* (pp. 799–803). Boston, MA: Springer US.
144. Sun, T., et al. (2014). Production of lycopene by metabolically-engineered Escherichia coli. *Biotechnology Letters, 36*(7), 1515–1522. https://doi.org/10.1007/s10529-014-1543-0
145. Papp, T., et al. (2013, June). Canthaxanthin production with modified Mucor circinelloides strains. *Appl MicrobiolBiotechnol, 97*(11), 4937–4950. https://doi.org/10.1007/s00253-012-4610-2

146. Beuttler, H., et al. (2011, February). Biosynthesis of zeaxanthin in recombinant pseudomonas putida. *Applied Microbiology and Biotechnology, 89*(4), 1137–1147. https://doi.org/10.1007/s00253-010-2961-0
147. Zhao, J., et al. (2013, May). Engineering central metabolic modules of Escherichia coli for improving B-carotene production. *Metabolic Engineering, 17*(1), 42–50. https://doi.org/10.1016/j.ymben.2013.02.002
148. Zhang, C., Chen, X., Lindley, N. D., & Too, H. P. (2018, January). A 'plug-n-play' modular metabolic system for the production of apocarotenoids. *Biotechnology and Bioengineering, 115*(1), 174–183. https://doi.org/10.1002/bit.26462
149. Mandelli, F., Miranda, V. S., Rodrigues, E., & Mercadante, A. Z. (2012, April). Identification of carotenoids with high antioxidant capacity produced by extremophile microorganisms. *World Journal of Microbiology and Biotechnology, 28*(4), 1781–1790. https://doi.org/10.1007/s11274-011-0993-y
150. Rivera, S. M., Christou, P., & Canela-Garayoa, R. (2014). Identification of carotenoids using mass spectrometry. Wiley. https://doi.org/10.1002/mas.21390.
151. Boujida, N., et al. (2018, October). Isolation and characterization of halophilic bacteria producing exopolymers with emulsifying and antioxidant activities. *Biocatalysis and Agricultural Biotechnology, 16*, 631–637. https://doi.org/10.1016/j.bcab.2018.10.015
152. Janda, J. M., & Abbott, S. L. (2007, September). 16S rRNA gene sequencing for bacterial identification in the diagnostic laboratory: Pluses, perils, and pitfalls. https://doi.org/10.1128/JCM.01228-07
153. Rani, A., et al. (2021). A review on microbial products and their perspective application as antimicrobial agents. https://doi.org/10.3390/biom11121860
154. Forman, H. J., & Zhang, H. (2021). Author correction: Targeting oxidative stress in disease: Promise and limitations of antioxidant therapy. *Nature Reviews. Drug Discovery, 20*(8). https://doi.org/10.1038/s41573-021-00267-5
155. Nishida, Y., Nawaz, A., Hecht, K., & Tobe, K. (2022). Astaxanthin as a novel mitochondrial regulator: A new aspect of carotenoids, beyond antioxidants. https://doi.org/10.3390/nu14010107
156. Pedro, A. C., et al. (2022). The importance of antioxidant biomaterials in human health and technological innovation: A Review. https://doi.org/10.3390/antiox11091644
157. Kaushal, J. B., Raut, P., & Kumar, S. (2023). Organic electronics in biosensing: A promising frontier for medical and environmental applications. https://doi.org/10.3390/bios13110976
158. Zandi, P., & Schnug, E. (2022). Reactive oxygen species, antioxidant responses and implications from a microbial modulation perspective. https://doi.org/10.3390/biology11020155
159. Kelkel, M., Jacob, C., Dicato, M., & Diederich, M. (2010). Potential of the dietary antioxidants resveratrol and curcumin in prevention and treatment of hematologic malignancies. https://doi.org/10.3390/molecules15107035
160. Salehi, B., et al. (2018). Resveratrol: A double-edged sword in health benefits. https://doi.org/10.3390/biomedicines6030091
161. Kamat, C. D., Gadal, S., Mhatre, M., Williamson, K. S., Pye, Q. N., & Hensley, K. (2008). Antioxidants in central nervous system diseases: Preclinical promise and translational challenges. https://doi.org/10.3233/JAD-2008-15314
162. Kamat, C. D., Gadal, S., Mhatre, M., Williamson, K. S., Pye, Q. N., & Hensley, K. (2008). Antioxidants in central nervous system diseases: Preclinical promise and translational challenges. *Journal of Alzheimer's Disease, 15*(3).
163. Ienco, E. C., et al. (2011). Oxidative stress treatment for clinical trials in neurodegenerative diseases. https://doi.org/10.3233/JAD-2011-110164
164. Addor, F. A. S. (2017). Antioxidants in dermatology. https://doi.org/10.1590/abd1806-4841.20175697
165. Baek, J., & Lee, M.-G. (2016). Communications in Free Radical Research Oxidative stress and antioxidant strategies in dermatology oxidative stress and antioxidant strategies in dermatology. *Redox Report, 21*(4).

166. Dahan, S., Khalouf, R., Salcedo, N., Taieb, C., & Gassia, V. (2024). Knowledge of free radicals and antioxidants in dermatology in the general population. *JEADV Clinical Practice, 3*(1). https://doi.org/10.1002/jvc2.274
167. García-Muñoz, A. M., Victoria-Montesinos, D., Ballester, P., Cerdá, B., & Zafrilla, P. (2024, October 01). A Descriptive Review of the Antioxidant Effects and Mechanisms of Action of Berberine and Silymarin. *Multidisciplinary Digital Publishing Institute (MDPI)*. https://doi.org/10.3390/molecules29194576
168. Singh, K. N., Patil, S., & Barkate, H. (2020). Protective effects of astaxanthin on skin: Recent scientific evidence, possible mechanisms, and potential indications. https://doi.org/10.1111/jocd.13019
169. Ye, Y., Ji, J., Sun, Z., Shen, P., & Sun, X. (2020). Recent advances in electrochemical biosensors for antioxidant analysis in foodstuff. https://doi.org/10.1016/j.trac.2019.115718
170. Huang, C. W., Lin, C., Nguyen, M. K., Hussain, A., Bui, X. T., & Ngo, H. H. (2023). A review of biosensor for environmental monitoring: Principle, application, and corresponding achievement of sustainable development goals. https://doi.org/10.1080/21655979.2022.2095089
171. Gavrilaş, S., Ştefan Ursachi, C., Perţa-Crişan, S., & Munteanu, F. D. (2022). Recent trends in biosensors for environmental quality monitoring. https://doi.org/10.3390/s22041513
172. Vigneshvar, S., Sudhakumari, C. C., Senthilkumaran, B., & Prakash, H. (2016). Recent advances in biosensor technology for potential applications—an overview. https://doi.org/10.3389/fbioe.2016.00011
173. Kucherenko, I. S., Soldatkin, O. O., Kucherenko, D. Y., Soldatkina, O. V., & Dzyadevych, S. V. (2019). Advances in nanomaterial application in enzyme-based electrochemical biosensors: a review. https://doi.org/10.1039/c9na00491b
174. Petrucci, R., Bortolami, M., Di Matteo, P., & Curulli, A. (2022). Gold nanomaterials-based electrochemical sensors and biosensors for phenolic antioxidants detection: Recent advances. https://doi.org/10.3390/nano12060959
175. Su, A., Yu, Q., Luo, Y., Yang, J., Wang, E., & Yuan, H. (2021, November 01). Metabolic engineering of microorganisms for the production of multifunctional non-protein amino acids: γ-aminobutyric acid and δ-aminolevulinic acid. *Wiley*. https://doi.org/10.1111/1751-7915.13783
176. Fernandez-Orozco, R., et al. (2009). Evaluation of bioprocesses to improve the antioxidant properties of chickpeas. *LWT, 42*(4), 885–892. https://doi.org/10.1016/j.lwt.2008.10.013
177. Šedík, P., Zagula, G., Ivanišová, E., Kňazovická, V., Horská, E., & Kačániová, M. (2018). Nutrition marketing of honey: Chemical, microbiological, antioxidant and antimicrobial profile. *Potravinarstvo Slovak Journal of Food Sciences, 12*(1), 767–774. https://doi.org/10.5219/988

Green Nanomedicine: Plant Nanoparticles for Targeting Oxidative Stress in Cancer

Mariaamalraj Samykannu, Poornasundari Baluchamy, and Arivoli Subramaniyan

Abstract Cancer is a significant health challenge worldwide, and oxidative stress plays a key role in its development and progression. Excessive oxidative stress can damage healthy cells, leading to mutations and uncontrolled cell growth. Traditional cancer treatments like chemotherapy and radiation can also increase oxidative stress, causing harmful side effects. To overcome these issues, researchers are exploring green nanomedicine, which uses plant-based biogenic nanoparticles as a natural and effective approach to cancer treatment. Biogenic nanoparticles are tiny particles synthesized using plant extracts. These nanoparticles are environmentally friendly, cost-effective, and rich in bioactive compounds with antioxidant and anticancer properties, enhancing the effectiveness of cancer therapies. Additionally, plant-based nanoparticles can be designed to target cancer cells, minimizing damage to healthy tissues specifically. Recent studies have shown that nanoparticles synthesized from medicinal plants like turmeric, green tea, and neem can inhibit cancer cell growth, induce cell death, and improve drug delivery. These nanoparticles can also enhance the body's natural defense mechanisms against oxidative stress. The use of biogenic nanoparticles in cancer treatment represents a promising, less toxic alternative to conventional therapies. This chapter will discuss the role of oxidative stress in cancer, the advantages of plant-based nanoparticles, and their potential in green nanomedicine. By integrating nanotechnology with natural plant compounds, we can develop safer and more efficient cancer treatments for the future.

Keywords Therapy-induced oxidative damage · Function of the immune system · Cancer therapy using metal nanoparticles · Biological compatibility and biodegradability · Maintaining healthy cells by reducing oxidative damage

---

M. Samykannu (✉)
Department of Biotechnology, Sethu Institute of Technology, Pulloor, Kariapatti, Tamil Nadu, India
e-mail: amalbiotech@gmail.com

P. Baluchamy · A. Subramaniyan
Department of Zoology, Thiruvalluvar University, Serkkadu, Vellore, Tamil Nadu, India

K. Kasinathan and A. Muthukumaran (eds.), *Recent Advances in Oxidative Stress Associated Chronic Diseases Volume 2*, https://doi.org/10.1007/978-981-95-3758-7_11

1 Introduction

Cancer is the most concerning health issue that has garnered global attention in recent decades. It ranks as the second major cause of mortality in developing nations. The cell division transpires in humans to generate new cells as the body requires. As cells age or sustain an injury, they perish, and new cells replace them. Occasionally, this systematic mechanism fails, resulting in the inappropriate proliferation of aberrant or damaged cells. These cells may develop into tumors, which are masses of tissue. Neoplasm or malignant is the common name of cancer. By 2030, the worldwide incidence of cancer is projected to reach 21.7 million cases, with 13 million fatalities attributed to the disease [1]. Tumors may be malignant or benign (non-cancerous), cancerous tumors infiltrate adjacent tissues and can disseminate to remote locations in the body, resulting in the formation of new tumors, a phenomenon known as metastasis. Cancerous tumors are sometimes referred to as malignant tumors. Numerous cancers develop solid tumors; however, hematological malignancies, such as leukemias, typically do not. Tobacco intake, obesity, poor diet, and alcohol consumption are 22% of cancer-causing agents. Inherited genetic defects cause 5–10%. Infection by some viruses, such as Human Papilloma Virus, HIV, Human T-lymphotropic virus, Sarcoma virus, bacteria, and parasites, constitutes an environmental component responsible for around 16–18% of global cancer cases. Errors in cell division, UV rays and DNA damage are other reasons. Oxidative stress, characterized by an imbalance favoring reactive oxygen species (ROS) over antioxidants, has been associated with neurodegenerative diseases, cardiovascular diseases, diabetes mellitus, and various other maladies [2]. These relationships underscore the necessity of achieving equilibrium between the relative concentrations of reactive oxygen species and antioxidants. In the last decades, antioxidant molecules have been recognized as one of the most effective alternative and complementary therapies, integrating a double therapeutic and preventive facet [3]. Cells have intricate biochemical and genetic systems to sustain this equilibrium, and their disruption can lead to significant pathogenic effects. Antioxidant systems require antioxidants to neutralize oxidants and mitigate free radicals, which are classified as either enzymatic or non-enzymatic. Superoxide dismutase, catalase, glutathione peroxidase, and the thioredoxin-thioredoxin dehydrogenase system are all examples of enzymatic antioxidants. In contrast, non-enzymatic antioxidants contain glutathione, vitamins C and E, selenium, and other metabolites [4, 5].

2 Type of Cancers

2.1 Carcinoma

Carcinomas represent the predominant category of cancer. Epithelial cells, including the body's internal and external sides, constitute their creation. Epithelial cells exhibit several kinds, frequently characterized by a columnar morphology when observed microscopically. Breast, colon and prostate are examples of adenocarcinoma. It forms fluids and mucous in epithelial cells. The basal layer of the epidermis is affected by basal carcinoma. Stomach, intestine, lungs, bladder, and kidney cells are affected by Squamosa cell carcinoma [6]. Transitional cell carcinoma is a malignancy that affects a particular form of epithelial tissue, the urothelium. The linings of the bladder, ureters, renal pelvis, and other organs are composed of this tissue, which is composed of numerous layers of epithelial cells that can expand and contract. Transitional cell carcinomas are a subset of malignancies that affect the bladder, ureters, and kidneys.

2.2 Sarcoma

The etiology of sarcoma remains ambiguous. Sarcomas are malignant tumors in bone and soft tissues, including muscle, adipose tissue, blood vessels, lymphatic vessels, and fibrous tissue. Osteosarcoma is the primary cancer of bone. Sarcoma occurs when cells undergo alterations in their DNA. The DNA within a cell contains the directives that instruct the cell's functions. DNA instructs the cells to proliferate at a specific rate in healthy cells. The DNA instructs the cells to undergo apoptosis at a predetermined time. In cancer cells, the alterations in DNA provide alternative instructions. The alterations instruct the cancer cells to proliferate rapidly. Cancer cells can persist when healthy cells would perish. This results in an excessive number of cells. Cancer cells may aggregate to create a mass known as a tumor. The tumor may proliferate to infiltrate and obliterate healthy bodily tissue. Cancer cells might eventually detach and metastasize to other regions of the body. When cancer disseminates, it is referred to as metastatic cancer.

2.3 Leukamia

Cancer that is formed in the bone marrow is called leukemia. Leukemia is generally believed to arise when some blood cells undergo alterations (*mutations*) in their genetic material or DNA. The DNA within a cell encodes the directives that govern its functions. Typically, DNA instructs the cell to proliferate at a predetermined rate and to undergo apoptosis at a specified period. In leukemia, the alterations instruct the blood cells to persist in proliferation and division. When this occurs, blood cell

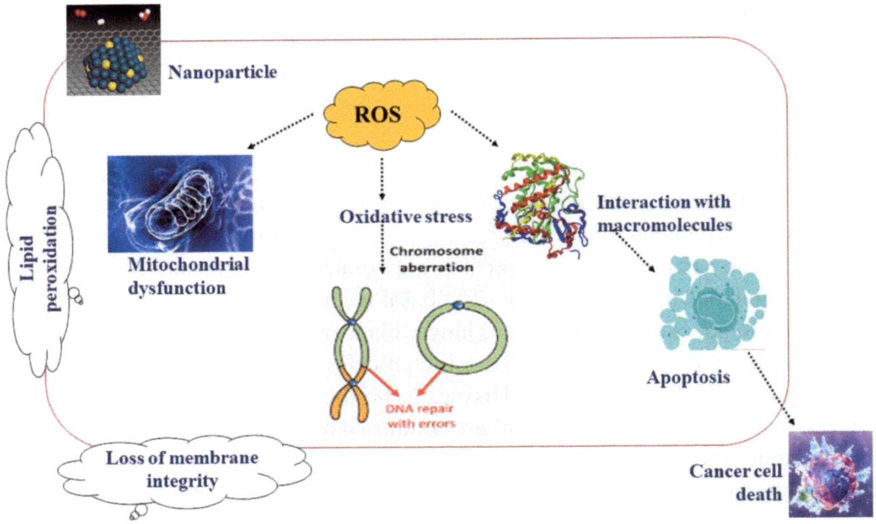

Fig. 1 Mechanism of cell death

production becomes unregulated. Eventually, these atypical cells can proliferate in the bone marrow, displacing healthy blood cells and resulting in a diminished quantity of white blood cells, red blood cells, and platelets, manifesting as the signs and symptoms of leukemia. The diminished quantity of regular blood cells can impede the body's ability to deliver oxygen to tissues, regulate hemorrhage, or combat infections. The leukemia classification is determined by the specific blood cell that undergoes malignant transformation and its proliferation rate as shown in Fig. 1. Leukemia predominantly affects adults over the age of 55. However, it remains the most prevalent malignancy among children under 15.

2.4 *Lymphoma*

Lymphoma is a malignancy in lymphocytes, specifically T cells or B cells. These are leukocytes that combat disease and are vital to the immune system. In lymphoma, aberrant lymphocytes accrue in lymph nodes, lymphatic arteries, and other body parts. There are two types of lymphoma reported: one is Hodgkin lymphoma, which is usually formed from B cells and the second one is non-Hodgkin lymphoma, which is from B and T cells. This large group of cancers starts in lymphocytes [7]. A compromised immune system due to medication or sickness may increase the risk of lymphoma. Individuals with a compromised immune system include individuals undergoing pharmacological treatment to modulate immunological function, such as post-organ transplantation patients [8]. Specific health disorders, like HIV infection, can compromise the immune system.

2.5 *Melanoma*

Skin cancer is the most prominent kind of malignancy. Cancer of the squamous cells, basal cells, and melanoma are the three major sorts of skin cancer. Melanoma is significantly less prevalent than the other varieties but is significantly more likely to invade adjacent cells and spread to other body regions. Melanoma is the primary cause of skin cancer-related fatalities. Atypical moles, ultraviolet exposure, and medical history can influence the likelihood of melanoma. This encompasses the arms, back, face, and legs dermis. Melanoma can also develop in the ocular region [9]. Occasionally, it may occur within the body, such as in the nasal cavity or pharynx. Melanomas can also occur in regions that are less exposed to sunlight. This encompasses the plantar surfaces, palmar surfaces, and nail beds. Melanoma can also occur internally. These concealed melanomas are more prevalent in individuals with brown or black skin.

3 Treatments for Cancer

Many treatments are available to treat cancer, such as surgery, chemotherapy, radiation, hormone therapy, immune therapy, and stem cell treatment. The treatments employed are contingent upon the cancer's type grade and the patient's health and preferences. Chemotherapy utilizes medications that target swiftly growing cells to eradicate tumors. The medications can also assist in the reduction of malignancies; however, the adverse effects may be severe. Hormone therapy entails the administration of medications that alter the function of specific hormones or impede the body's capacity to generate them. This is a prevalent method when hormones are a significant factor, as is the case with prostate and

breast malignancies. Immune therapy is a supplementary therapy that employs medications and other treatments to enhance the immune system's ability to combat malignant cells. High-dose radiation is employed in radiation therapy to eradicate malignant cells [7]. Additionally, a physician may suggest the use of radiation to reduce the symptoms associated with a tumor or to diminish the tumor before surgery. Individuals with blood-related malignancies, such as lymphoma or leukemia, may particularly benefit from stem cell treatment. It entails the removal of cells, such as red or white blood cells, that have been obliterated by chemotherapy or radiation. The cells are subsequently fortified by laboratory technicians and reintroduced into the body. Surgery frequently constitutes a component of a treatment regimen for individuals with a malignant tumor. A surgeon may excise lymph nodes to mitigate or avert the dissemination of the disease. However, these approaches are costly and associated with many adverse effects. However, as the issue continues, innovative strategies are required for illness management, particularly due to the inadequacy of traditional chemotherapeutic methods. Consequently, there is a necessity for innovative ways to prevent and treatment of cancer to manage the mortality rate associated

Table 1 Cancers associated with oxidative stress

Cancer type	Role of oxidative stress	References
Lung cancer	ROS from tobacco smoke and pollutants cause DNA damage (e.g., 8-oxoG), activating mutations in KRAS, EGFR, and TP53	Klaunig et al. [10]
Colorectal cancer	Chronic inflammation (e.g., IBD) increases ROS, activating APC , KRAS mutations, and Wnt/β-catenin	Grivennikov & Karin [11, 12]
Prostate cancer	Androgen receptor signaling elevates ROS, promoting genomic instability and PTEN loss	Kumar et al. [13]
Breast cancer	ROS drive HER2 and PI3K oncogenic pathways and EMT via HIF-1α	Gorrini et al. [7], Piskounovaet al. [14]
Liver cancer (HCC)	Oxidative stress from alcohol, HBV/HCV, and aflatoxins induces p53 mutations and NF-κB activation	Sies et al. [15]
Pancreatic cancer	KRAS mutations increase ROS, sustaining proliferation and chemoresistance	DeNicola et al. [16], Harris et al. [17]
Skin cancer (Melanoma)	UV-induced ROS damages DNA (e.g., BRAF V600E mutation) and suppresses immune surveillance	Piskounovaet al. [14]
Bladder cancer	Cigarette smoke and industrial toxins generate ROS, causing FGFR3 and TP53 mutations	Sabharwal & Schumacker [18], Khandrika et al. [19]

with this disease as shown in Table 1. Herbal medicine has emerged as a very safe, non-toxic, and readily accessible source of chemicals for cancer treatment as shown in Table 2. Herbs are believed to mitigate the impact of ailments in the body owing to their diverse qualities.

4 Plant-Derived Nanoparticles

The biosynthesis of nanoparticles represents a significant convergence of nanotechnology and biotechnology, garnering heightened interest due to the escalating demand for environmentally friendly material synthesis technologies. Biomolecules as reductants possess considerable advantages over chemical reductants owing to their non-biocompatible characteristics [26]. Conventional techniques for nanoparticle synthesis include chemical and physical processes. The chemical synthesis method encompasses chemical reduction, electrochemical methods, and photochemical reduction. The traditional chemical approach employs reducing agents, such as sodium borohydride and hydrazine, alongside radiation chemical methods produced by ionizing radiation to create nanoparticles. The physical synthesis processes encompass condensation, evaporation, and laser ablation. These technologies are

Table 2 Problems due to oxidative stress in cancer

Problem	Mechanism/Impact	References
DNA damage & mutations	ROS causes oxidative DNA lesions (8-oxoG, strand breaks), leading to oncogenic mutations.	Hoeijmakers [20], Klaunig et al. [10]
Oncogene activation & tumor Suppressor Inactivation	Oxidative stress promotes oncogene (e.g., RAS, MYC) activation and inhibits tumor suppressors (e.g., p53, PTEN).	Gorrini et al. [7]
Chronic inflammation	ROS activate NF-κB, STAT3, and COX-2, sustaining pro-tumorigenic inflammation	Grivennikov & Karin [11, 12]
Metastasis & EMT	ROS enhances epithelial-mesenchymal transition (EMT) and invasion via HIF-1α, MMPs, and TGF-β	Giannoni et al. [21]
Chemo/radio resistance	Elevated antioxidant systems (e.g., NRF2, SOD) protect cancer cells from therapy-induced oxidative damage.	Gorrini et al. [7], Diehn et al. [22]
Angiogenesis	ROS stabilize HIF-1α, promoting VEGF secretion and tumor vascularization.	Ushio-Fukai & Nakamura [23]
Altered metabolism (Warburg effect)	ROS stabilize HIF-1α, driving glycolytic metabolism even in normoxia.	DeBerardinis et al. [24]
Immune evasion	ROS suppress T-cell function and promote immunosuppressive TME (e.g., MDSCs, Tregs).	Schafer et al. [25]

costly, inefficient, and may emit hazardous waste into the environment; therefore, more environmentally sustainable approaches are favored. Green synthesis is a burgeoning domain within bio-nanotechnology, offering economic and environmental advantages as a substitute for chemical and physical methodologies as shown in Fig. 2. This method utilizes non-toxic, safe chemicals that are environmentally friendly and safe materials. Diverse natural resources, including plant extracts, algae, microorganisms, and their derivatives, such as polysaccharides like fucoidan and chitosan, can be utilized for the environmentally friendly synthesis of metal oxide nanoparticles [27].

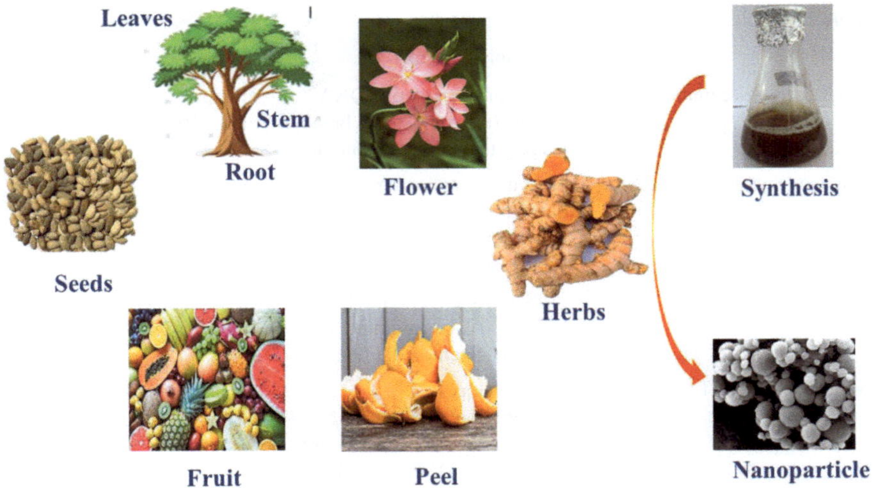

Fig. 2 Nanoparticle synthesis process from plants

5 Green Synthesis of Gold Nanoparticles

Gold nanoparticles possess extensive applications in DNA recognition, hereditary medicine, and nanocatalysis. The varying functionalities of these nanoparticles are regulated by their dimensions [28, 29]. The green synthesis of gold nanoparticles (AuNPs) from plant sources is an environmentally sustainable method. In the biosynthesis of gold nanoparticles (AuNPs) from plants, plant components such as leaves, bark, stems, and roots are utilized, cut into minute fragments, and cooked in distilled water to produce the extract. The extract can be refined using filtration and centrifugation. In the case of the metal salt solution $HAuCl_4$, $AgNO_3$ is often combined with plant extract at ambient temperature [30, 31]. Akhtar et al. [32] reported that AuNPs synthesized from hibiscus and curcumin exhibited activity against carcinoma HCT116 and Michigan Cancer Foundation MCF-7 cell lines, attributable to their diminutive size and homogeneous dispersion. Numerous studies have examined AuNPs as a potential agent for medication delivery to tumor locations through active or passive targeting. A burgeoning field of research employs nanotechnology methods for ocular cancer [33]. The cytotoxic effect of AuNPs synthesized from *Brazilian red Propolis* extract was evaluated toward bladder (T24) and prostate cancer (Pc-3) cell lines [34]. AuNPs synthesized from *Tecoma capensis* and its anticancer efficacy were tested on the human breast cancer cell line MCF-7. It triggers DNA breakage and nuclear condensation. Upon interaction with MCF-7 cancer cells, *T. capensis*-AuNPs induced cell shrinkage and apoptosis, primarily through the disruption of cell membrane integrity, inhibition of ATP synthesis, interference with metabolic and physiological methods, induction of oxidative stress, and obstruction of electron transfer [35]. Sun et al. [36] demonstrate the anticancer efficacy of AuNPs. The plant extracts *Marsdenia tenacissima* were utilized to illustrate its application as an in vitro anticancer agent in lung cancer cells A549 cells. AuNPs have numerous advantages compared to alternative nanoparticles in oncological treatment. Their biocompatibility, low toxicity, and adjustable size render them optimal candidates for drug delivery systems. Gold nanoparticles possess robust optical and physical characteristics, rendering them exceptionally efficient for diagnostic imaging and therapeutic applications, including photothermal therapy and radiosensitization [37].

According to Honsy et al. [35], gold nanoparticles mediated by *Tecoma capensis* (L.) induced apoptosis in malignant breast cells, leading to cellular contraction or apoptosis. The synthesized nanoparticles permeate the cell membrane, interact with, and disrupt diverse biomolecules, including proteins, nucleic acids, and other compounds. The gold nanoparticles synthesized using Cynodon dactylon aqueous extract demonstrate potential anticancer efficacy against human breast tumor cells while exhibiting non-toxicity with healthy cells. The preventive effect was dose-dependent, with green-synthesized AuNPs exhibiting an IC50 value of 31.34 μg/mL. Post-therapy, ROS levels increased, leading to programmed cell death in MCF-7 cancer cells [38]. In fact, [39] synthesized AuNPs from the flower extract of *Lonicera japonica.* The activity was seen in standard HEK293 cells, namely human embryonic kidney cells, demonstrating a non-toxic impact and dose-dependent anticancer activity against cervical cancer (HeLa) cells. Interestingly, [40] synthesized AuNPs from *Citrus macroptera* fruit extract shown greater activity against three different cancerous cells liver cancer (HepG2), alveolar basal epithelial cells (tA 549), and breast cancer cell line (MDA-MB 468). AuNPs synthesized from *Catharanthus roseus* and *Carica papaya* have potential against different cancer cells . The AuNPs from ethanolic extract of *Halodule uninervis* and shows more excellent activity toward varous human cancer lines inhibit cell viability and induce apoptosis [41].

6 Green Synthesis of Silver Nanoparticles

Silver nanoparticles (AgNPs) can be generated using physical, chemical, and biological techniques. Nonetheless, chemical and physical synthesis processes are far more detrimental than biological approaches, which employ an environmentally favorable and less toxic approach [42]. Phytochemicals significantly enhance the biological activity of AgNPs. Cell growth inhibition and DNA fragmentation of AgNPs were tested and synthesized from fruit extraction of *Azadiracta indica.* The cell viability assay was performed by A549 lung cancer cell lines using an MTT assay, and the outcome shows that the synthesized AgNPs showed more significant activity toward the tested cell lines [43] as shown in Table 3. AgNPs are prone to infiltrate cells and can specifically target organs or locations. They are safeguarded against physiological pH, moisture, and enzymes and can be delivered via oral, nasal, intraocular, or parenteral routes. Plant-synthesized AgNPs have been utilized for medication delivery and, in certain instances, for disease treatment [44]. AgNPs synthesized from

Ocimum *gratissimum* (gr-AgNPs) *O. tenuiflorum*, and *O. americanum* and their anticancer properties were evaluated against MCF-7 cell lines. The outcome shows all the synthesized AgNPs were effective against the treated cell lines [45]. Targeted medication therapy utilizing nanosized formulations can address this issue by targeting only the proliferating malignant cells for cytotoxicity. Nanosized formulations significantly advance in treating chronic diseases, including cancer. The AgNPs derived from *C. roseus* exhibit antimetastatic properties against cervical (HeLa229) cell lines by promoting apoptosis and halting the cell cycle [46]. AgNPs produced from *S. trilobatum* leaf extract can trigger apoptosis and subsequently cause cell death in oral cancer KB cell lines [47]. *Azadirachta indica* is a prevalent medicinal plant in South and Southeast Asia, and it is cultivated globally due to its health advantages. Silver nanoparticles synthesized from *A. indica* exhibit a significant inhibitory effect on cell proliferation in H1975 lung cancer and MCF-7 breast cancer cell lines. The treated cells were examined using an inverted microscope. Utilizing an inverted microscope, it was noted that treatment with AgNPs resulted in MCF-7 and H1975 cells losing their unique morphology, with numerous instances of cell shrinkage and necrosis identified post-exposure to AgNPs compared to the untouched control cells. The findings indicated a reduction in cell feasibility for both cancer cell lines as the quantities of AgNPs increased, implying the cytotoxic effects of the nanoparticles on cancer cells [48]. AgNPs synthesized from *Acacia nilotica* were evaluated against colorectal cancer, which ranks third among malignant tumors in humans, with a rising death rate worldwide and a persistently low survival rate. The primary treatment is chemotherapy with 5-fluorouracil, leucovorin, and oxaliplatin as the first-line regimen; however, chemoresistance and disease recurrence continue to pose significant challenges. The synthesized AgNPs were assessed against two cell lines, SW620 and SW480. The synthesized AgNPs exhibit significant cytotoxicity against both cell lines. Silver nanoparticles exert harmful effects via physiochemical interactions with intracellular proteins and DNA through phosphate groups and nitrogen bases. AgNPs modulate many signaling pathways in cell cycles by either upregulating or downregulating the expression levels of critical genes, hence diminishing cell proliferation and survival [49–51]. Therefore, [52] examine the impact of silver nanoparticles derived from Averrhoa bilimbi leaf extract on the A549 and MCF-7 cancer cell lines. The triggering of apoptosis in cancer cells occurs through the formation of reactive oxygen species (ROS). The making of reactive oxygen species is affected by the dimensions of the silver nanoparticles employed in the procedure. The AgNPs function as a tumor suppressor, regulating cell division and inhibiting uncontrolled growth. The disruption of the mitochondrial transmembrane potential. This results in the inhibition of cellular respiration, ultimately causing cell death.

Table 3 Green synthesis of AgNPs from plant sources and its activity tested against different cell lines

Plant	Synthesized nanoparticles	Size of nanoparticle	Cell lines tested	Ref.
Ossimum gratissium *O. tenuiflorum* *O. americanum*	AgNPs	46 nm 69 nm 58 nm	MCF-7	Alex et al. [45]
Olea europaea	AgNPs	23 nm	T47D & MCF-7	Felimban et al. [53]
Callisia fragrans	AgNPs	48 nm	MCF-7, HepG-2, KB, LU-1, & MKN	Nguyen et al. [54]
Ipomoea pes-caprae	AgNPs	116 nm	MCF-7	Veeramani et al. [55]
Averrhoa bilimbi	AgNPs	5 nm	A549 & MCF-7	Hublikar et al. [52]
Mentha pulegium	AgNPs	20 nm	HCT116, HepG2, & HeLa	Wang et al. [56]
Nepeta deflersiana	AgNPs	33 nm	HeLa	Al-Sheddi et al. [57]
Putranjiva roxburghii	AgNPs	69 nm	MCF-7	Nayaka et al. [58]
Conocarpus lancifolius	AgNPs	30 nm	MDA MB-231	Oves et al. [59]
Pueraria tuberosa	AgNPs	162 nm	MCF-7, MDA-MB-231, SKOV-3, U-87, & NCI/ADR	Satpathy et al. [60]
Artemisia turcomanica	AgNPs	22 nm	Gastric cancer AGS	Mousavi et al. [61]
Diospyros kaki	AgNPs	38 nm	Glioblastoma-U118, human colorectal adenocarcinoma-Caco-2, human ovarian sarcoma Skov-3	Keskin et al. [62]
Curcumin	AgNPs	45 nm	HT-29	Taher et al. [63]
Mukia maderaspatna	AgNPs	50 nm	MCF-7	Devi et al. [64]
Syzygium aromaticum	AgNPs	40 nm	MCF-7 & A549	Venugopal et al. [65]
Abutilon indicum	AgNPs	95 nm	MCF-7	Ramesh et al. [66]
Teucrium polium	AgNPs	15 nm	NALM-6	Amini et al. [67]

(continued)

Table 3 (continued)

Plant	Synthesized nanoparticles	Size of nanoparticle	Cell lines tested	Ref.
Cucumis prophetarum	AgNPs	50 nm	A549, MDA-MB-231, HepG2, & MCF-7	Hemlata et al. [68]
Guaiacum officinale	AgNPs	25 nm	SH-SY5Y cells	Rangavitala et al. [69, 70]
Aristolochia bracteolate	AgNPs	40 nm	MCF-7	Durairaju et al. [71]
Carica papaya	AgNPs	50 nm	MCF-7 & Hep-2	John et al. [72]
Cynara scolymus	AgNPs	223 nm	MCF-7	Erdogan et al. [73]
Tea & olive leaf extract		40 nm	Caco-2	Ozturk et al. [74]
Moullava spicata	AgNPs	20 nm	SH-SY5Y	Rangavitala et al. [69, 70]
Orange peel	AgNPs	40 nm	MCF-7	Ghoreishi et al. [75]
Taxus wallichiana	AgNPs	29 nm	Brain cancer cell line U251	Yousaf et al. [76]

7 Green Synthesis of Copper Nanoparticles

Copper nanoparticles (CuNPs) can be synthesized by environmentally friendly processes utilizing plant extracts. These approaches are more naturally sustainable and enhance the features of the generated nanoparticles regarding biocompatibility and functional capabilities. Copper is a vital trace mineral inherently found in certain foods. It is a cofactor in multiple metabolic pathways for various enzymes that facilitate energy production and manufacture connective tissue, neurotransmitters, collagen, and red blood cells. Copper significantly contributes to appropriate brain development, immunological activities, dismutation of superoxide radicals, skin regeneration, angiogenesis, and healing processes [77]. *Nilgirianthus ciliatus* leaf extract was used to synthesize CuO NPs and its anticancer action was assessed against breast cancer cells MCF-7, lung cancer A549, and normal human fibroblast—L929 cell lines. The result shows that the synthesized CuO NPs show more significant cytotoxicity against MCF-7 A549 cell lines and low toxicity on normal human fibroblast-L929 cell lines [78]. Jaculin et al. [79] assessed the cytotoxicity of CuONPs produced with A. tricolor extract against the MCF-7 breast cancer cell line, demonstrating a dose-dependent impact in the present investigation. The findings indicated that increased CuONP content corresponded with heightened cytotoxic effects on MCF-7 cells. CuONPs induce oxidative stress in cancer cells by generating ROS, leading to cellular damage and apoptosis. Furthermore, CuONPs may disrupt critical cellular processes, including DNA replication and repair, leading to cellular apoptosis. *Syzygium alternifolium*-mediated CuONPs, measuring 5–13 nm and exhibiting a spherical morphology, had significant anticancer activity against MDA-MB-231 human breast cancer cell lines. Additionally, phenols and proteins in the stem bark extract contributed to this anticancer efficacy [80]. 38 nm-sized CuONPs were synthesized from *Ficus racemosa*, and their anticancer activity was tested against lung cancer cell line A459. After 48 hrs of treatment, alterations in cell viability were detected in the cell lines treated with CuO NPs at the concentration of 75 μg/mL [81]. According to Rajagopal et al. [82] *Wrightia tinctoria*-mediated CuO NPs show more excellent activity against MCF-7 cells and Kidney epithelial cells. Copper-based chemicals exert significant biological effects, and copper, as a vital trace metal, is essential in the efficacy of anticancer medicines as shown in Table 4. The green production of CuONPs mediated by *Amomum subulatum* fruit extract exhibits cytotoxic effects against human breast cancer cells (MCF-7) and cervical cancer cells (HeLa). Multiple processes have been proposed to elucidate the deadly impact of metal nanoparticles, including the mitochondrial membrane and DNA damage resulting in the apoptotic demise of cancer cells. Nuclear pores of varying dimensions facilitate cellular entry through distinct methods [83].

8 Green Synthesis of Zinc Oxide Nanoparticles

The application of ZnO nanoparticles in oncological therapy has attracted significant attention from researchers, resulting in numerous publications illustrating the anticancer properties of ZnO nanoparticles. Zinc oxide (ZnO) nanoparticles, owing to their exceptional biocompatibility and biodegradability, can be considered effective nano-platforms for cancer therapy [92, 93], . *Punica granatum* fruit peel extract was used to synthesize ZnO NPs and tested against HCT116 colorectal, A549 lung cancer, and Hela cells. After 72 hrs of treatment, the synthesized NPs exhibited notable cytotoxic activity against all the tested cell lines [94]. Mongy et al. [95] synthesized ZnO NPs using *Rhus coriaria* fruit aqueous extract as a reducing and capping agent. The synthesized NPs possess excellent antitumor effects on MCF-7 and MDA-MB-231 breast cancer cells. ZnO NPs were observed to provoke S-phase arrest and inhibit colony-forming and metastatic potential by modulating apoptosis and metastasis-related genes. Green synthesis of zinc oxide nanoparticles (ZnO NPs) from the leaves of *Raphanus sativus var. Longipinnatus* and interpretation of its anticancer activity against Lung cancer cell line A549. The potential mechanisms encompass DNA damage pathways, paraptosis, autophagy, radiosensitization, overcoming chemoresistance and induction of apoptosis, or arresting the cancer cell cycle. The ZnO NPs synthesized from aqueous shilajit extract show greater activity against human cervical cancer cell lines [96]. Sana et al. [97] observed that the ZnO nanoparticles produced from *Crotalaria verrucosa* leaf extract had enhanced potency versus HeLa and DU145 cell lines as shown in Table 5. A further proliferation study was conducted to assess the direct impact of nanoparticles on metastasis, specifically their potential to inhibit tumor cell movement. The produced ZnO nanoparticles exhibit cytotoxicity, apoptosis-mediated DNA destruction, and antiproliferative

Table 4 Phytochemical-mediated synthesis of metal nanoparticles for cancer therapy

Phytochemical molecules	Plant source	Metal nanoparticle synthesized	Anticancer mechanism	Ref.
Curcumin	*Curcuma longa* (turmeric)	Gold NPs (AuNPs), silver NPs (AgNPs)	Induces ROS apoptosis via p53 activation; inhibits NF-κB	Mohan et al. [84]
Epigallocatechin gallate (EGCG)	*Camellia sinensis* (green tea)	Gold NPs (AuNPs), zinc oxide NPs (ZnO NPs)	Antioxidant-to-prooxidant shift; suppresses Bcl-2 and MMP-9	Zhou et al. [85]
Quercetin	*Allium cepa* (onion), *Ginkgo biloba*	Silver NPs (AgNPs), iron oxide NPs (Fe_2O_3 NPs)	DNA intercalation; downregulates PI3K/Akt pathway	Ghosh et al. [86], Ahmad et al. [87]
Resveratrol	*Vitis vinifera* (grapes), *Polygonum cuspidatum*	Gold NPs (AuNPs), selenium NPs (SeNPs)	Activates SIRT1; induces mitochondrial apoptosis	Siddiqui et al. [88]
Allicin	*Allium sativum* (garlic)	Silver NPs (AgNPs), copper NPs (CuNPs)	Generates ROS; inhibits thioredoxin reductase	Elegbede et al. [89]
Ellagic acid	*Punica granatum* (pomegranate)	Gold NPs (AuNPs), platinum NPs (PtNPs)	Arrests cell cycle (G1/S phase); upregulates p21	Kumar et al. [90]
Berberine	*Berberis vulgaris* (barberry)	Silver NPs (AgNPs), zinc oxide NPs (ZnO NPs)	Inhibits topoisomerase I/II; suppresses HIF-1α	Javed et al. [91]

potential in the examined carcinoma cells. According to ZnO [98] nanoparticles measuring 20 nm, synthesized from guava leaf extract, had considerable anticancer efficacy against lung cancer (A549), cervical cancer (HeLa), breast cancer (MCF-7), and normal human dermal fibroblast cells. The cytotoxicity of *Amaranthus dubius*-mediated ZnO nanoparticles was evaluated by a cell proliferation experiment on HCT116 colorectal human cancer cells and CCD112 normal human colon cells. In HCT116 cell lines, cell viability exhibited a modest rise, suggesting potential resistance or adaptability of the cancer cells at elevated concentrations. At a 15.63 µg/mL concentration, nanoparticles enhanced vitality in the normal cell line, suggesting non-toxic effects where minimal stresses elicit advantageous responses. *Sisymbrium irio*-mediated ZnO NPs show potent cytotoxic activity against BHK and HepG2 cell lines [99]. ZnO NPs mediated aqueous extract of neem leaves exhibits a strong anti-cancer effect against human colorectal cancer-HCT116 and lung cancer-A549 cell lines [87].

Table 5 Phytochemical compounds responsible for anticancer activity

Phytochemical	Source	Mechanism of action	References
Curcumin	Turmeric (*Curcuma longa*)	– ROS-mediated apoptosis – NF-κB inhibition – Bcl-2 downregulation	Aggarwal et al. [101]
Resveratrol	Grapes (*Vitis vinifera*)	– SIRT1 activation (senescence) – COX-2 suppression – p21/WAF1 upregulation	Jang et al. [102]
Vinblastine/ vincristine	*Catharanthus roseus*	– Microtubule disruption → mitotic arrest	Noble et al. [103]
Paclitaxel (Taxol)	*Taxus brevifolia*	– Microtubule stabilization → apoptosis	Wani et al. [104]
Artemisinin	*Artemisia annua*	– Iron-dependent ROS generation	Efferth et al. [66]
Quercetin	Onions, apples	– Topoisomerase II inhibition – Anti-angiogenic (VEGF suppression)	Russo et al. [105]
EGCG	Green tea (*Camellia sinensis*)	– Pro-oxidant (ROS) – EGFR/NF-κB modulation	Khan & Mukhtar [106]

9 Anticancer Activity of Palladium NPs

Palladium is a rare silvery-white metal part of the platinum group of metals. The biological features and therapeutic uses have been determined, especially regarding nanoparticles as anti-bacterial and anticancer agents [107]. Turmeric extract comprises many reactive chemicals that enhance its health advantages. The chemicals comprise curcuminoids, phenolic compounds, curcumin, proanthocyanidins, and flavonoids [108]. Curcumin, the principal ingredient in turmeric extract, has numerous health advantages, encompassing anti-inflammatory, antioxidant, and possibly anticancer characteristics. PdNPs synthesized from turmeric can prevent MCF-7 breast cancer cell line propagation. The viability of MCF-7 cells markedly diminished as the concentration of PdNPs synthesized from *Saudi propolis* escalated from 31.25 to 1000 g/mL [107]. Gondake et al. [109] reported that PdNPs synthesized from *Cassia absus* fruit extract shows decreased cell viability against the MCF-7 cell line. 5 nm–15 nm-sized PdNPs synthesized from fruit extract of *Couroupita guianensis.* UV spectra confirmed the synthesis at 420 nm and its anticancer potential was tested against A549 lung cancer cells after 24 hrs of treatment. The IC_{50} value was 121 mg/ml^{-1}. Palladium nanoparticles exhibit exceptional catalytic activity, extensive surface area, and chemical durability. They operate efficiently in various catalytic processes, including oxidation, dehydrogenation, and hydrogenation [110]. *Barleria prionitis* is a medicinal plant that is reported for its antifertility, antioxidant, antidiabetic, anti-inflammatory, anti-arthritic, and cytoprotective activity. The PdNPs synthesized from *Barleria prionitis* show moderate cell viability against the MCF-F cell line [111]. PdNPs synthesized from *Gloriosa superba* tuber extract demonstrate considerable efficacy against the MCF-7 cell line. After 48 hrs of treatment, the reducing viability was $36.26 \pm 0.91\%$, exhibiting significant cytotoxicity resulting in cell death [112]. The intricate and distinctive phytochemistry of *Dioscorea bulbifera* consists of a diverse array of bioactive natural compounds with significant therapeutic relevance. The phytochemical substances in the plant play a crucial role in the rapid synthesis and morphological evolution of unusual metal nanoparticles. PdNPs synthesized from *D. bulbifera* exhibited modest activity against HeLa cell lines, with a cytotoxic impact of 33.15% at a dosage of 10 $\mu g\ mL^{-1}$ [113]. *Bauhinia variegata* is a medicinal plant traditionally used in Unani, Folk, Ayurveda, Iranian, and Siddha medicines. The 9 nm-sized PdNPs synthesized from *Bauhinia variegata* have a potent cytotoxic effect against MCF-7; the IC_{50} value was 41.37 μg/ml. The cell viability was decreased with increasing concentration of the dosage from 0.05–1000 μg/mL it was compared with the standard drug doxorubicin used as a control [114]. The IC_{50} value for A549 lung cancer cells treated with *Madhuca longifolia*-mediated PdNPs was 29.20 μg/ml [115]. *Vaccinium macrocarpon*, generally referred to as Cranberry possesses a striking, brilliant red hue and a characteristic cranberry flavor attributed to its unique flavonoids, such as cyan pigments, proanthocyanidins, and flavanol glycosides. Several triterpenoids were discovered in cranberries, demonstrating significant medicinal properties against tumor growth. PdNPs derived from *V. macrocarpon* impede the growth of MCF-7 cell lines [116].

10 Applications of Nanoparticles

Nanoparticles (NPs) have emerged as versatile tools to modulate oxidative stress in cancer through multiple strategies. Antioxidant NPs, such as cerium oxide (CeO_2) and selenium (Se) NPs, scavenge reactive oxygen species (ROS) by mimicking endogenous enzymes like superoxide dismutase (SOD) and catalase, thereby protecting normal cells during radiotherapy [117]. Conversely, pro-oxidant NPs (e.g., iron oxide (Fe_3O_4) and gold (Au) NPs) exploit the Fenton reaction to generate cytotoxic ROS, selectively killing cancer cells and synergize with chemotherapy. For targeted therapy, liposomal and polymeric NPs (e.g., PLGA) deliver antioxidants (e.g., quercetin) specifically to tumors, minimizing off-target effects. Radiosensitizing NPs like hafnium oxide (HfO_2) enhance radiation-induced ROS in tumors while shielding healthy tissue. Hybrid NPs (e.g., Au-Fe_3O_4) combine dual functions—antioxidant protection for normal cells and pro-oxidant action for cancer cells. Diagnostic NPs (e.g., quantum dots, MnO_2 NPs) enable ROS imaging via fluorescence or MRI, facilitating early detection. Additionally, siRNA-loaded NPs silence oxidative stress-related genes (e.g., NOX4) to sensitize tumors to treatment. These approaches highlight the potential of nanotechnology to precisely balance oxidative stress for therapeutic and diagnostic gains in oncology.

11 Future Scope

The future of plant-derived nanoparticles (NPs) in cancer therapy lies in advancing precision, personalization, and clinical translation. One key direction is the development of smart, stimuli-responsive NPs that selectively release antioxidants or pro-oxidants based on tumor-specific conditions like pH, ROS levels, or enzyme activity. Combining plant NPs with immunotherapy could enhance antitumor immune responses while mitigating oxidative damage to healthy cells. Another promising avenue is multifunctional hybrid nanosystems, integrating plant NPs with conventional drugs, photothermal agents, or gene-editing tools for synergistic effects. Efforts must also focus on scaling up production using standardized, eco-friendly methods to ensure reproducibility and regulatory compliance. Additionally, AI-driven design could optimize NP formulations for specific cancer subtypes, improving targeting efficiency. Finally, rigorous clinical trials are needed to validate safety and efficacy, bridging the gap between lab research and real-world applications. Green nanomedicine could revolutionize oxidative stress modulation by addressing these challenges and offering safer, more effective cancer therapies.

12 Conclusion

Green nanomedicine, leveraging plant-derived nanoparticles (NPs), offers a revolutionary approach to target oxidative stress in cancer therapy. Unlike conventional treatments, these NPs exploit the dual role of reactive oxygen species (ROS) in cancer progression as tumor promoters and therapeutic targets while minimizing off-target toxicity [117]. Phytochemicals like curcumin, epigallocatechin gallate (EGCG), and resveratrol serve as natural reducing agents for synthesizing biocompatible Au, Ag, and Se NPs, which exhibit inherent antioxidant and pro-oxidant properties [84]. For instance, EGCG-functionalized AuNPs selectively induce ROS overload in cancer cells, triggering apoptosis, while CeO_2 NPs protect normal tissues by scavenging excess ROS during radiotherapy. Additionally, plant-based NPs enhance drug delivery, as seen in quercetin-loaded liposomes that improve bioavailability and reduce systemic toxicity. Despite their promise, challenges like batch variability and scalability must be addressed to advance clinical translation. Future directions include hybrid nanosystems combining plant NPs with immunotherapy and ligand-targeted delivery for precision medicine. By bridging traditional phytotherapy and nanotechnology, green nanomedicine provides a sustainable, multifunctional platform to modulate oxidative stress, offering safer and more effective cancer therapies. As research progresses, these innovations could redefine personalized oncology, emphasizing the synergy between nature and nanotechnology.

References

1. Parkin, D. M., Bray, F., Ferlay, J., & Pisani, P. (2005). Global cancer statistics, 2002. *CA: A Cancer Journal for Clinicians, 55*(2), 74–108.
2. Sies, H. (2015). Oxidative stress: a concept in redox biology and medicine. *Redox biology, 4*, 180–183. https://doi.org/10.1016/j.redox.2015.01.002
3. Kong, Q., Beel, J. A., & Lillehei, K. O. (2000). A threshold concept for cancer therapy. *Medical Hypotheses, 55*(1), 29–35.
4. Bansal, A., & Simon, M. C. (2018). Glutathione metabolism in cancer progression and treatment resistance. *The Journal of Cell Biology, 217*(7), 2291–2298.
5. Pei, J., Pan, X., Wei, G., & Hua, Y. (2023). Research progress of glutathione peroxidase family (GPX) in redoxidation. *Frontiers in Pharmacology, 14*, 1147414.
6. Grivennikov, S. I., Greten, F. R., & Karin, M. (2010). Immunity, inflammation, and cancer. *Cell, 140*(6), 883–899. https://doi.org/10.1016/j.cell.2010.01.025
7. Gorrini, C., Harris, I. S., & Mak, T. W. (2013). Modulation of oxidative stress as an anticancer strategy. *Nature Reviews Drug Discovery, 12*(12), 931–947.
8. Manzoor, S. I., Jabeen, F., Patel, R., Rizvi, M. M. A., Imtiyaz, K., Malik, M. A., & Dar, T. A. (2025). Green synthesis of biocompatible silver nanoparticles using Trillium govanianum rhizome extract: comprehensive biological evaluation and in silico analysis. *Materials Advances, 6*(2), 682–702. https://doi.org/10.1039/D4MA00959B
9. Schieber, M., & Chandel, N. S. (2014). ROS function in redox signaling and oxidative stress. Current biology, 24(10), R453-R462. https://doi.org/10.1016/j.cub.2014.03.034
10. Klaunig, J. E., Kamendulis, L. M., & Hocevar, B. A. (2010). Oxidative stress and oxidative damage in carcinogenesis. *Toxicologic pathology, 38*(1), 96–109. https://doi.org/10.1177/0192623309356453

11. Grivennikov, S. I., & Karin, M. (2010a). Dangerous liaisons: STAT3 and NF-κB collaboration in cancer. *Cancer Cell, 17*(2), 115–117.
12. Grivennikov, S. I., & Karin, M. (2010b). Inflammation and oncogenesis: A vicious connection. *Current Opinion in Genetics & Development, 20*(1), 65–71.
13. Kumar, B., Koul, S., Khandrika, L., Meacham, R. B., & Koul, H. K. (2008). Oxidative stress is inherent in prostate cancer cells and is required for aggressive phenotype. *Cancer Research, 68*(6), 1777–1785.
14. Piskounova, E., Agathocleous, M., Murphy, M. M., Hu, Z., Huddlestun, S. E., Zhao, Z., ... & Morrison, S. J. (2015). Oxidative stress inhibits distant metastasis by human melanoma cells. *Nature, 527*(7577), 186–191. https://doi.org/10.1038/nature15726
15. Sies, H., Berndt, C., & Jones, D. P. (2017). Oxidative stress. *Annual review of biochemistry, 86*(1), 715–748. https://doi.org/10.1146/annurev-biochem-061516-045037
16. DeNicola, G. M., Karreth, F. A., Humpton, T. J., Gopinathan, A., Wei, C., Frese, K., Mangal, D., Yu, K. H., Yeo, C. J., Calhoun, E. S., Scrimieri, F., Winter, J. M., Hruban, R. H., Iacobuzio-Donahue, C., Kern, S. E., Blair, I. A., & Tuveson, D. A. (2011). Oncogene-induced Nrf2 transcription promotes ROS detoxification and tumorigenesis. *Nature, 475*(7354), 106–109.
17. Harris, I. S., Treloar, A. E., Inoue, S., Sasaki, M., Gorrini, C., Lee, K. C., Yung, K. Y., Brenner, D., Knobbe-Thomsen, C. B., Cox, M. A., Elia, A., Berger, T., Cescon, D. W., Adeoye, A., Brustle, A., Molyneux, S. D., Mason, J. M., Li, W. Y., Yamamoto, K., Wakeham, A., Berman, H. K., Khokha, R., Done, S. J., Kavanagh, T. J., Lam, C. W., & Mak, T. W. (2015). Glutathione and thioredoxin antioxidant pathways synergize to drive cancer initiation and progression. *Cell, 160*(3), 389–401.
18. Sabharwal, S. S., & Schumacker, P. T. (2014). Mitochondrial ROS in cancer: initiators, amplifiers or an Achilles' heel?. *Nature Reviews Cancer, 14*(11), 709–721. https://doi.org/10.1038/nrc3803
19. Khandrika, L., Kumar, B., Koul, S., Maroni, P., & Koul, H. K. (2009). Oxidative stress in prostate cancer. *Cancer Letters, 282*(2), 125–136.
20. Hoeijmakers, J. H. (2009). DNA damage, aging, and cancer. *New England Journal of Medicine, 361*(15), 1475–1485.
21. Giannoni, E., Buricchi, F., Raugei, G., Ramponi, G., & Chiarugi, P. (2011). Intracellular reactive oxygen species activate Src tyrosine kinase during cell adhesion and anchorage-dependent cell growth. *Cancer Letters, 313*(2), 155–164.
22. Diehn, M., Cho, R. W., Lobo, N. A., Kalisky, T., Dorie, M. J., Kulp, A. N., Qian, D., Lam, J. S., Ailles, L. E., Wong, M., & Joshua, B. (2009). Association of reactive oxygen species levels and radioresistance in cancer stem cells. *Nature, 458*(7239), 780–783.
23. Ushio-Fukai, M., & Nakamura, Y. (2008). Reactive oxygen species and angiogenesis: NADPH oxidase as target for cancer therapy. *Cancer letters, 266*(1), 37–52. https://doi.org/10.1016/j.canlet.2008.02.044
24. DeBerardinis, R. J., Lum, J. J., Hatzivassiliou, G., & Thompson, C. B. (2008). The biology of cancer: Metabolic reprogramming fuels cell growth and proliferation. *Cell Metabolism, 7*(1), 11–20.
25. Schafer, M. J., White, T. A., Iijima, K., Haak, A. J., Ligresti, G., Atkinson, E. J., ... & LeBrasseur, N. K. (2017). Cellular senescence mediates fibrotic pulmonary disease. *Nature communications, 8*(1), 14532. https://doi.org/10.1038/ncomms14532
26. Huang, X. H., Jain, P. K., El-Sayed, I. H., & El-Sayed, M. A. (2007). Gold nanoparticles: Interesting optical properties and recent applications in cancer diagnostic and therapy. *Nanomedicine, 2*(5), 681–693.
27. Keabadile, O. P., Aremu, A. O., Elugoke, S. E., & Fayemi, O. E. (2020). Green and traditional synthesis of copper oxide nanoparticles—Comparative study. *Nanomaterials, 10*(12), 2502.
28. Alivisatos, A. P. (1996). Perspectives on the physical chemistry of semiconductor nanocrystals. *The Journal of Physical Chemistry, 100*(31), 13226–13239.
29. Jin, R. C., Cao, Y. W., Mirkin, C. A., Kelly, K. L., Schatz, G. C., & Zheng, J. G. (2001). Photoinduced conversion of silver nanospheres to nanoprisms. *Science, 294*(5548), 1901–1903.

30. Ahmed, S., Annu, I. S., & Yudha, S. S. (2016). Biosynthesis of gold nanoparticles: A green approach. *Journal of Photochemistry and Photobiology B: Biology, 161*, 141–153.
31. Al-Saadi, H., Abdulrasool, H., & Murphy, E. (2023). Evaluation of Clinical Assessment in Predicting Complicated Acute Diverticulitis. *Cureus, 15*(2). https://doi.org/10.7759/cureus.34709
32. Akhtar, S., Asiri, S. M., Khan, F. A., Gunday, S. T., Iqbal, A., Alrushaid, N., Labib, O. A., Deen, G. R., & Henari, F. Z. (2022). Formulation of gold nanoparticles with hibiscus and curcumin extracts induced anticancer activity. *Arabian Journal of Chemistry, 15*(2), 103594.
33. Bharadwaj, K. K., Rabha, B., Pati, S., Sarkar, T., Choudhury, B. K., Barman, A., Bhattacharjya, D., Srivastava, A., Baishya, D., Edinur, H. A., Kari, Z. A., & Noor, N. H. M. (2021). Green synthesis of gold nanoparticles using plant extracts as beneficial prospect for cancer theranostics. *Molecules, 26*(21), 6389.
34. Botteon, C. E. A., Silva, L. B., Ccana-Ccapatinta, G. V., Silva, T. S., Ambrosio, S. R., Veneziani, R. C. S., Bastos, J. K., & Marcato, P. D. (2021). Biosynthesis and characterization of gold nanoparticles using Brazilian red propolis and evaluation of its antimicrobial and anticancer activities. *Scientific Reports, 11*(1), 1974.
35. Hosny, M., Fawzy, M., El-Badry, Y. A., Hussein, E. E., & Eltaweil, A. S. (2022). Plant-assisted synthesis of gold nanoparticles for photocatalytic, anticancer, and antioxidant applications. *Journal of Saudi Chemical Society, 26*(2), 101419.
36. Sun, B., Hu, N., Han, L., Pi, Y., Gao, Y., & Chen, K. (2019). Anticancer activity of green synthesised gold nanoparticles from Marsdeniatenacissima inhibits A549 cell proliferation through the apoptotic pathway. *Artificial Cells, Nanomedicine, and Biotechnology, 47*(1), 4012–4019.
37. Sachan, N. R. S., Singh, A., Karnwal, A., Shidiki, A., & Kumar, G. (2024). Plant-mediated gold nanoparticles in cancer therapy: Exploring anticancer mechanisms, drug delivery applications, and future prospects. *Frontiers in Nanotechnology, 6*, 1490980.
38. Vinayagam, R., Santhoshkumar, M., Lee, K. E., David, E., & Kang, S. G. (2021). Bioengineered gold nanoparticles using Cynodon dactylon extract and its cytotoxicity and antibacterial activities. *Bioprocess and Biosystems Engineering, 44*, 1253–1262.
39. Patel, R., Ashcroft, J., Patel, A., Ashrafian, H., Woods, A. J., Singh, H., ... & Leff, D. R. (2019). The impact of transcranial direct current stimulation on upper-limb motor performance in healthy adults: a systematic review and meta-analysis. *Frontiers in Neuroscience, 13*, 1213. https://doi.org/10.3389/fnins.2019.01213
40. Majumdar, S., Mishra, V., Nandi, S., Abdullah, M., Barman, A., Raghavan, A., ... & Visweswariah, S. S. (2019). Correction for Majumdar et al.,"Absence of Receptor Guanylyl Cyclase C Enhances Ileal Damage and Reduces Cytokine and Antimicrobial Peptide Production during Oral Salmonella enterica Serovar Typhimurium Infection". *Infection and Immunity, 87*(9), 10–1128. https://doi.org/10.1128/IAI.00495-19
41. Wehbe, N., Mesmar, J. E., El Kurdi, R., Al-Sawalmih, A., Badran, A., Patra, D., & Baydoun, E. (2025). Halodule uninervis extract facilitates the green synthesis of gold nanoparticles with anticancer activity. *Scientific Reports, 15*(1), 4286.
42. Suganya, S., Ishwarya, R., Jayakumar, R., Govindarajan, M., Alharbi, N. S., Kadaikunnan, S., Khaled, J. M., Al-Anbr, M. N., & Vaseeharan, B. (2019). New insecticides and antimicrobials derived from Sargassum wightii and Halimeda gracillis seaweeds: Toxicity against mosquito vectors and antibiofilm activity against microbial pathogens. *South African Journal of Botany, 125*, 466–480.
43. Alharbi, N. S., & Alsubhi, N. S. (2022). Green synthesis and anticancer activity of silver nanoparticles prepared using fruit extract of Azadirachta indica. *Journal of Radiation Research and Applied Sciences, 15*(3), 335–345.
44. Lee, K. X., Shameli, K., Mohamad, S. E., Yew, Y. P., Isa, E. D., Yassin, M. T., Lim, W. L., & Teow, S. Y. (2019). Bio-mediated synthesis and characterisation of silver nanocarrier, and its potent anticancer action. *Nanomaterials, 9*(10), 1423.
45. Alex, A. M., Subburaman, S., Chauhan, S., Ahuja, V., Abdi, G., & Tarighat, M. A. (2024). Green synthesis of silver nanoparticle prepared with Ocimum species and assessment of anticancer potential. *Scientific Reports, 14*(1), 11707.

46. Hussein, H. S., Ngugi, C., Tolo, F. M., & Maina, E. N. (2024). Anticancer potential of silver nanoparticles biosynthesized using Catharanthus roseus leaves extract on cervical (HeLa229) cancer cell line. *Scientific African, 25*, e02268.
47. Ganesan, A., Kumar, G., & Natarajan, P. M. (2024). Anticancer activity of silver nanoparticles synthesized from aqueous leaf extract of Solanum trilobatum (purple fruit pea eggplant) on human oral cancer cells. *Journal of Herbmed Pharmacology, 13*(2), 260–268.
48. Alharbi, N. S., & Alsubhi, N. S. (2023). Silver nanoparticles biosynthesized using Azadirachta indica fruit and leaf extracts: Optimization, characterization, and anticancer activity. *Journal of Nanomaterials, 2023*(1), 9916777.
49. Alduraihem, N. S., Bhat, R. S., Al-Zahrani, S. A., Elnagar, D. M., Alobaid, H. M., & Daghestani, M. H. (2023). Anticancer and antimicrobial activity of silver nanoparticles synthesized from pods of Acacia nilotica. *PRO, 11*(2), 301.
50. Farah, M. A., Ali, M. A., Chen, S. M., Li, Y., Al-Hemaid, F. M., Abou-Tarboush, F. M., Al-Anazi, K. M., & Lee, J. (2016). Silver nanoparticles synthesized from Adenium obesum leaf extract induced DNA damage, apoptosis, and autophagy via generation of reactive oxygen species. *Colloids and Surfaces B: Biointerfaces, 141*, 158–169.
51. Yang, T., Yao, Q., Cao, F., Liu, Q., Liu, B., & Wang, X. H. (2016). Silver nanoparticles inhibit the function of hypoxia-inducible factor-1 and target genes: Insight into the cytotoxicity and antiangiogenesis. *International Journal of Nanomedicine, 11*, 6679–6692.
52. Hublikar, L. V., Ganachari, S. V., & Patil, V. B. (2023). Phytofabrication of silver nanoparticles using Averrhoa bilimbi leaf extract for anticancer activity. *Nanoscale Advances, 5*(16), 4149–4157.
53. Felimban, A. I., Alharbi, N. S., & Alsubhi, N. S. (2022). Optimization, characterization, and anticancer potential of silver nanoparticles biosynthesized using Olea europaea. *International Journal of Biomaterials, 2022*, 6859637.
54. Nguyen, L. A., Van Mai, B., Van Nguyen, D., Nguyen, N. Q., Van Pham, V., Pham, T. L., & Le, H. T. (2023). Green synthesis of silver nanoparticles using Callisia fragrans leaf extract and its anticancer activity against MCF-7, HepG2, KB, LU-1, and MKN-7 cell lines. *Green Processing and Synthesis, 12*(1), 20230024.
55. Veeramani, S., Ravindran, E., Ramadoss, P., Joseph, C., Shanmugam, K., & Renganathan, S. (2018). Silver nanoparticles green synthesis with Aq. Extract of stems Ipomoea pes-caprae, characterization, antimicrobial and anticancer potential. International journal of medical. *Nano Research, 5*(024).
56. Wang, Y., & Wei, S. (2021). Green fabrication of bioactive silver nanoparticles using Mentha pulegium extract under alkaline: An enhanced anticancer activity. *ACS Omega, 7*(1), 1494–1504.
57. Al-Sheddi, E. S., Farshori, N. N., Al-Oqail, M. M., Al-Massarani, S. M., Saquib, Q., Wahab, R., Musarrat, J., Al-Khedhairy, A. A., & Siddiqui, M. A. (2018). Anticancer potential of green synthesized silver nanoparticles using extract of Nepeta deflersiana against human cervical cancer cells (HeLA). *Bioinorganic Chemistry and Applications, 2018*(1), 9390784.
58. Nayaka, S., Bhat, M. P., Chakraborty, B., Pallavi, S. S., Airodagi, D., Muthuraj, R., Halaswamy, H. M., Dhanyakumara, S. B., Shashiraj, K. N., & Kupaneshi, C. (2020). Seed extract-mediated synthesis of silver nanoparticles from Putranjiva roxburghii wall., phytochemical characterization, antibacterial activity and anticancer activity against MCF-7 cell line. *Indian Journal of Pharmaceutical Sciences, 82*(2), 260–269.
59. Oves, M., Rauf, M. A., Aslam, M., Qari, H. A., Sonbol, H., Ahmad, I., Zaman, G. S., & Saeed, M. (2022). Green synthesis of silver nanoparticles by Conocarpus Lancifolius plant extract and their antimicrobial and anticancer activities. *Saudi Journal of Biological Sciences, 29*(1), 460–471.
60. Satpathy, S., Patra, A., Ahirwar, B., & Hussain, M. D. (2018). Antioxidant and anticancer activities of green synthesized silver nanoparticles using aqueous extract of tubers of Pueraria tuberosa. *Artificial Cells, Nanomedicine, and Biotechnology, 46*(sup3), 71–85.
61. Mousavi, B., Tafvizi, F., & Zaker Bostanabad, S. (2018). Green synthesis of silver nanoparticles using Artemisia turcomanica leaf extract and the study of anticancer effect and apoptosis

induction on gastric cancer cell line (AGS). *Artificial Cells, Nanomedicine, and Biotechnology, 46*(supl), 499–510.

62. Keskin, C., Ölçekçi, A., Baran, A., Baran, M. F., Eftekhari, A., Omarova, S., Khalilov, R., Aliyev, E., Sufianov, A., Beilerli, A., & Gareev, I. (2023). Green synthesis of silver nanoparticles mediated Diospyros kaki L. (persimmon): Determination of chemical composition and evaluation of their antimicrobials and anticancer activities. Frontiers. *Chemistry, 11*, 1187808.
63. Taher, N.G. & AL-Abodi, E.E., 2023. Anticancer activity of synthesized green silver nanoparticles against human colon cancer cell lines. Onkologia i Radioterapia, *17*(11)
64. Devi, G. K., & Sathishkumar, K. (2017). Synthesis of gold and silver nanoparticles using Mukia maderaspatna plant extract and its anticancer activity. *IET Nanobiotechnology, 11*(2), 143–151.
65. Venugopal, K., Rather, H. A., Rajagopal, K., Shanthi, M. P., Sheriff, K., Illiyas, M., Rather, R. A., Manikandan, E., Uvarajan, S., Bhaskar, M., & Maaza, M. (2017). Synthesis of silver nanoparticles (ag NPs) for anticancer activities (MCF 7 breast and A549 lung cell lines) of the crude extract of Syzygium aromaticum. *Journal of Photochemistry and Photobiology B: Biology, 167*, 282–289.
66. Efferth, T., Dunstan, H., Sauerbrey, A., Miyachi, H., & Chitambar, C. R. (2001). The anti-malarial artesunate is also active against cancer. *International Journal of Oncology, 18*(4), 767–773.
67. Amini, S. M., Salavati Pour, M. S., Vahidi, R., Kouhbananinejad, S. M., Bardsiri, M. S., Farsinejad, A., & Parsa, M. J. M. (2021). Green synthesis of stable silver nanoparticles using Teucrium polium extract: In-vitro anticancer activity on NALM-6. *Nanomedicine Research Journal, 6*(2), 170–178.
68. Hemlata, M., & P.R., Singh, A.P. and Tejavath, K.K. (2020). Biosynthesis of silver nanoparticles using Cucumis prophetarum aqueous leaf extract and their antibacterial and antiproliferative activity against cancer cell lines. *ACS Omega, 5*(10), 5520–5528.
69. Rangavitala, P., & Taranath, T. C. (2024a). Green synthesis of Guaiacum officinale L. leaf extract silver nanoparticles for potential antioxidant and anticancer activity. *Journal of Research in Nanoscience and Nanotechnology, 12*(1), 25–40.
70. Rangavitala, P., & Taranath, T. C. (2024b). Green synthesized Moullava spicata (Dalz.) Nicolson leaf extract mediated silver nanoparticles potentiate antioxidant and anticancer activity in human bone marrow neuroblastoma cancer cells. *International Journal of Pharmaceutical Sciences and Drug Research, 16*(4), 662–670.
71. Durairaju, P., Raja, G., Venkatesh, G., Govindasamy, C., Newehy, A. S. E., Haseena, S., Vennila, P., & Balasubramaniyan, S. (2025). Green synthesis, characterization, antimicrobial and anticancer activity of silver nanoparticles using Aristolochia bracteolate plant leaf extract: Experimental and theoretical calculations. *Bulletin of the Chemical Society of Ethiopia, 39*(2), 271–286.
72. John, T., Parmar, K. A., Kotval, S. C., & Jadhav, J. (2021). Synthesis, characterization, antibacterial and anticancer properties of silver nanoparticles synthesized from Carica papaya peel extract. *International Journal of Nanoscience and Nanotechnology, 17*(1), 23–32.
73. Erdogan, O., Abbak, M., Demirbolat, G. M., Birtekocak, F., Aksel, M., Pasa, S., & Cevik, O. (2019). Green synthesis of silver nanoparticles via Cynara scolymus leaf extracts: The characterization, anticancer potential with photodynamic therapy in MCF7 cells. *PLoS One, 14*(6), e0216496.
74. Ozturk, S. N., Tomak, A., & Karakus, C. O. (2024). Green synthesis of silver nanoparticles using plant extract blends and its impact on antibacterial and biological activity. *Nano, 19*(10), 2450076.
75. Ghoreishi, S. M., & Mortazavi-Derazkola, S. (2024). Biological evaluation of silver nanoparticles synthesized from bitter orange peel extract: Antibacterial and anticancer properties on human breast cancer cell line. *Journal of Nanostructures, 14*(2), 505–512.
76. Yousaf, A., Waseem, M., Javed, A., Baig, S., Ismail, B., Baig, A., Shahzadi, I., Nawazish, S., & Zaman, I. (2022). Augmented anticancer effect and antibacterial activity of silver nanoparticles synthesized by using Taxus wallichiana leaf extract. *PeerJ, 10*, e14391.

77. Sandoval, C., Ríos, G., Sepúlveda, N., Salvo, J., Souza-Mello, V., & Farías, J. (2022). Effectiveness of copper nanoparticles in wound healing process using in vivo and in vitro studies: A systematic review. *Pharmaceutics, 14*(9), 1838.
78. Rajamma, R., Gopalakrishnan Nair, S., Abdul Khadar, F., & Baskaran, B. (2020). Antibacterial and anticancer activity of biosynthesised CuO nanoparticles. *IET Nanobiotechnology, 14*(9), 833–838.
79. Jaculin, C., Viswanathan, D., & Govindasamy, R. (2024). Phytofabricated copper oxide nanoparticles from Amaranthus tricolor leaf extract: Assessing antioxidant and anticancer potentials. *International Journal of Pharmaceutical Investigation, 14*(2).
80. Yugandhar, P., Vasavi, T., Devi, U. M., Prasad, D., Narasimha, G., & Savithramma, N. (2017). Bioinspired green synthesis of copper oxide nanoparticles from Syzygium alternifolium (Wt.) Walp: Characterization and evaluation of its synergistic antimicrobial and anticancer activity. *Applied Nanoscience, 7*, 417–427.
81. Maurya, A. K., Varshney, S., Verma, V., Singh, A. K., Singh, A. K., Mishra, A., Singh, V. K., & Verma, S. K. (2024). Synthesis of silver and copper oxide nanoparticles using Ficus racemosa leaf extract: Characterization, anticancer potential, and dye degradation efficacy. *Journal of umm Al-Qura University for Applied Sciences.*
82. Rajagopal, G., Nivetha, A., Sundar, M., Panneerselvam, T., Murugesan, S., Parasuraman, P., Kumar, S., Ilango, S., & Kunjiappan, S. (2021). Mixed phytochemicals mediated synthesis of copper nanoparticles for anticancer and larvicidal applications. *Heliyon, 7*(6).
83. Dhir, S., Dutt, R., Singh, R. P., Chauhan, M., Virmani, T., Kumar, G., Alhalmi, A., Aleissa, M. S., Rudayni, H. A., & Al-Zahrani, M. (2023). Amomum subulatum fruit extract mediated green synthesis of silver and copper oxide nanoparticles: Synthesis, characterization, antibacterial and anticancer activities. *PRO, 11*(9), 2698.
84. Mohan, S., Oluwafemi, O. S., Kalarikkal, N., Thomas, S., & Songca, S. P. (2016). Biomaterials for phototherapeutic applications. *Journal of Drug Delivery Science and Technology, 33*, 134–143.
85. Zhou, P., Yang, X. L., Wang, X. G., Hu, B., Zhang, L., Zhang, W., ... & Shi, Z. L. (2020). A pneumonia outbreak associated with a new coronavirus of probable bat origin. *Nature, 579*(7798), 270–273. https://doi.org/10.1038/s41586-020-2012-7
86. Ghosh, S., Patil, S., Ahire, M., Kitture, R., Gurav, D. D., Jabgunde, A. M., Kale, S., Pardesi, K., Shinde, V., Bellare, J., & Chopade, B. A. (2019). Synthesis of gold nanoanisotrops using Dioscorea bulbifera tuber extract. *Journal of Nanomaterials, 2019*, 1–13.
87. Ahmad, N., Bhatnagar, S., Ali, S. S., Dutta, R., & Gupta, P. K. (2020). Phytochemical-functionalized gold nanoparticles and their biomedical applications. *Saudi Journal of Biological Sciences, 27*(1), 706–712.
88. Siddiqui, F., Salam, R. A., Lassi, Z. S., & Das, J. K. (2020). The intertwined relationship between malnutrition and poverty. *Frontiers in Public Health, 8*, 453. https://doi.org/10.3389/fpubh.2020.00453
89. Elegbede, J. A., Lateef, A., Azeez, M. A., Asafa, T. B., Yekeen, T. A., Oladipo, I. C., Adebayo, E. A., Beukes, L. S., & Gueguim-Kana, E. B. (2021). Biofabrication of gold nanoparticles using xylanases through valorization of corncob by aspergillus Niger and Trichoderma longibrachiatum: Antimicrobial, antioxidant, anticoagulant and thrombolytic activities. *Journal of Drug Delivery Science and Technology, 64*, 102366.
90. Kumar, B., Smita, K., Cumbal, L., & Debut, A. (2021). Fabrication of silver nanoplates using Nephelium lappaceum (Rambutan) peel: A sustainable approach. *Journal of Molecular Liquids, 323*, 114987.
91. Javed, B., Ikram, M., Farooq, F., Sultana, T., Raja, N. I., & Mashwani, Z. U. R. (2020). Biogenesis of silver nanoparticles to treat cancer, diabetes, and microbial infections: A mechanistic overview. *Applied Microbiology and Biotechnology, 105*(6), 2261–2275.
92. Costa, D., Savio, L., & Pradier, C. M. (2016). Adsorption of amino acids and peptides on metal and oxide surfaces in water environment: A synthetic and prospective review. *The Journal of Physical Chemistry B, 120*(29), 7039–7052.

93. Faisal, S., Jan, H., Shah, S. A., Shah, S., Khan, A., Akbar, M. T., Rizwan, M., Jan, F., Wajidullah, A. N., Khattak, A., & Syed, S. (2022). Paraclostridium benzoelyticum bacterium-mediated zinc oxide nanoparticles and their in vivo multiple biological applications. *Oxidative Medicine and Cellular Longevity, 2022*, 1–15.
94. Mohamad Sukri, S. N., Shameli, K., Teow, S. Y., Chew, J., Ooi, L. T., Lee-Kiun Soon, M., Ismail, N. A., & Moeini, H. (2023). Enhanced antibacterial and anticancer activities of plant extract mediated green synthesized zinc oxide-silver nanoparticles. *Frontiers in Microbiology, 14*, 1194292.
95. Mongy, Y., & Shalaby, T. (2024). Green synthesis of zinc oxide nanoparticles using Rhus coriaria extract and their anticancer activity against triple-negative breast cancer cells. *Scientific Reports, 14*, 13470.
96. Perumal, P., Sathakkathulla, N. A., Kumaran, K., Ravikumar, R., Selvaraj, J. J., Nagendran, V., Gurusamy, M., Shaik, N., Prabhakaran, S. G., Palanichamy, V. S., Ganesan, V., Thiraviam, P. P., Gunalan, S., & Rathinasamy, S. (2024). Green synthesis of zinc oxide nanoparticles using aqueous extract of shilajit and their anticancer activity against HeLa cells. *Scientific Reports, 14*, 2204.
97. Sana, S. S., Kumbhakar, D. V., Pasha, A., Pawar, S. C., Grace, A. N., Singh, R. P., Nguyen, V. H., Le, Q. V., & Peng, W. (2020). Crotalaria verrucosa leaf extract mediated synthesis of zinc oxide nanoparticles: Assessment of antimicrobial and anticancer activity. *Molecules, 25*(21), 4896.
98. Son, N. N., Thanh, V. M., & Huong, N. T. (2023). Anticancer activities of zinc oxide nanoparticles synthesized using guava leaf extract. *ChemistrySelect, 8*(45), e202303214.
99. Maher, S., Nisar, S., Aslam, S. M., Saleem, F., Behlil, F., Imran, M., Assiri, M. A., Nouroz, A., Naheed, N., Khan, Z. A., & Aslam, P. (2023). Synthesis and characterization of ZnO nanoparticles derived from biomass (Sisymbrium irio) and assessment of potential anticancer activity. *ACS Omega, 8*(18), 15920–15931.
100. El-Beltagi, H. S., Ragab, M., Osman, A., El-Masry, R. A., Alwutayd, K. M., Althagafi, H., Alqahtani, L. S., Alazragi, R. S., Alhajri, A. S., & El-Saber, M. M. (2024). Biosynthesis of zinc oxide nanoparticles via neem extract and their anticancer and antibacterial activities. *PeerJ, 12*, e17588.
101. Aggarwal, B. B., & Harikumar, K. B. (2009). Potential therapeutic effects of curcumin, the anti-inflammatory agent, against neurodegenerative, cardiovascular, pulmonary, metabolic, autoimmune and neoplastic diseases. *The International Journal of Biochemistry & Cell Biology, 41*(1), 40–59.
102. Jang, M., Cai, L., Udeani, G. O., Slowing, K. V., Thomas, C. F., Beecher, C. W., Fong, H. H., Farnsworth, N. R., Kinghorn, A. D., Mehta, R. G., Moon, R. C., & Pezzuto, J. M. (1997). Cancer chemopreventive activity of resveratrol, a natural product derived from grapes. *Science, 275*(5297), 218–220.
103. Noble, R. L., Beer, C. T., & Cutts, J. H. (1958). Role of chance observations in chemotherapy: Vinca rosea. *Annals of the New York Academy of Sciences, 76*(3), 882–894.
104. Wani, M. C., Taylor, H. L., Wall, M. E., Coggon, P., & McPhail, A. T. (1971). Plant antitumor agents. VI. Isolation and structure of taxol, a novel antileukemic and antitumor agent from Taxus brevifolia. *Journal of the American Chemical Society, 93*(9), 2325–2327.
105. Russo, M., Spagnuolo, C., Tedesco, I., Bilotto, S., & Russo, G. L. (2012). The flavonoid quercetin in disease prevention and therapy: Facts and fancies. *Biochemical Pharmacology, 83*(1), 6–15.
106. Khan, N., & Mukhtar, H. (2007). Tea polyphenols for health promotion. *Life Sciences, 81*(7), 519–533.
107. Al-Fakeh, M. S., Osman, S. O. M., Gassoumi, M., Rabhi, M., & Omer, M. (2021). Characterization, antimicrobial and anticancer properties of palladium nanoparticles biosynthesized optimally using Saudi propolis. *Nanomaterials, 11*(10), 2666.
108. Alobaidi, Y. M., Ali, M. M., & Mohammed, A. M. (2022). Synthesis of calcium oxide nanoparticles from waste eggshell by thermal decomposition and their applications. *Jordan Journal of Biological Sciences, 15*(2).

109. Gondake, S. P., Kshirsagar, S. R., & Shinde, S. I. (2022). Synthesis, characterization and antioxidant/anticancer activity of palladium nanoparticles using Cassia absus seed extract. *Bulletin of Environment, Pharmacology and Life Sciences, 1*, 772–777.
110. Vignesh, A., Amal, T. C., Kalaiyarasan, J., Selvakumar, S., & Vasanth, K. (2023). An effective bio-inspired synthesis of palladium nanoparticles using Crateva religiosa G. Forst. Leaf extract: A multifunctional approach for environmental and biomedical applications. *Biomass Conversion and Biorefinery*, 1–24.
111. Rokade, S. S., Joshi, K. A., Mahajan, K., Tomar, G., Dubal, D. S., Singh, V., Kitture, R., Bellare, J., & Ghosh, S. (2017). Novel anticancer platinum and palladium nanoparticles from Barleria prionitis. *Global journal of nanomedicine, 2*(5), 555600.
112. Rokade, S. S., Joshi, K. A., Mahajan, K., Patil, S., Tomar, G., Dubal, D. S., Parihar, V. S., Kitture, R., Bellare, J. R., & Ghosh, S. (2018). Gloriosa superba mediated synthesis of platinum and palladium nanoparticles for induction of apoptosis in breast cancer. *Bioinorganic Chemistry and Applications, 2018*(1), 4924186.
113. Ghosh, S., Nitnavare, R., Dewle, A., Tomar, G. B., Chippalkatti, R., More, P., Kitture, R., Kale, S., Bellare, J., & Chopade, B. A. (2015). Novel platinum–palladium bimetallic nanoparticles synthesized by Dioscorea bulbifera: Anticancer and antioxidant activities. *International Journal of Nanomedicine*, 7477–7490.
114. Vaghela, H., Shah, R., & Pathan, A. (2018). Palladium nanoparticles mediated through bauhinia variegata: Potent in vitro anticancer activity against mcf-7 cell lines and antimicrobial assay. *Current Nanomaterials, 3*(3), 168–177.
115. Gangwar, C., Yaseen, B., Nayak, R., Baker, A., Bano, N., Singh, N. K., & Naik, R. M. (2023). Madhuca longifolia leaves-mediated palladium nanoparticles synthesis via a sustainable approach to evaluate its biomedical application. *Chemical Papers, 77*(6), 3075–3091.
116. Tamizhdurai, P., Albakri, G. S., Khalid, M., Alreshidi, M. A., & Yadav, K. K. (2024). Bio fabricated palladium nano particles using phytochemicals from aqueous cranberry fruit extract for anti-bacterial, cytotoxic activities and photocatalytic degradation of anionic dyes. *RSC Advances, 14*(33), 23730–23743.
117. Das, S., Dowding, J. M., Klump, K. E., McGinnis, J. F., Self, W., & Seal, S. (2017). Cerium oxide nanoparticles: Applications and prospects in nanomedicine. *Nanomedicine, 8*(9), 1483–1508.

Recent Advances in Oxidative Stress Associated Rheumatoid Arthritis

B. Mythili Gnanamangai and K. Prabha

Abstract Rheumatoid arthritis (RA) is caused by oxidative stress, a biochemical process that negatively impacts cell growth, DNA, and cellular functions. The evidence in the cellular, molecular, and clinical domains strongly supports this theory. Several oxidative damage markers, including malondialdehyde (MDA), protein carbonyls, and DNA oxidation products (8-OHdG), are strongly associated with disease activity, treatment resistance, and progression of radiological imaging. It appears that OS is central to RA's pathophysiology beyond its role as a secondary outcome of inflammation. It is important to note that OS also contributes to extra-articular manifestations of RA, such as atherosclerosis, one of the biggest causes of excess morbidity. According to Romero-Bueno, redox imbalances in RA are independently associated with premature vascular aging, proving that redox management could be integrated into cardiovascular risk reduction plans. Furthermore, researchers have found that oxidative stress may predict flare risk and drug response, further supporting its role in stratified care. There has been evidence that MDA and antioxidant enzyme levels fluctuate prior to changes in clinical scores, allowing for early therapeutic adjustments as discussed in this chapter.

Keywords Cytokine-driven signaling cascades · Cellular macromolecules · Redox-based biomarkers · Synovial fibroblasts and endothelial cells · Anti-citrullinated protein antibodies

B. M. Gnanamangai (✉)
Department of Biotechnology, K.S. Rangasamy College of Technology, Tiruchengode, Tamil Nadu, India
e-mail: mythilignanamangai@ksrct.ac.in

K. Prabha
Department of Food Technology, K.S. Rangasamy College of Technology, Tiruchengode, Tamil Nadu, India

K. Kasinathan and A. Muthukumaran (eds.), *Recent Advances in Oxidative Stress Associated Chronic Diseases Volume 2*, https://doi.org/10.1007/978-981-95-3758-7_12

1 Introduction (Overview of Rheumatoid Arthritis)

Rheumatoid arthritis (RA) is a chronic, systemic autoimmune disorder characterized by persistent synovial inflammation, which progressively leads to cartilage degradation, bone erosion, and joint deformity. It predominantly affects small joints symmetrically but may also involve large joints as the disease advances. The global prevalence of RA ranges from 0.5 to 1%, with significant variation across populations and regions. Women are disproportionately affected, particularly in postmenopausal age groups, with a female-to-male ratio of nearly 3:1. The onset of RA typically occurs between the ages of 40 and 60 years. However, elderly-onset RA is increasingly recognized and may follow a distinct clinical course [1, 2]. The pathogenesis of RA involves a complex interplay of genetic susceptibility and environmental triggers. Genetic associations are well-established, particularly with HLA-DRB1 alleles encoding the shared epitope. However, external factors such as cigarette smoking, air pollution, heavy metal exposure (e.g., cadmium), microbiota imbalance, and diet have been shown to interact with genetic predispositions to initiate the autoimmune process. These exposures can promote post-translational modifications such as citrullination of self-proteins, leading to the generation of anti-citrullinated protein antibodies (ACPAs), which precede the clinical onset of RA and contribute to its pathophysiology [1–4]. Clinically, RA manifests with swelling, tenderness, stiffness (especially in the morning), and functional limitation in affected joints. Extra-articular involvement is common and may include cardiovascular disease, interstitial lung disease, vasculitis, and neurocognitive impairment. Cardiovascular comorbidity is a leading cause of mortality in RA patients and is driven by chronic systemic inflammation and oxidative stress, which accelerate atherosclerosis and myocardial remodeling [5–10]. The diagnosis of RA is based on the 2010 ACR/EULAR classification criteria, which consider joint involvement, serology (rheumatoid factor and ACPAs), acute-phase reactants (CRP, ESR), and symptom duration. A score of six or more points is required for classification. These criteria are sensitive for early diagnosis, which is critical because irreversible joint damage may begin within months of symptom onset. Seropositive RA, marked by the presence of RF and/or ACPAs, tends to exhibit a more aggressive disease course with rapid radiographic progression and more severe systemic features [1, 6, 7]. On a molecular level, RA is driven by dysregulated immune responses involving T-helper cells, B cells, macrophages, and fibroblast-like synoviocytes. These cells orchestrate an inflammatory environment by producing cytokines like TNF-α, IL-1β, and IL-6, chemokines, and matrix metalloproteinases. These inflammatory mediators stimulate osteoclast differentiation, promoting bone resorption and cartilage damage. ACPAs may directly activate macrophages and osteoclasts, further amplifying the inflammatory loop and contributing to pain via interaction with sensory neurons [1, 2, 7]. Oxidative stress is increasingly recognized as a central contributor to RA pathology. Reactive oxygen and nitrogen species (ROS/RNS) generated in the inflamed synovium damage proteins, DNA, and lipids, exacerbating joint destruction. Moreover, RA patients often impair endogenous antioxidant defense systems such as superoxide dismutase, catalase, and glutathione peroxidase. This imbalance reinforces inflammatory cascades and contributes to systemic complications, including cardiovascular dysfunction and endothelial damage [4, 11–13]. Recent research also points to the early involvement of oxidative stress even before clinical manifestations appear, suggesting that biomarkers like malondialdehyde (MDA), 8-isoprostane, and reduced antioxidant enzyme levels could serve in early diagnosis and prognosis. Evidence indicates that oxidative mechanisms underlie joint damage and systemic effects, such as myocardial remodeling and cognitive decline seen in RA [9, 14].

1.1 Pathophysiology of RA

RA is mediated by a dysregulated immune response wherein innate and adaptive immune cells coordinate a chronic inflammatory attack on the synovial membrane. Antigen-presenting cells activate autoreactive CD4 + T cells, which then stimulate B cells to produce autoantibodies, most notably RF and ACPAs as shown in Fig. 1. These autoantibodies form immune complexes that deposit in the synovium, triggering complement activation and cytokine release that attract more immune cells and sustain inflammation [1, 6, 13]. Fibroblast-like synoviocytes (FLS), activated by this proinflammatory environment, undergo a phenotypic transformation, becoming aggressive, invasive cells that perpetuate joint destruction. These cells produce matrix metalloproteinases (MMPs) and cytokines like TNF-α and IL-6 that degrade cartilage and support osteoclastogenesis. Osteoclasts, activated by RANKL (expressed by FLS and T cells), resorb bone, contributing to erosions observed in imaging and pathology [5, 6]. A defining molecular feature of RA pathophysiology is the role of oxidative stress. Elevated ROS levels in RA joints arise from mitochondrial dysfunction, NOX2 activity, and oxidative bursts from infiltrating neutrophils and macrophages. These reactive species damage synovial components, including lipids, proteins, and nucleic acids, fostering chronic inflammation and structural damage. ROS-mediated oxidative post-translational modifications can generate neoantigens, which the immune system recognizes and drive further autoantibody production, particularly ACPAs [15–17]. The inflamed synovium is hypoxic due to rapid cellular proliferation and increased metabolic demands, further enhancing ROS generation through hypoxia/reperfusion injury and mitochondrial electron transport dysfunction. The resulting vicious cycle of hypoxia-induced ROS generation and inflammation promotes angiogenesis, immune activation, and persistent pannus formation [16, 18]. In T cells, oxidative stress induces somatic mutations, particularly in the p53 gene, and impairs DNA repair mechanisms. These defects contribute to apoptosis resistance and senescence, accumulating dysfunctional T cells with a pro-inflammatory phenotype. Chronic oxidative DNA damage in CD4 + T cells, for example, is linked to premature immunosenescence and skewing of the T cell receptor repertoire toward autoreactivity [16, 19]. Moreover, ROS serves not only as damaging agents but also as signaling molecules. They modulate key immune pathways by influencing cytokine production, antigen presentation, and T cell tolerance. NADPH oxidase-derived ROS can suppress or enhance immune responses depending on the context. Dysregulation of this balance, as observed in RA, shifts ROS functions toward promoting inflammation, angiogenesis, and matrix degradation [15, 20]. Concurrently, endogenous antioxidant defense mechanisms are impaired in RA. Levels and activities of critical antioxidants such as superoxide dismutase (SOD), catalase (CAT), and glutathione peroxidase (GPx) are significantly reduced in RA patients. This imbalance between oxidants and antioxidants contributes to sustained oxidative stress, amplifying joint damage and systemic manifestations, including cardiovascular dysfunction and fatigue [2, 11, 14, 21]. Lipid peroxidation products like malondialdehyde (MDA), protein carbonyls, and 8-isoprostane, along with markers of DNA oxidation, are elevated in RA and correlate strongly with disease activity indices. This correlation underscores the potential of oxidative stress biomarkers as tools for disease monitoring and stratification. Their role in early diagnosis and prognostic evaluation is increasingly recognized [11, 21, 22]. The cumulative evidence positions oxidative stress not as a mere byproduct but as a central pathogenic mechanism in RA. It intersects with immune dysregulation, genetic vulnerability, cellular senescence, and metabolic dysfunction to sustain chronic inflammation and drive joint destruction. Understanding these pathways is essential for developing novel biomarkers and redox-modulating therapies. RA is mediated by a dysregulated immune response wherein innate and adaptive immune cells coordinate a chronic inflammatory attack on the synovial membrane. Antigen-presenting cells activate autoreactive CD4 + T cells, which stimulate B cells to produce autoantibodies, most notably RF and ACPAs. These autoantibodies form immune complexes that deposit in the synovium, triggering complement activation and cytokine release that attract more immune cells and sustain inflammation [1, 6, 13]. Fibroblast-like synoviocytes (FLS), activated by this proinflammatory environment, undergo a phenotypic transformation, becoming aggressive, invasive cells that perpetuate joint destruction. These cells produce matrix metalloproteinases and cytokines like TNF-α and IL-6 that degrade cartilage and support osteoclastogenesis. Osteoclasts, activated by RANKL (expressed by FLS and T cells), resorb bone, contributing to erosions observed in imaging and pathology [1, 6]. A defining molecular feature of RA pathophysiology is the role of oxidative stress. Elevated ROS levels in RA joints arise from mitochondrial dysfunction, NOX2 activity, and oxidative bursts from infiltrating neutrophils and macrophages. These reactive species damage synovial components, including lipids, proteins, and nucleic acids, fostering chronic inflammation and structural damage. ROS-mediated oxidative post-translational modifications can generate neoantigens, which the immune system recognizes and drive further autoantibody production, particularly ACPAs [15–17]. The inflamed synovium is hypoxic due to rapid cellular proliferation and increased metabolic demands, further enhancing ROS generation through hypoxia/reperfusion injury and mitochondrial electron transport dysfunction. The resulting vicious cycle of hypoxia-induced ROS generation and inflammation promotes angiogenesis, immune activation, and persistent pannus formation [16, 18]. In T cells,

Fig. 1 Rheumatoid arthritis pathogenesis

oxidative stress induces somatic mutations, particularly in the p53 gene, and impairs DNA repair mechanisms. These defects contribute to apoptosis resistance and senescence, accumulating dysfunctional T cells with a pro-inflammatory phenotype [18, 19]. Moreover, ROS serves not only as damaging agents but also as signaling molecules. They modulate key immune pathways such as NF-κB and MAPKs. Dysregulation of this balance in RA leads to inflammation, angiogenesis, and matrix degradation [15, 20]. Concurrently, antioxidant defense mechanisms are impaired in RA. Critical antioxidant enzymes such as SOD, CAT, and GPx are significantly reduced, compounding oxidative damage. Numerous studies report elevated lipid peroxidation markers like MDA and protein carbonyls in RA patients, correlating with disease activity scores such as DAS28 and CRP [11, 13, 21, 22].

1.2 Concept and Mechanisms of Oxidative Stress

Oxidative stress arises from an imbalance between pro-oxidants and antioxidants, favoring excessive production or insufficient clearance of reactive species. ROS such as superoxide ($O_2^{\cdot -}$), hydroxyl (·OH), hydrogen peroxide (H_2O_2), and peroxynitrite ($ONOO^-$) are continuously generated by mitochondrial oxidative phosphorylation and various enzyme systems. Under homeostatic conditions, these molecules play physiological roles in signaling, cell growth, and immune response. However, under pathological states like RA, their accumulation leads to oxidative damage to lipids, proteins, and DNA [23–25]. Mitochondria are primary sources of ROS. In RA, mitochondrial dysfunction results in the leakage of electrons from the respiratory chain, forming superoxide radicals. Normally, these are detoxified by antioxidant enzymes, including SOD and GPx, but in RA, this protective mechanism is overwhelmed [1, 11, 26]. Excess ROS activates NF-κB and MAPK signaling pathways, perpetuating inflammation and joint degradation [24, 27]. Lipid peroxidation, primarily measured via MDA and 4-HNE, disrupts membrane integrity and generates electrophilic aldehydes that propagate further damage. In RA synovial fibroblasts, 4-HNE impairs mitochondrial respiration and induces proangiogenic responses, promoting pannus formation [28]. Moreover, oxidative stress shifts energy metabolism from oxidative phosphorylation to glycolysis, mimicking the Warburg effect and fueling inflammatory cell proliferation [28]. Multiple studies have confirmed that RA patients exhibit significantly higher ROS levels and reduced antioxidant enzyme activities than healthy controls [2, 20, 25, 26, 28]. This redox imbalance is reflected in increased oxidative stress biomarkers and is now considered a hallmark of RA progression and a potential therapeutic target.

1.3 Link Between Oxidative Stress and Autoimmune Inflammation

The autoimmune inflammation characteristic of RA is intimately linked with oxidative stress through mutual amplification of redox imbalance and immune cell dysregulation. Activated immune cells, particularly neutrophils, macrophages, and T cells, are key sources of ROS and RNS in the synovial and peripheral compartments. These reactive species inflict oxidative damage and activate inflammatory pathways, notably NF-κB and MAPK signaling, reinforcing cytokine synthesis and tissue degradation [1, 15, 20, 29]. T-helper cell polarization plays a pivotal role in this process. Th1 cells produce pro-inflammatory cytokines (e.g., IFN-γ, TNF-α, IL-2) that potentiate oxidative stress by stimulating NADPH oxidase and inducible nitric oxide synthase (iNOS) while impairing antioxidant defenses. Conversely, Th2 cytokines (e.g., IL-4, IL-10) offer some protection by suppressing these pro-oxidant pathways, though their activity is often diminished in active RA [29]. NF-κB activation is central to the redox-inflammation loop. ROS act as upstream signals that promote NF-κB nuclear translocation, thereby driving the expression of pro-inflammatory cytokines, adhesion molecules, and MMPs. Once induced, these cytokines, including TNF-α and IL-1β, further enhance ROS production via mitochondrial disruption and inflammatory enzyme activation, particularly NOX and COX-2 [24, 29, 30]. Autoantibodies such as RF and ACPA, which are hallmarks of RA, are generated in a redox-sensitive environment. ROS-mediated post-translational modifications (e.g., citrullination, carbamylation) yield neoantigens that break immune tolerance and promote B cell activation. B cells and plasma cells further contribute to inflammation by releasing additional cytokines and complement-activating immune complexes [1, 2, 30]. Inflammatory monocytes and macrophages infiltrating the synovium produce large quantities of superoxide, peroxynitrite, and nitric oxide, particularly through iNOS and xanthine oxidase. These RNS oxidize DNA, protein thiols, and lipids, contributing to tissue injury and further immune activation. Neutrophils in RA synovial fluid exhibit a respiratory burst response, releasing ROS that damage cartilage and stimulate synoviocytes and chondrocytes to release MMPs and cytokines [29, 31]. Mitochondrial ROS also act as signaling molecules in T cell differentiation and survival, favoring a proinflammatory phenotype. Oxidative stress impairs apoptotic signaling, contributing to the survival of autoreactive immune cells. This apoptotic resistance is reinforced by redox-dependent stabilization of anti-apoptotic proteins like Bcl-2 and degradation of p53, facilitating chronic immune activation [15, 19]. The therapeutic benefit of antioxidants in modulating disease activity further illustrates the tight interdependence between oxidative stress and autoimmune inflammation. Antioxidant compounds reduce ROS burden, downregulate NF-κB activity, and attenuate cytokine production. Clinical and experimental evidence shows that dietary antioxidants like curcumin, flavonoids, and green tea polyphenols modulate redox-sensitive inflammatory pathways in RA [20, 29]. Thus, oxidative stress is not a secondary byproduct of inflammation but a critical pathogenic partner that sustains autoimmunity, damages synovial tissue, and amplifies systemic immune responses in RA.

1.4 Rationale for Focusing on Oxidative Stress in RA

Targeting oxidative stress in rheumatoid arthritis (RA) is not merely an adjunctive approach but a potentially pivotal strategy in both understanding and managing the disease. As an established driver of inflammation, immune activation, and joint destruction, oxidative stress represents a pathophysiological axis that interacts closely with classical immune pathways to perpetuate chronic synovitis and systemic complications [15, 17, 20, 30]. Numerous studies have demonstrated that RA patients exhibit elevated levels of oxidative stress markers such as malondialdehyde (MDA), thiobarbituric acid reactive substances (TBARS), and protein carbonyls, along with diminished antioxidant enzyme activity, including superoxide dismutase (SOD), catalase (CAT), and glutathione peroxidase (GPx). These changes are closely correlated with clinical parameters of disease activity, including DAS28 scores and levels of inflammatory markers like CRP and ESR [1, 2, 11, 13, 20, 32, 33]. The oxidative environment in RA promotes post-translational modifications of proteins (e.g., citrullination, carbamylation), creating neoantigens that breach immune tolerance and initiate or amplify autoreactive responses. Oxidative stress also disrupts DNA repair pathways and contributes to gene mutations like p53, facilitating apoptosis resistance and synovial hyperplasia [17, 19, 34]. Experimental models support the therapeutic relevance of redox modulation. Nrf2-knockout mice with induced arthritis exhibit more severe joint destruction and oxidative damage than wild-type controls, underlining the protective role of endogenous antioxidant pathways in arthritis [34]. Additionally, antioxidant supplementation has been shown to reduce ROS levels and improve inflammatory and clinical indices in both animal models and human studies, with promising effects observed for natural antioxidants, Nrf2 activators, and specific vitamins [29, 35]. From a clinical standpoint, oxidative stress parameters may serve as biomarkers of disease activity and as prognostic tools and indicators of therapeutic response. This utility is underscored by the demonstrated association between oxidative stress levels and disease severity, fatigue, cardiovascular risk, and even depression in RA patients [13, 14, 18]. Thus, focusing on oxidative stress in RA offers multifaceted benefits: it deepens our mechanistic understanding of disease progression, provides novel targets for therapeutic intervention, and enables risk stratification through redox-based biomarkers. Oxidative pathways contribute to both joint and systemic manifestations, forming a critical and modifiable axis in the RA pathophysiological network.

2 Role of Oxidative Stress in RA Pathogenesis

2.1 Sources of Reactive Oxygen Species (ROS) in RA

Reactive oxygen species (ROS) in rheumatoid arthritis (RA) arise from multiple intracellular and extracellular sources within the inflamed joint microenvironment, contributing substantially to disease initiation, progression, and tissue damage as shown in Fig. 2. Key cellular contributors include activated immune cells—particularly neutrophils, macrophages, and lymphocytes—as well as synovial fibroblasts and endothelial cells. These cells produce ROS through several enzymatic systems, including NADPH oxidase (NOX), mitochondrial respiratory chain enzymes, xanthine oxidase, and myeloperoxidase [1, 15, 17]. Neutrophils are prominent in the RA synovium and represent a significant source of ROS. Their oxidative burst response, triggered by immune complex stimulation via Fcγ receptors, leads to superoxide ($O_2^{\cdot -}$) and downstream hydrogen peroxide (H_2O_2) generation. This response is amplified by NOX2, a membrane-bound complex that transfers electrons from NADPH to oxygen, forming superoxide [1, 29, 36].

Mitochondrial dysfunction in synovial fibroblasts and T cells is another critical source of ROS. Under hypoxic and inflammatory conditions characteristic of RA synovia, the mitochondrial electron transport chain becomes less efficient, leading to electron leakage and ROS formation. Dysfunctional mitochondria also impair cellular antioxidant defenses, exacerbating oxidative stress and contributing to the survival of autoreactive T cells [16, 17, 37]. Monocytes and macrophages contribute to the ROS pool through NOX and inducible nitric oxide synthase (iNOS) activity. These enzymes produce superoxide and nitric oxide (NO), which can combine to form peroxynitrite ($ONOO^-$), a highly reactive and damaging nitrogen species [19, 29, 38]. Additionally, myeloperoxidase released from neutrophils and monocytes catalyzes the formation of hypochlorous acid (HOCl), another potent oxidant found at elevated levels in RA joints [37]. Endothelial cells and fibroblast-like synoviocytes (FLS) within the inflamed joint also generate ROS in response to cytokines

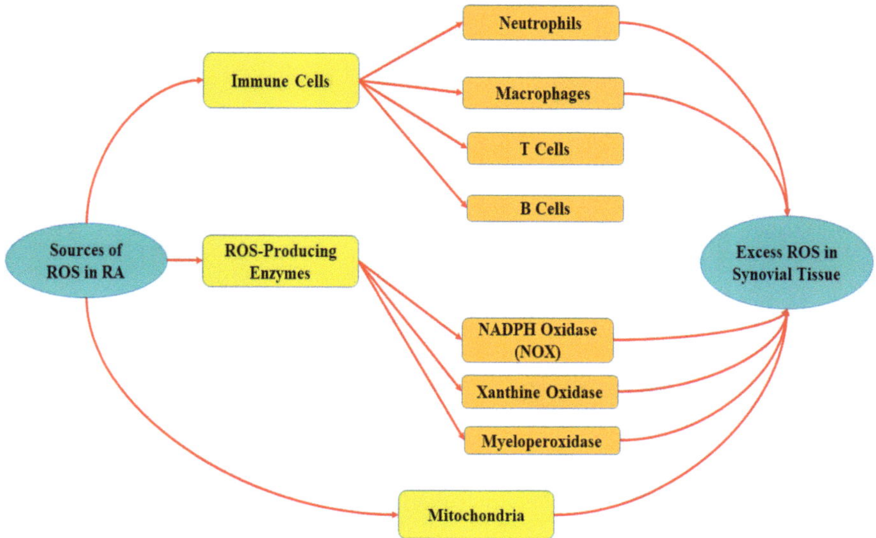

Fig. 2 ROS and its sources

such as IL-1β and TNF-α. ROS derived from these cells amplify the inflammatory loop by activating redox-sensitive transcription factors like NF-κB and AP-1, which regulate genes involved in inflammation, angiogenesis, and matrix degradation [15, 20, 36]. Oxidative stress is further perpetuated by impaired clearance of ROS due to reduced levels of antioxidant enzymes such as SOD, CAT, and GPx, as discussed in earlier sections. The imbalance favors a sustained oxidative environment, fostering DNA damage, protein carbonylation, lipid peroxidation, and generation of neoantigens—all reinforcing autoimmune inflammation in RA [11, 13, 25, 26]. Together, these findings underscore that ROS in RA are not a single-source product but originate from a dynamic interplay of immune cell activation, metabolic dysfunction, and cytokine-driven signaling cascades. This multifactorial origin supports the rationale for targeting ROS production and antioxidant defense mechanisms in RA therapeutics.

2.2 ROS-Induced Molecular Damage (DNA, Lipids, Proteins)

The overproduction of reactive oxygen species (ROS) in rheumatoid arthritis (RA) leads to extensive molecular damage that significantly contributes to disease progression as shown in Fig. 3. Excess ROS can induce a variety of deleterious modifications to cellular macromolecules, notably affecting DNA, lipids, and proteins.

DNA Damage: Excessive ROS generation in RA results in oxidative modifications to DNA, including base oxidation, single- and double-strand breaks, and

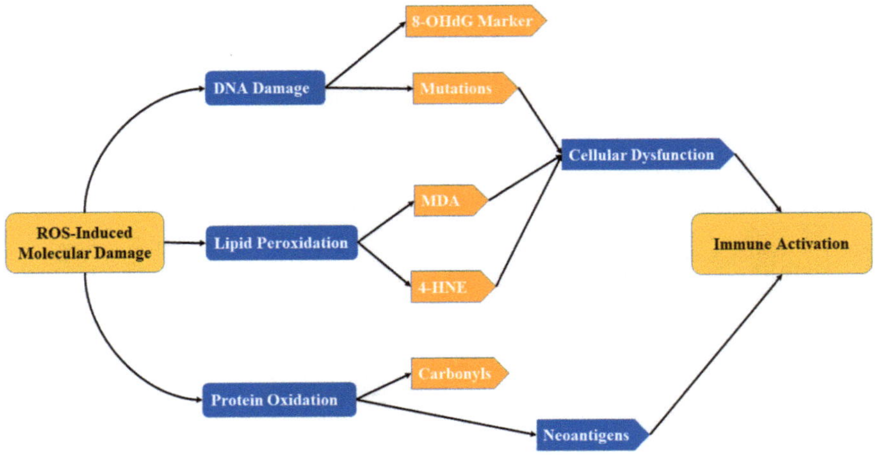

Fig. 3 ROS-induced molecular damage

the formation of DNA adducts. Such damage undermines genomic integrity, triggering mutations that may further exacerbate autoimmunity. Studies have shown that increased DNA damage in peripheral blood neutrophils and synovial cells correlates closely with higher disease activity and joint inflammation [39]. Lipid Peroxidation: ROS readily attack polyunsaturated fatty acids within cellular membranes, initiating lipid peroxidation. This process produces toxic aldehydes such as malondialdehyde (MDA) and 4-hydroxy-2-nonenal (4-HNE). These reactive by-products form adducts with lipids and proteins, thereby propagating inflammatory responses and contributing to tissue degradation. Elevated levels of lipid peroxidation products in the synovial tissue and plasma of RA patients underscore the role of oxidative stress in the disease's pathogenesis [40]. Protein Oxidation: Proteins are susceptible to oxidative modifications, including carbonylation, nitration, and disulfide bond alterations. These modifications can compromise protein function and structure, leading to the formation of neo-epitopes. In RA, oxidatively modified proteins may break immune tolerance and stimulate the production of autoantibodies such as rheumatoid factor (RF) and anti-citrullinated protein antibodies (ACPA), thereby sustaining chronic inflammation [41]. Alteration of Signaling Pathways: In addition to causing direct macromolecular damage, ROS also modulates intracellular signaling pathways. Oxidative modifications of kinases, phosphatases, and transcription factors (e.g., NF-κB, MAPK) alter cell signaling, further amplifying pro-inflammatory responses. Mitochondrial-derived ROS can disrupt apoptotic pathways, promoting synovial hyperplasia and chronic inflammation [42]. The molecular damage induced by ROS—through DNA strand breaks, lipid peroxidation, and protein oxidation—drives tissue injury and contributes to the dysregulation of immune responses in RA. These processes underscore the pivotal role of oxidative stress in RA pathogenesis and highlight the potential of targeting ROS-induced damage as a therapeutic strategy [43, 44].

2.3 Synovial Inflammation and Joint Destruction via Oxidative Pathways

Synovial inflammation in rheumatoid arthritis (RA) is tightly linked to the overproduction of reactive oxygen species (ROS), which act as both cytotoxic and signaling molecules within the inflamed joint environment as shown in Fig. 4. Oxidative stress contributes not only to tissue damage but also perpetuates the inflammatory cycle in RA, thereby enhancing joint destruction. The inflamed synovial tissue in RA shows signs of hypoxia-reoxygenation, a process that activates hypoxia-inducible factor-1α (HIF-1α) and nuclear factor-κB (NF-κB). These transcription factors upregulate inflammation, angiogenesis, and cell survival genes, contributing to synovial hyperplasia and pannus formation [45]. ROS generated during mitochondrial dysfunction and NADPH oxidase activation promotes synoviocyte proliferation and inflammatory cytokine release, including TNF-α, IL-1β, and IL-6, further exacerbating synovial inflammation [46]. ROS also contributes directly to the activation of matrix metalloproteinases (MMPs), particularly MMP-1 and MMP-3, leading to the degradation of collagen and proteoglycans in cartilage. Oxidative damage to synovial fibroblasts alters their function, supporting a more aggressive, invasive phenotype that promotes bone and cartilage erosion [47]. Furthermore, studies have reported increased levels of reactive aldehydes like methylglyoxal (MG) in RA synovial fluid, which form advanced glycation end products (AGEs). These AGEs interact with RAGE receptors on synovial cells and macrophages, initiating ROS production and inflammatory signaling [48].

Inflammatory cells such as neutrophils and macrophages release ROS and reactive nitrogen species (RNS) in response to cytokines. This leads to activation of cyclooxygenase-2 (COX-2), phospholipase A2, and inducible nitric oxide synthase (iNOS), all of which contribute to the maintenance of inflammation and oxidative injury in the synovium [49]. Moreover, oxidative modification of proteins in the synovial microenvironment promotes the generation of autoantigens, stimulating autoreactive T and B cells. This autoimmune cascade sustains synovitis and drives joint damage through continuous immune complex deposition and complement activation [50]. Taken together, oxidative stress orchestrates a multifaceted attack on joint integrity, combining inflammatory activation, enzymatic tissue degradation, and autoimmunity. Targeting oxidative pathways, therefore, holds promise as a therapeutic strategy in controlling synovial inflammation and halting joint destruction in RA.

2.4 Crosstalk Between Immune Cells and Oxidative Pathways

In rheumatoid arthritis (RA), chronic inflammation and joint destruction are driven by intricate interactions between immune cells and oxidative stress mechanisms. Beyond their cytotoxic effects, reactive oxygen species (ROS) act as critical signaling

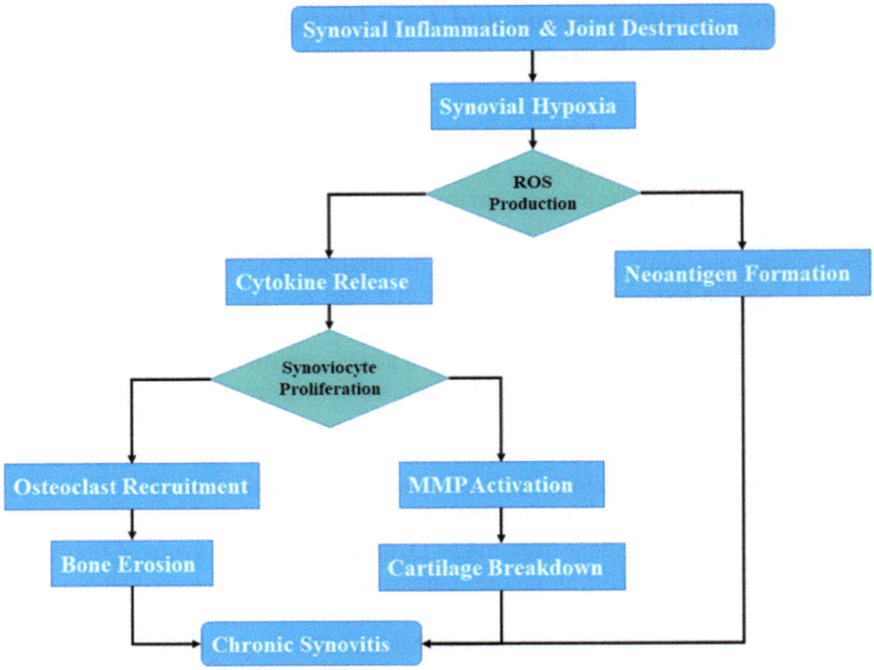

Fig. 4 Synovial inflammation and joint destruction

molecules that modulate immune cell behavior and communication in the synovial environment. Macrophages and T lymphocytes, the dominant immune cells in RA synovium, are key sources and targets of ROS. Oxidative stress shapes macrophage polarization, pushing them toward an M1-like phenotype characterized by high production of pro-inflammatory cytokines such as TNF-α and IL-1β. This redox-sensitive phenotype promotes the activation of NF-κB and MAPK signaling cascades, amplifying inflammatory responses [51]. Meanwhile, ROS influences T cell differentiation, promoting Th1 and Th17 responses, which are implicated in RA pathology. The inhibition of ROS-generating enzymes like NADPH oxidase has been shown to dampen Th17 expansion, indicating the regulatory role of ROS in T cell programming [52]. Synovial fibroblasts also contribute to this oxidative-immune crosstalk. These cells interact directly with T cells and macrophages, promoting ROS production through cell–cell contact mechanisms and cytokine stimulation. CTLA-4Ig, a co-stimulatory blocker, has been found to reduce ROS generation in T cells by preserving Rap1 activity, a redox-sensitive GTPase involved in cell activation, thereby interrupting the inflammatory cycle [53]. Neutrophils, abundant in RA synovial fluid, exhibit an altered oxidative phenotype that enhances antigen presentation. Recent evidence suggests that under high ROS conditions, neutrophils transdifferentiate into dendritic cell-like phenotypes (N-DCs), expressing HLA-DR and co-stimulatory molecules. These cells bridge innate and adaptive immunity, promoting autoimmunity in a redox-dependent manner [54]. ROS also regulates immune checkpoints and adhesion molecule expression, facilitating immune cell retention and activation in synovial tissues. Through redox-sensitive modulation of integrins and chemokine receptors, immune cells remain anchored in the inflamed joint, sustaining chronic inflammation [55]. The bidirectional communication between immune cells and oxidative stress mechanisms perpetuates immune activation, tissue damage, and joint degradation in RA. Disrupting this crosstalk represents a promising avenue for therapeutic intervention.

2.5 Role of Mitochondrial Dysfunction in ROS Production

Mitochondrial dysfunction is increasingly recognized as a key contributor to the oxidative and inflammatory milieu characteristic of rheumatoid arthritis (RA). Mitochondria are the primary source of intracellular reactive oxygen species (ROS), especially under stress conditions, and alterations in mitochondrial function can amplify ROS production, contributing to synovial inflammation, immune dysregulation, and joint destruction. RA patients exhibit profound mitochondrial alterations in peripheral blood mononuclear cells (PBMCs), including decreased mitochondrial membrane potential ($\Delta\Psi m$), reduced ATP levels, and elevated mitochondrial ROS generation. The increased expression of mitochondrial regulators such as ROMO1 and OMA1 further supports the link between mitochondrial dysfunction and oxidative stress in RA immune cells [56]. These impairments in energy metabolism correlate negatively with antioxidant capacity and are strongly associated with disease severity. Within synovial fibroblasts, mitochondrial dysfunction enhances resistance to apoptosis and sustains pro-inflammatory phenotypes. It has been shown that TL1A, a TNF-like cytokine, promotes mitochondrial dysfunction in fibroblast-like synoviocytes (FLS) via TNFR2 signaling. This leads to impaired respiration, loss of $\Delta\Psi m$, and excessive mitochondrial ROS production, ultimately increasing the secretion of inflammatory cytokines such as IL-6 and IL-8. Inhibition of mitochondrial ROS via Mito-TEMPO reverses these inflammatory features, highlighting a direct role of mitochondrial ROS in amplifying synovial inflammation [57]. The mitochondrial calcium uniporter (MCU) also plays a key role in regulating mitochondrial ROS in RA. Increased MCU expression in RA-FLS leads to enhanced ROS generation, disrupted mitochondrial dynamics, and aggressive cell migration. Pharmacological inhibition of MCU normalizes mitochondrial function, reduces ROS production, and impairs the invasive behavior of FLS both in vitro and in vivo [58]. Mitochondrial DNA (mtDNA) mutations, frequently detected in RA synovial tissues, are another indicator of oxidative damage. These mutations correlate with increased TNF-α levels, synovial vascularity, and immune cell infiltration. Notably, treatment with TNF-α inhibitors reduces mtDNA mutation rates and mitochondrial ROS levels, underscoring the reversible nature of mitochondrial damage under anti-inflammatory therapy [59]. Moreover, studies show impaired mitophagy (selective degradation of damaged mitochondria) allows dysfunctional mitochondria to persist, leading to chronic ROS release and inflammation. Experimental models utilizing ultrasound-activated piezoelectric nanosheets that induce mitophagy have successfully suppressed mitochondrial ROS and alleviated arthritis symptoms, pointing to novel therapeutic strategies targeting mitochondrial quality control [60]. Altogether, mitochondrial dysfunction in RA drives ROS overproduction and reinforces inflammatory circuits that maintain and exacerbate disease pathology. Restoring mitochondrial health or curbing mitochondria-derived ROS represents a promising frontier in RA treatment.

3 Oxidative Biomarkers in Rheumatoid Arthritis

3.1 Lipid Peroxidation Markers (e.g., Malondialdehyde, TBARS)

Lipid peroxidation is a hallmark of oxidative stress and a significant mechanism contributing to tissue damage in rheumatoid arthritis (RA) as shown in Fig. 4. This process primarily involves the oxidative degradation of polyunsaturated fatty acids in cell membranes, generating a cascade of reactive aldehyde by-products, most notably malondialdehyde (MDA). MDA has emerged as one of the most widely used and reliable biomarkers for assessing lipid peroxidation and systemic oxidative stress in RA patients. Elevated serum and synovial fluid levels of MDA have been consistently reported in RA patients and are positively correlated with disease activity scores, including DAS28 and inflammatory markers like C-reactive protein (CRP) and erythrocyte sedimentation rate (ESR). A recent study demonstrated that MDA levels above 16.21 nmol/mL could discriminate RA patients from healthy individuals with 100% sensitivity and specificity, highlighting its diagnostic potential [61]. Similar associations were found in earlier studies that reported significantly increased thiobarbituric acid reactive substances (TBARS), indicating enhanced lipid peroxidation in both plasma and synovial samples [62, 63]. The link between MDA and disease pathophysiology is not limited to oxidative insult alone. MDA and related lipid aldehydes can react with proteins and nucleic acids to form advanced lipoxidation end products (ALEs), contributing to neo-antigen formation, immune activation, and joint destruction. Moreover, lipid peroxidation products potentiate redox-sensitive inflammatory pathways, particularly those regulated by NF-κB and MAPKs, thus establishing a vicious cycle of oxidative stress and inflammation in RA [48]. Emerging studies further underscore the role of MDA as a prognostic marker. Longitudinal assessments have shown that anti-inflammatory treatments like tocilizumab and infliximab significantly reduce MDA and TBARS levels alongside clinical improvement, suggesting their utility as markers for therapeutic efficacy [64]. Given their diagnostic and prognostic value, lipid peroxidation markers such as MDA and TBARS are gaining recognition as biomarkers of oxidative damage and as dynamic indicators of disease activity and treatment response in RA. Their inclusion in routine monitoring could improve disease management and help guide antioxidant-based adjunct therapies.

3.2 *Protein Oxidation and Nitrosative Stress Markers (e.g., Nitric Oxide)*

Oxidative stress in rheumatoid arthritis (RA) extends beyond lipid peroxidation and encompasses protein oxidation and nitrosative stress, both pivotal in the disease's progression. These processes involve modifying proteins through reactive oxygen and nitrogen species, resulting in functional alterations, immunogenicity, and sustained inflammation. Protein oxidation is characterized by the accumulation of protein carbonyls and advanced oxidation protein products (AOPPs), which are significantly elevated in RA patients compared to healthy controls. These markers correlate inversely with thiol group levels, suggesting that compromised antioxidant defense mechanisms lead to protein damage. Elevated protein carbonyls reflect oxidative insult to serum and joint proteins and have been associated with higher disease activity score (DAS28) and CRP levels [65, 66]. Nitrosative stress arises from excessive nitric oxide (NO) and its reactive derivatives like peroxynitrite. NO is synthesized in large quantities in RA by inducing inducible nitric oxide synthase (iNOS), particularly in macrophages, endothelial cells, and synoviocytes within the inflamed synovium. Studies show that RA patients have significantly higher serum NO and nitrite/nitrate levels than healthy individuals [67, 68]. This increase occurs even in early-stage RA and may precede detectable structural damage, suggesting a role for NO in initiating inflammation [69]. Despite these elevated levels, the relationship between NO and traditional disease activity markers (CRP, ESR, RF) is inconsistent. Some studies show weak or no correlations, while others report associations with disease severity or functional scores [70, 71]. This variability may reflect the compartmentalized production of NO within tissues versus systemic circulation. Increased nitrotyrosine levels—a marker of peroxynitrite-mediated protein nitration—have been detected in RA synovium and serum, particularly within macrophages and endothelial cells. This supports the role of NO-derived species in direct protein damage and immune modulation [72]. Additionally, S-nitrosoproteins (SNP), a more stable form of NO interaction with thiol-containing proteins, are found in elevated concentrations in RA serum and synovial fluid. These modifications affect signaling, gene regulation, and immune responses, and their levels decline after corticosteroid therapy, supporting their clinical relevance [73]. Protein oxidation and nitrosative stress markers provide crucial insights into the biochemical disturbances in RA. Their assessment may complement lipid peroxidation markers in evaluating oxidative stress burden, disease activity, and therapeutic responses (Fig. 5).

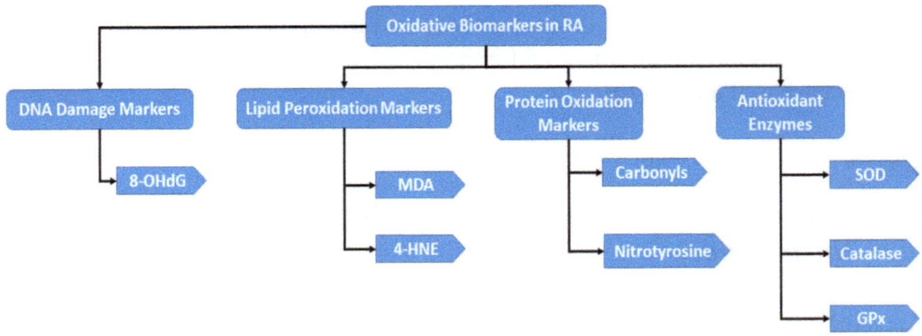

Fig. 5 Oxidative biomarkers in RA

3.3 DNA Oxidation and Damage Biomarkers

Oxidative DNA damage is an emerging hallmark of rheumatoid arthritis (RA), reflecting both the systemic oxidative stress burden and localized inflammatory responses within the synovium. Accumulating evidence highlights the critical role of DNA oxidation markers in disease pathogenesis, progression, and potential therapeutic targeting. One of the most studied oxidative DNA lesions is 8-hydroxy-2′-deoxyguanosine (8-OHdG), a mutagenic base modification caused by ROS. Elevated levels of 8-OHdG have been consistently reported in the serum, plasma, and synovial fluid of RA patients compared to healthy controls, indicating significant oxidative DNA damage in RA [74, 75]. These levels correlate with disease activity and markers of systemic inflammation, such as CRP and DAS28, supporting their clinical utility [76]. In addition to 8-OHdG, RA patients demonstrate increased DNA strand breaks and abasic sites as detected by comet assays and specific oxidative damage markers. These alterations are observed in immune cells like PBMCs and neutrophils and in fibroblast-like synoviocytes, suggesting widespread oxidative damage [77, 78]. Furthermore, age-dependent increases in lipid peroxidation-derived DNA adducts, such as heptanone-etheno-2′-deoxycytidine (HεdC), have been shown to correlate with joint swelling and RA severity, providing a novel class of biomarkers linked to both lipid peroxidation and DNA damage [79]. A growing body of research also implicates deficient DNA repair capacity in RA pathogenesis. RA patients display reduced efficiency in global genome repair and altered chromatin structure, which facilitate the persistence of oxidative DNA lesions [80]. These impairments appear reversible with anti-inflammatory treatment, suggesting their dynamic nature and potential as pharmacodynamic markers. More recently, circulating cell-free DNA (cfDNA) has been explored as a non-invasive biomarker. Elevated plasma cfDNA levels have been associated with disease activity indices, seropositivity, and joint erosion scores, reinforcing their relevance as prognostic and diagnostic tools in RA [81]. Overall, oxidative DNA damage biomarkers such as 8-OHdG, HεdC, DNA strand breaks, and cfDNA offer significant potential in reflecting oxidative injury and, stratifying disease activity and predicting therapeutic response. Incorporating these markers into RA diagnostics may provide a more comprehensive view of disease status and underlying redox imbalance.

3.4 Antioxidant Enzymes (SOD, CAT, GPx)

Antioxidant enzymes are pivotal in neutralizing the oxidative stress contributing to inflammation and tissue damage in rheumatoid arthritis (RA). Among these, superoxide dismutase (SOD), catalase (CAT), and glutathione peroxidase (GPx) are the most studied in RA patients. SOD catalyzes the conversion of superoxide radicals into hydrogen peroxide, which CAT and GPx then neutralize. Several studies have shown elevated SOD activity in RA, likely as a compensatory response to increased ROS production. However, persistent oxidative stress often arises despite this elevation due to insufficient CAT and GPx activities. In one study, RA patients showed significantly higher SOD but lower CAT and glutathione reductase (GR) activities compared to controls, indicating incomplete ROS detoxification and potential hydrogen peroxide accumulation [13, 22]. CAT activity is frequently reduced in RA and this deficit has been linked with increased disease activity. Lower CAT levels correlate with enhanced inflammatory burden, possibly due to its suppression by inflammatory mediators or oxidative inactivation during chronic inflammation (70). GPx, another hydrogen peroxide-neutralizing enzyme, is also diminished in RA patients, particularly those with high disease activity. Its deficiency is associated with lower total glutathione levels, suggesting exhaustion of the glutathione system in RA pathology [13]. Combined deficits in GPx and CAT exacerbate oxidative stress, resulting in sustained tissue damage and immune dysregulation. Erythrocyte-based assessments consistently show reduced activities of these enzymes, underscoring systemic oxidative imbalance beyond local joint inflammation [22]. These antioxidant enzymes reflect disease severity and may serve as potential targets for antioxidant therapy in RA.

3.5 Non-enzymatic Antioxidants (Albumin, Glutathione, Bilirubin)

Non-enzymatic antioxidants form a complementary defense against oxidative stress in RA by directly neutralizing free radicals and supporting enzymatic systems. The key molecules in this group include albumin, glutathione (GSH), and bilirubin. Albumin, the most abundant plasma protein, possesses antioxidant properties through its free thiol groups and ability to bind transition metals. In RA, serum albumin levels are frequently reduced due to chronic inflammation and nutritional impairment. Decreased albumin levels inversely correlate with disease activity scores and inflammatory biomarkers, making it a valuable negative acute-phase reactant and antioxidant indicator [67]. GSH is the central intracellular antioxidant, acting directly on ROS and as a substrate for GPx. RA patients exhibit consistently lower GSH levels in both plasma and synovial fluid, especially during active disease. These reductions are associated with increased lipid peroxidation and lower total antioxidant capacity, indicating severe redox imbalance [13, 22]. Bilirubin, a product of heme catabolism, also functions as a potent endogenous antioxidant. Studies show that RA patients have significantly reduced serum bilirubin levels, which inversely correlate with disease activity scores such as DAS28 and biomarkers like CRP and IL-6. Notably, the bilirubin/albumin ratio has emerged as a novel indicator for RA disease status, with lower ratios predicting higher disease activity [66]. Together, these non-enzymatic antioxidants are important contributors to oxidative defense and are increasingly recognized for their diagnostic and prognostic relevance in RA.

3.6 Recent Research on Oxidative Stress Marker Profiling in RA Patients

Recent studies have emphasized the diagnostic, prognostic, and therapeutic value of oxidative stress biomarkers in rheumatoid arthritis (RA). These markers are profiled to clarify the disease mechanism, guide early detection, and assess treatment response. A 2024 observational study evaluated oxidative stress markers in untreated RA patients and found elevated levels of SOD, non-protein thiols (NP-SH), and TBARS, with reduced catalase activity. These changes were associated with higher body mass index and disease activity scores, suggesting their utility as early indicators of oxidative burden in RA [82]. A major systematic review analyzed 16 cross-sectional case–control studies and confirmed that oxidative stress markers such as MDA, nitric oxide, and SOD are significantly altered in RA patients. The review further emphasized that profiling these markers could improve early diagnosis, particularly during preclinical disease stages, and enhance monitoring of progression and treatment outcomes [44]. One of the most promising emerging biomarkers is plasma heme oxygenase-1 (HO-1). Its expression has been shown to correlate strongly with disease activity and radiographic progression. HO-1 levels were elevated in both early and established RA, and significantly declined following anti-inflammatory therapy. This makes HO-1 a compelling candidate as a first-in-class biomarker of synovial oxidative stress [83]. Peripheral endothelial dysfunction—a predictor of cardiovascular comorbidities in RA—has also been linked with oxidative markers. In one case–control study, RA patients demonstrated higher plasma MDA levels and significantly lower paroxonase-1 (PON-1) and protein-SH (PSH) concentrations than controls. These findings support their use in profiling RA's vascular risk and systemic oxidative burden [84]. Additionally, nutritional factors have gained attention in recent research. A 2021 study linked low zinc status with reduced activities of antioxidant enzymes such as SOD and GPx, and increased oxidative stress in RA patients. These results suggest that integrating trace element analysis with oxidative biomarker profiling may provide a more comprehensive assessment of disease risk and activity [85]. Altogether, oxidative stress profiling is evolving into a powerful diagnostic and prognostic tool in RA. Incorporating conventional and novel markers into clinical practice may enable earlier detection, better monitoring, and personalized treatment strategies.

4 Clinical Significance and Disease Correlation

4.1 Association of Oxidative Stress with Disease Activity (DAS28 Scores)

Oxidative stress markers have demonstrated a strong association with disease activity in patients with rheumatoid arthritis (RA), as measured by the disease activity score-28 (DAS28). This correlation is central to understanding the pathophysiological role of redox imbalance in RA and its potential use as a monitoring tool. Multiple studies have reported that higher oxidative stress levels—reflected through elevated ROS, TBARS, MDA, and protein carbonylation—correlate with higher DAS28 scores. For instance, Datta et al. found that all oxidative stress damage markers, including hydroxyl radicals and protein carbonylation in synovial fluid neutrophils, were significantly elevated in patients with high DAS28 scores, showing a direct association between redox status and disease severity [85]. Similarly, Mukhopadhyay et al. reported a significant positive correlation between ROS levels in peripheral blood neutrophils and DAS28 at both baseline ($r = 0.63$) and follow-up ($r = 0.75$), reinforcing that ROS measurements mirror clinical disease activity [86]. Pradhan et al., further emphasized oxidative stress's mechanistic role by showing that RA patients' neutrophils exhibited enhanced ROS generation, which correlated positively with DAS28 and protein oxidation. This effect was compounded by increased NADPH oxidase and myeloperoxidase activity, highlighting a direct molecular link between oxidative stress and disease activity [87]. Another recent study by Hosi [88] confirmed significantly elevated MDA and reduced GSH in RA patients, which are positively associated with moderate to severe DAS28 categories, supporting the reliability of oxidative stress biomarkers in tracking disease progression [88]. These findings collectively support incorporating oxidative stress parameters into routine clinical evaluation as surrogate biomarkers for disease activity, particularly in patients with high inflammatory burden or poor clinical control.

4.2 Correlation with Joint Erosion and Radiological Damage

Oxidative stress contributes significantly to the structural damage observed in rheumatoid arthritis (RA), including joint erosion and radiographic progression.

Recent investigations have clarified how reactive oxygen species (ROS) interact with cartilage and bone tissue to promote degradation. One of the most direct associations was identified in a 2024 clinical study, which demonstrated that elevated plasma heme oxygenase-1 (HO-1)—marker of oxidative stress—was strongly correlated with radiographic progression in RA patients. Baseline HO-1 levels were significantly higher in patients with early and established RA compared to controls, and levels decreased with treatment. Moreover, baseline HO-1 levels outperformed traditional serological markers in predicting long-term joint erosion and radiological damage over 2 years [86]. Animal model studies also support this oxidative damage pathway. In Nrf2-knockout mice, which lack a key antioxidant transcription factor, RA-induced cartilage destruction was significantly more severe than in wild-type controls. These mice exhibited higher VEGF expression and more significant oxidative burden, further supporting the idea that impaired redox defense mechanisms exacerbate joint degradation [87]. Another clinical study showed that the redox state of albumin-thiol—a sensitive oxidative stress marker—correlated positively with serum matrix metalloproteinase-3 (MMP-3), a key enzyme involved in cartilage breakdown. The oxidized albumin-thiol ratio was also associated with DAS28 and CRP, reinforcing the dual role of oxidative stress in inflammation and tissue degradation [88]. Moreover, the salivary oxidative stress index (OSI), which includes total antioxidant and oxidant balance, has been positively linked with radiographic bone alterations in RA patients' temporomandibular joint (TMJ). Lower TAS and higher OSI were observed in RA subjects with condylar erosion compared to healthy controls, suggesting peripheral tissue oxidative profiling may reflect joint damage status [82]. These converging data from systemic and local biomarker profiles highlight the importance of oxidative stress in predicting and monitoring radiographic progression in RA. This also supports the integration of redox biomarkers into composite scoring systems for joint health assessment.

4.3 Role in Extra-Articular Manifestations and Comorbidities (e.g., Cardiovascular Disease)

Rheumatoid arthritis (RA), though primarily an articular disease, is frequently associated with a range of extra-articular manifestations (EAMs) and comorbidities, many of which are mediated or exacerbated by oxidative stress. These include cardiovascular disease (CVD), interstitial lung disease, vasculitis, and sarcopenia—conditions that significantly contribute to the morbidity and mortality of RA patients. Oxidative stress plays a central mechanistic role in many of these non-articular complications. For instance, chronic inflammation and increased ROS production contribute to accelerated atherosclerosis, a hallmark cardiovascular manifestation in RA. Endothelial dysfunction driven by systemic oxidative imbalance promotes plaque formation and vascular inflammation. Patients with RA exhibit a twofold higher risk of myocardial infarction and up to a 50% increase in cardiovascular mortality compared to the general population [89]. HO-1, an oxidative stress-related biomarker, has been shown to correlate with subclinical myocardial dysfunction and radiographic damage. Its levels are significantly higher in patients with cardiovascular manifestations, providing a molecular link between oxidative stress and vascular comorbidity [82]. Severe extra-articular disease was also independently associated with the risk of first-ever cardiovascular events in RA. A longitudinal cohort study showed that patients with extra-articular manifestations had over a threefold increased risk of new-onset coronary artery disease, even after controlling for traditional risk factors like age, sex, smoking, and RF status [90]. Comorbidities such as diabetes, hypertension, and dyslipidemia—collectively known as cardiometabolic disorders—are also more prevalent in RA patients and have been associated with greater oxidative stress and disease severity. These comorbidities not only elevate cardiovascular risk but are also linked to higher disease activity and lower remission rates [91]. Furthermore, systemic oxidative stress markers, such as oxidized albumin and advanced oxidation protein products, correlate with endothelial dysfunction and organ involvement in RA. Their monitoring may aid in predicting the development of complications such as interstitial lung disease, rheumatoid vasculitis, and neurovascular events [92]. These insights collectively emphasize the need for redox biomarker profiling for assessing joint disease and early detection and management of extra-articular manifestations and comorbid conditions, particularly cardiovascular diseases.

4.4 Relationship with Patient-Reported Outcomes (Pain, Fatigue)

Patient-reported outcomes (PROs) such as pain and fatigue are key indicators of disease burden in rheumatoid arthritis (RA), often reflecting patients' quality of life more accurately than laboratory or imaging findings. Growing evidence suggests that oxidative stress may underlie or modulate these symptoms. A randomized controlled trial evaluating the antioxidant effects of garlic supplementation in RA patients found significant reductions in serum oxidative stress markers (e.g., MDA and increased total antioxidant capacity) and notable improvements in patient-reported fatigue and physical function. This demonstrates that reducing oxidative stress can directly enhance PROs [93]. Fatigue in RA is a multifactorial phenomenon, and oxidative imbalance is increasingly recognized as a biological contributor. A metabolomics-based study linked high fatigue levels in untreated RA patients to altered profiles of urea cycle metabolites and tocopherols—biochemical signatures consistent with redox stress. The study reported that several fatigue-associated metabolites correlated with increased oxidative stress, supporting the hypothesis that systemic redox imbalance is a metabolic driver of fatigue symptoms [94]. In parallel, a multidimensional model of fatigue in RA showed that pain and sleep disturbances were the most substantial direct contributors to fatigue, but oxidative stress-related inflammation was also noted as a possible mediating pathway. In this model, mood, functional capacity, and sense of control were indirectly associated with oxidative stress and influenced perceived fatigue severity [95]. Furthermore, fatigue severity has shown a strong correlation with pain intensity and disease activity, independently associated with ROS production and systemic inflammation. An extensive observational study confirmed that patient-reported fatigue scores were highly correlated with VAS pain, DAS28, and CRP levels—all redox-sensitive variables in RA [96]. Importantly, even non-pharmacologic factors such as sleep and chronic stress—which modulate ROS generation—have been shown to exacerbate fatigue and pain perception. This suggests oxidative stress may be a biological bridge between external psychosocial triggers and internal symptom perception [97]. Together, these studies underscore the physiological relevance of oxidative stress in mediating PROs like pain and fatigue. Integrating antioxidant strategies may provide dual benefits—ameliorating redox imbalance and improving the lived experience of RA.

5 Genetic and Epigenetic Insights

5.1 Genetic Polymorphisms in Oxidative Stress-Related Genes (e.g., HIF1A, NOS2A, p53)

Environmental and inflammatory triggers modulate oxidative stress and host genetic factors influencing redox balance. Several genetic polymorphisms in oxidative stress-related genes—including HIF1A, NOS2A, p53, and others—have been associated with rheumatoid arthritis (RA) susceptibility, severity, and progression. A comprehensive review by Carmona et al. identified HIF1A, MSRA, HO-1, p53, NCF1, and NOS2A as key nuclear and mitochondrial genes implicated in oxidative regulation relevant to RA. These genes govern cellular responses to hypoxia, nitric oxide synthesis, DNA repair, and ROS generation—core mechanisms in RA pathophysiology. Mutations or single nucleotide polymorphisms (SNPs) in these genes may impair antioxidant defense or exaggerate pro-oxidative pathways, thereby enhancing inflammation and joint destruction [98]. Specific SNPs in NOS2A, which encodes inducible nitric oxide synthase (iNOS), have been associated with RA disease activity. In an extensive genotyping study, the NOS2A (CCTTT)n repeat polymorphism showed a relationship with elevated DAS28 scores and inflammatory markers, suggesting a functional consequence in NO-mediated tissue damage [99, 100]. Mutations in the p53 tumor suppressor gene—critical for apoptosis and DNA damage response—have been extensively examined in RA. Chronic oxidative stress in RA joints may trigger permanent p53 mutations, altering cell-cycle regulation and contributing to synovial hyperplasia. Zvaifler et al. proposed that oxidative stress-induced p53 mutations could shift transient inflammation toward chronic joint disease [101]. Polymorphisms in the p53 codon 72 (Arg/Pro) variant were also shown to influence disease severity rather than susceptibility. Italian RA patients with the Pro/Pro genotype demonstrated significantly more radiographic erosions and higher

Sharp/van der Heijde scores at five years, indicating its potential as a marker for disease progression [102]. Beyond these, polymorphisms in NCF1 and NCF4—genes involved in NADPH oxidase function—were associated with altered ROS generation and RA severity in both animal models and human populations. The rs729749 SNP in NCF4, for instance, was linked to autoantibody-negative RA in a Swedish cohort, supporting a role for impaired ROS signaling in seronegative disease subsets [100]. These findings underscore the importance of genetic variants in modulating oxidative stress in RA and open avenues for personalized medicine approaches targeting redox-sensitive pathways.

5.2 *Epigenetic Modifications and ROS Regulation*

Epigenetic mechanisms such as DNA methylation, histone modification, and non-coding RNAs are critical in regulating genes involved in oxidative stress responses in rheumatoid arthritis (RA). These changes can affect cellular redox homeostasis, inflammatory activation, and synovial hyperplasia—hallmarks of RA pathogenesis. DNA methylation changes have been identified in RA patients' synovial fibroblasts and immune cells, leading to altered expression of pro-oxidant and antioxidant genes. For instance, hypomethylation of genes encoding inflammatory mediators and ROS-producing enzymes like NOX family members has been associated with synovial tissue hyperactivity and sustained oxidative stress [103]. Recent integrative studies of the RA epigenome have shown global alterations in histone modification marks, including H3K27ac, H3K4me1, and H3K36me3, in fibroblast-like synoviocytes (FLS). These marks are associated with enhancer activation and chromatin accessibility in oxidative metabolism and mitochondrial function genes. Aberrant histone landscapes may promote ROS production by enhancing the transcription of NADPH oxidase subunits and repressing antioxidant regulators [104]. MicroRNAs (miRNAs) also contribute significantly to the epigenetic regulation of redox pathways. Downregulation of miR-146a and miR-155—both negative regulators of inflammatory signaling—have been observed in RA patients and is linked with increased ROS generation and TNF-α signaling. The loss of these miRNAs may promote oxidative injury by derepressing pathways that activate iNOS and NOX enzymes [105]. Moreover, proinflammatory cytokines such as TNF-α and IL-1β have been shown to epigenetically repress HDAC5, a histone deacetylase that regulates interferon response genes. Loss of HDAC5 expression leads to enhanced nuclear IRF1 activity and increased ROS-related gene transcription. This feedback loop suggests that inflammatory signaling in RA can epigenetically amplify oxidative stress [106]. Emerging data also point to MeCP2, a methyl-DNA-binding protein, as a critical modulator of oxidative stress response in RA. MeCP2 interacts with DNA methylation and histone modifications to silence antioxidant genes. Its upregulation in RA synoviocytes has been associated with the downregulation of ROS-scavenging enzymes like GPx and CAT, promoting redox imbalance [107]. Collectively, these epigenetic alterations converge to dysregulate ROS production and clearance in RA.

Targeting such reversible modifications offer a promising avenue for redox-based therapies that can restore oxidative balance and modulate inflammatory responses.

5.3 Interplay Between Genetic Susceptibility and Environmental Triggers

The pathogenesis of rheumatoid arthritis (RA) reflects a complex interplay between genetic predisposition and environmental exposures that together influence the immune response, redox status, and eventual clinical manifestation of disease. Among the most well-established genetic-environment interactions is the synergy between HLA-DRB1 shared epitope (SE) alleles and smoking. Individuals with SE-positive alleles who are exposed to tobacco smoke demonstrate significantly elevated risk of developing seropositive RA. Smoking induces citrullination of proteins in the lungs, and in genetically susceptible individuals, this triggers immune responses that result in anti-citrullinated protein antibody (ACPA) formation—often years before joint symptoms begin [108]. Beyond HLA-DRB1, polymorphisms in genes such as PTPN22, STAT4, and MSRA have been linked to increased oxidative stress susceptibility and heightened immune reactivity in response to environmental factors like infections, air pollution, and psychological stress. These environmental cues may act as epigenetic modifiers, influencing DNA methylation and histone modifications that skew immune tolerance and redox regulation [109]. Newer data also support the role of airborne pollutants, including particulate matter and VOCs, in exacerbating oxidative stress and enhancing the expression of oxidized LDL and protein damage products in individuals with preclinical or genetic risk for RA. These environmental stressors have been shown to account for up to 40% of anti-LDG antibody production in genetically predisposed groups [110]. This convergence of gene-environment interaction was also evident in a recent study analyzing HLA-DRB1 and smoking status in a Pakistani cohort, where smoking increased RA risk 5.8-fold in carriers of the SE alleles compared to 2.2-fold in non-smokers. This reinforces the idea of cumulative risk and offers a rationale for integrating genetic screening with lifestyle interventions to reduce RA onset and progression [111]. The interaction between inherited susceptibility and modifiable environmental insults forms the cornerstone of RA risk stratification. Recognizing and mitigating these environmental influences in high-risk populations may represent a crucial early intervention and prevention strategy.

6 Therapeutic Implications of Targeting Oxidative Stress

6.1 Conventional RA Therapies and Their Effects on Oxidative Status

Conventional therapies for rheumatoid arthritis (RA), including non-steroidal anti-inflammatory drugs (NSAIDs), glucocorticoids (GCs), and disease-modifying anti-rheumatic drugs (DMARDs) such as methotrexate (MTX), not only reduce inflammation but also exert significant effects on oxidative stress pathways. Methotrexate, the cornerstone DMARD in RA therapy, has shown a dual role—while it effectively reduces inflammation and ROS production by dampening immune cell activation, it also induces oxidative damage through mitochondrial toxicity and folate depletion. A 2018 study comparing treatment-naive RA patients and those on MTX monotherapy found elevated oxidative stress markers (e.g., TBARS) and reduced antioxidant enzyme activity (SOD, CAT, GR) in both groups but significantly lower in the MTX-treated cohort, suggesting a partial amelioration of oxidative imbalance [112]. NSAIDs such as diclofenac and ibuprofen have been observed to lower ROS indirectly by reducing COX-2-mediated prostaglandin synthesis and neutrophil activity. However, prolonged NSAID use may impair mitochondrial respiration and glutathione cycling, leading to paradoxical oxidative injury in hepatic and renal tissues [113, 114]. Glucocorticoids exhibit strong antioxidant effects by suppressing iNOS and COX-2 gene expression and scavenging free radicals. Yet chronic GC use is linked to oxidative injury in bone, skin, and vascular tissues. These opposing outcomes necessitate dosage control and antioxidant co-therapy in long-term steroid regimens [115]. Interestingly, ceruloplasmin levels—a positive acute-phase reactant and ROS scavenger—were significantly elevated in patients on MTX therapy, suggesting an adaptive upregulation of antioxidant defenses. In the same patients, erythrocyte antioxidant enzymes were still suppressed, reinforcing the need for nutritional or pharmacological antioxidant supplementation during therapy [116]. Furthermore, studies suggest that antioxidant vitamins (e.g., C and E) or plant-derived flavonoids could enhance the redox-modulatory effects of conventional treatments. This synergy between DMARDs and antioxidant supplementation is being explored as a potential approach to minimize oxidative tissue injury while preserving anti-inflammatory efficacy [114]. While conventional RA therapies mitigate inflammation and indirectly reduce ROS production, they do not fully normalize oxidative stress parameters. These insights support adjunct antioxidant therapy to counterbalance residual oxidative damage and improve treatment tolerability.

6.2 Antioxidant Supplementation: Current Evidence

Given the established role of oxidative stress in rheumatoid arthritis (RA) pathogenesis, numerous studies have explored the efficacy of antioxidant supplementation in managing disease activity and oxidative burden. While findings remain mixed, several trials suggest potential clinical benefits under specific conditions. In a randomized controlled trial, Pourhabibi-Zarandi et al. reported that curcumin supplementation (500 mg/day) for 8 weeks significantly reduced DAS28 scores, tender/swollen joint counts, and visual analog pain scores. It also increased total antioxidant capacity (TAC) and reduced malondialdehyde (MDA), although the change in MDA was not statistically significant compared to placebo [76] (Fig. 6).

Similarly, Jalili et al. administered a multi-antioxidant supplement (selenium, zinc, vitamins A, C, and E) for 12 weeks in female RA patients. The intervention led to significant improvements in erythrocyte antioxidant enzymes (GPx, SOD, CAT) and total antioxidant capacity, as well as reductions in high-sensitivity CRP. However, joint count parameters remained unchanged, suggesting antioxidant therapy may primarily benefit systemic redox status rather than local inflammation [117]. Qamar et al. highlighted the emerging role of selenium in RA management. As a cofactor for glutathione peroxidase and thioredoxin reductase, selenium modulates ROS levels and immune cell signaling. Preclinical and early clinical data suggest that selenium supplementation may reduce synovial inflammation and slow disease progression [118]. On the contrary, a Mendelian randomization study by Zhou et al. found no causal relationship between most antioxidant levels (*vitamin C, β-carotene, tocopherols*) and RA risk. The only exception was circulating retinol metabolites, which were inversely associated with seropositive RA, suggesting that retinol may have a unique protective role in disease initiation [119]. Pattison and Winyard reviewed existing interventional trials and concluded that while antioxidant-rich diets may lower the risk of developing inflammatory arthritis, clinical trial data remain inconclusive regarding their therapeutic use in established RA. This inconsistency was attributed to heterogeneity in dosage, bioavailability, trial duration,

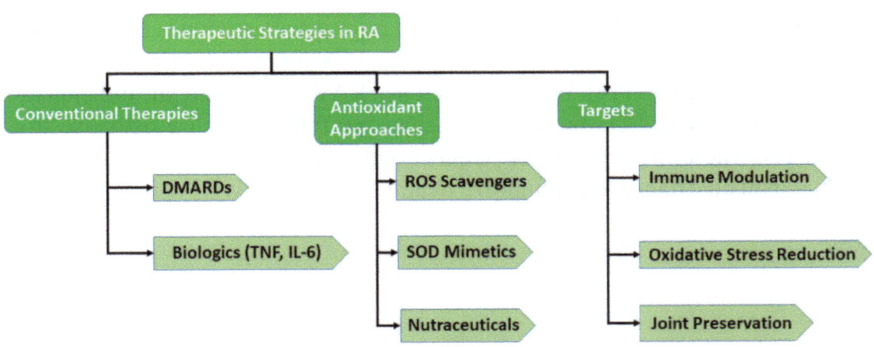

Fig. 6 Therapeutic strategies in RA

and patient characteristics [120]. Therefore, although antioxidant supplementation shows promise—particularly with compounds like curcumin, selenium, and retinol—its clinical application remains nuanced. More large-scale, standardized trials are necessary to establish specific guidelines for antioxidant use in RA.

6.2.1 Garlic, Vitamin E, NAC, and Polyphenols

Several natural antioxidant compounds, such as garlic, Vitamin E, N-acetylcysteine (NAC), and polyphenols, have been extensively studied for their potential therapeutic roles in rheumatoid arthritis (RA) due to their ability to modulate oxidative stress and inflammation. Garlic, rich in organosulfur compounds, exhibits potent antioxidant and anti-inflammatory properties. Clinical trials have shown that garlic supplementation significantly reduced serum levels of inflammatory biomarkers (CRP and TNF-α) and improved patient-reported outcomes, such as fatigue, pain, and joint tenderness, in RA patients [96]. Similar findings were noted in studies where garlic, in combination with olive and sesame oils, led to marked reductions in markers like ESR, RF, and anti-CCP, suggesting a broader anti-inflammatory benefit. Vitamin E, a lipid-soluble antioxidant, has significantly improved RA's oxidative status and disease activity scores when used as adjunct therapy. High-dose α-tocopherol treatment in animal models led to reduced joint swelling, lower oxidative damage, and decreased pain sensitivity. Clinical studies have similarly shown that Vitamin E and C supplementing reduces inflammatory markers, improves joint scores, and enhances antioxidant enzyme activity [63]. N-acetylcysteine (NAC), a precursor to glutathione, is known for replenishing intracellular antioxidants and scavenging ROS. Though direct evidence in RA is limited, studies in related autoimmune conditions suggest that NAC can modulate immune responses and reduce oxidative stress, warranting its exploration in RA. Polyphenols, naturally found in fruits and vegetables like broccoli, berries, and tea, offer robust antioxidant capabilities. Experimental models of arthritis treated with broccoli extract, rich in polyphenols, showed reduced malondialdehyde (MDA) levels and a tendency to improve joint inflammation scores. Additionally, broader literature supports the role of polyphenol-rich diets in decreasing oxidative stress and modulating immune response in RA. Collectively, these compounds have shown promise as adjunct therapies to reduce oxidative burden and inflammation in RA patients, thereby potentially improving quality of life and clinical outcomes.

6.2.2 Clinical Trials and Human Studies

Human studies investigating antioxidant supplementation in rheumatoid arthritis (RA) provide mixed but encouraging evidence regarding its impact on oxidative stress and disease activity. A recent double-blind RCT by Pourhabibi-Zarandi et al., evaluated curcumin (500 mg/day) in 48 women with RA. The results showed significant improvements in disease activity (DAS28), reductions in pain, swollen/tender joint counts, and increases in serum total antioxidant capacity (TAC). Although malondialdehyde (MDA) levels declined, changes were not statistically significant between groups, indicating a mild but beneficial antioxidant effect [121]. Saffron supplementation (100 mg/day for 12 weeks) was similarly shown to significantly improve clinical parameters, including DAS28, pain visual analog scale, and inflammatory markers such as ESR and hs-CRP. Though total antioxidant capacity and pro-oxidant markers like TNF-α and IFN-γ improved, intergroup differences were non-significant [122]. Jalili et al., conducted a 12-week intervention using a multi-antioxidant supplement (Selenium, Zinc, Vitamins A, C, and E) in RA patients. This study revealed improvements in erythrocyte antioxidant enzymes (SOD, CAT, GPx) and reductions in hs-CRP and oxidative stress biomarkers. However, clinical joint count parameters remained unchanged [123]. A trial by Abdollahzad et al. using Coenzyme Q10 (100 mg/day) for 2 months showed a significant reduction in MDA levels and suppression of TNF-α expression, indicating antioxidant and anti-inflammatory effects. However, total antioxidant capacity and IL-6 levels remained unaffected [119, 124]. While some trials report benefits, others provide more cautious interpretations. A systematic review and meta-analysis by Zeng et al. [125] evaluated 24 RCTs on antioxidant therapy in RA. It confirmed significant reductions in MDA and modest improvements in clinical parameters (DAS28, VAS, CRP) but showed no consistent benefit for TAC, SOD, or GPx. The authors emphasized heterogeneity in study quality and called for standardized large-scale trials [126]. Moreover, the systematic review by Nelson et al. analyzed 27 RCTs and identified moderate evidence for specific antioxidants—quercetin, saffron, and ubiquinone—while reporting inconclusive or null effects for vitamin E and some spice-based compounds. They concluded that dietary interventions may benefit select patient subsets but are unlikely to replace conventional therapy [120]. Clinical trials affirm that antioxidant supplementation may improve redox balance and inflammation in RA, particularly when used as an adjunct. However, heterogeneity in dosing, patient selection, and endpoints limits definitive conclusions. Future studies should aim for methodological consistency and clinical outcome relevance.

6.3 Chinese Herbal Medicines and Natural Products with Anti-ROS Activity

Chinese herbal medicines (CHMs) and natural plant-derived products are gaining recognition as adjunct therapies in rheumatoid arthritis (RA), particularly due to their anti-inflammatory and antioxidant properties. These traditional formulations often modulate oxidative stress by enhancing endogenous antioxidant defense and reducing reactive oxygen species (ROS) production, thereby influencing disease activity and progression. A comprehensive review by Wang et al. [127] emphasized that many traditional Chinese herbs—such as *Tripterygium wilfordii, Curcuma longa*, and *Paeonia lactiflora*—possess bioactive constituents (e.g., flavonoids, terpenoids, and phenolic acids) with proven anti-ROS activity. These compounds attenuate oxidative stress by suppressing NADPH oxidase expression, enhancing superoxide dismutase (SOD) activity, and modulating mitochondrial ROS production [124]. A classical herbal formula, Qianghuo Shengshi decoction, was demonstrated through systems biology to contain active compounds like notopterol and glycyrrhizic acid that exert antioxidant effects by targeting key inflammatory and ROS-related pathways, such as MAPK and PI3K-Akt. These herbs reduce oxidative damage in synovial tissue and interfere with downstream inflammatory cascades [125]. Epimedii Herba (Yinyanghuo) is another extensively studied TCM compound with polyphenolic constituents like icariin, which exerts ROS-scavenging and anti-inflammatory effects. It has been shown to inhibit NOX2-mediated superoxide generation and enhance antioxidant enzyme expression in RA animal models, supporting its role in redox modulation [128]. In a network pharmacology analysis, Huo-Luo-Xiao-Ling Dan, a classic four-herb formula, demonstrated the ability to suppress oxidative injury by modulating leukocyte trafficking and reducing oxidative protein damage in an adjuvant-induced arthritis rat model. The modified herbal blend significantly reduced ROS accumulation and inhibited joint degradation pathways [118]. Moreover, a cohort study found that long-term CHM use was associated with a reduced risk of stroke in RA patients, implying systemic antioxidant protection beyond joint tissues. Commonly used herbs in this context included Dan-Shen and Gui-Zhi-Shao-Yao-Zhi-Mu-Tang, known for ROS suppression and endothelial function improvement [127]. While the variability in TCM formulations and practitioner diagnoses presents challenges in standardization, the growing pharmacological evidence supports their redox-regulatory roles in RA management. Integrating CHMs with conventional therapies may provide a synergistic approach to reduce oxidative burden and control disease symptoms.

6.4 Emerging Pharmacological Agents Targeting Oxidative Pathways

In recent years, significant progress has been made in identifying novel pharmacological agents that specifically modulate oxidative stress pathways in rheumatoid arthritis (RA). These targeted interventions represent a promising frontier in RA therapy, aimed at restoring redox homeostasis alongside immune modulation. Among the most well-characterized redox-sensitive therapeutic targets is the Nuclear factor erythroid 2-related factor 2 (Nrf2) pathway. Nrf2 orchestrates cellular antioxidant defenses by regulating genes involved in glutathione synthesis, detoxification enzymes, and redox-buffering proteins. Dysregulation of Nrf2 in RA contributes to sustained oxidative injury and joint degradation. Pharmacologic Nrf2 activators, such as dimethyl fumarate (DMF) and novel triterpenoids, have shown promise in preclinical RA models by suppressing ROS production, inhibiting NF-κB signaling, and reducing synovial inflammation [129]. Another innovative approach involves targeting nitric oxide synthases (NOS). Elevated inducible NOS (iNOS) activity in RA joints contributes to the overproduction of nitric oxide, which reacts with superoxide to form peroxynitrites—a highly toxic oxidant. iNOS inhibitors, currently under evaluation in clinical trials, show the potential to dampen inflammatory oxidative stress without impairing physiological NO functions [130]. MicroRNA-based therapies are also under investigation. miR-30a-3p, for instance, has been shown to reduce oxidative damage in fibroblast-like synoviocytes (FLS) by activating the Keap1/Nrf2 pathway, inhibiting ROS-induced proliferation and promoting apoptosis of aberrant synoviocytes. Its therapeutic mimic or agomir has demonstrated effectiveness in preclinical arthritis models [131, 132]. Natural flavonoids like isoliquiritigenin (ISL) have emerged as Nrf2 activators with antioxidant and anti-apoptotic properties. ISL treatment in RA chondrocyte cultures and animal models significantly decreased oxidative injury and pro-inflammatory cytokine expression by enhancing Nrf2/HO-1 signaling. This suggests that plant-derived molecules may function as targeted redox modulators [133]. Finally, nanocarrier systems such as liposomal encapsulation of antioxidant enzymes (e.g., superoxide dismutase, SOD) are being developed for improved delivery and retention at inflamed joint sites. PEGylated SOD liposomes have demonstrated prolonged circulation time and efficient ROS scavenging in arthritic rat models, indicating their potential to enhance site-specific antioxidant therapy [134]. These emerging agents, spanning synthetic drugs, RNA-based therapeutics, phytochemicals, and nanomedicine platforms, offer targeted strategies to rebalance oxidative stress in RA. While many remain in preclinical stages, their mechanistic specificity and dual anti-inflammatory properties position them as strong candidates for future RA management.

6.5 *Role of Diet, Lifestyle, and Microbiome in Modulating Oxidative Stress*

The role of diet, lifestyle, and the gut microbiome in regulating oxidative stress and inflammation in rheumatoid arthritis (RA) has gained increasing attention as evidence supports their combined influence on immune function, redox balance, and disease outcomes. A comprehensive review by Gioia et al. emphasized that dietary habits can modulate both inflammation and oxidative stress in RA, with the Mediterranean Diet (MD) emerging as particularly beneficial due to its richness in antioxidants, omega-3 polyunsaturated fatty acids (PUFAs), polyphenols, and fiber. These nutrients directly scavenge reactive oxygen species (ROS) and indirectly reduce oxidative damage by modulating gene expression and gut microbiota composition [135]. The gut microbiome, in particular, is a critical intermediary between diet and immune-mediated oxidative processes. Dourado et al. reported that MD and specific dietary components—such as n-3 PUFAs and polyphenols—support microbial diversity and strengthen intestinal barrier integrity, thereby reducing systemic endotoxemia and oxidative burden [136]. Similarly, data from the MADEIRA trial demonstrated that adherence to a Mediterranean Diet and physical activity significantly reduced RA disease activity scores and improved oxidative stress-related cardiometabolic outcomes. Improvements were noted in inflammatory markers and 1,25(OH)2D levels, suggesting synergistic effects between diet, nutrient metabolism, and immune modulation [137]. Lifestyle interventions—especially smoking cessation, physical activity, and weight control—also play pivotal roles. In a narrative review, Shekhar et al. highlighted that obesity and smoking are associated with more significant oxidative stress and poorer treatment responses. Conversely, exercise improves mitochondrial function and endogenous antioxidant enzyme activity (SOD, CAT), reducing RA severity [138]. Recent mechanistic insights by Ma et al. showed that dietary timing and microbiota-driven diurnal rhythms influence the oxidative inflammatory axis in RA. Specifically, microbial β-glucosidase activity from *Parabacteroides distasonis* was shown to release antioxidant compounds like glycitein, modulating NF-κB signaling through the SIRT5 pathway—emphasizing the potential of chrono-nutrition in RA management [139]. These findings suggest that nutritional and lifestyle modifications, especially those promoting microbiome health, are essential adjuncts to pharmacological interventions in RA. They offer sustainable, low-risk strategies to attenuate oxidative stress and improve disease control.

7 Biomarkers as Diagnostic and Prognostic Tools

7.1 Utility of Oxidative Stress Biomarkers in Clinical Practice

The application of oxidative stress biomarkers in rheumatoid arthritis (RA) has garnered increasing interest due to their potential to aid diagnosis, monitor disease progression, and predict therapeutic response. As redox imbalance is deeply implicated in RA pathogenesis, various biomarkers reflecting lipid, protein, and DNA oxidation and antioxidant capacity are being explored in clinical settings. Markers such as malondialdehyde (MDA), thiobarbituric acid reactive substances (TBARS), protein carbonyls, 8-hydroxy-2′-deoxyguanosine (8-OHdG), and enzymatic antioxidants (e.g., SOD, CAT, GPx) have shown consistent alterations in RA patients compared to healthy controls. A 2016 systematic review emphasized that these biomarkers reflect active disease processes and could potentially support disease staging or response evaluation, though heterogeneity in findings limits standardization [140]. Recent prospective data by Andersen et al. identified plasma heme oxygenase-1 (HO-1) as a promising first-in-class oxidative stress biomarker in early RA (eRA), correlating strongly with disease activity and radiographic progression over two years. Plasma HO-1 levels outperformed conventional inflammatory markers in predicting long-term joint damage, making it a strong candidate for clinical use [141]. Another clinical study showed that SOD and TBARS levels were significantly elevated in untreated RA patients and decreased upon pharmacological intervention, implying their potential as early indicators of disease activity and treatment efficacy [114]. Protein carbonyls and oxidized LDL are associated with higher disease activity score-28 (DAS28) and inflammatory markers. A study by Mititelu et al. found that the integration of oxidative and inflammatory biomarkers—such as neutrophil-to-lymphocyte ratio (NLR), monocyte-to-lymphocyte ratio (MLR), and TBARS—enhanced predictive capacity for disease outcomes [142]. While these biomarkers show promise, a lack of standardization in assay techniques, threshold values, and interpretive criteria currently limits clinical utility. Nonetheless, oxidative stress markers can enrich RA's diagnostic and prognostic landscape when used alongside clinical scores (e.g., DAS28), serological markers (e.g., CRP, anti-CCP), and imaging tools.

7.2 Prognostic Value for Predicting Flares or Treatment Response

The identification of reliable prognostic biomarkers is crucial for the individualized management of rheumatoid arthritis (RA). Among various candidates, oxidative stress biomarkers have emerged as valuable tools for predicting disease flares, gauging treatment response, and identifying patients at risk of poor outcomes. A notable marker in this context is plasma heme oxygenase-1 (HO-1). As previously noted in Sect. 7.1, HO-1 levels correlate with disease activity, radiographic progression, and synovial inflammation. Despite conventional treatment, its early elevation in untreated patients was associated with persistent disease activity, suggesting prognostic utility in identifying individuals less likely to achieve remission [141]. Malondialdehyde (MDA) and TBARS levels have been shown to predict subclinical inflammation and symptom resurgence in clinically stable patients. In longitudinal studies, sustained high MDA levels were associated with poor methotrexate responsiveness and more frequent flares [114]. In a trial analyzing antioxidant enzyme responses to therapy, improvement in SOD and GPx activity within the first 8 weeks of treatment was predictive of favorable DAS28 reduction at 6 months. Patients lacking this antioxidant rebound were more likely to exhibit incomplete response or relapse after glucocorticoid tapering [123]. Protein oxidation markers, such as protein carbonyl content and oxidized albumin, have also been associated with poor therapeutic response. Patients with elevated carbonyl stress at baseline were less likely to respond to TNF-α inhibitors, suggesting that oxidative imbalance could interfere with cytokine-targeted therapy efficacy [142]. Furthermore, a systematic review by Zeng et al., confirmed that changes in oxidative markers—specially MDA, SOD, and CAT—can serve as early indicators of pharmacological responsiveness. The meta-analysis concluded that antioxidant dynamics could aid clinicians in differentiating responders from non-responders even before clinical scores change significantly [126]. Incorporating these markers into routine practice could help tailor therapy intensity, select appropriate agents, and anticipate the need for escalation. Although none are currently standardized for clinical use, accumulating evidence supports their inclusion in prospective predictive algorithms.

7.3 Integration into Personalized Medicine Approaches

Oxidative stress biomarkers offer promising avenues for incorporation into personalized medicine frameworks in rheumatoid arthritis (RA). Their ability to reflect individual variations in redox status, immune response, and treatment susceptibility makes them attractive tools for tailoring disease management. Plasma levels of heme oxygenase-1 (HO-1), which correlate with disease activity and long-term joint damage, exemplify the potential of oxidative biomarkers in stratifying patients by risk and optimizing therapeutic regimens. As demonstrated by Andersen et al., patients with elevated baseline HO-1 may benefit from early aggressive treatment, whereas those with lower levels might follow a more conservative approach [141]. The variability in oxidative stress responses across patients—highlighted by divergent SOD, GPx, and MDA levels—suggests that redox profiling could assist in defining therapeutic responders. In this context, Jalili et al. proposed that the trajectory of antioxidant enzyme recovery could forecast DMARD responsiveness, forming the basis for early adaptive therapy models [123]. Mititelu et al. introduced the idea of multiparametric panels that integrate oxidative stress markers (e.g., protein carbonyls, TBARS), systemic inflammation indices (e.g., NLR, MLR), and traditional clinical scores to improve predictive power. These multimodal approaches align well with personalized medicine principles, allowing nuanced risk stratification [142]. Recent advances in machine learning and transcriptomic profiling also support the inclusion of oxidative biomarkers in digital diagnostics. Wang et al. identified a mitochondrial oxidative stress-related gene signature (CDKN1A, GADD45B, MAFF) that accurately predicts RA risk and immune infiltration status. Their diagnostic model represents a step toward precision-based interventions targeting oxidative pathways [143]. Quiñonez-Flores et al. emphasized the interconnection between oxidative biomarkers and classical RA pathogenic elements, advocating for their inclusion in clinical algorithms alongside CRP, ESR, and anti-CCP, especially in seronegative or early RA presentations where traditional markers may be insufficient [140]. While implementation challenges remain—particularly around standardization and cost—integrating oxidative stress assessment with genomics, imaging, and immunophenotyping can personalize care at every disease stage. Such integrated approaches may guide therapy choices and reduce trial-and-error prescribing, improving outcomes and reducing toxicity.

8 Limitations and Controversies

8.1 *Variability in Biomarker Assessment Techniques*

The clinical application of oxidative stress biomarkers in rheumatoid arthritis (RA) is hindered by substantial variability in assessment techniques. This inconsistency arises from differences in assay protocols, biospecimen types, detection methods, and patient-related factors, limiting comparability and reproducibility across studies. As highlighted by Quiñonez-Flores et al., the current evidence supporting oxidative biomarkers in RA is heterogeneous and lacks uniformity in sample handling and quantification strategies. Variability in methods to assess MDA, TBARS, SOD, and protein carbonyls has resulted in conflicting conclusions regarding their association with disease activity [140]. Further complicating interpretation, Hermann et al. found that enzymatic antioxidant activity (e.g., CAT and SOD) showed wide inter-individual variability in untreated RA patients, with values influenced by body mass index and other metabolic confounders. Such findings reinforce the importance of considering demographic and lifestyle variables in biomarker assessment [114]. A study by Mullins et al. demonstrated that oxidative stress markers exhibit intra- and inter-individual variability even in healthy individuals, with coefficients of variation ranging from 130 to 240% depending on the marker and physiological state. These fluctuations raise questions about using single-point measurements for clinical decision-making [144]. Pre-analytical and analytical factors—such as storage temperature, anticoagulant use, and time between collection and analysis—further affect market stability. As shown in urinary biomarker studies, normalization techniques (e.g., creatinine adjustment) can enhance reproducibility, but standardized protocols remain underdeveloped in rheumatology contexts. Despite promising results from several trials, the lack of assay harmonization has prevented these biomarkers from being adopted into routine RA clinical workflows. Consensus on biomarker panels, assay validation, and reference ranges is essential for integrating oxidative stress profiles into diagnostic and prognostic algorithms.

8.2 *Conflicting Data in Clinical Studies*

The clinical application of oxidative stress biomarkers in rheumatoid arthritis (RA) is hindered by substantial variability in assessment techniques. This inconsistency arises from differences in assay protocols, biospecimen types, detection methods, and patient-related factors, limiting comparability and reproducibility across studies. As highlighted by Quiñonez-Flores et al., the current evidence supporting oxidative biomarkers in RA is heterogeneous and lacks uniformity in sample handling and quantification strategies. Variability in methods to assess MDA, TBARS, SOD, and protein carbonyls has resulted in conflicting conclusions regarding their association with disease activity [140]. Further complicating interpretation, Hermann et al. found that enzymatic antioxidant activity (e.g., CAT and SOD) showed wide inter-individual variability in untreated RA patients, with values influenced by body mass index and other metabolic confounders. Such findings reinforce the importance of considering demographic and lifestyle variables in biomarker assessment [114]. A study by Mullins et al. demonstrated that oxidative stress markers exhibit intra- and inter-individual variability even in healthy individuals, with coefficients of variation ranging from 130 to 240% depending on the marker and physiological state. These fluctuations raise questions about using single-point measurements for clinical decision-making [144]. Pre-analytical and analytical factors—such as storage temperature, anticoagulant use, and time between collection and analysis—further affect market stability. As shown in urinary biomarker studies, normalization techniques (e.g., creatinine adjustment) can enhance reproducibility, but standardized protocols remain underdeveloped in rheumatology contexts. Despite promising results from several trials, the lack of assay harmonization has prevented these biomarkers from being adopted into routine RA clinical workflows. Consensus on biomarker panels, assay validation, and reference ranges is essential for integrating oxidative stress profiles into diagnostic and prognostic algorithms.

8.3 Gaps in Longitudinal and Interventional Research

Despite significant advances in understanding oxidative stress in rheumatoid arthritis (RA), substantial gaps remain in the design and execution of longitudinal and interventional studies exploring its diagnostic, prognostic, and therapeutic relevance. These limitations hinder the translation of oxidative stress findings into clinical practice. One of the most prominent issues is the scarcity of long-term prospective cohort studies that evaluate changes in oxidative markers over the full disease course or concerning structured therapy timelines. Most existing studies are cross-sectional or short-term interventions, which limits the understanding of dynamic biomarker fluctuations during remission, relapse, or treatment tapering. A 2016 systematic review highlighted the heterogeneity of oxidative stress biomarker outcomes and concluded that a lack of longitudinal designs significantly restricted interpretability and clinical translatability [140]. Similarly, Ahmed et al. [145] underscored that while oxidative biomarkers like MDA, protein carbonyls, and antioxidant enzyme levels are consistently altered in RA, their longitudinal association with clinical disease activity or flare prediction remains inadequately investigated. They called for standardized study designs and the inclusion of oxidative endpoints in routine RA research protocols [146]. Furthermore, despite growing numbers, interventional studies using antioxidant therapies have been underpowered, short-term, or lacking in mechanistic clarity. For example, while propolis, curcumin, and vitamin E supplementation have shown promise in reducing oxidative stress and inflammation, most trials have used small cohorts and variable dosages, precluding generalized conclusions [106, 121]. Additionally, linking oxidative stress dynamics to treatment outcomes remains an unmet need. While a few trials have correlated oxidative changes with DAS28 score trajectories, others fail to integrate conventional or biological drug regimens with redox profiles comprehensively. This reflects a disconnect between redox biology and therapeutic pharmacology in RA clinical trials. Lastly, few studies address sex-based or age-related differences in oxidative stress trajectories or therapeutic responses despite growing evidence of their relevance. Future research must adopt inclusive, multicentric, and long-term frameworks integrating redox biology into predictive and therapeutic models.

9 Future Perspectives

9.1 *Need for Standardized Biomarker Panels*

While oxidative stress biomarkers hold strong potential in rheumatoid arthritis (RA) diagnosis and prognosis, their widespread clinical use remains constrained by a lack of standardized panels and validated thresholds. Currently, studies use a wide range of isolated markers—such as malondialdehyde (MDA), thiobarbituric acid reactive substances (TBARS), protein carbonyls, and antioxidant enzyme levels (SOD, CAT, GPx)—with inconsistent methodologies and endpoints. Quiñonez-Flores et al. highlighted this heterogeneity, reporting that no consensus exists on assay standardization, sample processing, or clinical interpretation for oxidative biomarkers in RA [140]. Similarly, Hermann et al. emphasized the influence of demographic and metabolic factors on oxidative stress readouts, stressing the importance of reproducibility and adjustment in biomarker reporting [114]. Mititelu et al. proposed a multi-marker approach integrating TBARS, protein carbonyls, and inflammatory indices (e.g., NLR, MLR), which improves predictive accuracy. This type of composite panel may be better suited to reflect the multifactorial nature of oxidative imbalance in RA [142]. Additionally, Mukhopadhyay et al. demonstrated that reactive oxygen species (ROS) levels in neutrophils, assessed via flow cytometry, significantly correlated with DAS28 scores at multiple time points. This supports their potential as part of a serial monitoring panel when assessed with high-sensitivity techniques [147]. Despite this growing body of evidence, there is no universal guideline for biomarker selection or cut-off values. Ahmed et al. called for international consensus on which markers should be validated, how they should be measured, and what reference ranges are suitable for routine use [146]. Future initiatives should focus on harmonizing biomarker measurement platforms (e.g., ELISA vs. HPLC), establishing multi-biomarker scoring systems, and integrating oxidative stress data into clinical decision-support tools alongside conventional inflammatory markers.

9.2 Integrating Omics Approaches (Proteomics, Metabolomics)

The application of omics technologies—particularly proteomics and metabolomics—has opened new horizons in understanding the pathogenesis, stratification, and treatment response in rheumatoid arthritis (RA). These high-throughput approaches enable system-wide profiling of oxidative stress-related pathways and the identification of robust, multidimensional biomarkers that could augment or replace traditional clinical parameters. A comprehensive review by Puentes-Osorio et al. systematically classified over 190 omics-derived biomarkers in RA, including 78 proteomic and 17 metabolomic markers. Many of these markers are linked to oxidative processes, such as SOD1, glutathione peroxidase, carbonylated proteins, and lipid peroxidation by-products. These markers correlate with disease activity and treatment responsiveness, providing mechanistic insight into redox imbalance [145]. Hui et al., conducted a large-scale polygenic risk score analysis using transcriptomic, proteomic, and metabolomic datasets from over 58,000 participants, identifying key oxidative stress-related genes and proteins—such as AGER, TFF3, and HLA-DQA2—that are significantly associated with RA. These findings were enriched in pathways including NF-κB signaling and cytokine receptor interactions, central to redox regulation and immune activation [148]. Tasaki et al., employed multi-omics monitoring during RA drug treatment and discovered that successful therapy was associated with a molecular shift toward a healthy state across transcriptomic, proteomic, and immunophenotypic profiles. Specific redox-associated proteins and inflammatory markers tracked closely with disease remission, suggesting their utility in defining molecular response rather than relying solely on clinical scores [149, 150]. Moreover, metabolomics studies have identified disease-specific metabolic shifts, such as glutathione depletion, increased ROS-linked metabolites (e.g., succinate, pyruvate), and dysregulated lipid metabolism in RA patients compared to healthy controls. These findings point to novel oxidative stress-linked pathways that could be exploited for diagnostics and targeted interventions [151]. Cassotta et al., emphasized the future role of omics in nutritional and lifestyle modulation of oxidative stress in RA. By integrating dietary data with proteomic and metabolomic outputs, they proposed a nutrigenomic framework for oxidative stress management through precision nutrition and microbiome-oriented therapy [152]. As omics technologies become more accessible and affordable, their integration into RA research is expected to produce particular redox signatures. These tools promise to improve early diagnosis, predict flares, guide drug selection, and monitor disease remission from a molecular perspective.

9.3 Development of Oxidative Stress-Targeted Therapies

Therapeutic targeting of oxidative stress pathways in rheumatoid arthritis (RA) is an evolving frontier, leveraging the mechanistic insights into redox imbalance to mitigate disease progression. Over the past decade, various interventions, from conventional antioxidants to advanced nanotherapeutics and biologics, have been evaluated for their efficacy in reducing oxidative damage in RA. A 2021 systematic review and meta-analysis encompassing 24 randomized controlled trials found that antioxidant therapies, particularly those targeting malondialdehyde (MDA) reduction effectively decreased oxidative stress and improved clinical parameters like DAS28 and C-reactive protein. However, the results for antioxidant enzymes (e.g., SOD, GPx) and total antioxidant capacity (TAC) were mixed, indicating that therapeutic benefit may depend on the oxidative target and patient phenotype [126]. Recent innovations include nanocatalytic agents such as ferrihydrite nanoparticles, which simultaneously scavenge hydrogen peroxide and generate oxygen in inflamed synovial tissue, alleviating hypoxia and oxidative injury. These nanomedicines exhibit catalase-mimetic activity, suppress inflammatory cytokines, and promote synovial repair in animal models, representing a promising next-generation therapy [153]. On the biologics front, therapies such as tocilizumab and anti-TNF-α agents have significantly reduced serum and synovial oxidative stress markers, including ROMs and lipid peroxidation by-products. These agents exert indirect antioxidant effects by modulating upstream cytokine cascades that drive ROS production [154]. Moreover, multi-targeted botanical agents like curcumin, N-acetylcysteine (NAC), and Chinese herbal decoctions have demonstrated antioxidant properties by activating the Nrf2 pathway, inhibiting NADPH oxidase, and improving mitochondrial redox buffering. These compounds are increasingly evaluated in conjunction with conventional DMARDs in clinical trials [124]. Despite these advances, challenges persist in translating oxidative therapies into mainstream care. These include the lack of biomarker-based patient stratification, variable bioavailability of natural compounds, and limited long-term safety data. Addressing these gaps through biomarker-guided trials and mechanistic studies will be critical to optimizing redox-based therapy in RA.

9.4 Opportunities in Predictive, Preventive, and Personalized Rheumatology

The evolving field of precision medicine offers transformative opportunities to reshape rheumatoid arthritis (RA) care into a predictive, preventive, and personalized model. Oxidative stress (OS) biomarkers are emerging as a critical component of this transition, given their mechanistic relevance and measurable alterations in preclinical and clinical RA stages. Predictive strategies increasingly leverage OS-related signatures to identify high-risk individuals before overt clinical onset. A

2025 study by Wang et al. developed a diagnostic model using mitochondrial oxidative stress-associated genes (CDKN1A, GADD45B, MAFF) to stratify RA risk and characterize immune infiltration, providing a feasible approach for early detection and tailored surveillance [143]. Preventive medicine in RA also draws from OS biology. Ruscitti et al. emphasized the value of predictive testing in first-degree relatives of RA patients to initiate ultra-early lifestyle and antioxidant interventions. Genetic susceptibility and redox imbalance may help guide risk reduction efforts before inflammation becomes clinically apparent [155]. Personalized rheumatology, where interventions are adapted to molecular and oxidative profiles, is facilitated by advances in biomarker discovery. Multi-omics tools (Sect. 9.2) have shown promise in capturing individual redox dynamics, while algorithms integrating clinical, serologic, and oxidative markers have enhanced the prediction of flares and treatment nonresponse [149]. Mahler et al. highlighted how stratified intervention trials—based on autoantibodies, genetics, lifestyle, and potentially oxidative parameters—could refine the clinical trial design and improve the chances of successful RA prevention. They stressed that combining modifiable and non-modifiable risk factors, including OS markers, will be key to personalized preventive strategies [156]. Furthermore, patient acceptability of predictive testing is growing. In a multinational study, Mosor et al. reported that individuals at risk of RA were willing to adopt preventive measures if provided with accurate predictive biomarker information. This underscores the importance of communicating OS-related risk in a patient-centric manner [157]. Integrating OS biomarkers into predictive frameworks may allow earlier diagnosis, enable preventive interventions, and improve individualized treatment strategies, ultimately altering the trajectory of RA at both individual and population levels.

10 Conclusion

10.1 Summary of Key Insights

The accumulated research on oxidative stress (OS) in rheumatoid arthritis (RA) underscores its pivotal role in the disease's onset, progression, and therapeutic response. Across experimental, clinical, and omics-based studies, OS emerges as a byproduct of inflammation and a central driver of immune dysregulation, tissue damage, and systemic manifestations. The pathophysiological mechanisms of ROS overproduction in RA were outlined, detailing mitochondrial dysfunction, synovial oxidative injury, and the crosstalk between immune cells and redox pathways. ROS-induced damage to lipids, proteins, and DNA reinforces a pro-inflammatory environment that perpetuates joint erosion and systemic complications [4, 20, 39, 55, 158]. Oxidative markers correlate with disease severity, including DAS28 scores, radiological progression, and extra-articular involvement. Genetic and epigenetic predispositions affecting redox enzymes (e.g., SOD2, GPx, HIF1A) further explain patient-specific variability in OS burden and treatment response [48, 68, 79]. Therapeutic implications were outlined, showcasing conventional drugs, antioxidant supplementation, and herbal compounds with Nrf2-activating or ROS-scavenging properties. Both clinical and preclinical studies show that redox modulation can attenuate inflammation and improve outcomes, though challenges in standardization and evidence consistency remain [106, 121, 124]. The clinical relevance of oxidative biomarkers, emphasizing their utility in diagnosis, prognosis, and personalization of therapy, was further addressed. While several markers (e.g., MDA, HO-1, protein carbonyls) show strong correlations with disease activity, standardization and longitudinal validation are needed for clinical integration [114, 140, 143]. The methodological and translational gaps were sourced out, including variability in biomarker assays, the need for multicentric longitudinal studies, and limitations in integrating oxidative metrics into clinical workflows. These obstacles highlight the urgency of developing standardized panels and predictive algorithms. Future-focused strategies, such as multi-omics profiling, predictive prevention, and nanotechnology-based therapies, were addressed. The integration of OS signatures into precision medicine frameworks represents a promising direction for personalized rheumatology, with a clear potential to inform early diagnosis, risk stratification, and individualized intervention [145, 153, 155]. Overall, oxidative stress is not just a pathological consequence of RA but a modifiable axis central to understanding and managing the disease. Continued efforts to refine biomarker panels, harmonize detection protocols, and tailor therapies based on oxidative signatures will significantly advance research and patient care.

10.2 Importance of Addressing Oxidative Stress in RA Management

More importantly, OS has been implicated in extra-articular manifestations such as atherosclerosis, which significantly contribute to excess morbidity in RA [140]. Romero-Bueno demonstrated that redox imbalance is independently associated with premature vascular aging in RA, supporting the integration of redox management into cardiovascular risk mitigation strategies [159]. Recent studies also suggest that oxidative stress is a potential predictor of flare risk and drug response, furthering its value in stratified care. For example, MDA and antioxidant enzyme levels have been shown to fluctuate ahead of changes in clinical scores, offering a window for early therapeutic adjustment [114]. Moreover, the downstream effects of OS on genetic integrity and epigenetic modulation, including mutations in p53 and regulation via miRNAs, highlight the role of oxidative processes in disease perpetuation and therapy resistance [160]. Oxidative stress remains underrepresented in routine clinical practice despite its significance, partly due to variability in biomarker assay techniques and the absence of standardized reference ranges. Nevertheless, as discussed in Sect. 9, growing evidence supports the development of validated biomarker panels and redox modulation strategies in targeted therapy. Incorporating oxidative stress assessment into routine RA care can improve disease monitoring, optimize treatment selection, and address systemic risk. Recognizing OS not merely as a bystander but as a therapeutic target will enrich the current inflammation-centric management paradigm and align RA care with modern personalized medicine principles.

10.3 Directions for Ongoing Research and Innovation

Future research in oxidative stress (OS) and rheumatoid arthritis (RA) must bridge the gap between mechanistic understanding and translational applicability. While substantial progress has been made in characterizing oxidative biomarkers and elucidating ROS-mediated pathology, several priority areas remain for innovation and standardization. Firstly, there is a clear need for longitudinal studies that evaluate oxidative stress markers across different disease phases, from preclinical autoimmunity through flare and remission. As emphasized by Ahmed et al. and Quiñonez-Flores et al., most studies are cross-sectional, providing only a snapshot of redox balance without capturing its dynamic nature over time [140, 146]. Secondly, interventional studies targeting oxidative stress require a more rigorous design. While NAC, vitamin E, polyphenols, and Chinese herbal extracts have shown antioxidant benefits in small trials, few have undergone long-term evaluation in randomized, biomarker-guided settings [103]. Future trials must incorporate redox-based inclusion criteria, dose standardization, and composite oxidative endpoints to validate clinical utility. Third, integrating omics platforms—including redox proteomics, metabolomics, and transcriptomics—offers the potential to identify redox subtypes of RA. These tools can unveil patient-specific oxidative pathways, enabling molecular classification and stratification in line with precision medicine principles [145]. Fourth, the development of standardized oxidative biomarker panels remains essential. This includes defining validated assays, establishing normative ranges, and integrating these into existing disease activity algorithms (e.g., alongside DAS28, CRP, and anti-CCP). Tools such as multi-analyte panels and redox scoring indices are in the early stages of development but hold promise for widespread clinical deployment [142]. Finally, future innovation must also explore neuroimmune-redox interactions, given emerging evidence linking oxidative pathways with fatigue, depression, and cognitive dysfunction in RA. This underexplored area could pave the way for holistic, symptom-targeted therapies that address quality-of-life outcomes beyond joint damage. Altogether, these directions signal a shift toward multidisciplinary, biomarker-driven research that positions oxidative stress as a disease mediator and a tangible target for intervention, monitoring, and personalization of care in RA.

Acknowledgement The authors are thankful to the extensive support by DST-FIST (SR/FST/College-235/2014), DST-FIST (SR/FST/College/2023/1423) for their support.

References

1. Zamudio-Cuevas, Y., et al. (2022). Rheumatoid arthritis and oxidative stress: A review of a decade. *Cellular and Molecular Biology, 68*(6), 174–184. https://doi.org/10.14715/cmb/2022.68.6.28
2. Ahmed, A. B., Chetia, P., & Kalita, H. (2024). A systematic review on association of oxidative stress in rheumatoid arthritis. *Indian Journal of Pharmaceutical Education and Research, 58*(3), 709–721. https://doi.org/10.5530/ijper.58.3.89
3. Baasu, S., et al. (2020). Evaluation of oxidative stress biomarkers in rheumatoid arthritis: Potential diagnostic markers. *Journal of Medical Biochemistry, 39*(3), 300–308.
4. Duarte, G., et al. (2021). Oxidative stress and antioxidant profile in rheumatoid arthritis. *Clinical Rheumatology, 40*(4), 1503–1511.
5. Del Buono, M., Abbate, A., & Toldo, S. (2018). Interplay of inflammation, oxidative stress and cardiovascular disease in rheumatoid arthritis. *Heart*. https://doi.org/10.1136/heartjnl-2018-313313
6. Smolen, J. S., Aletaha, D., & McInnes, I. B. (2016). Rheumatoid arthritis. *Lancet, 388*(10055), 2023–2038. https://doi.org/10.1016/S0140-6736(16)30173-8
7. Firestein, G. S., & McInnes, I. B. (2017). Immunopathogenesis of rheumatoid arthritis. *Immunity, 46*(2), 183–196. https://doi.org/10.1016/j.immuni.2017.02.006
8. Meune, C., Touzé, E., Trinquart, L., et al. (2009). Trends in cardiovascular mortality in patients with rheumatoid arthritis: a meta-analysis. *Rheumatology, 48*(10), 1309–1313. https://doi.org/10.1093/rheumatology/kep252
9. Pironti, G., et al. (2018). Cardiomyopathy, oxidative stress and impaired contractility in a rheumatoid arthritis mouse model. *Heart*.
10. Ikonomidis, I., et al. (2008). IL-1 inhibition improves vascular and ventricular function in RA. *Circulation, 117*, 2662–2669.
11. Hussain, Q., et al. (2020). Cadmium and oxidative stress in rheumatoid arthritis. *Toxicology Mechanisms and Methods, 30*(7), 527–535. https://doi.org/10.1080/15376516.2020.1784425
12. Al-Jawadi, W. A., et al. (2021). Lipid peroxidation and antioxidant activity in rheumatoid arthritis patients. *Journal of Clinical and Diagnostic Research, 15*(3), BC01–BC04. https://doi.org/10.7860/JCDR/2021/47625.14676
13. Nawaz, H., et al. (2021). Comparative study of antioxidant enzyme activity in rheumatoid arthritis. *Pakistan Journal of Pharmaceutical Sciences, 34*(4), 1601–1607.
14. Tiwari, V., et al. (2020). Neurocognitive deficits and oxidative stress in rheumatoid arthritis. *Neurobiology of Diseases, 143*, 104996.
15. Quiñonez-Flores, C. M., et al. (2016). Oxidative stress relevance in the pathogenesis of rheumatoid arthritis. *BioMed Research International., 2016*, 6097417. https://doi.org/10.1155/2016/6097417
16. Yoo, S. J., et al. (2016). Roles of reactive oxygen species in rheumatoid arthritis pathogenesis. *Journal of Rheumatic Diseases, 23*(6), 340–347. https://doi.org/10.4078/jrd.2016.23.6.340
17. Tasset, I., et al. (2019). Redox regulation of inflammation in rheumatoid arthritis. *Clinical and Experimental Immunology, 197*(3), 204–214.
18. Lee, S. Y., et al. (2011). Oxidative DNA damage in CD4+ T cells linked to inflammation in RA. *Arthritis Research & Therapy, 13*(6), R211.
19. Hitchon, C. A., & El-Gabalawy, H. S. (2004). Oxidation in rheumatoid arthritis. *Arthritis Research & Therapy, 6*(6), 265–278. https://doi.org/10.1186/ar1447
20. Mukhopadhyay, K., et al. (2021). Evaluation of oxidative stress in patients with rheumatoid arthritis. *Journal of Family Medicine and Primary Care, 10*(5), 1981–1986. https://doi.org/10.4103/jfmpc.jfmpc_2412_20
21. Nakajima, A., et al. (2014). Serum ROM correlates with CRP and disease activity in RA. *Modern Rheumatology, 24*(1), 1–5.
22. Balogh, E., et al. (2018). Oxidative stress impairs energy metabolism in synovial tissue. *Arthritis Research & Therapy, 20*, 95. https://doi.org/10.1186/s13075-018-1592-1

23. Szabo-Taylor, K. E. et al. (2013). Oxidative stress in rheumatoid arthritis. In *Studies on arthritis and joint disorders* (pp. 145–159). Springer. https://doi.org/10.1007/978-1-4614-6166-1_8
24. Phull, A. R., et al. (2018). Oxidative stress and ROS-mediated signaling in RA. *Chemico-Biological Interactions, 281*, 121–136. https://doi.org/10.1016/j.cbi.2017.12.024
25. Fonseca, L. J. S., et al. (2019). Oxidative stress in rheumatoid arthritis. *Oxidative Medicine and Cellular Longevity, 2019*, 7536805. https://doi.org/10.1155/2019/7536805
26. Desai, P. B., et al. (2010). Oxidative stress and enzymatic antioxidant status in RA. *European Review for Medical and Pharmacological Sciences, 14*(12), 959–967. PMID: 21284345
27. Ahmed, A. B., et al. (2024). Systematic review of oxidative stress in RA. *Indian Journal of Pharmaceutical Education and Research, 58*(3), 709–721.
28. Mahdi, J. K., et al. (2019). Oxidative stress status in RA patients. *Thi-Qar Medical Journal, 17*(1), 531–538. https://doi.org/10.32792/jmed.v16i2.7
29. Bala, A., et al. (2017). Oxidative stress in inflammatory cells in RA. *Inflammopharmacology, 25*(5), 595–610. https://doi.org/10.1007/s10787-017-0397-1
30. Del Buono, M., et al. (2018). Inflammation and oxidative stress in RA. *Heart*. https://doi.org/10.1136/heartjnl-2018-313313
31. Datta, S., et al. (2014). Oxidant status and tissue damage in RA. *Clinical Rheumatology, 33*(11), 1557–1564. https://doi.org/10.1007/s10067-014-2597-z
32. Wang, X., et al. (2022). ROS in the RA synovial microenvironment. *Antioxidants, 11*(6). https://doi.org/10.3390/antiox11061153
33. Bauerová, K, & Bezek, Å. (1999). ROS/RNS in RA pathogenesis. *General Physiology and Biophysics, 18*(Spec No), 15–20. PMID: 10703714
34. Tiwari, V., et al. (2020). Oxidative stress and neurocognitive decline in RA. *Neurobiology of Diseases, 143*, 104996.
35. Al-Jawadi, W. A., et al. (2021). Antioxidant activity in RA patients. *Journal of Clinical and Diagnostic Research, 15*(3), BC01–BC04.
36. Mirshafiey, A., & Mohsenzadegan, M. (2008). The role of reactive oxygen species in immunopathogenesis of rheumatoid arthritis. *Iranian Journal of Allergy, Asthma, and Immunology, 7*(4), 195–202. PMID: 19052348
37. Wang, X., Fan, D., Cao, X., et al. (2022). The role of reactive oxygen species in the rheumatoid arthritis-associated synovial microenvironment. *Antioxidants, 11*(6). https://doi.org/10.3390/antiox11061153
38. Bauerová K, & Bezek Š. (1999). Role of reactive oxygen and nitrogen species in etiopathogenesis of rheumatoid arthritis. *General Physiology and Biophysics, 18*(Spec No), 15–20.
39. Martelli-Palomino, G., Paoliello-Paschoalato, A. B., Crispim, J., et al. (2016). DNA damage in peripheral neutrophils from patients with rheumatoid arthritis correlates with disease activity. *Clinical and Experimental Rheumatology, 34*(2), 247–254. https://doi.org/10.1007/s10067-014-2597-z
40. Xie, Z., Hou, H., Luo, D., et al. (2020). Ferroptosis and lipid peroxidation in rheumatoid arthritis: A potential therapeutic target. *Inflammation, 44*, 35–47. https://doi.org/10.1007/s10753-020-01338-2
41. Datta, S., Kundu, S., Ghosh, P., De, S., et al. (2014). Correlation of oxidant status with oxidative tissue damage in patients with rheumatoid arthritis. *Clinical Rheumatology, 33*(11), 1557–1564. https://doi.org/10.1007/s10067-014-2636-1
42. Phull, A.-R., Nasir, B., Haq, I. U., & Kim, S.-J. (2018). Oxidative stress, consequences and ROS mediated cellular signaling in rheumatoid arthritis. *Chemico-Biological Interactions, 281*, 121–136. https://doi.org/10.1016/j.cbi.2017.12.024
43. Ogawa, M., Matsuda, T., Ogata, A., et al. (2013). Increased formation of oxidative DNA adducts in rheumatoid arthritis: Implications for pathogenesis. *Autoimmunity Reviews, 12*(8), 800–805. https://doi.org/10.1016/j.autrev.2013.01.001
44. Mao, H., Xie, S., Yu, C., et al. (2025). Identification and integrated analysis of oxidative stress-related genes as potential biomarkers for diagnosis and treatment of rheumatoid arthritis using bioinformatics methods. *Research Square*. Preprint. https://doi.org/10.21203/rs.3.rs-6176622/v1

45. Hitchon, C. A., & El-Gabalawy, H. S. (2004). Oxidation in rheumatoid arthritis. *Arthritis Research & Therapy, 6*(6), 265–278. https://doi.org/10.1186/ar1447
46. Szabó-Taylor, K. É., Nagy, G., Eggleton, P., & Winyard, P. G. (2013). Oxidative stress in rheumatoid arthritis. In: M. J. Alcaraz et al. (Eds.), *Studies on arthritis and joint disorders*. Springer. https://doi.org/10.1007/978-1-4614-6166-1_8
47. Bala, A., Mondal, C., Haldar, P. K., & Khandelwal, B. (2017). Oxidative stress in inflammatory cells of patient with rheumatoid arthritis: Clinical efficacy of dietary antioxidants. *Inflammopharmacology, 25*(5), 595–604. https://doi.org/10.1007/s10787-017-0397-1
48. Mukhopadhyay, S., Sen, S., Majhi, B., Das, K. P., & Kar, M. (2007). Methyl glyoxal elevation is associated with oxidative stress in rheumatoid arthritis. *Free Radical Research, 41*(5), 507–514. https://doi.org/10.1080/10715760601148285
49. Navarro-Compán, V., Melguizo-Madrid, E., Hernández-Cruz, B., et al. (2013). Interaction between oxidative stress and smoking is associated with an increased risk of rheumatoid arthritis: A case–control study. *Rheumatology, 52*(3), 487–493. https://doi.org/10.1093/rheumatology/kes286
50. Steinz, M. M., Santos-Alves, E., & Lanner, J. T. (2020). Skeletal muscle redox signaling in rheumatoid arthritis. *Clinical Science, 134*(21), 2835–2850. https://doi.org/10.1042/CS20190728
51. Ibrahim, S., & Huttunen, K. (2021). Orchestrated modulation of rheumatoid arthritis via crosstalking intracellular signaling pathways. *Inflammopharmacology, 29*, 965–974. https://doi.org/10.1007/s10787-021-00800-3
52. López-Armada, M. J., Fernández-Rodríguez, J., & Blanco, F. (2022). Mitochondrial dysfunction and oxidative stress in rheumatoid arthritis. *Antioxidants, 11*(6), 1151. https://doi.org/10.3390/antiox11061151
53. Remans, P. H., Wijbrandts, C. A., Sanders, M. E., et al. (2006). CTLA-4Ig suppresses reactive oxygen species by preventing synovial adherent cell-induced inactivation of Rap1, a Ras family GTPase mediator of oxidative stress in rheumatoid arthritis T cells. *Arthritis & Rheumatism, 54*(10), 3135–3143. https://doi.org/10.1002/art.22139
54. Bagchi, A., Ghosh, P., Ghosh, A., & Chatterjee, M. (2022). Role of oxidative stress in induction of trans-differentiation of neutrophils in patients with rheumatoid arthritis. *Free Radical Research, 56*(4), 290–302. https://doi.org/10.1080/10715762.2022.2089567
55. Tanaka, Y. (2000). Crosstalking between chemokines and adhesion molecules in the pathogenesis of rheumatoid arthritis. *Japanese Journal of Inflammation, 20*, 107–117. https://doi.org/10.2492/JSIR1981.20.107
56. Abbasifard, A., Solgi, G., Khodadadi, I., et al. (2024). Peripheral blood mononuclear cells show markers of mitochondrial dysfunction in rheumatoid arthritis. *Cell Biochemistry and Function, 42*. https://doi.org/10.1002/cbf.4025
57. Al-Azab, M., Qaed, E., Ouyang, X., et al. (2020). TL1A/TNFR2-mediated mitochondrial dysfunction of fibroblast-like synoviocytes increases inflammatory response in patients with rheumatoid arthritis via reactive oxygen species generation. *The FEBS Journal, 287*. https://doi.org/10.1111/febs.15181
58. Promila, L., Sarkar, K., Guleria, S., et al. (2024). Mitochondrial calcium uniporter regulates human fibroblast-like synoviocytes invasion via altering mitochondrial dynamics and dictates rheumatoid arthritis pathogenesis. bioRxiv (preprint). https://doi.org/10.1101/2024.08.13.607857
59. Harty, L. C., Biniecka, M., Fox, E., et al. (2011). The mutation of mitochondrial DNA is central to the pathogenesis of rheumatoid and psoriatic arthritis. *Annals of the Rheumatic Diseases, 70*, A20. https://doi.org/10.1136/ard.2010.148965.18
60. Li, B., Yang, C., Guo, M., et al. (2022). Ultrasound-remote selected activation mitophagy for precise treatment of rheumatoid arthritis by two-dimensional piezoelectric nanosheets. *ACS Nano*. https://doi.org/10.1021/acsnano.2c09834
61. Nabih, G. A., El Sheshtawy, N. E., El Mikkawy, D. M. E., & Kamel, M. A. (2024). Serum malondialdehyde as a marker of oxidative stress in rheumatoid arthritis. *Egyptian Rheumatology and Rehabilitation, 51*(43). https://doi.org/10.1186/s43166-024-00275-4

62. Hirao, M., Yamasaki, N., Oze, H., et al. (2012). Serum level of oxidative stress marker is dramatically low in patients with rheumatoid arthritis treated with tocilizumab. *Rheumatology International, 32*, 4041–4045. https://doi.org/10.1007/s00296-011-2135-0
63. Zamudio-Cuevas, Y., Martínez-Flores, K., Martínez-Nava, G. A., et al. (2022). Rheumatoid arthritis and oxidative stress: a review of a decade. *Cellular and Molecular Biology, 68*(6), 174–184. https://doi.org/10.14715/cmb/2022.68.6.28
64. Mukhopadhyay, K., De, S., Kundu, S., et al. (2021). Evaluation of levels of oxidative stress as a potential biomarker in patients with rheumatoid arthritis. *Journal of Family Medicine and Primary Care, 10*(5), 1981–1986. https://doi.org/10.4103/jfmpc.jfmpc_2412_20
65. Mahmoud, A. A., & Ismail, M. A. (2011). Serum protein carbonyl content, total thiol and nitric oxide in patients with rheumatoid arthritis. *Journal of American Science, 7*(5), 683–686.
66. Lemaréchal, H., Allanore, Y., Chenevier-Gobeaux, C., et al. (2006). Serum protein oxidation in patients with rheumatoid arthritis and effects of infliximab therapy. *Clinica Chimica Acta, 372*(1–2), 147–153. https://doi.org/10.1016/J.CCA.2006.04.002
67. Choi, J. (2003). Nitric oxide production is increased in patients with rheumatoid arthritis but does not correlate with laboratory parameters of disease activity. *Clinica Chimica Acta, 336*(1–2), 83–87. https://doi.org/10.1016/S0009-8981(03)00324-3
68. Khallaf, H., Nosair, N., Alashkar, D., & Abdella, D. H. M. (2016). Assessment of serum nitrite level in patients with rheumatoid arthritis. *Tanta Medical Journal, 44*, 12–15. https://doi.org/10.4103/1110-1415.180544
69. Kaur, H., & Halliwell, B. (1994). Evidence for nitric oxide-mediated oxidative damage in chronic inflammation: Nitrotyrosine in serum and synovial fluid from rheumatoid patients. *FEBS Letters, 350*, 9–12. https://doi.org/10.1016/0014-5793(94)00722-5
70. Devecı, H., & Özmen, Z. (2020). Evaluation of serum nitric oxide level and its relationship with disease activity parameters in patients with rheumatoid arthritis. *Journal of Contemporary Medicine*. https://doi.org/10.16899/JCM.769098
71. Guzel, S., Seven, A., Güzel, E., & Hamuryudan, V. (2012). Nitric oxide and superoxide dismutase in rheumatoid arthritis: Correlation with disease activity. *Turkish Journal of Family Medicine and Primary Care, 6*(2), 84–88. https://doi.org/10.21763/TJFMPC.90457
72. Sandhu, J., Robertson, S., Birnboim, H., & Goldstein, R. (2003). Distribution of protein nitrotyrosine in synovial tissues of patients with rheumatoid arthritis and osteoarthritis. *The Journal of Rheumatology, 30*(6), 1173–1181.
73. Hilliquin, P., Borderie, D., Hernvann, A., Menkès, C., & Ekindjian, O. (1997). Nitric oxide as S-nitrosoproteins in rheumatoid arthritis. *Arthritis & Rheumatism, 40*(8), 1512–1517. https://doi.org/10.1002/ART.1780400820
74. Salman, S. A., & Al-Jumaily, R. (2024). Evaluation of some immunological and DNA damage parameters among patients with rheumatoid arthritis. *Iraqi Journal of Science, 65*(10). https://doi.org/10.24996/ijs.2024.65.10.9
75. Seven, A., Güzel, S., Aslan, M., & Hamuryudan, V. (2008). Lipid, protein, DNA oxidation and antioxidant status in rheumatoid arthritis. *Clinical Biochemistry, 41*(7–8), 538–543. https://doi.org/10.1016/j.clinbiochem.2008.01.029
76. Altindag, O., Karakoç, M., Koçyiğit, A., Çelik, H., & Soran, N. (2007). Increased DNA damage and oxidative stress in patients with rheumatoid arthritis. *Clinical Biochemistry, 40*(3–4), 167–171. https://doi.org/10.1016/j.clinbiochem.2006.10.006
77. Souliotis, V., Vlachogiannis, N., Pappa, M., Argyriou, A., & Sfikakis, P. (2019). DNA damage accumulation, defective chromatin organization and deficient DNA repair capacity in patients with rheumatoid arthritis. *Clinical Immunology, 203*, 28–36. https://doi.org/10.1016/j.clim.2019.03.009
78. Martelli-Palomino, G., Paoliello-Paschoalato, A. B., Crispim, J., et al. (2016). DNA damage increase in peripheral neutrophils from patients with rheumatoid arthritis is associated with disease activity and the presence of shared epitope. *Clinical and Experimental Rheumatology, 35*(2), 247–254.

79. Ogawa, M., Matsuda, T., Ogata, A., et al. (2013). DNA damage in rheumatoid arthritis: An age-dependent increase in the lipid peroxidation-derived DNA adduct, heptanone-etheno-2′-deoxycytidine. *Autoimmune Diseases, 2013* (Article ID 183487). https://doi.org/10.1155/2013/183487
80. Pappa, M., Vlachogiannis, N., Argyriou, A., Souliotis, V., & Sfikakis, P. (2019). SAT0058: deficient DNA damage response and repair in active rheumatoid arthritis and the effect of treatment. *Annals of the Rheumatic Diseases, 78*(Suppl 2), 1093. https://doi.org/10.1136/annrheumdis-2019-eular.5129
81. Taha, S. I., Samaan, S. F., Hawash, S. S., et al. (2025). Circulating cell-free DNA as a potential biomarker for prediction of disease activity and prognosis among Egyptian rheumatoid arthritis patients. *International Journal of Immunopathology and Pharmacology, 39*. https://doi.org/10.1177/03946320251315036
82. Andersen, B., Greisen, S., Stengaard-Pedersen, K., et al. (2024). Plasma haem oxygenase-1 may represent a first-in-class biomarker of oxidative stress in rheumatoid arthritis. *Scandinavian Journal of Rheumatology, 53*(1), 1–5. https://doi.org/10.1080/03009742.2024.2392364
83. Bassu, S., Zinellu, A., Sotgia, S., et al. (2020). Oxidative stress biomarkers and peripheral endothelial dysfunction in rheumatoid arthritis: A monocentric cross-sectional case-control study. *Molecules, 25*(17). https://doi.org/10.3390/molecules25173855
84. Duarte, G. B. S., Callou, K. R. A., Almondes, K., et al. (2021). Evaluation of biomarkers related to zinc nutritional status, antioxidant activity and oxidative stress in rheumatoid arthritis patients. *Nutrition and Health, 28*(3), 257–264. https://doi.org/10.1177/02601060211015594
85. Datta, S., Kundu, S., Ghosh, P., et al. (2014). Correlation of oxidant status with oxidative tissue damage in patients with rheumatoid arthritis. *Clinical Rheumatology, 33*, 1557–1564. https://doi.org/10.1007/s10067-014-2597-z
86. Mukhopadhyay, K., De, S., Kundu, S., et al. (2021). Evaluation of levels of oxidative stress as a potential biomarker in patients with rheumatoid arthritis. *Journal of Family Medicine and Primary Care, 10*, 1981–1986. https://doi.org/10.4103/jfmpc.jfmpc_2412_20
87. Pradhan, A., Bagchi, A., De, S., et al. (2019). Role of redox imbalance and cytokines in mediating oxidative damage and disease progression of patients with rheumatoid arthritis. *Free Radical Research, 53*(7), 768–779. https://doi.org/10.1080/10715762.2019.1629586
88. Hosi, S. S. (2024). Investigating the role of oxidative stress and immune response in rheumatoid arthritis. *International Journal of Medical Toxicology and Legal Medicine, 27*(3 Suppl). https://doi.org/10.47059/ijmtlm/v27i3s/771/530
89. Wruck, C., Fragoulis, A., Gurzynski, A., et al. (2010). Role of oxidative stress in rheumatoid arthritis: Insights from the Nrf2-knockout mice. *Annals of the Rheumatic Diseases, 70*, 844–850. https://doi.org/10.1136/ard.2010.132720
90. Kizaki, K., Yoshizumi, Y., Takahashi, T., et al. (2015). Elevated oxidative stress monitored via the albumin-thiol redox state is correlated with matrix metalloproteinase-3 elevation in patients with rheumatoid arthritis. *Clinical Laboratory, 61*(1–2), 175–178. https://doi.org/10.7754/CLIN.LAB.2014.140822
91. Yaman, D., Bulut, D. G., Ustaoğlu, G., et al. (2021). Dental and temporomandibular joint alterations in rheumatoid arthritis patients and their association with salivary oxidative stress. *Turkish Journal of Medical Sciences, 51*, 2073–2080. https://doi.org/10.3906/sag-2102-260
92. Figus, F., Piga, M., Azzolin, I., McConnell, R., & Iagnocco, A. (2021). Rheumatoid arthritis: Extra-articular manifestations and comorbidities. *Autoimmunity Reviews, 20*(5), 102776. https://doi.org/10.1016/j.autrev.2021.102776
93. Turesson, C., McClelland, R. L., Christianson, T. J. H., & Matteson, E. L. (2006). Severe extra-articular disease manifestations are associated with an increased risk of first ever cardiovascular events in patients with rheumatoid arthritis. *Annals of the Rheumatic Diseases, 66*, 70–75. https://doi.org/10.1136/ard.2006.052506
94. Ruscitti, P., Conforti, A., Di Muzio, C., et al. (2022). Cardiometabolic comorbidities may identify a more severe subset of rheumatoid arthritis, results from a "real-life" study. *Annals of the Rheumatic Diseases*. https://doi.org/10.1136/annrheumdis-2022-eular.1665

95. Giles, J. T. (2019). Extra-articular manifestations and comorbidity in rheumatoid arthritis: Potential impact of pre-rheumatoid arthritis prevention. *Clinical Therapeutics, 41*(8), 1402–1416. https://doi.org/10.1016/j.clinthera.2019.04.018
96. Moosavian, S. P., Paknahad, Z., & Habibagahi, Z. (2020). A randomized, double-blind, placebo-controlled clinical trial evaluating the garlic supplement effects on some serum biomarkers of oxidative stress and quality of life in women with rheumatoid arthritis. *International Journal of Clinical Practice, 74*(10), e13498. https://doi.org/10.1111/ijcp.13498
97. Surowiec, I., Gjesdal, C., Jonsson, G., et al. (2016). Metabolomics study of fatigue in patients with rheumatoid arthritis naïve to biological treatment. *Rheumatology International, 36*, 703–711. https://doi.org/10.1007/s00296-016-3426-2
98. Rongen-van Dartel, S. A. A., Repping-Wuts, H., Donders, R., et al. (2016). A multidimensional 'path analysis' model of factors explaining fatigue in rheumatoid arthritis. *Clinical and Experimental Rheumatology, 34*(2), 200–206.
99. Garip, Y., Eser, F., Aktekin, L., & Bodur, H. (2011). Fatigue in rheumatoid arthritis: Association with severity of pain, disease activity and functional status. *Acta Reumatológica Portuguesa, 36*(4), 364–369.
100. Bohanec Grabar, P., Logar, D., Tomšič, M., Rozman, B., & Dolžan, V. (2009). Genetic polymorphisms modifying oxidative stress are associated with disease activity in rheumatoid arthritis patients. *Disease Markers, 26*(1), 41–48. https://doi.org/10.3233/DMA-2009-0603
101. Davis, M. C., Zautra, A., Younger, J., et al. (2008). Chronic stress and regulation of cellular markers of inflammation in rheumatoid arthritis: Implications for fatigue. *Brain, Behavior, and Immunity, 22*(1), 24–32. https://doi.org/10.1016/j.bbi.2007.06.013
102. Carmona, F., Martín, J., & Javier Martín, J. (2013). Genetic component of oxidative stress in rheumatoid arthritis. In *Oxidative stress in applied basic research and clinical practice.* Springer. https://doi.org/10.1007/978-1-4614-6166-1_7
103. Tak, P. P., Zvaifler, N. J., Green, D. R., & Firestein, G. S. (2000). Rheumatoid arthritis and p53: How oxidative stress might alter the course of inflammatory diseases. *Immunology Today, 21*(2), 78–82. https://doi.org/10.1016/S0167-5699(99)01552-2
104. Macchioni, P., Nicoli, D., Casali, B., et al. (2007). The codon 72 polymorphic variants of p53 in Italian rheumatoid arthritis patients. *Clinical and Experimental Rheumatology, 25*(3), 416–421.
105. Olsson, L., Lindqvist, A. K., Källberg, H., et al. (2007). A case-control study of rheumatoid arthritis identifies an associated single nucleotide polymorphism in the NCF4 gene. *Arthritis Research & Therapy, 9*, R98. https://doi.org/10.1186/ar2299
106. Strietholt, S., Maurer, B., Peters, M. A., Pap, T., & Gay, S. (2008). Epigenetic modifications in rheumatoid arthritis. *Arthritis Research & Therapy, 10*(1), 219. https://doi.org/10.1186/ar2500
107. Ai, R., Laragione, T., Hammaker, D., et al. (2018). Comprehensive epigenetic landscape of rheumatoid arthritis fibroblast-like synoviocytes. *Nature Communications, 9*, 1921. https://doi.org/10.1038/s41467-018-04310-9
108. Payet, M., Dargai, F., Gasque, P., & Guillot, X. (2021). Epigenetic regulation (including micro-RNAs, DNA methylation and histone modifications) of rheumatoid arthritis: A systematic review. *International Journal of Molecular Sciences, 22*(22), 12170. https://doi.org/10.3390/ijms222212170
109. Angiolilli, C., Grabiec, A. M., Ferguson, B. S., et al. (2016). Inflammatory cytokines epigenetically regulate rheumatoid arthritis fibroblast-like synoviocyte activation by suppressing HDAC5 expression. *Annals of the Rheumatic Diseases, 75*(2), 430–438. https://doi.org/10.1136/annrheumdis-2014-205635
110. Miao, C., Yang, Y. Y., He, X., & Li, J. (2013). New advances of DNA methylation and histone modifications in rheumatoid arthritis, with special emphasis on MeCP2. *Cellular Signalling, 25*(4), 875–882. https://doi.org/10.1016/j.cellsig.2012.12.017
111. Klareskog, L., Padyukov, L., Lorentzen, J., & Alfredsson, L. (2006). Mechanisms of disease: Genetic susceptibility and environmental triggers in the development of rheumatoid arthritis. *Nature Clinical Practice Rheumatology, 2*, 425–433. https://doi.org/10.1038/ncprheum0249

112. Arleevskaya, M., Takha, E., Petrov, S., et al. (2022). Interplay of Environmental, Individual and Genetic Factors in Rheumatoid Arthritis Provocation. *International Journal of Molecular Sciences, 23*(15), 8140. https://doi.org/10.3390/ijms23158140
113. Takha, E., Larionova, R., Petrov, S. V., et al. (2022). Possible mechanism of the implementing the trigger role of air pollution in rheumatoid arthritis (preliminary data). *Hygiene and Sanitation, 101*(2), 139–145. https://doi.org/10.47470/0016-9900-2022-101-2-139-145
114. Quiñonez-Flores, C., González-Chávez, S., Del Río Nájera, D., & Pacheco-Tena, C. (2016). Oxidative stress relevance in the pathogenesis of the rheumatoid arthritis: A systematic review. *BioMed Research International, 2016*, 6097417. https://doi.org/10.1155/2016/6097417
115. Eman, N. T., Rafiq, S., Hayat, A., et al. (2024). Investigating the role of genetic predisposition in developing rheumatoid arthritis. *Innovative Research in Applied, Biological and Chemical Sciences, 69*. https://doi.org/10.62497/irabcs.2024.69
116. Premkuma, B. (2018). Antioxidant defense and disease activity in rheumatoid arthritis. *Research Journal of Pharmacy and Technology, 11*, 1810–1814. https://doi.org/10.5958/0974-360X.2018.00336.0
117. Saxena, R., Suneja, S., Sharma, D., et al. (2015). Cumulative effect of systemic inflammation and oxidative stress in 40 known cases of active rheumatoid arthritis. *International Journal of Research in Orthopaedics, 1*, 7–10. https://doi.org/10.18203/ISSN.2455-4510.INTJRESORTHOP20160341
118. Wang, X., Kong, Y., & Li, Z. (2024). Advantages of Chinese herbal medicine in treating rheumatoid arthritis: A focus on its anti-inflammatory and anti-oxidative effects. *Frontiers in Medicine, 11* (Article 1371461). https://doi.org/10.3389/fmed.2024.1371461
119. Pourhabibi-Zarandi, F., Rafraf, M., Zayeni, H., et al. (2024). The efficacy of curcumin supplementation on serum total antioxidant capacity, malondialdehyde, and disease activity in women with rheumatoid arthritis. *Phytotherapy Research*. https://doi.org/10.1002/ptr.8225
120. Jalili, M., Kolahi, S., Aref-Hosseini, S. R., et al. (2014). Beneficial role of antioxidants on clinical outcomes and erythrocyte antioxidant parameters in rheumatoid arthritis patients. *International Journal of Preventive Medicine, 5*, 835–840.
121. Qamar, N., John, P., & Bhatti, A. (2021). Emerging role of selenium in treatment of rheumatoid arthritis: An insight on its antioxidant properties. *Journal of Trace Elements in Medicine and Biology, 66*, 126737. https://doi.org/10.1016/j.jtemb.2021.126737
122. Zhou, X., Mi, J., & Liu, Z. (2022). Causal association of diet-derived circulating antioxidants with the risk of rheumatoid arthritis: A Mendelian randomization study. *Seminars in Arthritis and Rheumatism, 56*, 152079. https://doi.org/10.2139/ssrn.4116094
123. Pattison, D. J., & Winyard, P. G. (2008). Dietary antioxidants in inflammatory arthritis: Do they have any role in etiology or therapy? *Nature Clinical Practice Rheumatology, 4*(11), 590–596. https://doi.org/10.1038/ncprheum0920
124. Abdollahzad, H., Aghdashi, M., Asghari Jafarabadi, M., & Alipour, B. (2015). Effects of coenzyme Q10 supplementation on inflammatory cytokines (TNF-α, IL-6) and oxidative stress in rheumatoid arthritis patients: A randomized controlled trial. *Archives of Medical Research, 46*(7), 527–533. https://doi.org/10.1016/j.arcmed.2015.08.006
125. Zeng, L., Yu, G., Yang, K., et al. (2021). The efficacy of antioxidative stress therapy on oxidative stress levels in rheumatoid arthritis: A systematic review and meta-analysis. *Oxidative Medicine and Cellular Longevity, 2021*, 3302886. https://doi.org/10.1155/2021/3302886
126. Hamidi, Z., Aryaeian, N., Abolghasemi, J., et al. (2020). The effect of saffron supplement on clinical outcomes and metabolic profiles in patients with active rheumatoid arthritis. *Phytotherapy Research, 34*, 1650–1658. https://doi.org/10.1002/ptr.6633
127. Wang, J., Tao, C., Xu, G., et al. (2023). A Q-marker screening strategy based on ADME studies and systems biology for Chinese herbal medicine, taking Qianghuo Shengshi decoction in treating rheumatoid arthritis as an example. *Molecular Omics*. https://doi.org/10.1039/d3mo00029j
128. Nelson, J., Sjöblom, H., Gjertsson, I., et al. (2020). Do interventions with diet or dietary supplements reduce the disease activity score in rheumatoid arthritis? A systematic review of randomized controlled trials. *Nutrients, 12*(10), 2991. https://doi.org/10.3390/nu12102991

129. Zhang, L., Yan, Y., He, J., et al. (2022). Epimedii herba: An ancient Chinese herbal medicine in the prevention and treatment of rheumatoid arthritis. *Frontiers in Chemistry, 10* (Article 1023779). https://doi.org/10.3389/fchem.2022.1023779
130. Nanjundaiah, S., Lee, D., Ma, Z., et al. (2012). Traditional Chinese herbal formula HLXL controls ongoing arthritis by regulating immune mediators of leukocyte trafficking and joint damage. *BMC Complementary and Alternative Medicine, 12*(S1), P26. https://doi.org/10.1186/1472-6882-12-S1-P26
131. Lai, N., Livneh, H., Fan, Y. H., et al. (2019). Use of Chinese herbal medicines by rheumatoid arthritis patients was associated with lower risk of stroke: A retrospective cohort study. *Complementary Therapies in Medicine, 45*, 124–129. https://doi.org/10.1016/j.ctim.2019.05.029
132. Saha, S. (2024). An overview of therapeutic targeting of Nrf2 signaling pathway in rheumatoid arthritis. *ACS Omega, 9*, 10049–10057. https://doi.org/10.1021/acsomega.4c00163
133. Huang, J. B., Chen, Z. R., Yang, S. L., & Hong, F. F. (2023). Nitric oxide synthases in rheumatoid arthritis. *Molecules, 28*(11), 4414. https://doi.org/10.3390/molecules28114414
134. Lv, X., Huang, J., & Wang, H. (2021). MiR-30a-3p ameliorates oxidative stress in rheumatoid arthritis synovial fibroblasts via activation of Nrf2-ARE signaling pathway. *Immunology Letters, 234*, 16–24. https://doi.org/10.1016/j.imlet.2021.01.004
135. Hung, S. Y., Chen, J. L., Tu, Y. K., et al. (2023). Isoliquiritigenin inhibits apoptosis and ameliorates oxidative stress in rheumatoid arthritis chondrocytes through the Nrf2/HO-1-mediated pathway. *Biomedicine & Pharmacotherapy, 170*, 116006. https://doi.org/10.1016/j.biopha.2023.116006
136. Corvo, M. L., Boerman, O., Oyen, W., et al. (1999). Intravenous administration of superoxide dismutase entrapped in long circulating liposomes. II. In vivo fate in a rat model of adjuvant arthritis. *Biochimica et Biophysica Acta 1419*(2), 325–334. https://doi.org/10.1016/S0005-2736(99)00081-4
137. Gioia, C., Lucchino, B., Tarsitano, M., et al. (2020). Dietary habits and nutrition in rheumatoid arthritis: Can diet influence disease development and clinical manifestations? *Nutrients, 12*(5), 1456. https://doi.org/10.3390/nu12051456
138. Dourado, E., Ferro, M., Guerreiro, C. S., & Fonseca, J. E. (2020). Diet as a modulator of intestinal microbiota in rheumatoid arthritis. *Nutrients, 12*(11), 3504. https://doi.org/10.3390/nu12113504
139. Papandreou, P., Gioxari, A., Daskalou, E., et al. (2023). Mediterranean diet and physical activity nudges versus usual care in women with rheumatoid arthritis: Results from the MADEIRA randomized controlled trial. *Nutrients, 15*(3), 676. https://doi.org/10.3390/nu15030676
140. Shekhar, K. V., Pathak, M. M., & Pisulkar, G. (2023). Diet and lifestyle impact on rheumatoid arthritis: A comprehensive review. *Cureus, 15*(11), e48625. https://doi.org/10.7759/cureus.48625
141. Ma, F., Li, Z., Liu, H., et al. (2024). Dietary-timing-induced gut microbiota diurnal oscillations modulate inflammatory rhythms in rheumatoid arthritis. *Cell Metabolism*. https://doi.org/10.1016/j.cmet.2024
142. Andersen, B., Greisen, S., Stengaard-Pedersen, K., et al. (2024). Plasma haem oxygenase-1 may represent a first-in-class biomarker of oxidative stress in rheumatoid arthritis. *Scandinavian Journal of Rheumatology, 1–5*. https://doi.org/10.1080/03009742.2024.2392364
143. Hermann, E. F., Kleibert, K. R. U., Nunes, P. L., et al. (2024). Assessment of oxidative stress biomarkers in patients before pharmacological treatment for rheumatoid arthritis. *Acta Scientiarum Health Sciences, 46*(1). https://doi.org/10.4025/actascihealthsci.v46i1.65536
144. Mititelu, R. R., Pădureanu, R., Bacanoiu, M., et al. (2020). Inflammatory and oxidative stress markers—Mirror tools in rheumatoid arthritis. *Biomedicines, 8*(5), 125. https://doi.org/10.3390/biomedicines8050125
145. Ahmed, A.B., Chetia, P., & Kalita, H. (2024). A systematic review on association of oxidative stress in rheumatoid arthritis based on cross-sectional case-control studies. *Indian Journal of Pharmaceutical Education and Research, 58*(3). https://doi.org/10.5530/ijper.58.3.79

146. Wang, D., Li, Q., Diao, X., & Wang, Q. (2025). Mitochondrial oxidative stress-related diagnostic model accurately assesses rheumatoid arthritis risk stratification and immune infiltration characterization. *Biotechnology Journal, 20*(2), e202400615. https://doi.org/10.1002/biot.202400615
147. Mullins, A. L., van Rosendal, S. V., Briskey, D., Fassett, R., Wilson, G., & Coombes, J. (2013). Variability in oxidative stress biomarkers following a maximal exercise test. *Biomarkers, 18*(6), 446–454. https://doi.org/10.3109/1354750X.2013.810668
148. Mukhopadhyay, K., De, S., Kundu, S., Ghosh, P., Chatterjee, S., & Chatterjee, M. (2021). Evaluation of levels of oxidative stress as a potential biomarker in patients with rheumatoid arthritis. *Journal of Family Medicine and Primary Care, 10*(5), 1981–1986. https://doi.org/10.4103/jfmpc.jfmpc_2412_20
149. Puentes-Osorio, Y., Amariles, P., Calleja, M., et al. (2021). Potential clinical biomarkers in rheumatoid arthritis with an omic approach. *Autoimmunity Highlights, 12*(1). https://doi.org/10.1186/s13317-021-00152-6
150. Hui, J., He, D., Liu, C., et al. (2024). A large-scale multi-omics polygenic risk score analysis identified candidate risk locus associated with rheumatoid arthritis. *Joint Bone Spine, 91*, 105841. https://doi.org/10.1016/j.jbspin.2024.105841
151. Tasaki, S., Suzuki, K., Kassai, Y., et al. (2018). Multi-omics monitoring of drug response in rheumatoid arthritis in pursuit of molecular remission. *Nature Communications, 9*, 2755. https://doi.org/10.1038/s41467-018-05044-4
152. Priori, R., Scrivo, R., Brandt, J. E., et al. (2013). Metabolomics in rheumatic diseases: The potential of an emerging methodology for improved patient diagnosis, prognosis, and treatment efficacy. *Autoimmunity Reviews, 12*(10), 1022–1030. https://doi.org/10.1016/j.autrev.2013.04.002
153. Cassotta, M., Forbes-Hernández, T., Cianciosi, D., et al. (2021). Nutrition and rheumatoid arthritis in the 'Omics' era. *Nutrients, 13*(3), 763. https://doi.org/10.3390/nu13030763
154. Yang, B., & Shi, J. (2023). Ferrihydrite nanoparticles alleviate rheumatoid arthritis by nanocatalytic antioxidation and oxygenation. *Nano Letters*. https://doi.org/10.1021/acs.nanolett.3c02743
155. Hirao, M., Yamasaki, N., Oze, H., et al. (2011). Serum level of oxidative stress marker is dramatically low in patients with rheumatoid arthritis treated with tocilizumab. *Rheumatology International, 32*, 4041–4045. https://doi.org/10.1007/s00296-011-2135-0
156. Ruscitti, P., Nunziato, M., Caso, F., et al. (2024). Prevention of rheumatoid arthritis using a familial predictive medicine approach. *Autoimmunity Reviews*, 103653. https://doi.org/10.1016/j.autrev.2024.103653
157. Mahler, M., Martinez-Prat, L., Sparks, J., & Deane, K. (2020). Precision medicine in the care of rheumatoid arthritis: Focus on prediction and prevention of future clinically-apparent disease. *Autoimmunity Reviews, 19*(8), 102506. https://doi.org/10.1016/j.autrev.2020.102506
158. Mirshafiey, A., & Mohsenzadegan, M. (2008). ROS in immunopathogenesis of RA. *Iranian Journal of Allergy, Asthma, and Immunology, 7*(4), 195–202.
159. Mosor, E., Stoffer, M., Steiner, G., et al. (2017). "And suddenly you are a person at risk of developing rheumatoid arthritis!" Different perspectives of individuals on predictive testing—Results of an international qualitative interview study. *Annals of the Rheumatic Diseases, 76*, 1511. https://doi.org/10.1136/annrheumdis-2017-eular.4705
160. Romero-Bueno, F. (2013). Oxidative stress and premature atherosclerosis in rheumatoid arthritis. In *Advances in pathobiology and management of Paget's disease of bone* (pp. 169–181). https://doi.org/10.1007/978-1-4614-6166-1_9
161. Altindag, O., et al. (2007). Increased DNA damage and oxidative stress in RA. *Clinical Biochemistry, 40*(3–4), 167–171.
162. Hussain, W. F., & Al-Hashemi, W. (2018). Oxidative stress in Iraqi RA patients. *Journal of Al-Nahrain University Science, 21*(2), 1–6.
163. Wruck, C. J., et al. (2010). Role of oxidative stress in RA: Nrf2-knockout mice. *Annals of the Rheumatic Diseases, 70*, 844–850.

164. Bauerova, K., et al. (2015). Oxidative stress markers in adjuvant arthritis. *Autoimmunity, 48*(1), 46–56.
165. Veselinovic, M., et al. (2014). Oxidative stress and disease activity in RA. *Molecular and Cellular Biochemistry, 391*(1–2), 225–232.
166. da Fonseca, L. J. S., Nunes-Souza, V., Goulart, M., & Rabelo, L. (2019). Oxidative stress in rheumatoid arthritis: what the future might hold regarding novel biomarkers and add-on therapies. *Oxidative Medicine and Cellular Longevity, 2019*, 7536805. https://doi.org/10.1155/2019/7536805

An Overview of Natural Antioxidant Pharmacodynamics and Pharmacokinetics

Sivakumar S. Moni

Abstract Oxidative stress is a key factor in chronic diseases such as cancer, cardiovascular, neurodegenerative and metabolic disorders such as diabetes. Antioxidants, whether endogenous or derived from natural sources, neutralize Reactive oxygen species (ROS), mitigate cell damage, and promote health. This review examines the role of natural antioxidants, including phenolic compounds, vitamins, carotenoids, and omega-3 fatty acids, in preventing and treating disease. Phenolic compounds such as flavonoids and anthocyanins have significant antioxidant and anti-inflammatory potential and modulate various signaling pathways. Vitamins A, C, and E have unique antioxidant mechanisms that improve immune function and reduce oxidative damage. Carotenoids such as β-carotene and lycopene offer protective effects against various chronic diseases. The eicosapentaenoic acid (EPA) and docosahexaenoic acid (DHA) significantly reduce inflammation and oxidative damage. Despite the promising benefits, limitations in bioavailability and stability hinder their therapeutic potential, so advances in delivery systems, such as nanotechnology-based formulations, are needed. By integrating natural antioxidants into clinical and dietary therapies, their potential to attenuate oxidative stress and manage chronic diseases opens up opportunities for innovative treatment strategies.

Keywords Oxidative stress · Antioxidants · Reactive oxygen species (ROS) · Reactive nitrogen species (RNS) · Natural antioxidants · Pharmacodynamics · Pharmacokinetic

1 Introduction

Oxidative stress is an imbalance between reactive oxygen species (ROS) production and the body's antioxidant defenses. It is closely linked to cellular respiration, the metabolic process that generates energy through ATP [1]. ROS are inevitably

S. S. Moni (✉)
Department of Pharmaceutics, College of Pharmacy and Health Research Centre, Jazan University, Jazan 45142, Kingdom of Saudi Arabia
e-mail: drsmsivakumar@gmail.com

K. Kasinathan and A. Muthukumaran (eds.), *Recent Advances in Oxidative Stress Associated Chronic Diseases Volume 2*, https://doi.org/10.1007/978-981-95-3758-7_13

produced as by-products, particularly during mitochondrial oxidative phosphorylation. Free radicals include the superoxide anion (O_2^-), produced during mitochondrial respiration and enzymatic reactions, and the hydroxyl radical (OH·), formed through processes like the Fenton reaction or radiation exposure [2, 3]. The superoxide anion is moderately reactive, serving as a signaling molecule and a precursor to more harmful ROS, such as hydrogen peroxide. The hydroxyl radical, the most reactive ROS, causes severe damage to DNA, proteins, and lipids. While free radicals are vital for cellular signaling, immune defense, and tissue repair, their excessive accumulation disrupts cellular integrity, leading to DNA mutations, lipid peroxidation, and protein misfolding [4, 5]. It is implicated in diseases such as cancer, cardiovascular disorders, neurodegenerative conditions, and aging. ROS support essential physiological roles in a controlled state, but various modern factors exacerbate oxidative stress, tipping the balance toward cellular damage and disease [6]. Antioxidants prevent oxidation, a chemical reaction that produces free radicals that can damage cells, proteins and DNA [7]. This damage, known as oxidative stress, is associated with aging and various chronic diseases as shown in Fig. 1. Antioxidants counteract oxidative stress by neutralizing free radicals, preventing cell damage and promoting overall health [8–12].

2 Natural Antioxidants

Natural antioxidants are bioactive substances found in plants, fruits, vegetables and other natural sources that help neutralize harmful free radicals in the body [13]. The reactive oxygen species (ROS) and reactive nitrogen species (RNS) primary source for oxidative stress and damage DNA, proteins and lipids [14, 15]. Oxidative stress is associated with aging, cancer, cardiovascular disease, and neurodegenerative disorders. Natural antioxidants prevent oxidative damage, modulate critical biological signaling pathways, and boost the immune system and promote overall health as shown in Fig. 2 [16]. Phenolic compounds are a diverse group of phytochemicals characterized by one or more hydroxyl groups attached to aromatic rings. These compounds, abundant in plants, are considered some of the most effective natural antioxidants. They include subgroups such as flavonoids, phenolic acids, tannins, lignans, and stilbenes, each with unique structural properties and antioxidant mechanisms. The antioxidant activity of phenolic compounds primarily neutralizes free radicals, thus interrupting the chain reactions of lipid peroxidation and other oxidative processes.

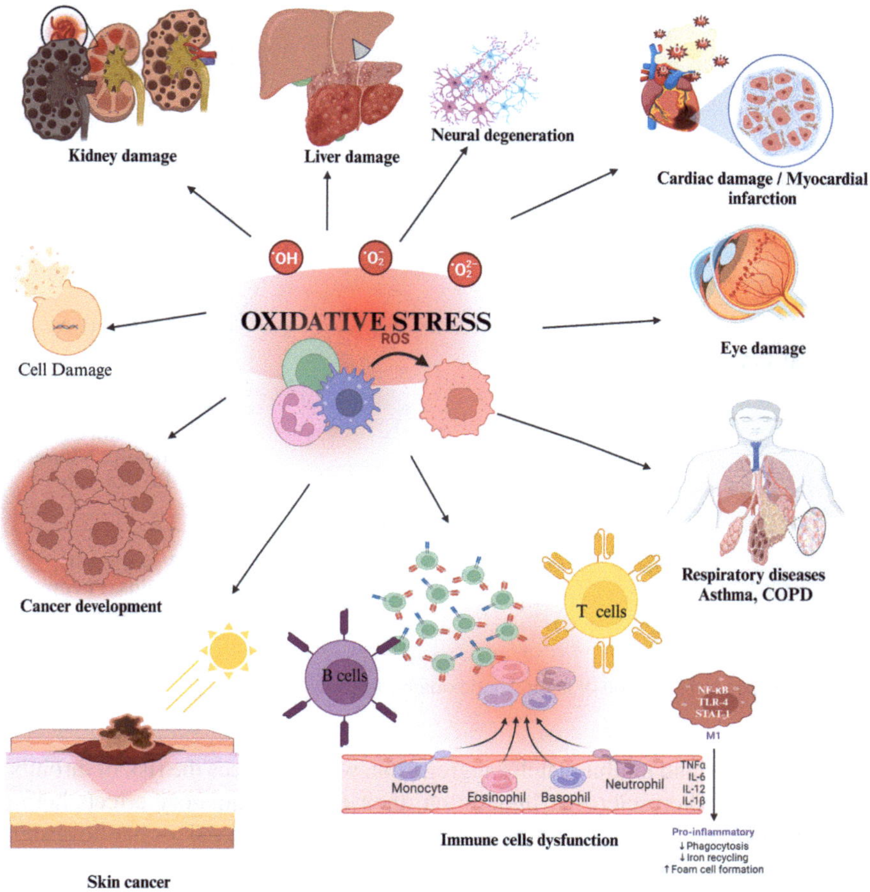

Fig. 1 Schematic representation of oxidative stress. This figure was created with BioRender.com, Bio Render, Canada

3 Classification of Natural Antioxidants

3.1 Phenolic Compounds

Phenolic compounds exert notable anti-inflammatory, anticancer, and cardioprotective effects by modulating key intracellular signaling pathways (Table 1). For example, they activate the nuclear factor erythroid 2-related factor 2 (Nrf2-), antioxidant response elements (AREs) signaling pathway, which enhances the production of superoxide dismutase (SOD), catalase and glutathione peroxidase [17, 18]. Phenolic acids such as caffeic acid and gallic acid, commonly found in coffee, tea and berries, are potent free radical scavengers with proven cardiovascular benefits. These acids reduce oxidative damage to endothelial cells, improve the lipid profile by lowering

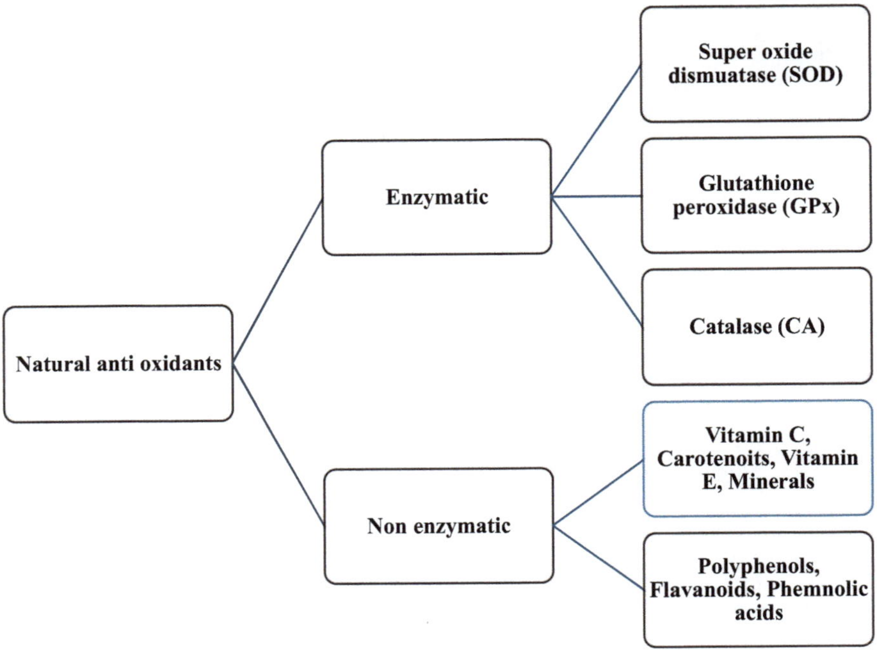

Fig. 2 Scheme representation of natural antioxidant categories

LDL cholesterol and inhibit platelet aggregation, which reduces the risk of thrombus formation and atherosclerosis. In addition, their anti-inflammatory effects help to lower markers such as CRP and NF-κB, further reducing the risk of cardiovascular disease [19, 20]. The neuroprotective effects of flavonoids and the cardioprotective properties of phenolic acids highlight their importance in dietary measures to combat diseases caused by oxidative stress, making foods such as berries, tea, and coffee essential components of a health-promoting diet. Despite their extensive biological activities, the bioavailability of phenolic compounds is often limited due to poor absorption, rapid metabolism and excretion. Besides, the metabolism of phenolic compounds by the gut microbiota can generate metabolites with different and sometimes enhanced biological activities that contribute to their systemic effects. These compounds have attracted considerable attention in developing functional foods, dietary supplements and pharmaceuticals designed to mitigate oxidative stress and improve human health. The multiple mechanisms and health benefits of phenolic compounds underscore their importance as natural antioxidants in preventive and therapeutic settings [21, 22].

Table 1 Classification of phenolic compounds as natural antioxidants: sources and functional properties

Class	Molecules	Source and properties	References
Simple phenols	Hydroquinone, resorcinol	1. Natural sources—hydroquinone and resorcinol are phenolic antioxidants found in vegetables such as potatoes, onions, broccoli, carrots, spinach, cabbage, peppers and tomatoes, and fruits such as berries, pears, apples, and grapes 2. Free radical scavengers—these compounds have a strong antioxidant effect by donating hydrogen atoms to unstable free radicals, preventing cell damage caused by oxidative stress 3. Metal chelation—hydroquinone and resorcinol have metal-chelating properties that help reduce metal-catalyzed oxidation reactions critical in lipid peroxidation and aging processes 4. Health benefits—due to their strong antioxidant potential, they contribute to cardiovascular protection, anti-inflammatory effects, and potentially neuroprotective properties, supporting overall well-being 5. Nutritional importance—regularly consuming fruits and vegetables rich in hydroquinone and resorcinol is essential for maintaining cellular health, preventing oxidative damage, and promoting longevity	[23–25]

(continued)

Table 1 (continued)

Class	Molecules	Source and properties	References
Phenolic acids	Ferulic acid, caffeic acid, and gallic acid	1. Natural sources—phenolic acids such as ferulic, caffeic, and gallic acids are abundant in whole grains, coffee, berries, and grapes and contribute to their potent antioxidant properties 2. Inhibition of lipid peroxidation—these compounds prevent lipid peroxidation, which is responsible for cell membrane damage, aging, and the progression of various chronic diseases 3. Neutralization of reactive oxygen species (ROS) and reducing inflammation-related diseases 4. Health benefits—ferulic acid improves skin protection, caffeic acid has neuroprotective effects, while gallic acid has antimicrobial, anti-cancer, and cardioprotective properties 5. Nutritional Importance—regularly consuming foods rich in phenolic acids strengthens the body's defenses against oxidative damage and promotes overall health and longevity	[26–28]
Flavonoids	Quercetin, catechins, rutin	1. Natural sources—flavonoids such as quercetin, catechins, and rutin are abundant in green tea, citrus fruits, onions, and berries and are essential to a healthy diet 2. Anti-inflammatory properties associated with cardiovascular disease, arthritis, and neurodegenerative disorders 3. Cell membrane protection—helping cell membranes from oxidative damage by stabilizing the lipid bilayers and preventing cellular destruction caused by free radicals 4. Health benefits—immune-boosting and antihistamine properties. Catechins (found in green tea) contribute to heart health and metabolic regulation, while rutin supports vascular health and strengthens capillaries 5. Nutritional importance—regular consumption of flavonoid-rich foods strengthens the body's defenses against oxidative stress, promotes cell longevity and reduces the risk of chronic disease	[28–31]

(continued)

Table 1 (continued)

Class	Molecules	Source and properties	References
Isoflavones	Genistein, daidzein	1. Natural sources—genistein and daidzein are mainly found in soybeans and legumes and essential to a plant-based diet 2. Reducing oxidative stress—isoflavones are essential in neutralizing free radicals, thereby reducing oxidative stress and protecting cells from damage that leads to aging and chronic disease 3. Health benefits—genistein has been linked to bone health, cardiovascular protection and cancer prevention, while daidzein improves metabolic health 4. Nutritional importance—regular consumption of isoflavone helps with hormonal balance, heart health, and cell protection, making it essential for overall well-being and longevity	[32–34]
Anthocyanins	Cyanidin, delphinidin	1. Natural sources—anthocyanins such as cyanidin and delphinidin are abundant in berries, red grapes, and eggplants and give these foods bright red, blue, and purple colors 2. Powerful antioxidant action—these compounds are potent antioxidants that effectively neutralize free radicals and reduce oxidative stress, contributing to aging and chronic disease 3. DNA protection—anthocyanins help protect DNA from damage and prevent mutations that can lead to cancer and other degenerative diseases 4. Health benefits—cyanidin supports cardiovascular health and has anti-inflammatory effects, while delphinidin is associated with neuroprotective properties and improved cognitive function 5. Nutritional Importance—regularly consuming anthocyanin-containing foods boosts cellular defense mechanisms, improves immune function, and promotes longevity and disease prevention	[35–38]

(continued)

Table 1 (continued)

Class	Molecules	Source and properties	References
Stilbenes	Resveratrol and piceatannol	1. Natural sources—stilbenes such as resveratrol and piceatannol occur naturally in grapes, red wine, peanuts, certain berries, and other edible plants 2. Anti-aging properties—these compounds are widely known for their anti-aging effects, helping slow cellular aging, and improving mitochondrial role 3. Health benefits—resveratrol is known for its cardioprotective and neuroprotective effects, while piceatannol contributes to cancer prevention 4. Nutritional importance—regular consumption of stilbene-rich foods supports heart health, cognitive function and overall longevity, making them an important part of a balanced and health-promoting diet	[39–41]

(continued)

Table 1 (continued)

Class	Molecules	Source and properties	References
Tannins	Ellagitannins, proanthocyanidins	1. Natural sources—tannins, including ellagitannins and proanthocyanidins, are abundant in tea, pomegranate, berries, and nuts and contribute to their astringent flavor and health benefits 2. Powerful antioxidant action—these compounds have free radical scavenging properties, protecting cells from oxidative stress and preventing lipid peroxidation 3. Metal chelation—tannins can bind and neutralize metal ions, reducing metal-catalyzed oxidation reactions that contribute to aging and cell damage 4. Antimicrobial properties—ellagitannins and proanthocyanidins have a strong antimicrobial activity that inhibits the growth of harmful bacteria and fungi, positively affecting gut and oral health 5. Nutritional Importance—regular consumption of tannin-rich foods supports antioxidant defense, cardiovascular health and microbial balance, contributing to overall health and longevity	[42, 43]
Coumarins	Scopoletin, umbelliferone	1. Natural sources—coumarins such as scopoletin and umbelliferone are found in citrus fruits, carrots, celery, and various herbs and medicinal plants 2. Anti-inflammatory properties—these compounds have a strong anti-inflammatory effect and help to reduce swelling, pain, and tissue damage associated with chronic inflammation 3. Health benefits—scopoletin is known for its immunomodulatory and neuroprotective properties, while umbelliferone plays a role in UV protection, antimicrobial defense, and liver health 4. Nutritional importance—regular consumption of foods rich in coumarin contributes to anti-aging, cardiovascular protection and improved metabolic health, making it a valuable part of a balanced diet	[44–46]

3.2 Flavonoids

Flavonoids are abundant in foods such as green tea, berries and dark chocolate and play a crucial role in scavenging free radicals. Flavonoids are highly regarded for their neuroprotective effects because they help reduce oxidative stress in Alzheimer's and Parkinson's [47]. Flavonoids modulate key cellular signaling pathways, such as the Nrf2 pathway that activates the production of superoxide dismutase (SOD) and glutathione peroxidase (GPx) [48, 49]. Subclasses of flavonoids such as quercetin found in apples and onions, anthocyanins found in berries and epigallocatechin-3-gallate (EGCG) found in green tea have been shown to have significant potential to reduce the accumulation of β-amyloid plaques, improve synaptic plasticity and prevent mitochondrial dysfunction, which are essential factors in the progression of neurodegenerative diseases [50]. The catechins, a flavonoid found in green tea, have been linked to reducing inflammation, supporting brain health, and improving cardiovascular function. In addition, the flavonoids in green tea can regulate blood pressure and improve lipid profiles, which promotes overall heart health [51, 52]. Berries help reduce inflammation and improve blood flow, benefiting brain and heart health. The flavanols in dark chocolate have been shown to improve cardiovascular functions. Regular consumption of flavonoid-rich foods as part of a healthy diet offers a natural approach to relieving inflammation, supporting brain function, and maintaining a robust cardiovascular system. The synergistic effect of these bioactive compounds underscores the importance of including flavonoid-rich foods in the daily diet to promote long-term health and well-being as shown in Table 2 [53–55]. Quercetin is a naturally occurring flavonoid widely distributed in plants and is known for its strong antioxidant, anti-inflammatory, and therapeutic properties. Quercetin is abundant in fruits such as apples and berries, vegetables such as onions and kale, and beverages such as green tea and red wine, and it is an essential part of a balanced diet. Its antioxidant effect is based on its ability to neutralize ROS, protect cellular structures from oxidative damage and chelate transition metals that catalyze the formation of free radicals. In addition, quercetin helps the SOD and GPx, contributing to cellular defense mechanisms. Beyond its antioxidant role, quercetin modulates the Nuclear Factor Kappa B (NF-κB) signaling pathway and suppresses histamine release [56–60]. These properties make it helpful in treating various chronic diseases. In addition, quercetin has been studied for its anti-cancer potential, as it can induce apoptosis, inhibit angiogenesis, and interrupt the proliferation of cancer cells. Due to its wide range of biological activities and safety profile, quercetin is often used in dietary supplements and functional foods to promote health and prevent oxidative stress-related diseases.

Table 2 Classification of vitamins as natural antioxidants: sources and functional properties

Class	Name	Source and properties	References
Fat-soluble vitamins	Vitamin A (retinol, beta-carotene)	1. Natural sources—vitamin A, including retinol and beta-carotene, is found in carrots, sweet potatoes, spinach, liver, and eggs and provides essential nutrients for overall health 2. Free radical protection—prevent cell aging and disease development 3. Supporting eye health—beta-carotene is crucial for maintaining good vision, protecting the retina from light-induced damage and reducing the risk of age-related macular degeneration 4. Improving the immune system 5. Nutritional importance—regular consumption of vitamin A-rich foods ensures optimal vision, a strong immune response and protection against oxidative damage, contributing to overall well-being	[83–85]
Fat-soluble vitamins	Vitamin D (cholecalciferol, ergocalciferol)	1. Natural sources—vitamin D, including cholecalciferol (D_3) and ergocalciferol (D_2), is derived from sunlight, fatty fish, egg yolk, and fortified dairy products and is crucial in maintaining overall health 2. Regulation of calcium metabolism—this vitamin is essential for calcium absorption and bone mineralization, helping to prevent osteoporosis and ensure strong skeletal development 3. Modulation of the immune system—vitamin D improves immune function by regulating inflammatory responses and supporting immune cell activity, thereby reducing autoimmunity 4. Nutritional and lifestyle importance—regular sun exposure and consuming vitamin D-rich foods are critical for maintaining bone strength, immune system resilience, and overall metabolic health	[86–88]

(continued)

Table 2 (continued)

Class	Name	Source and properties	References
Fat-soluble vitamins	Vitamin E (tocopherols, tocotrienols)	1. Natural sources—vitamin E, including tocopherols and tocotrienols, is abundant in nuts, seeds, vegetable oils and green leafy vegetables, making it an essential part of a balanced diet 2. Protecting cell membranes—vitamin E is critical in protecting cell membranes by acting as a potent lipid-soluble antioxidant that prevents lipid peroxidation. As a key component of cellular protection, it neutralizes reactive oxygen species (ROS) and stabilizes polyunsaturated fatty acids (PUFAs) in cell membranes, thereby preserving cell integrity and function. By reducing oxidative stress, vitamin E helps prevent membrane damage, inflammation and the progression of chronic disease, making it essential for overall health 3. Prevention of oxidative stress—preventing aging and various chronic diseases 4. Health benefits—tocopherols are known for their cardioprotective and neuroprotective effects, while tocotrienols have anti-inflammatory and anti-cancer properties 5. Nutritional importance—regular consumption of vitamin E-rich foods supports skin health, immune function and cardiovascular well-being, making it essential for longevity and disease prevention	[89–91]

(continued)

Table 2 (continued)

Class	Name	Source and properties	References
Fat-soluble vitamins	Vitamin K (phylloquinone, menaquinones)	1. Natural sources—vitamin K, including phylloquinone (K_1) and menaquinone (K_2), acts as an indirect antioxidant by reducing oxidative stress and preventing lipid peroxidation. Vitamin K is found in leafy vegetables (spinach, kale, broccoli), soybeans and fermented foods and is crucial in protecting blood vessels from oxidative damage, reducing inflammation and supporting cardiovascular health. It also helps maintain cellular redox balance by influencing the activity of antioxidant enzymes, contributing to overall cell protection and longevity 2. Regulation of blood clotting—helping to synthesize clotting factors that prevent excessive bleeding and support wound healing 3. Supporting bone health—vitamin K is essential for calcium metabolism, promoting bone mineralization and reducing the risk of osteoporosis and fractures 4. Cardiovascular benefits—menaquinones (K_2) help prevent vascular calcification, reduce arterial stiffness and support heart health by maintaining proper blood vessel function 5. Nutritional importance—regular consumption of vitamin K-rich foods is essential for healthy blood clotting, strong bones, and the protection of the cardiovascular system, contributing to overall well-being and longevity	[92, 93]

(continued)

Table 2 (continued)

Class	Name	Source and properties	References
Water-soluble vitamins	Vitamin C (ascorbic acid)	1. Natural Sources—Vitamin C is widely found in fresh fruits and vegetables, especially citrus fruits like oranges, lemons, and grapefruits, as well as kiwi, strawberries, guava, papaya, and mango. Vegetables such as broccoli, bell peppers, spinach, kale, and tomatoes are also excellent sources. 2. Scavenging free radicals and thus prevents cell damage caused by oxidative stress. By neutralizing ROS, proteins, lipids, and DNA from oxidative changes caused by environmental pollutants. This antioxidant effect reduces inflammation, supports immune defense and slows down the aging process, making vitamin C essential for maintaining cellular health and overall well-being 3. Collagen synthesis and skin health—vitamin C is vital for collagen production and supports skin elasticity, wound healing, joint health and connective tissue strength, contributing to overall structural integrity 4. Nutritional importance—provides antioxidant protection, vital for disease prevention and longevity	[94–96]
Water-soluble vitamins	Vitamin B2 (riboflavin)	1. Natural sources—vitamin B2 (Riboflavin) is found in dairy products, eggs, lean meat, and green vegetables and is essential to a nutrient-rich diet 2. Role in redox reactions—it is an important player in redox reactions and facilitates electron transfer in metabolic pathways that generate cellular energy (ATP). Supports oxidation–reduction reactions essential for mitochondrial function 3. Supports glutathione recycling—this vitamin contributes to the regeneration of glutathione, an important intracellular antioxidant that protects cells from oxidative stress and detoxifies harmful compounds 4. Health benefits—riboflavin contributes to healthy skin, vision, and nervous system function while reducing inflammation and oxidative damage 5. Nutritional importance—regular consumption of foods rich in vitamin B2 supports energy metabolism, antioxidant protection and overall cellular health, making it essential for optimal physiological function	[97–99]

(continued)

Table 2 (continued)

Class	Name	Source and properties	References
Water-soluble vitamins	Vitamin B3 (niacin, nicotinamide)	1. Natural sources—vitamin B3, including niacin and nicotinamide, is found in meat, fish, whole grains, and legumes and is essential to a balanced diet 2. Reducing oxidative damage—this vitamin is key in neutralizing free radicals and reducing oxidative stress, protecting cells from damage that contributes to aging and chronic disease 3. Improving energy metabolism—cellular respiration and supports sustained energy production 4. Health Benefits—niacin supports blood circulation and lowers cholesterol levels, while nicotinamide contributes to DNA repair and nerve protection 5. Nutritional importance—regular consumption of vitamin B3-rich foods promotes cellular energy efficiency, antioxidant protection, and age-related diseases	[100–102]

3.3 Anthocyanins

Anthocyanins, abundant in fruits, vegetables, and flowers, contribute to the visual appeal of many plants and their health-promoting potential. The antioxidant potential of anthocyanins results from reducing oxidative stress and preventing cell damage. They effectively neutralize ROS and RNS, preventing chain reactions that lead to lipid peroxidation, protein oxidation and DNA damage. In addition, anthocyanins are strong metal chelators that bind to pro-oxidant metals such as iron and copper, thus inhibiting the formation of highly reactive hydroxyl radicals via the Fenton and Haber–Weiss reactions. Together, these mechanisms protect biological systems from oxidative damage, allowing anthocyanins to effectively mitigate the progression of diseases caused by oxidative stress [35]. Anthocyanins have shown neuroprotective effects in studies on neurodegenerative diseases such as Alzheimer's and Parkinson's, primarily by protecting neurons from oxidative damage. They also contribute to cancer prevention by reducing oxidative DNA damage, inhibiting tumor cell proliferation, and inducing apoptosis. Their anti-inflammatory effect complements their antioxidant effect and provides additional protection against diseases such as diabetes, obesity, and metabolic syndrome [36, 37]. Despite their broad therapeutic potential, anthocyanins are hampered by issues related to stability and bioavailability. In addition, anthocyanins are extensively metabolized in the gastrointestinal tract, resulting in low systemic bioavailability. Their mechanisms of action improve their pharmacokinetic properties, and anthocyanins are emerging against oxidative stress-related diseases [61]. The therapeutic potential of flavonoids, despite their essential antioxidant and anti-inflammatory properties, is often influenced by their bioavailability and pharmacokinetics. Flavonoids generally have low bioavailability due to their poor solubility, extensive metabolism, and rapid elimination from the body. After ingestion, these compounds undergo enzymatic changes in the small intestine and are metabolized by the intestinal microbiota in the large intestine, producing various metabolites that may have different biological activities. Absorbed flavonoids undergo extensive first-pass metabolism in the liver, where they are converted to glucuronidated, sulfated, or methylated forms, which may affect their bioactivity and systemic distribution. These metabolic processes often lead to reduced bioavailability of the parent compounds, limiting their direct physiological effects [61, 62]. Improving the pharmacokinetic profile of flavonoids is a vital area of research to increase their efficacy. Factors such as the food matrix, co-consumption with other bioactive compounds and individual differences in the gut microbiota composition significantly influence the absorption and metabolism of flavonoids. For example, the catechins in green tea or the anthocyanins in berries may have improved bioavailability when consumed with dietary fats that promote their absorption. While their metabolites may contribute to their overall health benefits, the exact understanding of these effects requires further investigation. Despite these challenges, the widespread presence of flavonoids in the diet and their synergistic interaction with other nutrients underscore their importance in promoting health and mitigating diseases caused by oxidative stress [63, 64].

3.4 *Chalcones*

Chalcones are widely distributed in various plant species. This unique structural configuration, particularly the conjugated double bonds and hydroxyl groups on the aromatic rings, contributes significantly to their ability to neutralize ROS and RNS. Chalcones are abundant in various natural resources, including fruits, vegetables, spices and medicinal plants. Notable examples include isoliquiritigenin from licorice, xanthohumol from hops, and cardamonin from cardamom, all exhibiting vigorous antioxidant activity. Their natural occurrence and diverse structural modifications make chalcones versatile agents that neutralize oxidative damage and maintain cellular redox balance. The antioxidant effect of chalcones is based on their ability to act as electron donors, effectively scavenging free radicals and inhibiting lipid peroxidation. In addition, their phenolic hydroxyl groups help to chelate metal ions such as iron and copper, which catalyze the formation of hydroxyl radicals in biological systems. As natural antioxidants, chalcones are a promising starting point for developing novel interventions against diseases caused by oxidative stress [65, 66]. Despite their robust antioxidant and therapeutic properties, the bioavailability of chalcones is often limited by factors such as poor water solubility, rapid metabolism, and low systemic absorption. After ingestion, chalcones undergo phase I and phase II metabolism in the liver, including oxidation, glucuronidation, and sulfation, which can significantly reduce the bioactive concentration of the parent compound in the bloodstream. Additionally, chalcones are subject to degradation in the gastrointestinal tract, further limiting their absorption. These pharmacokinetic issues affect their effectiveness in reaching target tissues and maintaining therapeutic concentrations. However, the metabolites of chalcones may also have biological activity that contributes to their overall antioxidant and health benefits. Addressing these limitations through dietary formulations, co-administration with absorption enhancers, or chemical modifications could improve their pharmacokinetic profile and unlock their full therapeutic potential in combating diseases related to oxidative stress [67, 68].

3.5 *Resveratrol*

Resveratrol is in grapes, peanuts, and berries and is recognized for its remarkable antioxidant properties. It has gained much attention due to its potential role in promoting longevity and mitigating age-related diseases. Resveratrol protects cells and tissues from damage that contributes to aging by neutralizing free radicals and reducing oxidative stress. This antioxidant action is complemented by resveratrol's ability to activate sirtuins, particularly SIRT1, a group of proteins associated with cell health and lifespan extension. Through this mechanism, resveratrol improves DNA repair, supports mitochondrial function and delays cellular senescence, making it a promising compound in anti-aging research [69, 70]. In addition to its anti-aging effects, resveratrol is crucial in heart health and inflammation control. It supports cardiovascular wellness by improving endothelial function and reducing arterial stiffness and blood pressure. Resveratrol also inhibits the oxidation of LDL cholesterol, a critical step in the development of atherosclerosis, while promoting HDL cholesterol, which is known for its protective effect on the heart. Its anti-inflammatory properties further enhance cardiovascular benefits by lowering pro-inflammatory markers such as C-reactive protein and cytokines. These combined effects help to lower the risk of heart attacks and strokes. With regular and moderate consumption of foods or beverages containing resveratrol, individuals can benefit from its protective effects on the heart and its role in reducing inflammation, paving the way for healthier aging and overall well-being [71–73]. Innovations such as nanotechnology-based formulations, including nanoemulsions, liposomes, and polymeric nanoparticles, significantly improve resveratrol's bioavailability and therapeutic potential. These advanced delivery systems improve resveratrol's stability, absorption, and targeted delivery, and they address the limitations resulting from its rapid metabolism and poor solubility. By optimizing the pharmacokinetics of resveratrol, these innovations expand their therapeutic applications in the treatment of various diseases [74, 75].

3.6 Vitamins

Vitamins, essential micronutrients, play a central role as antioxidants in mitigating oxidative stress by neutralizing ROS and RNS, thus protecting cell structures and promoting overall health [76]. Among these vitamins, vitamins A, C, E and specific B-complex vitamins (especially B6, B9, and B12) are well-known for their antioxidant potential, each with different pharmacokinetic and pharmacodynamic properties [77]. Vitamin A, in its forms, such as beta-carotene, acts as scavenging singlet oxygen and peroxyl radicals while influencing gene expression and immune response. Vitamin A metabolites such as retinaldehyde and retinoic acid, which are absorbed in the small intestine and mainly stored in the liver, play a crucial role in cell differentiation and apoptosis [78]. Vitamin C, a water-soluble antioxidant, scavenges many free radicals and regenerates vitamin E, increasing its antioxidant capacity. It is absorbed via sodium-dependent vitamin C transporters, widely distributed in the tissue, metabolized to dehydroascorbic acid and excreted mainly via the urine, which requires regular intake due to its short plasma half-life [79, 80]. Vitamin E, consisting of tocopherols and tocotrienols, protects polyunsaturated fatty acids in cell membranes from lipid peroxidation and is absorbed via dietary fats, distributed via lipoproteins, metabolized in the liver and excreted as carboxy ethyl hydroxy chroman derivatives. Its antioxidant effect stabilizes cell membranes and modulates inflammatory reactions [81, 82].

The B-complex vitamins, including B6, B9, and B12, have a synergistic antioxidant effect by supporting mitochondrial function, lowering homocysteine levels and protecting DNA from oxidative damage. These vitamins are absorbed via specific transport mechanisms, stored in limited quantities (B12 mainly in the liver), and converted into active coenzymes such as pyridoxal phosphate, tetrahydrofolate, methylcobalamin and excreted in the urine [103, 104]. The pharmacokinetic profiles of these vitamins highlight their bioavailability, tissue distribution and metabolic activity, which influence their efficacy in preventing oxidative damage associated with various chronic diseases. Vitamins (A and E) have an extended half-life and storage capacity, allowing for sustained antioxidant action, while vitamins (C and B-complex) require a continuous supply for optimal plasma levels [77, 105]. Therapeutically, these vitamins are critical for reducing oxidative stress, boosting immune function, improving mitochondrial activity and attenuating age-related and chronic disease progression. While vitamin A supports vision and epithelial health, vitamin C promotes collagen synthesis and immune defense, vitamin E reduces lipid peroxidation in neurodegeneration, and B-complex vitamins combat oxidative stress and inflammation [106].

3.6.1 Vitamin C

Vitamin C, a crucial nutrient for maintaining health, is found abundantly in citrus fruits like oranges, lemons, grapefruits, strawberries, peppers, broccoli, and kiwis [107, 108]. Unlike most animals, humans cannot produce vitamin C without a specific enzyme, making dietary intake essential. It helps in preventing chronic conditions linked to oxidative damage, such as cancer, cardiovascular disease, and neurodegenerative disorders. Vitamin C enhances immune defense by improving the activity of white blood cells, including neutrophils, lymphocytes, and phagocytes, which combat infections [109]. It also supports antibody and cytokine production, ensuring a robust immune response. Regular intake has been shown to shorten the duration and severity of colds and respiratory infections, particularly for individuals under physical stress or in harsh environments. Its role in reducing inflammation is valuable in managing autoimmune diseases and chronic inflammatory conditions. Beyond its immune-boosting effects, vitamin C plays a vital role in skin health and cardiovascular protection. It is essential for collagen production and maintaining skin structure, blood vessels, tendons, and bones. By supporting collagen synthesis, vitamin C aids in wound healing and mitigates skin aging by improving elasticity and reducing wrinkles. Its antioxidant activity shields the skin from damage caused by ultraviolet (UV) rays and environmental pollutants. In cardiovascular health, vitamin C lowers blood pressure by relaxing blood vessels and reducing strain on the heart [110–112]. It also inhibits the oxidation of low-density lipoprotein (LDL) cholesterol, a critical factor in the development of arteriosclerosis. Studies suggest that higher levels of vitamin C are associated with a reduced risk of heart attacks and strokes, highlighting its importance for heart health and longevity. Vitamin C, enhancing immune function to promote healthy skin and protecting the heart, underscores its significance in a balanced diet. Its role in combating oxidative stress, supporting structural proteins like collagen, and preventing chronic diseases makes it indispensable for overall well-being. Including vitamin C-rich foods regularly can help maintain health, boost resilience to illnesses, and support long-term wellness [113].

3.6.2 Vitamin E

Vitamin E is an essential fat-soluble nutrient with strong antioxidant properties that help protect the body's cells from damage caused by oxidative stress. This vitamin is crucial in neutralizing free radicals, unstable molecules produced during normal metabolism, or by environmental toxins such as pollution and ultraviolet (UV) radiation. By combating oxidative damage, vitamin E protects cell membranes, proteins and DNA, helping to maintain cellular integrity and overall health. Its functions go beyond antioxidation, including strengthening the immune system, repairing the skin, and protecting the cardiovascular system. This broad spectrum of benefits underscores the importance of including adequate vitamin E in the daily diet [114, 115]. Natural food sources of vitamin E are abundant and varied and offer a convenient way to achieve the recommended intake. Nuts such as almonds and sunflower seeds are among the wealthiest sources as they are high in tocopherols and tocotrienols, the two primary chemical forms of vitamin E [116]. In addition, vegetable oils such as sunflower, safflower and wheat germ oil serve as excellent carriers of this nutrient and are often used in cooking and salad dressings. These foods provide vitamin E and supplement it with other vital nutrients, making them a valuable part of a balanced diet. By consuming these foods, individuals can easily reap the many health benefits of this essential vitamin [117]. One of the outstanding benefits of vitamin E lies in its ability to protect and rejuvenate the skin. It also supports skin hydration by strengthening the lipid barrier, promoting elasticity and helping to repair damaged tissue. In addition, with its anti-inflammatory properties, vitamin E helps to alleviate conditions such as eczema and psoriasis, which is why it is often included in dermatological treatments and cosmetic products. Its benefits also extend to wound healing, where it accelerates healing and minimizes scarring, cementing its reputation as a cornerstone of skin care [118]. In addition to skin health, vitamin E is crucial in reducing inflammation and boosting overall immune function. Influencing inflammatory processes helps alleviate chronic inflammatory conditions such as arthritis and cardiovascular disease. Its ability to prevent the oxidation of low-density lipoprotein (LDL) cholesterol is critical in reducing the risk of atherosclerosis, one of the leading causes of heart attacks and strokes. In addition, new research highlights the neuroprotective properties of vitamin E, which show promise in slowing cognitive decline and reducing the risk of neurodegenerative diseases such as Alzheimer's. As a nutrient with multiple benefits, vitamin E is a cornerstone of health, emphasizing the need for its consistent inclusion in the daily diet through natural sources [119, 120].

3.6.3 β-Carotene

β-Carotene is a vital nutrient and precursor of vitamin A, known for its many health benefits and essential role in maintaining bodily functions. This natural pigment belongs to the carotenoid family, giving many fruits and vegetables bright orange, yellow, and green hues. As a powerful antioxidant, beta-carotene helps neutralize free radicals that can cause oxidative damage to cells, contributing to its protective effect against various diseases [121]. Its conversion into vitamin A in the body is fundamental for various biological processes, including vision, immune system function, and cellular health, which underlines its indispensability in a balanced diet. With their deep orange flesh, sweet potatoes provide significant amounts of beta-carotene and various other nutrients, such as fiber and potassium, making them a versatile addition to meals [122, 123]. Spinach, which looks green due to its chlorophyll, is another excellent source of beta-carotene that supports overall health through its synergistic combination of vitamins, minerals, and antioxidants. Including these foods in your daily meals ensures a steady supply of this beneficial compound [124]. The health benefits of beta-carotene are far-reaching and have profound effects on various systems in the body. Its role in supporting eye health is particularly noteworthy as it supports retina formation, which is crucial for vision, especially in dim light. In addition, beta-carotene boosts immune function by boosting the production of white blood cells, allowing the body to fight infection better. New research also suggests that beta-carotene may reduce the risk of certain cancers thanks to its antioxidant properties that protect DNA and cell structures from damage. With its multiple benefits, beta-carotene is a cornerstone of overall wellness and underscores the importance of a diet rich in colorful vegetables and leafy greens [125].

3.7 Alkaloids as Potent Antioxidants

Alkaloids, primarily abundant in various vegetables and fruits, are critical in neutralizing ROS and mitigating oxidative stress, a key factor in numerous chronic diseases [126]. Alkaloids such as solanine from potatoes, tomatine from tomatoes, and berberine from fruits like barberry exhibit robust free radical scavenging activities. Their molecular structure, often characterized by heterocyclic nitrogen atoms, lends itself to electron donation, stabilizing reactive molecules. These alkaloids can upregulate endogenous antioxidant defense mechanisms [127]. Including alkaloid-rich fruits and vegetables in the diet bolsters antioxidant defense and prevents conditions like cardiovascular diseases, neurodegeneration, and certain cancers. While alkaloids are potent antioxidants, their bioavailability significantly influences their therapeutic potential. Many alkaloids exhibit poor bioavailability due to low solubility, rapid metabolism, and limited intestinal absorption. For instance, despite its potent antioxidant effects, berberine suffers from low oral bioavailability owing to extensive first-pass metabolism in the liver and efflux by P-glycoprotein in the intestines. Advances in pharmacokinetic research have sought to overcome these limitations through innovative pharmaceutical formulations [128, 129].

3.8 Selenium

Selenium is a key nutrient that enhances the activity of glutathione peroxidase. This enzyme helps neutralize harmful ROS and protects cells from oxidative damage, a factor linked to aging, cancer, cardiovascular diseases, and neurodegenerative disorders. Selenium's antioxidant role is indispensable in preventing cellular damage, supporting DNA repair, and reducing inflammation caused by free radicals. It contributes to the production of thyroid hormones by supporting enzymes that convert the inactive hormone thyroxine (T4) into its active form, triiodothyronine (T3), ensuring metabolic processes run efficiently. Adequate selenium levels help prevent thyroid disorders such as hypothyroidism and autoimmune conditions like Hashimoto's thyroiditis. Selenium also enhances immune responses by promoting the proliferation and activity of white blood cells, making it critical for fighting infections. Its ability to modulate inflammation further supports the immune system, reducing the risk of chronic inflammatory diseases. Regular consumption of selenium-rich foods is essential for sustaining these vital functions, highlighting its importance in a balanced diet for overall health and disease prevention [130, 131]. Its bioavailability and pharmacokinetics vary depending on the form (organic, such as selenomethionine, or inorganic, such as sodium selenite) and route of administration. The organic forms are more bioavailable as they are efficiently absorbed in the small intestine and incorporated into body proteins. Selenium is distributed in the bloodstream and is bound to selenoproteins such as glutathione peroxidase and selenoprotein P, which mediate its physiological effects. During liver metabolism, selenium is converted into bioactive metabolites, which are mainly excreted in the urine. The pharmacokinetic profile shows a biphasic excretion with a rapid initial phase followed by a slower tissue turnover. Dietary composition and health status influence its absorption and utilization [132]. Pharmaceutical formulation innovations aim to improve selenium's bioavailability and therapeutic potential, particularly in treating deficiency and oxidative stress-related diseases [133].

3.9 Lycopene

Lycopene is a naturally occurring carotenoid found in bright red and pink foods such as tomatoes, watermelons, and grapefruit. It is a powerful antioxidant known for mitigating oxidative stress and protecting cells from damage caused by ROS. Antioxidant activity reduces the incidence of various cancers, including prostate cancer, by preventing DNA damage, slowing the proliferation of malignant cells, and modifying the signaling pathways of tumor formation. The bioavailability of lycopene is increased when it is taken in processed forms such as tomato paste or cooked tomatoes, which increases its effectiveness for health [134, 135]. In addition to its anti-cancer properties, lycopene promotes cardiovascular health. It promotes cardiovascular health by reducing the oxidation of low-density lipoprotein (LDL), a key element in the progression of atherosclerosis. Lycopene improves vascular function by reducing inflammation and promoting endothelial health, leading to improved blood flow and a lower risk of heart attacks and cerebrovascular accidents. Research suggests that a diet rich in lycopene correlates with improved cholesterol levels and regulation of blood pressure. Consistent intake of lycopene-rich fruits and vegetables is an effective and natural way to promote sustainable cardiovascular and overall health [136, 137]. Innovations such as lycopene and nanotechnology-based formulations enhance lycopene's bioavailability and therapeutic potential by improving its stability, absorption, and targeted delivery. These advancements offer more effective solutions for managing deficiencies and oxidative stress-related disorders. They also optimize lycopene's pharmacokinetics, broadening its therapeutic applications [138].

3.9.1 Omega-3 Fatty Acids

Omega-3 fatty acids are known for their powerful antioxidant properties and wide-ranging health benefits. They are mainly divided into three types: Alpha-linolenic acid (ALA), which is found in plant sources such as flaxseed, chia seeds, and walnuts; Eicosapentaenoic acid (EPA) and Docosahexaenoic acid (DHA); both of which are found primarily in fatty fish such as salmon, mackerel, and sardines [139]. Algae-based supplements are also an excellent source of DHA, especially for vegetarians and vegans. Their ability to inhibit lipid peroxidation protects cell membranes, particularly in the brain and cardiovascular system, from oxidative damage. It plays a role in curbing inflammation by regulating the production of cytokines and prostaglandins. EPA and DHA are particularly effective in reducing the risk of chronic diseases such as cardiovascular disease, diabetes, and neurodegenerative disorders [140, 141]. They also support brain health by improving cognitive function and protecting against age-related mental decline. In addition, omega-3 fatty acids contribute to skin health by maintaining water balance, reducing inflammation, and fighting oxidative damage caused by environmental factors such as UV radiation and pollution. In addition to their antioxidant and anti-inflammatory effects, omega-3 fatty acids support overall cardiovascular well-being. They lower triglyceride levels, improve the elasticity of blood vessels, and help maintain healthy blood pressure. These fats also reduce the risk of atherosclerosis by preventing the oxidation of low-density lipoprotein (LDL) cholesterol, a critical factor in plaque formation. Omega-3 fatty acids are also associated with improved immune function, promoting a balanced immune response and protecting against excessive inflammation. By incorporating omega-3-rich foods or supplements into a regular diet, individuals can harness the antioxidant potential of omega-3 fatty acids and reap comprehensive health benefits for multiple systems in the body [142–144]. Innovations such as nanotechnology-based formulations significantly improve the bioavailability and therapeutic potential of omega-3 fatty acids by improving their stability, absorption, and targeted delivery. These advancements address challenges in managing deficiencies and oxidative stress-related disorders more effectively. By optimizing omega-3 pharmacokinetics, they expand their therapeutic applications in areas like cardiovascular health, inflammation, and neuroprotection [145, 146].

3.9.2 Palmitic Acid

Palmitic acid is one of nature's most abundant fatty acids and contributes significantly to the lipid profile of plants, animals, and microorganisms. While traditionally associated with its role in metabolic energy storage and membrane structure, recent research has investigated its potential as an antioxidant [147, 148]. The antioxidant properties of palmitic acid are primarily attributed to its ability to modulate cellular mechanisms, including regulating lipid peroxidation and stabilizing cell membranes. By reducing lipid peroxidation, palmitic acid helps prevent the degradation of polyunsaturated fatty acids in cell membranes, thereby maintaining cell integrity and function. The role of palmitic acid as an antioxidant also overlaps with its involvement in lipid signaling pathways that influence the production of ROS and the activation of antioxidant enzymes. Palmitic acid has been shown to act synergistically with other lipid-based antioxidants to enhance the protective effect against oxidative damage, especially in the neuronal and cardiovascular systems. However, the dual role of palmitic acid as a protective and potentially pro-oxidant molecule under certain metabolic conditions highlights the importance of balanced concentrations for therapeutic efficacy [149–151]. Palmitic acid is ubiquitous and readily available in various foods and natural sources. It is a significant component of palm oil, which derives its name from this fatty acid and is widely used in cooking, cosmetics, and pharmaceuticals. Other critical natural sources include animal fats, dairy products, and plant-derived oils such as coconut, cocoa butter, and sunflower. In addition to dietary sources, palmitic acid is synthesized endogenously in humans via the de novo lipogenesis pathway, where excess carbohydrates are converted into fatty acids for storage. Its presence in plants such as olives and seeds highlights its ecological importance as an energy-rich molecule for growth and reproduction [151–154]. Marine fish and microalgae are becoming alternative natural palmitic acid reservoirs, especially in sustainable production. In addition to its antioxidant potential, palmitic acid serves as a precursor for the biosynthesis of other biologically active lipids, including palmitoylated proteins, which play a crucial role in cell signaling and immunomodulation. Exploration of palmitic acid from underutilized natural resources such as desert plants and marine algae holds promise for its use in dietary supplements and pharmaceuticals, potentially enhancing its role in therapeutic interventions related to oxidative stress [155, 156]. Nanotechnology-based formulations, including nanoemulsions, liposomes, and solid lipid nanoparticles, significantly improve palmitic acid's bioavailability and therapeutic potential. These advanced delivery systems overcome challenges such as poor solubility and limited bio efficacy by improving stability, absorption and targeted delivery. By optimizing the pharmacokinetics of palmitic acid, these innovations expand their therapeutic applications, particularly in energy metabolism, skin health, and the treatment of lipid metabolism disorders [157, 158].

3.9.3 Stigmasterol

Stigmasterol, a phytosterol commonly found in plant lipids, has attracted attention for its potent antioxidant properties and role in attenuating oxidative stress. Stigmasterol, structurally like cholesterol but with a distinct double bond in its side chain, is a functional bioactive compound with various health benefits. As an antioxidant, stigmasterol scavenges free radicals, especially ROS, and protects cell structures from oxidative damage. Stigmasterol's ability to integrate into lipid bilayers further stabilizes cell membranes, reducing susceptibility to lipid peroxidation and preserving cellular integrity. This mechanism protects neurons, cardiovascular tissue and other organs susceptible to oxidative damage. Recent studies suggest that stigmasterol also exerts indirect antioxidant effects by modulating various signaling pathways. In addition, its role in attenuating inflammation, a process often associated with oxidative stress, enhances its antioxidant efficacy, making it a promising candidate for therapeutic applications in diseases such as neurodegeneration, atherosclerosis and cancer [159–161]. Stigmasterol is widely distributed in nature, especially in plants, where it is an essential component of cell membranes and a precursor for the biosynthesis of steroid compounds. It is abundant in vegetable oils such as soybean, rapeseed, and sunflower, among the richest dietary sources of this phytosterol. Legumes, nuts, seeds, and certain fruits such as avocados also contain significant amounts of stigmasterol, making these foods valuable for a health-conscious diet. In addition, medicinal plants such as *Centella asiatica*, *Withania somnifera* (ashwagandha), and ginseng are rich in stigmasterol, whose antioxidant properties contribute to their therapeutic effects. Seaweeds and algae have been identified as alternative stigmasterol sources, offering sustainable extraction and utilization opportunities in the nutraceutical and pharmaceutical industries [162, 163]. Stigmasterol has low bioavailability due to its poor solubility and limited absorption in the gastrointestinal tract. Its pharmacokinetics involve minimal intestinal absorption and rapid hepatic metabolism in which it is conjugated to glucuronides or sulfates. Excretion is primarily via bile and feces, with negligible excretion via urine. Nanotechnology formulations, including nanoemulsions and liposomes, have been developed to improve their solubility, stability, and targeted delivery, thereby increasing their bioavailability and expanding their therapeutic potential in treating cholesterol, inflammation, and chronic diseases [162–167].

3.9.4 β-Sitosterol

β-Sitosterol is a naturally occurring plant sterol structurally related to cholesterol but functionally different in its effect on human health. It is found in various plants and its therapeutic importance has been known for centuries [168, 169]. Sources include vegetable oils such as olive, soybean, and corn oil, as well as nuts such as almonds and seeds such as sesame and sunflower. It is also found in fruits such as avocados and bananas and in leafy vegetables such as spinach. Traditional medicinal plants, including saw palmetto and *Pygeum africanum*, are rich in β-sitosterol and are used to treat various health problems. Its role as a potent bioactive agent goes beyond cholesterol treatment and includes properties that promote immune modulation, reduction of inflammation and tissue repair, making it invaluable in treating a wide range of diseases. The ability of β-sitosterol to interact with and inhibit cholesterol absorption in the gut is one of its best-documented properties, effectively helping to improve lipid profiles in patients with hypercholesterolemia. Apart from its lipid-lowering benefits, β-sitosterol exhibits remarkable anti-inflammatory and anti-cancer activities. It interferes with inflammatory pathways by modulating the release of cytokines, thus alleviating chronic inflammation associated with diseases such as arthritis and inflammatory bowel disease. In oncology, its ability to induce apoptosis and suppress tumor growth has been studied extensively in breast, prostate and colon cancer models. In addition, β-sitosterol is an essential factor in treating symptoms of benign prostatic hyperplasia, where it helps reduce the prostate's size and improve urinary function. However, the compound's therapeutic potential is limited by its low bioavailability due to minimal absorption in the gastrointestinal tract and extensive excretion via bile. Current advances in drug delivery, such as nanoencapsulation and lipid-based formulations, aim to overcome these limitations and pave the way for more effective clinical applications of β-sitosterol in cardiovascular diseases, immune disorders and chronic diseases [170].

3.9.5 Limonene

Limonene, a naturally occurring monoterpene, is the main component of citrus essential oils and contributes to the characteristic aroma of fruits such as lemons, oranges, and grapefruits. It has gained much attention due to its strong antioxidant properties and wide range of health benefits. Limonene significantly prevents oxidative stress by scavenging ROS and followed by free radical damage [171, 172]. It also modulates the body's antioxidant defenses by increasing the activity of enzymes such as SOD and CAT, which are critical for neutralizing oxidative damage. These antioxidant mechanisms make limonene an important player in preventing oxidative damage associated with chronic diseases such as cardiovascular disease, neurodegenerative diseases and age-related cell deterioration. Limonene may inhibit the development and progression of certain cancers by inducing apoptosis, suppressing tumor growth and modulating detoxification pathways. In addition, limonene supports digestive health, helps relieve acid reflux and promotes intestinal motility. Due to its influence on neurotransmitters, its mood-enhancing properties also contribute to its therapeutic value. It is used in dietary supplements, aromatherapy, and in functional foods. Limonene is a versatile compound with significant antioxidant and health-promoting potential, making it a valuable addition to preventative and integrative health approaches. Limonene has a low bioavailability due to its poor water solubility and rapid metabolism. Its pharmacokinetics involve efficient absorption in the gastrointestinal tract, followed by extensive hepatic metabolism into oxygenated metabolites such as perillic acid and dihydroperillic acid, which are mainly excreted in the urine. To solve the problem of bioavailability, nanotechnological formulations such as nanoemulsions, liposomes, and polymeric nanoparticles have been developed. These systems improve the solubility, stability, and targeted delivery of limonene and optimize its therapeutic potential in treating inflammation, oxidative stress, and cancer [173–176].

3.9.6 Eugenol

Eugenol, a phenolic compound known for its strong antioxidant properties, is found primarily in the essential oils of cloves (*Syzygium aromaticum*), cinnamon (*Cinnamomum zeylanicum*), nutmeg (*Myristica fragrans*), basil (*Ocimum basilicum*), and bay leaves (*Laurus nobilis*). Clove oil is the richest source of eugenol and contains up to 85% of this bioactive compound. Its antioxidant effect is based on its phenolic structure, enabling it to scavenge ROS and RNS effectively. In addition to being a direct free radical scavenger, eugenol inhibits lipid peroxidation, a process in which free radicals attack polyunsaturated fatty acids in cell membranes, leading to cellular dysfunction [177, 178]. This inhibition helps to maintain membrane integrity and cellular stability. Eugenol increases the activity of key endogenous antioxidant enzymes, including SOD, CAT, and GPx, which together protect cells from oxidative damage. This potent antioxidant property makes eugenol a promising candidate for preventing diseases related to oxidative stress. Its abundant natural and favorable bioavailability further underlines its potential for therapeutic and nutraceutical applications. Pharmacokinetically, eugenol is absorbed in the gastrointestinal tract, metabolized in the liver to glucuronide and sulfate conjugates, and excreted primarily in the urine. However, its therapeutic use is hampered by problems such as limited solubility and stability. Advanced nanotechnology delivery systems have been developed to overcome these limitations, including nanoemulsions, liposomes, and polymeric nanoparticles. These innovative formulations improve eugenol's solubility, stability, and targeted delivery, maximizing its therapeutic efficacy in treating pain, inflammation, microbial infections, and oxidative stress [179–181].

3.9.7 α-Linolenic Acid

α-Linolenic acid (ALA) found abundantly in flaxseed, chia seeds, walnuts, and canola oil, ALA neutralizes free radicals by halting oxidative damage to vital biomolecules. Its role in inhibiting lipid peroxidation protects cell membranes from oxidative degradation. ALA increases the activity of SOD and GPx, which are essential for the neutralization of ROS. In addition, it reduces the production of harmful pro-oxidant molecules such as hydrogen peroxide and malondialdehyde, thus maintaining cellular redox balance. The compound modulates inflammatory pathways by downregulating markers such as NF-κB and interleukins, reducing oxidative stress-induced inflammation. These multifaceted mechanisms underscore the central role of ALA in preventing oxidative damage that contributes to chronic diseases such as cardiovascular disease, neurodegenerative diseases and cancer [182–186]. The antioxidant potential of ALA leads to comprehensive health benefits. It protects against cardiovascular disease by preventing the oxidation of LDL, an essential step in atherosclerosis, and reduces the risk of heart attacks and strokes. To protect the nervous system, ALA crosses the blood–brain barrier and alleviates oxidative stress in nervous tissue, which could delay the progression of Alzheimer's and Parkinson's disease. Its ability to inhibit DNA damage and modulate pro-apoptotic signaling pathways makes ALA a promising anti-cancer agent. In metabolic diseases like diabetes, ALA improves insulin sensitivity and reduces oxidative stress in pancreatic beta cells, improving overall metabolic health. It also benefits skin care by mitigating UV-induced oxidative damage, reducing the signs of aging and promoting wound healing. However, issues like ALA's instability under heat and light and its limited conversion to longer-chain omega-3 fatty acids such as DHA and EPA may compromise its systemic efficacy. Despite these limitations, ALA remains a valuable dietary antioxidant with immense therapeutic potential that warrants further research to optimize its clinical applications. ALA has low bioavailability due to its poor water solubility and susceptibility to oxidation, which may limit its absorption and therapeutic efficacy. Improving its pharmacokinetic profile through nanotechnology delivery systems, such as nanoemulsions or liposomes, improves its stability, solubility, and absorption, thus maximizing its health benefits [187–191].

3.9.8 Zingerone

Zingerone, a phenolic compound from ginger (*Zingiber officinale*), is considered one of the richest sources of natural antioxidants. It is mainly formed during the cooking or drying of fresh ginger, where gingerols are converted, enhancing its antioxidant effect. Structurally, zingerone is characterized by a vanillyl group attached to a ketone, contributing to its ability to reduce oxidative stress. Its antioxidant properties are based on its phenolic hydroxyl group, which donates hydrogen atoms to neutralize ROS and RNS, prevent cell damage and maintain the redox balance in biological systems [192, 193]. Zingerone has several antioxidant mechanisms of action. It protects lipids from peroxidation and thus protects cell membranes from oxidative degradation. This protection prevents damage to essential cellular components such as phospholipids, proteins and nucleic acids. Zingerone also enhances the activity of the body's antioxidant enzymes, including SOD, CAT, and GPx, thus strengthening the body's natural defense system against oxidative stress. Its ability to chelate metal ions such as iron and copper prevents the formation of highly reactive hydroxyl radicals through Fenton-like reactions, further reducing oxidative damage. In addition, zingerone reduces inflammatory markers often associated with oxidative stress, highlighting its dual role as an antioxidant and anti-inflammatory [194, 195]. Zingerone's bioavailability and natural occurrence make it a versatile compound with therapeutic potential. It is found in ginger and in ginger-derived products such as dried powder, tea, and essential oils, so it can be consumed in various ways through food.

The diverse antioxidant properties of zingerone and its availability as a widely used spice emphasize its importance for preventive health care and its potential application in functional foods, nutraceuticals, and pharmaceuticals. Zingerone faces challenges of low bioavailability, which restricts its therapeutic effectiveness. Nanotechnology-based delivery systems such as nanoemulsions, liposomes, and polymeric nanoparticles have been developed to overcome these limitations. These advanced formulations improve zingerone's solubility, stability, and absorption, enhancing its pharmacokinetic properties and enabling precise targeting. These innovations address the drawbacks of conventional delivery methods and significantly enhance zingerone's potential for treating conditions like inflammation, oxidative stress, and microbial infections [196–198].

4 Conclusion

The consumption of antioxidants is a promising approach to combating and preventing diseases related to oxidative stress, highlighting their important role in maintaining cellular integrity and overall health. Natural antioxidants, such as phenolic compounds, vitamins, carotenoids, and omega-3 fatty acids, offer various mechanisms of action, including protection against free radicals, inhibition of lipid peroxidation and modulation of inflammatory processes. These bioactive compounds neutralize ROS and strengthen the body's immune functions. Despite their proven benefits, problems such as poor bioavailability, rapid metabolism, and stability limit their clinical application. Advances in nanotechnology delivery systems, including nanoemulsions, liposomes, and polymeric nanoparticles, have shown that these obstacles can be overcome. These innovations improve stability, solubility, and targeted delivery of antioxidants, optimizing their therapeutic efficacy. Future research should focus on refining these technologies, exploring synergistic combinations of antioxidants and conducting large-scale clinical trials to validate their efficacy. From a nutritional perspective, consuming antioxidant-rich foods such as fruits, vegetables, nuts, and fatty fish can significantly reduce oxidative stress. A personalized diet based on genetic and metabolic profiles offers the opportunity to maximize the benefits of antioxidants tailored to individual needs. Focusing on the future, integrating antioxidants into functional foods, pharmaceuticals, and nutraceuticals could revolutionize preventive healthcare. In summary, while the therapeutic potential of antioxidants is enormous, a multidisciplinary approach encompassing advanced drug delivery, personalized nutrition and translational research is essential to unlock their full potential. This comprehensive strategy promises to transform the treatment of oxidative stress-related diseases, improve quality of life and pave the way for innovative, sustainable health solutions.

References

1. Juan, C. A., Pérez de la Lastra, J. M., Plou, F. J., & Pérez-Lebeña, E. (2021). The chemistry of reactive oxygen species (ROS) revisited: Outlining their role in biological macromolecules (DNA, lipids and proteins) and induced pathologies. *International Journal of Molecular Sciences, 22*(9), 4642. https://doi.org/10.3390/ijms22094642
2. Chaudhary, P., Janmeda, P., Docea, A. O., Yeskaliyeva, B., Abdull Razis, A. F., Modu, B., et al. (2023). Oxidative stress, free radicals and antioxidants: Potential crosstalk in the pathophysiology of human diseases. *Frontiers in Chemistry, 11*, 1158198. https://doi.org/10.3389/fchem.2023.1158198
3. Hong, Y., Boiti, A., Vallone, D., & Foulkes, N. S. (2024). Reactive oxygen species signaling and oxidative stress: Transcriptional regulation and evolution. *Antioxidants, 13*(3), 312. https://doi.org/10.3390/antiox13030312
4. Jomova, K., Raptova, R., Alomar, S. Y., Alwasel, S. H., Nepovimova, E., Kuca, K., & Valko, M. (2023). Reactive oxygen species, toxicity, oxidative stress, and antioxidants: Chronic diseases and aging. *Archives of Toxicology, 97*(10), 2499–2574. https://doi.org/10.1007/s00204-023-03562-9
5. Martemucci, G., Costagliola, C., Mariano, M., D'andrea, L., Napolitano, P., & D'Alessandro, A. G. (2022). Free radical properties, source and targets, antioxidant consumption and health. *Oxygen, 2*(2), 48–78. https://doi.org/10.3390/oxygen2020006
6. Afzal, S., Abdul Manap, A. S., Attiq, A., Albokhadaim, I., Kandeel, M., & Alhojaily, S. M. (2023). From imbalance to impairment: The central role of reactive oxygen species in oxidative stress-induced disorders and therapeutic exploration. *Frontiers in Pharmacology, 14*, 1269581. https://doi.org/10.3389/fphar.2023.1269581
7. Parcheta, M., Świsłocka, R., Orzechowska, S., Akimowicz, M., Choińska, R., & Lewandowski, W. (2021). Recent developments in effective antioxidants: The structure and antioxidant properties. *Materials, 14*(8), 1984. https://doi.org/10.3390/ma14081984
8. Maldonado, E., Morales-Pison, S., Urbina, F., & Solari, A. (2023). Aging hallmarks and the role of oxidative stress. *Antioxidants, 12*(3), 651. https://doi.org/10.3390/antiox12030651
9. Reddy, V. P. (2023). Oxidative stress in health and disease. *Biomedicines, 11*(11), 2925. https://doi.org/10.3390/biomedicines11112925
10. Nediani, C., Ruzzolini, J., & Dinu, M. (2024). Oxidative stress and inflammation as targets for novel preventive and therapeutic approaches in non-communicable diseases III. *Antioxidants, 13*(11), 1404. https://doi.org/10.3390/antiox13111404
11. Muscolo, A., Mariateresa, O., Giulio, T., & Mariateresa, R. (2024). Oxidative stress: The role of antioxidant phytochemicals in the prevention and treatment of diseases. *International Journal of Molecular Sciences, 25*(6), 3264. https://doi.org/10.3390/ijms25063264
12. Zhang, Y.-J., Gan, R.-Y., Li, S., Zhou, Y., Li, A.-N., Xu, D.-P., & Li, H.-B. (2015). Antioxidant phytochemicals for the prevention and treatment of chronic diseases. *Molecules, 20*(12), 21138–21156. https://doi.org/10.3390/molecules201219753
13. Rahaman, M. M., Hossain, R., Herrera-Bravo, J., Islam, M. T., Atolani, O., Adeyemi, O. S., et al. (2023). Natural antioxidants from some fruits, seeds, foods, natural products, and associated health benefits: An update. *Food Science & Nutrition, 11*(4), 1657–1670. https://doi.org/10.1002/fsn3.3217
14. Nakai, K., & Tsuruta, D. (2021). What are reactive oxygen species, free radicals, and oxidative stress in skin diseases? *International Journal of Molecular Sciences, 22*(19), 10799. https://doi.org/10.3390/ijms221910799
15. Aranda-Rivera, A. K., Cruz-Gregorio, A., Arancibia-Hernández, Y. L., Hernández-Cruz, E. Y., & Pedraza-Chaverri, J. (2022). RONS and oxidative stress: An overview of basic concepts. *Oxygen, 2*(4), 437–478. https://doi.org/10.3390/oxygen2040030
16. Dash, U. C., Bhol, N. K., Swain, S. K., Samal, R. R., Nayak, P. K., Raina, V., et al. (2024). Oxidative stress and inflammation in the pathogenesis of neurological disorders: Mechanisms and implications. *Acta Pharmaceutica Sinica B*. https://doi.org/10.1016/j.apsb.2024.10.004

17. Sharifi-Rad, J., Seidel, V., Izabela, M., Monserrat-Mequida, M., Sureda, A., Ormazabal, V., et al. (2023). Phenolic compounds as Nrf2 inhibitors: Potential applications in cancer therapy. *Cell Communication and Signaling: CCS, 21*(1), 89. https://doi.org/10.1186/s12964-023-01109-0
18. Kostenko, V., Akimov, O., Gutnik, O., Kostenko, H., Kostenko, V., Romantseva, T., et al. (2023). Modulation of redox-sensitive transcription factors with polyphenols as pathogenetically grounded approach in therapy of systemic inflammatory response. *Heliyon, 9*(5), e15551. https://doi.org/10.1016/j.heliyon.2023.e15551
19. Rahman, M. M., Rahaman, M. S., Islam, M. R., Rahman, F., Mithi, F. M., Alqahtani, T., Almikhlafi, M. A., Alghamdi, S. Q., Alruwaili, A. S., Hossain, M. S., Ahmed, M., Das, R., Emran, T. B., & Uddin, M. S. (2022). Role of phenolic compounds in human disease: Current knowledge and future prospects. *Molecules, 27*(1), 233. https://doi.org/10.3390/molecules27010233
20. Ziółkiewicz, A., Kasprzak-Drozd, K., Rusinek, R., Markut-Miotła, E., & Oniszczuk, A. (2023). The influence of polyphenols on atherosclerosis development. *International Journal of Molecular Sciences, 24*(8), 7146. https://doi.org/10.3390/ijms24087146
21. Ali Redha, A., Kodikara, C., & Cozzolino, D. (2024). Does encapsulation improve the bioavailability of polyphenols in humans? A concise review based on in vivo human studies. *Nutrients, 16*(21), 3625. https://doi.org/10.3390/nu16213625
22. Grabska-Kobyłecka, I., Szpakowski, P., Król, A., Książek-Winiarek, D., Kobyłecki, A., Głąbiński, A., & Nowak, D. (2023). Polyphenols and their impact on the prevention of neurodegenerative diseases and development. *Nutrients, 15*(15), 3454. https://doi.org/10.3390/nu15153454
23. Chiorcea-Paquim, A.-M. (2023). Electrochemistry of flavonoids: A comprehensive review. *International Journal of Molecular Sciences, 24*(21), 15667. https://doi.org/10.3390/ijms242115667
24. Giner, R. M., Ríos, J. L., & Máñez, S. (2022). Antioxidant activity of natural hydroquinones. *Antioxidants, 11*(2), 343. https://doi.org/10.3390/antiox11020343
25. Yang, H., Furutani, Y., Kudo, S., Hayashi, J.-I., & Norinaga, K. (2016). Experimental investigation of thermal decomposition of dihydroxybenzene isomers: Catechol, hydroquinone, and resorcinol. *Journal of Analytical and Applied Pyrolysis, 120*, 321–329. https://doi.org/10.1016/j.jaap.2016.05.019
26. Kiokias, S., & Oreopoulou, V. (2021). A review of the health protective effects of phenolic acids against a range of severe pathologic conditions (including coronavirus-based infections). *Molecules, 26*(17), 5405. https://doi.org/10.3390/molecules26175405
27. GutiErrez-Grijalva, E. P., Ambriz-Pere, D. L., Leyva-Lopez, N., Castillo-Lopez, R. I., & Heiedia, J. B. (2016). Review: Dietary phenolic compounds, health benefits and bioaccessibility. *Archivos Latinoamericanos de Nutrición, 66*(2), 87–100 (PMID: 29737665).
28. Sun, W., & Shahrajabian, M. H. (2023). Therapeutic potential of phenolic compounds in medicinal plants—Natural health products for human health. *Molecules, 28*(4), 1845. https://doi.org/10.3390/molecules28041845
29. Cione, E., La Torre, C., Cannataro, R., Caroleo, M. C., Plastina, P., & Gallelli, L. (2020). Quercetin, epigallocatechin gallate, curcumin, and resveratrol: From dietary sources to human microRNA modulation. *Molecules, 25*(1), 63. https://doi.org/10.3390/molecules25010063
30. Hu, L., Luo, Y., Yang, J., & Cheng, C. (2025). Botanical flavonoids: Efficacy, absorption, metabolism and advanced pharmaceutical technology for improving bioavailability. *Molecules, 30*(5), 1184. https://doi.org/10.3390/molecules30051184
31. Zahra, M., Abrahamse, H., & George, B. P. (2024). Flavonoids: Antioxidant powerhouses and their role in nanomedicine. *Antioxidants, 13*(8), 922. https://doi.org/10.3390/antiox13080922
32. Pejčić, T., Zeković, M., Bumbaširević, U., Kalaba, M., Vovk, I., Bensa, M., Popović, L., & Tešić, Ž. (2023). The role of isoflavones in the prevention of breast cancer and prostate cancer. *Antioxidants, 12*(2), 368. https://doi.org/10.3390/antiox12020368

33. Gómez-Zorita, S., González-Arceo, M., Fernández-Quintela, A., Eseberri, I., Trepiana, J., & Portillo, M. P. (2020). Scientific evidence supporting the beneficial effects of isoflavones on human health. *Nutrients, 12*(12), 3853. https://doi.org/10.3390/nu12123853
34. Yu, L., Rios, E., Castro, L., Liu, J., Yan, Y., & Dixon, D. (2021). Genistein: Dual role in women's health. *Nutrients, 13*(9), 3048. https://doi.org/10.3390/nu13093048
35. Mattioli, R., Francioso, A., Mosca, L., & Silva, P. (2020). Anthocyanins: A comprehensive review of their chemical properties and health effects on cardiovascular and neurodegenerative diseases. *Molecules, 25*(17), 3809. https://doi.org/10.3390/molecules25173809
36. Panchal, S. K., John, O. D., Mathai, M. L., & Brown, L. (2022). Anthocyanins in chronic diseases: The power of purple. *Nutrients, 14*(10), 2161. https://doi.org/10.3390/nu14102161
37. Speer, H., D'Cunha, N. M., Alexopoulos, N. I., McKune, A. J., & Naumovski, N. (2020). Anthocyanins and human health—A focus on oxidative stress, inflammation and disease. *Antioxidants, 9*(5), 366. https://doi.org/10.3390/antiox9050366
38. Luiza Koop, B., et al. (2022). Flavonoids, anthocyanins, betalains, curcumin, and carotenoids: Sources, classification and enhanced stabilization by encapsulation and adsorption. *Food Research International (Ottawa, Ont.), 153*, 110929. https://doi.org/10.1016/j.foodres.2021.110929
39. Martínez-Márquez, A., Selles-Marchart, S., Nájera, H., Morante-Carriel, J., Martínez-Esteso, M. J., & Bru-Martínez, R. (2024). Biosynthesis of piceatannol from resveratrol in grapevine can be mediated by cresolase-dependent *ortho*-hydroxylation activity of polyphenol oxidase. *Plants, 13*(18), 2602. https://doi.org/10.3390/plants13182602
40. Hu, W.-H., Dai, D. K., Zheng, B.Z.-Y., Duan, R., Dong, T.T.-X., Qin, Q.-W., & Tsim, K.W.-K. (2020). Piceatannol, a natural analog of resveratrol, exerts anti-angiogenic efficiencies by blockage of vascular endothelial growth factor binding to its receptor. *Molecules, 25*(17), 3769. https://doi.org/10.3390/molecules25173769
41. Duta-Bratu, C.-G., Nitulescu, G. M., Mihai, D. P., & Olaru, O. T. (2023). Resveratrol and other natural oligomeric stilbenoid compounds and their therapeutic applications. *Plants, 12*(16), 2935. https://doi.org/10.3390/plants12162935
42. Smeriglio, A., Barreca, D., Bellocco, E., & Trombetta, D. (2017). Proanthocyanidins and hydrolysable tannins: Occurrence, dietary intake and pharmacological effects. *British Journal of Pharmacology, 174*(11), 1244–1262. https://doi.org/10.1111/bph.13630
43. Cosme, F., Aires, A., Pinto, T., Oliveira, I., Vilela, A., & Gonçalves, B. (2025). A comprehensive review of bioactive tannins in foods and beverages: Functional properties, health benefits, and sensory qualities. *Molecules, 30*(4), 800. https://doi.org/10.3390/molecules30040800
44. Olennikov, D. N. (2023). Coumarins of lovage roots (*Levisticum officinale* W.D.J.Koch): LC-MS profile, quantification, and stability during postharvest storage. *Metabolites, 13*(1), 3. https://doi.org/10.3390/metabo13010003
45. Lončar, M., Jakovljević, M., Šubarić, D., Pavlić, M., Buzjak Služek, V., Cindrić, I., & Molnar, M. (2020). Coumarins in food and methods of their determination. *Foods, 9*(5), 645. https://doi.org/10.3390/foods9050645
46. Kornicka, A., Balewski, Ł, Lahutta, M., & Kokoszka, J. (2023). Umbelliferone and Its synthetic derivatives as suitable molecules for the development of agents with biological activities: A review of their pharmacological and therapeutic potential. *Pharmaceuticals, 16*(12), 1732. https://doi.org/10.3390/ph16121732
47. Calderaro, A., Patanè, G. T., Tellone, E., Barreca, D., Ficarra, S., Misiti, F., & Laganà, G. (2022). The neuroprotective potentiality of flavonoids on Alzheimer's disease. *International Journal of Molecular Sciences, 23*(23), 14835. https://doi.org/10.3390/ijms232314835
48. Behl, T., Kaur, G., Sehgal, A., Zengin, G., Singh, S., Ahmadi, A., & Bungau, S. (2022). Flavonoids, the family of plant-derived antioxidants making inroads into novel therapeutic design against ionizing radiation-induced oxidative stress in Parkinson's disease. *Current Neuropharmacology, 20*(2), 324–343. https://doi.org/10.2174/1570159X19666210524152817
49. Jomova, K., Alomar, S. Y., Alwasel, S. H., Eugenie, N., Kuca, K., & Valko, M. (2024). Several lines of antioxidant defense against oxidative stress: Antioxidant enzymes, nanomaterials with

multiple enzyme-mimicking activities, and low-molecular-weight antioxidants. *Archives of Toxicology, 98*, 1323–1367. https://doi.org/10.1007/s00204-024-03696-4

50. Meng-Zhen, S., Ju, L., Lan-Chun, Z., Cai-Feng, D., Shu-da, Y., Hao-Fei, Y., & Wei-Yan, H. (2022). Potential therapeutic use of plant flavonoids in AD and PD. *Heliyon, 8*(11), e11440. https://doi.org/10.1016/j.heliyon.2022.e11440
51. Panche, A. N., Diwan, A. D., & Chandra, S. R. (2016). Flavonoids: An overview. *Journal of Nutritional Science, 5*(e47), e47. https://doi.org/10.1017/jns.2016.41
52. Chagas, M. S. S., Behrens, M. D., Moragas-Tellis, C. J., Penedo, G. X. M., Silva, A. R., & Gonçalves-de-Albuquerque, C. F. (2022). Flavonols and flavones as potential anti-inflammatory, antioxidant, and antibacterial compounds. *Oxidative Medicine and Cellular Longevity, 2022*, 9966750. https://doi.org/10.1155/2022/9966750
53. Samanta, S., Sarkar, T., Chakraborty, R., Rebezov, M., Shariati, M. A., Thiruvengadam, M., & Rengasamy, K. R. R. (2022). Dark chocolate: An overview of its biological activity, processing, and fortification approaches. *Current Research in Food Science, 5*, 1916–1943. https://doi.org/10.1016/j.crfs.2022.10.017
54. García-Cordero, J., Martinez, A., Blanco-Valverde, C., Pino, A., Puertas-Martín, V., San Román, R., & de Pascual-Teresa, S. (2023). Regular consumption of cocoa and red berries as a strategy to improve cardiovascular biomarkers via modulation of microbiota metabolism in healthy aging adults. *Nutrients, 15*(10), 2299. https://doi.org/10.3390/nu15102299
55. Martin, M. Á., & Ramos, S. (2021). Impact of cocoa flavanols on human health. *Food and Chemical Toxicology: An International Journal Published for the British Industrial Biological Research Association, 151*, 112121. https://doi.org/10.1016/j.fct.2021.112121
56. Aghababaei, F., & Hadidi, M. (2023). Recent advances in potential health benefits of quercetin. *Pharmaceuticals, 16*(7), 1020. https://doi.org/10.3390/ph16071020
57. Deepika, & Maurya, P. K. (2022). Health benefits of quercetin in age-related diseases. *Molecules, 27*(8), 2498. https://doi.org/10.3390/molecules27082498
58. Ullah, A., Munir, S., Badshah, S. L., Khan, N., Ghani, L., Poulson, B. G., Emwas, A.-H., & Jaremko, M. (2020). Important flavonoids and their role as a therapeutic agent. *Molecules, 25*(22), 5243. https://doi.org/10.3390/molecules25225243
59. Mirza, M. A., Mahmood, S., Hilles, A. R., Ali, A., Khan, M. Z., Zaidi, S. A. A., Iqbal, Z., & Ge, Y. (2023). Quercetin as a therapeutic product: Evaluation of its pharmacological action and clinical applications—A review. *Pharmaceuticals, 16*(11), 1631. https://doi.org/10.3390/ph16111631
60. Bernini, R., & Velotti, F. (2021). Natural polyphenols as immunomodulators to rescue immune response homeostasis: Quercetin as a research model against severe COVID-19. *Molecules, 26*(19), 5803. https://doi.org/10.3390/molecules26195803
61. Luiza Koop, B., Nascimento da Silva, M., Diniz da Silva, F., Thayres Dos Santos Lima, K., Santos Soares, L., José de Andrade, C., et al. (2022). Flavonoids, anthocyanins, betalains, curcumin, and carotenoids: Sources, classification and enhanced stabilization by encapsulation and adsorption. *Food Research International (Ottawa, Ont.), 153*, 110929. https://doi.org/10.1016/j.foodres.2021.110929
62. Ysrafil, Y., Sapiun, Z., Slamet, N. S., Mohamad, F., Hartati, H., Damiti, S. A., et al. (2023). Anti-inflammatory activities of flavonoid derivates. *ADMET & DMPK, 11*(3), 331–359. https://doi.org/10.5599/admet.1918
63. Thilakarathna, S. H., & Rupasinghe, H. P. V. (2013). Flavonoid bioavailability and attempts for bioavailability enhancement. *Nutrients, 5*(9), 3367–3387. https://doi.org/10.3390/nu5093367
64. Xiong, H.-H., Lin, S.-Y., Chen, L.-L., Ouyang, K.-H., & Wang, W.-J. (2023). The interaction between flavonoids and intestinal microbes: A review. *Foods, 12*(2), 320. https://doi.org/10.3390/foods12020320
65. Jasim, H. A., Nahar, L., Jasim, M. A., Moore, S. A., Ritchie, K. J., & Sarker, S. D. (2021). Chalcones: Synthetic chemistry follows where nature leads. *Biomolecules, 11*(8), 1203. https://doi.org/10.3390/biom11081203
66. Samota, M. K., Yadav, D. K., Koli, P., Kaur, M., Kaur, M., Rani, H., et al. (2024). Exploring natural chalcones: Innovative extraction techniques, bioactivities, and health potential. *Sustainable Food Technology*. https://doi.org/10.1039/d4fb00126e

67. Mojzis, J., Varinska, L., Mojzisova, G., Kostova, I., & Mirossay, L. (2008). Antiangiogenic effects of flavonoids and chalcones. *Pharmacological Research, 57*, 259–265. https://doi.org/10.1016/j.phrs.2008.02.005
68. Salehi, B., Quispe, C., Chamkhi, I., El Omari, N., Balahbib, A., Sharifi-Rad, J., et al. (2020). Pharmacological properties of chalcones: A review of preclinical including molecular mechanisms and clinical evidence. *Frontiers in Pharmacology, 11*, 592654. https://doi.org/10.3389/fphar.2020.592654
69. Koushki, M., Amiri-Dashatan, N., Ahmadi, N., Abbaszadeh, H.-A., & Rezaei-Tavirani, M. (2018). Resveratrol: A miraculous natural compound for diseases treatment. *Food Science & Nutrition, 6*(8), 2473–2490. https://doi.org/10.1002/fsn3.855
70. Farhan, M., & Rizvi, A. (2023). The pharmacological properties of red grape polyphenol resveratrol: Clinical trials and obstacles in drug development. *Nutrients, 15*(20), 4486. https://doi.org/10.3390/nu15204486
71. Vikal, A., Maurya, R., Bhowmik, S., Khare, S., Raikwar, S., Patel, P., & Das Kurmi, B. (2024). Resveratrol: A comprehensive review of its multifaceted health benefits, mechanisms of action, and potential therapeutic applications in chronic disease. *Pharmacological Research—Natural Products, 3*, 100047. https://doi.org/10.1016/j.prenap.2024.100047
72. Meng, Q., Li, J., Wang, C., & Shan, A. (2023). Biological function of resveratrol and its application in animal production: A review. *Journal of Animal Science and Biotechnology, 14*(1), 25. https://doi.org/10.1186/s40104-022-00822-z
73. Bartra, C., Yuan, Y., Vuraić, K., Valdés-Quiroz, H., Garcia-Baucells, P., Slevin, M., Pastorello, Y., Suñol, C., & Sanfeliu, C. (2024). Resveratrol activates antioxidant protective mechanisms in cellular models of Alzheimer's disease inflammation. *Antioxidants, 13*(2), 177. https://doi.org/10.3390/antiox13020177
74. Wang, S., Su, R., Nie, S., Sun, M., Zhang, J., Wu, D., & Moustaid-Moussa, N. (2014). Application of nanotechnology in improving bioavailability and bioactivity of diet-derived phytochemicals. *The Journal of Nutritional Biochemistry, 25*(4), 363–376. https://doi.org/10.1016/j.jnutbio.2013.10.002
75. Kumar, A., Kurmi, B. D., Singh, A., & Singh, D. (2022). Potential role of resveratrol and its nano-formulation as anti-cancer agent. *Exploration of Targeted Anti-Tumor Therapy, 3*(5), 643–658. https://doi.org/10.37349/etat.2022.00105
76. Kozlov, A. V., Javadov, S., & Sommer, N. (2024). Cellular ROS and antioxidants: physiological and pathological role. *Antioxidants, 13*(5), 602. https://doi.org/10.3390/antiox13050602
77. Sugandhi, V. V., Pangeni, R., Vora, L. K., Poudel, S., Nangare, S., Jagwani, S., et al. (2024). Pharmacokinetics of vitamin dosage forms: A complete overview. *Food Science & Nutrition, 12*(1), 48–83. https://doi.org/10.1002/fsn3.3787
78. Miazek, K., Beton, K., Śliwińska, A., & Brożek-Płuska, B. (2022). The effect of β-carotene, tocopherols and ascorbic acid as anti-oxidant molecules on human and animal in vitro/in vivo studies: A review of research design and analytical techniques used. *Biomolecules, 12*(8), 1087. https://doi.org/10.3390/biom12081087
79. Chambial, S., Dwivedi, S., Shukla, K. K., John, P. J., & Sharma, P. (2013). Vitamin C in disease prevention and cure: An overview. *Indian Journal of Clinical Biochemistry: IJCB, 28*(4), 314–328. https://doi.org/10.1007/s12291-013-0375-3
80. Kaźmierczak-Barańska, J., Boguszewska, K., Adamus-Grabicka, A., & Karwowski, B. T. (2020). Two faces of vitamin C—Antioxidative and pro-oxidative agent. *Nutrients, 12*(5), 1501. https://doi.org/10.3390/nu12051501
81. Sen, C. K., Khanna, S., Rink, C., & Roy, S. (2007). Tocotrienols: The emerging face of natural vitamin E. In *Vitamins and hormones. Vitamin E* (pp. 203–261). https://doi.org/10.1016/s0083-6729(07)76008-9
82. Ahsan, H., Ahad, A., Iqbal, J., & Siddiqui, W. A. (2014). Pharmacological potential of tocotrienols: A review. *Nutrition & Metabolism (London), 11*, 52. https://doi.org/10.1186/1743-7075-11-52
83. Andrès, E., Lorenzo-Villalba, N., Terrade, J.-E., & Méndez-Bailon, M. (2024). Fat-soluble vitamins A, D, E, and K: Review of the literature and points of interest for the clinician. *Journal of Clinical Medicine, 13*(13), 3641. https://doi.org/10.3390/jcm13133641

84. Steinbauer, S., Wallner, M., Karl, L.-M., Gramatte, T., Essl, K., Iken, M., Weghuber, J., Blank-Landeshammer, B., & Röhrl, C. (2025). Differential enhancement of fat-soluble vitamin absorption and bioefficacy via micellization in combination with selected plant extracts in vitro. *Nutrients, 17*(2), 359. https://doi.org/10.3390/nu17020359
85. Siener, R., Machaka, I., Alteheld, B., Bitterlich, N., & Metzner, C. (2020). Effect of fat-soluble vitamins A, D, E and K on vitamin status and metabolic profile in patients with fat malabsorption with and without urolithiasis. *Nutrients, 12*(10), 3110. https://doi.org/10.3390/nu12103110
86. Park, C. Y., Shin, S., & Han, S. N. (2024). Multifaceted roles of vitamin D for diabetes: From immunomodulatory functions to metabolic regulations. *Nutrients, 16*(18), 3185. https://doi.org/10.3390/nu16183185
87. Condoleo, V., Pelaia, C., Armentaro, G., Severini, G., Clausi, E., Cassano, V., Miceli, S., Fiorentino, T. V., Succurro, E., Arturi, F., Andreozzi, F., Sesti, G., & Sciacqua, A. (2021). Role of vitamin D in cardiovascular diseases. *Endocrines, 2*(4), 417–426. https://doi.org/10.3390/endocrines2040037
88. Aiello, G., Lombardo, M., & Baldelli, S. (2024). Exploring vitamin D synthesis and function in cardiovascular health: A narrative review. *Applied Sciences, 14*(11), 4339. https://doi.org/10.3390/app14114339
89. Ungurianu, A., Zanfirescu, A., Nițulescu, G., & Margină, D. (2021). Vitamin E beyond its antioxidant label. *Antioxidants, 10*(5), 634. https://doi.org/10.3390/antiox10050634
90. Szewczyk, K., Chojnacka, A., & Górnicka, M. (2021). Tocopherols and tocotrienols—Bioactive dietary compounds; what is certain, what is doubt? *International Journal of Molecular Sciences, 22*(12), 6222. https://doi.org/10.3390/ijms22126222
91. Regner-Nelke, L., Nelke, C., Schroeter, C. B., Dziewas, R., Warnecke, T., Ruck, T., & Meuth, S. G. (2021). Enjoy carefully: The multifaceted role of vitamin E in neuro-nutrition. *International Journal of Molecular Sciences, 22*(18), 10087. https://doi.org/10.3390/ijms221810087
92. Aaseth, J. O., Finnes, T. E., Askim, M., & Alexander, J. (2024). The importance of vitamin K and the combination of vitamins K and D for calcium metabolism and bone health: A review. *Nutrients, 16*(15), 2420. https://doi.org/10.3390/nu16152420
93. Regulska-Ilow, B., Różańska, D., Zatońska, K., & Szuba, A. (2022). Estimation of vitamin K content and its sources in the diet of the polish participants of the PURE study. *Nutrients, 14*(9), 1917. https://doi.org/10.3390/nu14091917
94. Alberts, A., Moldoveanu, E.-T., Niculescu, A.-G., & Grumezescu, A. M. (2025). Vitamin C: A comprehensive review of its role in health, disease prevention, and therapeutic potential. *Molecules, 30*(3), 748. https://doi.org/10.3390/molecules30030748
95. Grădinaru, A. C., & Popa, S. (2025). Vitamin C: From self-sufficiency to dietary dependence in the framework of its biological functions and medical implications. *Life, 15*(2), 238. https://doi.org/10.3390/life15020238
96. Doseděl, M., Jirkovský, E., Macáková, K., Krčmová, L. K., Javorská, L., Pourová, J., Mercolini, L., Remião, F., Nováková, L., Mladěnka, P., & on behalf of The OEMONOM. (2021). Vitamin C—Sources, physiological role, kinetics, deficiency, use, toxicity, and determination. *Nutrients, 13*(2), 615. https://doi.org/10.3390/nu13020615
97. Aragão, M. Â., Pires, L., Santos-Buelga, C., Barros, L., & Calhelha, R. C. (2024). Revitalising riboflavin: Unveiling its timeless significance in human physiology and health. *Foods, 13*(14), 2255. https://doi.org/10.3390/foods13142255
98. Suwannasom, N., Kao, I., Pruß, A., Georgieva, R., & Bäumler, H. (2020). Riboflavin: The health benefits of a forgotten natural vitamin. *International Journal of Molecular Sciences, 21*(3), 950. https://doi.org/10.3390/ijms21030950
99. Frost, Z., Bakhit, S., Amaefuna, C. N., Powers, R. V., & Ramana, K. V. (2025). Recent advances on the role of B vitamins in cancer prevention and progression. *International Journal of Molecular Sciences, 26*(5), 1967. https://doi.org/10.3390/ijms26051967
100. Jahan-Mihan, A., Stevens, P., Medero-Alfonso, S., Brace, G., Overby, L. K., Berg, K., & Labyak, C. (2024). The role of water-soluble vitamins and vitamin D in prevention and

treatment of depression and seasonal affective disorder in adults. *Nutrients, 16*(12), 1902. https://doi.org/10.3390/nu16121902
101. Hossain, K. S., Amarasena, S., & Mayengbam, S. (2022). B vitamins and their roles in gut health. *Microorganisms, 10*(6), 1168. https://doi.org/10.3390/microorganisms10061168
102. Mosegaard, S., Dipace, G., Bross, P., Carlsen, J., Gregersen, N., & Olsen, R. K. J. (2020). Riboflavin deficiency—Implications for general human health and inborn errors of metabolism. *International Journal of Molecular Sciences, 21*(11), 3847. https://doi.org/10.3390/ijms21113847
103. Franco, C. N., Seabrook, L. J., Nguyen, S. T., Leonard, J. T., & Albrecht, L. V. (2022). Simplifying the B complex: How vitamins B6 and B9 modulate one carbon metabolism in cancer and beyond. *Metabolites, 12*(10), 961. https://doi.org/10.3390/metabo12100961
104. Abbaszadeh, F., Javadpour, P., Mousavi Nasab, M. M., & Jorjani, M. (2024). The role of vitamins in spinal cord injury: Mechanisms and benefits. *Oxidative Medicine and Cellular Longevity, 2024*(1), 4293391. https://doi.org/10.1155/2024/4293391
105. Mohd Zaffarin, A. S., Ng, S.-F., Ng, M. H., Hassan, H., & Alias, E. (2020). Pharmacology and pharmacokinetics of vitamin E: Nanoformulations to enhance bioavailability. *International Journal of Nanomedicine, 15*, 9961–9974. https://doi.org/10.2147/IJN.S276355
106. Pisoschi, A. M., Pop, A., Iordache, F., Stanca, L., Geicu, O. I., Bilteanu, L., & Serban, A. I. (2022). Antioxidant, anti-inflammatory and immunomodulatory roles of vitamins in COVID-19 therapy. *European Journal of Medicinal Chemistry, 232*, 114175. https://doi.org/10.1016/j.ejmech.2022.114175
107. Ali, A., Riaz, S., Khalid, W., Fatima, M., Mubeen, U., Babar, Q., et al. (2024). Potential of ascorbic acid in human health against different diseases: An updated narrative review. *International Journal of Food Properties, 27*(1), 493–515. https://doi.org/10.1080/10942912.2024.2327335
108. Mieszczakowska-Frąc, M., Celejewska, K., & Płocharski, W. (2021). Impact of Innovative technologies on the content of vitamin C and its bioavailability from processed fruit and vegetable products. *Antioxidants, 10*(1), 54. https://doi.org/10.3390/antiox10010054
109. Carr, A. C., & Frei, B. (1999). Toward a new recommended dietary allowance for vitamin C based on antioxidant and health effects in humans. *The American Journal of Clinical Nutrition, 69*(6), 1086–1107. https://doi.org/10.1093/ajcn/69.6.1086
110. Pullar, J. M., Carr, A. C., & Vissers, M. C. M. (2017). The roles of vitamin C in skin health. *Nutrients, 9*(8), 866. https://doi.org/10.3390/nu9080866
111. Moser, M. A., & Chun, O. K. (2016). Vitamin C and heart health: A review based on findings from epidemiologic studies. *International Journal of Molecular Sciences, 17*(8), 1328. https://doi.org/10.3390/ijms17081328
112. Morelli, M. B., Gambardella, J., Castellanos, V., Trimarco, V., & Santulli, G. (2020). Vitamin C and cardiovascular disease: An update. *Antioxidants, 9*(12), 1227. https://doi.org/10.3390/antiox9121227
113. McRae, M. P. (2008). Vitamin C supplementation lowers serum low-density lipoprotein cholesterol and triglycerides: A meta-analysis of 13 randomized controlled trials. *Journal of Chiropractic Medicine, 7*(2), 48–58. https://doi.org/10.1016/j.jcme.2008.01.002
114. Traber, M. G., & Stevens, J. F. (2011). Vitamins C and E: Beneficial effects from a mechanistic perspective. *Free Radical Biology & Medicine, 51*(5), 1000–1013. https://doi.org/10.1016/j.freeradbiomed.2011.05.017
115. Esfandiari, M., Bagheri, H., Mirarab-Razi, V., Mirarab Razi, S., Rashidinejad, A. (2023). Vitamin E. In S. M. Jafari, A. Rashidinejad, & J. Simal-Gandara (Eds.), *Handbook of food bioactive ingredients*. Springer. https://doi.org/10.1007/978-3-031-28109-9_24
116. Liao, S., Omage, S. O., Börmel, L., Kluge, S., Schubert, M., Wallert, M., & Lorkowski, S. (2022). Vitamin E and metabolic health: Relevance of interactions with other micronutrients. *Antioxidants (Basel, Switzerland), 11*(9), 1785. https://doi.org/10.3390/antiox11091785
117. Rahim, M. A., Ayub, H., Sehrish, A., Ambreen, S., Khan, F. A., Itrat, N., Nazir, A., Shoukat, A., Shoukat, A., Ejaz, A., Özogul, F., Bartkiene, E., & Rocha, J. M. (2023). Essential components from plant source oils: A review on extraction, detection, identification, and quantification. *Molecules, 28*(19), 6881. https://doi.org/10.3390/molecules28196881

118. Michalak, M., Pierzak, M., Kręcisz, B., & Suliga, E. (2021). Bioactive compounds for skin health: A review. *Nutrients, 13*(1), 203. https://doi.org/10.3390/nu13010203
119. Lewis, E. D., Meydani, S. N., & Wu, D. (2019). Regulatory role of vitamin E in the immune system and inflammation. *IUBMB Life, 71*(4), 487–494. https://doi.org/10.1002/iub.1976
120. Iddir, M., Brito, A., Dingeo, G., Fernandez Del Campo, S. S., Samouda, H., La Frano, M. R., & Bohn, T. (2020). Strengthening the immune system and reducing inflammation and oxidative stress through diet and nutrition: Considerations during the COVID-19 crisis. *Nutrients, 12*(6), 1562. https://doi.org/10.3390/nu12061562
121. Grune, T., Lietz, G., Palou, A., Ross, A. C., Stahl, W., Tang, G., et al. (2010). Beta-carotene is an important vitamin A source for humans. *The Journal of Nutrition, 140*(12), 2268S-2285S. https://doi.org/10.3945/jn.109.119024
122. Harari, A., Melnikov, N., Kandel Kfir, M., Kamari, Y., Mahler, L., Ben-Amotz, A., Harats, D., Cohen, H., & Shaish, A. (2020). Dietary β-carotene rescues vitamin A deficiency and inhibits atherogenesis in apolipoprotein E-deficient mice. *Nutrients, 12*(6), 1625. https://doi.org/10.3390/nu12061625
123. Kim, J. A., Jang, J.-H., & Lee, S.-Y. (2021). An updated comprehensive review on vitamin A and carotenoids in breast cancer: Mechanisms, genetics, assessment, current evidence, and future clinical implications. *Nutrients, 13*(9), 3162. https://doi.org/10.3390/nu13093162
124. Roberts, J. L., & Moreau, R. (2016). Functional properties of spinach (*Spinacia oleracea* L.) phytochemicals and bioactives. *Food & Function, 7*(8), 3337–3353. https://doi.org/10.1039/C6FO00051G
125. Tufail, T., Bader Ul Ain, H., Noreen, S., Ikram, A., Arshad, M. T., & Abdullahi, M. A. (2024). Nutritional benefits of lycopene and beta-carotene: A comprehensive overview. *Food Science & Nutrition, 12*(11), 8715–8741. https://doi.org/10.1002/fsn3.4502
126. Llauradó Maury, G., Méndez Rodríguez, D., Hendrix, S., Escalona Arranz, J. C., Fung Boix, Y., Pacheco, A. O., et al. (2020). Antioxidants in plants: A valorization potential emphasizing the need for the conservation of plant biodiversity in cuba. *Antioxidants, 9*(11), 1048. https://doi.org/10.3390/antiox9111048
127. Delbrouck, J. A., Desgagné, M., Comeau, C., Bouarab, K., Malouin, F., & Boudreault, P.-L. (2023). The therapeutic value of solanum steroidal (glyco)alkaloids: A 10-year comprehensive review. *Molecules, 28*(13), 4957. https://doi.org/10.3390/molecules28134957
128. Aune, D. (2019). Plant foods, antioxidant biomarkers, and the risk of cardiovascular disease, cancer, and mortality: A review of the evidence. *Advances in Nutrition (Bethesda, Md.), 10*(Suppl_4), S404–S421. https://doi.org/10.1093/advances/nmz042
129. Pandey, K. B., & Rizvi, S. I. (2009). Plant polyphenols as dietary antioxidants in human health and disease. *Oxidative Medicine and Cellular Longevity, 2*(5), 270–278. https://doi.org/10.4161/oxim.2.5.9498
130. Genchi, G., Lauria, G., Catalano, A., Sinicropi, M. S., & Carocci, A. (2023). Biological activity of selenium and its impact on human health. *International Journal of Molecular Sciences, 24*(3), 2633. https://doi.org/10.3390/ijms24032633
131. Kieliszek, M., & Serrano Sandoval, S. N. (2023). The importance of selenium in food enrichment processes. A comprehensive review. *Journal of Trace Elements in Medicine and Biology: Organ of the Society for Minerals and Trace Elements (GMS), 79*, 127260. https://doi.org/10.1016/j.jtemb.2023.127260
132. Zhang, F., Li, X., & Wei, Y. (2023). Selenium and selenoproteins in health. *Biomolecules, 13*(5), 799. https://doi.org/10.3390/biom13050799
133. Barchielli, G., Capperucci, A., & Tanini, D. (2022). The role of selenium in pathologies: An updated review. *Antioxidants, 11*(2), 251. https://doi.org/10.3390/antiox11020251
134. Shafe, M. O., Gumede, N. M., Nyakudya, T. T., & Chivandi, E. (2024). Lycopene: A potent antioxidant with multiple health benefits. *Journal of Nutrition and Metabolism, 2024*(1), 6252426. https://doi.org/10.1155/2024/6252426
135. Bin-Jumah, M. N., Nadeem, M. S., Gilani, S. J., Mubeen, B., Ullah, I., Alzarea, S. I., Ghoneim, M. M., Alshehri, S., Al-Abbasi, F. A., & Kazmi, I. (2022). Lycopene: A natural arsenal in the war against oxidative stress and cardiovascular diseases. *Antioxidants, 11*(2), 232. https://doi.org/10.3390/antiox11020232

136. Przybylska, S., & Tokarczyk, G. (2022). Lycopene in the prevention of cardiovascular diseases. *International Journal of Molecular Sciences, 23*(4), 1957. https://doi.org/10.3390/ijms23041957
137. Mozos, I., Stoian, D., Caraba, A., Malainer, C., Horbańczuk, J. O., & Atanasov, A. G. (2018). Lycopene and vascular health. *Frontiers in Pharmacology, 9*, 521. https://doi.org/10.3389/fphar.2018.00521
138. Falsafi, S. R., Rostamabadi, H., Babazadeh, A., Tarhan, Ö., Rashidinejad, A., Boostani, S., et al. (2022). Lycopene nanodelivery systems; recent advances. *Trends in Food Science & Technology, 119*, 378–399. https://doi.org/10.1016/j.tifs.2021.12.016
139. Banaszak, M., Dobrzyńska, M., Kawka, A., Górna, I., Woźniak, D., Przysławski, J., & Drzymała-Czyż, S. (2024). Role of omega-3 fatty acids eicosapentaenoic (EPA) and docosahexaenoic (DHA) as modulatory and anti-inflammatory agents in noncommunicable diet-related diseases—Reports from the last 10 years. *Clinical Nutrition ESPEN, 63*, 240–258. https://doi.org/10.1016/j.clnesp.2024.06.053
140. García-Maldonado, E., Alcorta, A., Zapatera, B., & Vaquero, M. P. (2023). Changes in fatty acid levels after consumption of a novel docosahexaenoic supplement from algae: A crossover randomized controlled trial in omnivorous, lacto-ovo vegetarians and vegans. *European Journal of Nutrition, 62*(4), 1691–1705. https://doi.org/10.1007/s00394-022-03050-3
141. Djuricic, I., & Calder, P. C. (2021). Beneficial outcomes of omega-6 and omega-3 polyunsaturated fatty acids on human health: An update for 2021. *Nutrients, 13*(7), 2421. https://doi.org/10.3390/nu13072421
142. Ibrahim, D., Arisha, A. H., Khater, S. I., Gad, W. M., Hassan, Z., Abou-Khadra, S. H., Mohamed, D. I., Ahmed Ismail, T., Gad, S. A., Eid, S. A. M., Abd El-Wahab, R. A., & Kishawy, A. T. Y. (2022). Impact of omega-3 fatty acids nano-formulation on growth, antioxidant potential, fillet quality, immunity, autophagy-related genes and aeromonas hydrophila resistance in Nile Tilapia (*Oreochromis niloticus*). *Antioxidants, 11*(8), 1523. https://doi.org/10.3390/antiox11081523
143. Chitre, N. M., Moniri, N. H., & Murnane, K. S. (2019). Omega-3 Fatty Acids as druggable therapeutics for neurodegenerative disorders. *CNS & Neurological Disorders Drug Targets, 18*(10), 735–749. https://doi.org/10.2174/1871527318666191114093749
144. Vardanyan, R., & Hruby, V. (2016). Hypolipidemic and antihyperlipidemic drugs. In *Synthesis of best-seller drugs* (pp. 285–315). https://doi.org/10.1016/B978-0-12-411492-0.00020-1
145. Karageorgou, D., Rova, U., Christakopoulos, P., Katapodis, P., Matsakas, L., & Patel, A. (2023). Benefits of supplementation with microbial omega-3 fatty acids on human health and the current market scenario for fish-free omega-3 fatty acid. *Trends in Food Science & Technology*. https://doi.org/10.1016/j.tifs.2023.04.018
146. Burns-Whitmore, B., Froyen, E., Heskey, C., Parker, T., & San Pablo, G. (2019). Alpha-linolenic and linoleic fatty acids in the vegan diet: Do they require dietary reference intake/adequate intake special consideration? *Nutrients, 11*(10), 2365. https://doi.org/10.3390/nu11102365
147. Murru, E., Manca, C., Carta, G., & Banni, S. (2022). Impact of dietary palmitic acid on lipid metabolism. *Frontiers in Nutrition, 9*, 861664. https://doi.org/10.3389/fnut.2022.861664
148. Mancini, A., Imperlini, E., Nigro, E., Montagnese, C., Daniele, A., Orrù, S., & Buono, P. (2015). Biological and nutritional properties of palm oil and palmitic acid: Effects on health. *Molecules, 20*(9), 17339–17361. https://doi.org/10.3390/molecules200917339
149. Ceja-Galicia, Z. A., Cespedes-Acuña, C. L. A., & El-Hafidi, M. (2025). Protection strategies against palmitic acid-induced lipotoxicity in metabolic syndrome and related diseases. *International Journal of Molecular Sciences, 26*(2), 788. https://doi.org/10.3390/ijms26020788
150. Carta, G., Murru, E., Banni, S., & Manca, C. (2017). Palmitic acid: Physiological role, metabolism and nutritional implications. *Frontiers in Physiology, 8*. https://doi.org/10.3389/fphys.2017.00902

151. Afifi, E. H., John Martin, J. J., Wang, Q., Li, X., Liu, X., Zhou, L., Li, R., Fu, D., Li, Q., Ye, J., & Cao, H. (2025). Fatty acid and lipid metabolism in oil palm: from biochemistry to molecular mechanisms. *International Journal of Molecular Sciences, 26*(6), 2531. https://doi.org/10.3390/ijms26062531
152. Gesteiro, E., Guijarro, L., Sánchez-Muniz, F. J., Vidal-Carou, M. C., Troncoso, A., Venanci, L., Jimeno, V., Quilez, J., Anadón, A., & González-Gross, M. (2019). Palm oil on the edge. *Nutrients, 11*(9), 2008. https://doi.org/10.3390/nu11092008
153. Albrahim, T., Alotaibi, M. H. M., Altamimi, N. M. M., Albariqi, A. M. A., Alqarni, L. A. O., Alassaf, S. N. A., Aloudah, H. S., Alahmed, M., Almnaizel, A. T., Aldraihem, M. R., & Alonazi, M. (2022). The impact of dietary consumption of palm oil and olive oil on lipid profile and hepatocyte injury in hypercholesterolemic rats. *Pharmaceuticals, 15*(9), 1103. https://doi.org/10.3390/ph15091103
154. van Rooijen, M. A., & Mensink, R. P. (2020). Palmitic acid versus stearic acid: Effects of interesterification and intakes on cardiometabolic risk markers—A systematic review. *Nutrients, 12*(3), 615. https://doi.org/10.3390/nu12030615
155. Conde, T. A., Neves, B. F., Couto, D., Melo, T., Neves, B., Costa, M., Silva, J., Domingues, P., & Domingues, M. R. (2021). Microalgae as sustainable bio-factories of healthy lipids: Evaluating fatty acid content and antioxidant activity. *Marine Drugs, 19*(7), 357. https://doi.org/10.3390/md19070357
156. Santin, A., Russo, M. T., Ferrante, M. I., Balzano, S., Orefice, I., & Sardo, A. (2021). Highly valuable polyunsaturated fatty acids from microalgae: Strategies to improve their yields and their potential exploitation in aquaculture. *Molecules, 26*(24), 7697. https://doi.org/10.3390/molecules26247697
157. Ashfaq, R., Rasul, A., Asghar, S., Kovács, A., Berkó, S., & Budai-Szűcs, M. (2023). Lipid nanoparticles: An effective tool to improve the bioavailability of nutraceuticals. *International Journal of Molecular Sciences, 24*(21), 15764. https://doi.org/10.3390/ijms242115764
158. Krishnaswami, V., Mohanan, D. P., Raja, S. A. J., Natarajan, B., & Velusamy, S. (2024). Nanotechnology-based advancements for effective delivery of phytoconstituents for ocular diseases. *Nano TransMed*, 100056. https://doi.org/10.1016/j.ntm.2024.100056
159. Liao, H., Zhu, D., Bai, M., Chen, H., Yan, S., Yu, J., et al. (2020). Stigmasterol sensitizes endometrial cancer cells to chemotherapy by repressing Nrf2 signal pathway. *Cancer Cell International, 20*(1), 480. https://doi.org/10.1186/s12935-020-01470-x
160. Lifsey, H. C., Kaur, R., Thompson, B. H., Bennett, L., Temel, R. E., & Graf, G. A. (2020). Stigmasterol stimulates transintestinal cholesterol excretion independent of liver X receptor activation in the small intestine. *The Journal of Nutritional Biochemistry, 76*, Article 108263. https://doi.org/10.1016/j.jnutbio.2019.108263
161. Bakrim, S., Benkhaira, N., Bourais, I., Benali, T., Lee, L.-H., El Omari, N., Sheikh, R. A., Goh, K. W., Ming, L. C., & Bouyahya, A. (2022). Health benefits and pharmacological properties of stigmasterol. *Antioxidants, 11*(10), 1912. https://doi.org/10.3390/antiox11101912
162. Zhang, X., Wang, J., Zhu, L., Wang, X., Meng, F., Xia, L., & Zhang, H. (2022). Advances in Stigmasterol on its anti-tumor effect and mechanism of action. *Frontiers in Oncology, 12*, 1101289. https://doi.org/10.3389/fonc.2022.1101289
163. Valitova, J., Renkova, A., Beckett, R., & Minibayeva, F. (2024). Stigmasterol: An enigmatic plant stress sterol with versatile functions. *International Journal of Molecular Sciences, 25*(15), 8122. https://doi.org/10.3390/ijms25158122
164. Shen, M., Yuan, L., Zhang, J., Wang, X., Zhang, M., Li, H., Jing, Y., Zeng, F., & Xie, J. (2024). Phytosterols: Physiological functions and potential application. *Foods, 13*(11), 1754. https://doi.org/10.3390/foods13111754
165. Evtyugin, D. D., Evtuguin, D. V., Casal, S., & Domingues, M. R. (2023). Advances and challenges in plant sterol research: Fundamentals, analysis. *Applications and Production. Molecules, 28*(18), 6526. https://doi.org/10.3390/molecules28186526
166. Bhalani, D. V., Nutan, B., Kumar, A., & Singh Chandel, A. K. (2022). Bioavailability enhancement techniques for poorly aqueous soluble drugs and therapeutics. *Biomedicines, 10*(9), 2055. https://doi.org/10.3390/biomedicines10092055

167. Goswami, M., Priya, Jaswal, S., Gupta, G. D., & Verma, S. K. (2023). A comprehensive update on phytochemistry, analytical aspects, medicinal attributes, specifications and stability of stigmasterol. *Steroids, 196*, 109244. https://doi.org/10.1016/j.steroids.2023.109244
168. Aguirre-Crespo, F. J., Aragón-Gastélum, J. L., Gutiérrez-Alcántara, E. J., Zamora-Crescencio, P., Gómez-Galicia, D. L., Alatriste-Kurzel, D. R., Alvarez, G., & Hernández-Núñez, E. (2024). β-Sitosterol mediates gastrointestinal smooth muscle relaxation induced by *Coccoloba uvifera* via muscarinic acetylcholine receptor subtype 3. *Scientia Pharmaceutica, 92*(2), 19. https://doi.org/10.3390/scipharm92020019
169. Khan, Z., Nath, N., Rauf, A., Emran, T. B., Mitra, S., & Islam, F. (2022). Multifunctional roles and pharmacological potential of β-sitosterol: Emerging evidence toward clinical applications. *Chemico-Biological Interactions, 365*, 110117. https://doi.org/10.1016/j.cbi.2022.110117
170. Babu, S., & Jayaraman, S. (2020). An update on β-sitosterol: A potential herbal nutraceutical for diabetic management. *Biomedicine & Pharmacotherapy, 131*, 110702. https://doi.org/10.1016/j.biopha.2020.110702
171. Lin, H., Li, Z., Sun, Y., Zhang, Y., Wang, S., Zhang, Q., Cai, T., Xiang, W., Zeng, C., & Tang, J. (2024). D-Limonene: Promising and sustainable natural bioactive compound. *Applied Sciences, 14*(11), 4605. https://doi.org/10.3390/app14114605
172. Eddin, L. B., Jha, N. K., Meeran, M. F. N., Kesari, K. K., Beiram, R., & Ojha, S. (2021). Neuroprotective potential of limonene and limonene containing natural products. *Molecules, 26*(15), 4535. https://doi.org/10.3390/molecules26154535
173. Kamran, S., Sinniah, A., Abdulghani, M. A. M., & Alshawsh, M. A. (2022). Therapeutic potential of certain terpenoids as anticancer agents: A scoping review. *Cancers, 14*(5), 1100. https://doi.org/10.3390/cancers14051100
174. Narayanankutty, A., Visakh, N. U., Sasidharan, A., Pathrose, B., Olatunji, O. J., Al-Ansari, A., Alfarhan, A., & Ramesh, V. (2022). Chemical composition, antioxidant, anti-bacterial, and anti-cancer activities of essential oils extracted from *Citrus limetta* Risso peel waste remains after commercial use. *Molecules, 27*(23), 8329. https://doi.org/10.3390/molecules27238329
175. Ibáñez, M. D., Sanchez-Ballester, N. M., & Blázquez, M. A. (2020). Encapsulated limonene: A pleasant lemon-like aroma with promising application in the agri-food industry. A review. *Molecules, 25*(11), 2598. https://doi.org/10.3390/molecules25112598
176. Andriotis, E. G., Papi, R. M., Paraskevopoulou, A., & Achilias, D. S. (2021). Synthesis of D-limonene loaded polymeric nanoparticles with enhanced antimicrobial properties for potential application in food packaging. *Nanomaterials, 11*(1), 191. https://doi.org/10.3390/nano11010191
177. Nisar, M. F., Khadim, M., Rafiq, M., Chen, J., Yang, Y., & Wan, C. C. (2021). Pharmacological properties and health benefits of eugenol: A comprehensive review. *Oxidative Medicine and Cellular Longevity, 2021*(1), 2497354. https://doi.org/10.1155/2021/2497354
178. de Araújo Lopes, A., da Fonseca, F. N., Rocha, T. M., de Freitas, L. B., Araújo, E. V. O., Wong, D. V. T., et al. (2018). Eugenol as a promising molecule for the treatment of dermatitis: Antioxidant and anti-inflammatory activities and its nano formulation. *Oxidative Medicine and Cellular Longevity, 2018*(1), 1–13. https://doi.org/10.1155/2018/8194849
179. Ulanowska, M., & Olas, B. (2021). Biological properties and prospects for the application of eugenol—A review. *International Journal of Molecular Sciences, 22*(7), 3671. https://doi.org/10.3390/ijms22073671
180. Ahmad, N., Ahmad, F. J., Bedi, S., Sharma, S., Umar, S., & Ansari, M. A. (2019). A novel nano formulation development of eugenol and their treatment in inflammation and periodontitis. *Saudi Pharmaceutical Journal, 27*(6), 778–790. https://doi.org/10.1016/j.jsps.2019.04.014
181. Fu, X., Gao, Y., Yan, W., Zhang, Z., Sarker, S., Yin, Y., et al. (2022). Preparation of eugenol nano emulsions for antibacterial activities. *RSC Advances, 12*(6), 3180–3190. https://doi.org/10.1039/D1RA08184E
182. Blondeau, N., Lipsky, R. H., Bourourou, M., Duncan, M. W., Gorelick, P. B., & Marini, A. M. (2015). Alpha-linolenic acid: An omega-3 fatty acid with neuroprotective properties-ready for use in the stroke clinic? *BioMed Research International, 2015*, 519830. https://doi.org/10.1155/2015/519830

183. Simopoulos, A. P. (2004). Omega-3 fatty acids and antioxidants in edible wild plants. *Biological Research, 37*(2), 263–277. https://doi.org/10.4067/S0716-97602004000200013
184. Superti, F., & Russo, R. (2024). Alpha-lipoic acid: biological mechanisms and health benefits. *Antioxidants, 13*(10), 1228. https://doi.org/10.3390/antiox13101228
185. Bertoni, C., Pini, C., Mazzocchi, A., Agostoni, C., & Brambilla, P. (2024). The role of alpha-linolenic acid and other polyunsaturated fatty acids in mental health: A narrative review. *International Journal of Molecular Sciences, 25*(22), 12479. https://doi.org/10.3390/ijms252212479
186. Tomczyk, M. (2024). Omega-3 fatty acids and muscle strength—Current state of knowledge and future perspectives. *Nutrients, 16*(23), 4075. https://doi.org/10.3390/nu16234075
187. Takic, M., Pokimica, B., Petrovic-Oggiano, G., & Popovic, T. (2022). Effects of dietary α-linolenic acid treatment and the efficiency of its conversion to eicosapentaenoic and docosahexaenoic acids in obesity and related diseases. *Molecules, 27*(14), 4471. https://doi.org/10.3390/molecules27144471
188. Casas, A., Perotti, C., Saccoliti, M., Sacca, P., Fukuda, H. C., & Batlle, A. M. (2002). ALA and ALA hexyl ester in free and liposomal formulations for the photosensitisation of tumour organ cultures. *British Journal of Cancer, 86*, 837–842. https://doi.org/10.1038/sj.bjc.6600144
189. Cassano, R., Serini, S., Curcio, F., Trombino, S., & Calviello, G. (2022). Preparation and study of solid lipid nanoparticles based on curcumin, resveratrol and capsaicin containing linolenic acid. *Pharmaceutics, 14*(8), 1593. https://doi.org/10.3390/pharmaceutics14081593
190. Kumar, R., Kumar, A., Kumar, D., Yadav, S., Shrivastava, N. K., Singh, J., Sonkar, A. B., Verma, P., Arya, D. K., Kaithwas, G., Agrarwal, A. K., & Singh, S. (2024). Harnessing potential of ω-3 polyunsaturated fatty acid with nanotechnology for enhanced breast cancer therapy: A comprehensive investigation into ALA-based liposomal PTX delivery. *Pharmaceutics, 16*(7), 913. https://doi.org/10.3390/pharmaceutics16070913
191. Subongkot, T., Ngawhirunpat, T., & Opanasopit, P. (2021). Development of ultradeformable liposomes with fatty acids for enhanced dermal rosmarinic acid delivery. *Pharmaceutics, 13*(3), 404. https://doi.org/10.3390/pharmaceutics13030404
192. Mao, Q.-Q., Xu, X.-Y., Cao, S.-Y., Gan, R.-Y., Corke, H., Beta, T., & Li, H.-B. (2019). Bioactive compounds and bioactivities of ginger (*Zingiber officinale* Roscoe). *Foods, 8*(6), 185. https://doi.org/10.3390/foods8060185
193. Ayustaningwarno, F., Anjani, G., Ayu, A. M., & Fogliano, V. (2024). A critical review of ginger's (*Zingiber officinale*) antioxidant, anti-inflammatory, and immunomodulatory activities. *Frontiers in Nutrition, 11*, 1364836. https://doi.org/10.3389/fnut.2024.1364836
194. Geng, X., Liu, H., Yuwen, Q., Wang, J., Zhang, S., Zhang, X., & Sun, J. (2021). Protective effects of zingerone on high cholesterol diet-induced atherosclerosis through lipid regulatory signaling pathway. *Human & Experimental Toxicology, 40*(10), 1732–1745. https://doi.org/10.1177/09603271211006170
195. Ballester, P., Cerdá, B., Arcusa, R., García-Muñoz, A. M., Marhuenda, J., & Zafrilla, P. (2023). Antioxidant activity in extracts from Zingiberaceae family: Cardamom, turmeric, and ginger. *Molecules, 28*(10), 4024. https://doi.org/10.3390/molecules28104024
196. Min, M., Egli, C., Bartolome, R. A., & Sivamani, R. K. (2024). Ex vivo evaluation of a liposome-mediated antioxidant delivery system on markers of skin photoaging and skin penetration. *Clinical, Cosmetic and Investigational Dermatology, 17*, 1481–1494. https://doi.org/10.2147/CCID.S461753
197. Kung, M.-L., Lin, P.-Y., Huang, S.-T., Tai, M.-H., Hsieh, S.-L., Wu, C.-C., et al. (2019). Zingerone nanotetramer strengthened the polypharmacological efficacy of zingerone on human hepatoma cell lines. *ACS Applied Materials & Interfaces, 11*(1), 137–150. https://doi.org/10.1021/acsami.8b14559
198. Yang, C.-M., Chu, T.-H., Tsai, K.-W., Hsieh, S., & Kung, M.-L. (2022). Phytochemically derived zingerone nanoparticles inhibit cell proliferation, invasion and metastasis in human oral squamous cell carcinoma. *Biomedicines, 10*(2), 320. https://doi.org/10.3390/biomedicines10020320

Properties, Molecular Effects and Signalling Pathways of Reactive Oxygen Species

Devipriya Anbumani, Kayal Vizhi Dhandapani, and Babujanarthanam Ranganathan

Abstract The main component of reactive oxygen species, free radicals are very reactive molecules with unpaired electrons that harm cells and cause oxidative stress. They have a substantial impact on conditions like cancer, neurodegeneration and cardiovascular disorders. Different characteristics and reactivities are displayed by free radicals, including superoxides, hydroxyl radicals, nitric oxide and peroxyl radicals. Every kind of free radical affects inflammation, oxidative stress and the aetiology of illness in a different way. Deciphering the specific characteristics and effects of different free radicals can inform the development of targeted therapeutic strategies to mitigate oxidative damage and promote cellular health. Reactive oxygen species exhibit dual roles as both deleterious and beneficial molecules, influencing various cellular processes. Their properties and molecular effects modulate signalling pathways, impacting cell growth, differentiation and survival. It activates the key signalling pathways, like MAPK/ERK, PI3K/Akt, NF-κB, Jak-STAT, PKC, P53 and Nrf2/Keap1, influencing cell fate decisions. These pathways orchestrate responses to oxidative stress, regulating inflammation, survival and antioxidant defences, with implications for disease mechanisms and treatment strategies.

Keywords Free radicals · Reactive oxygen species · Molecular target · Signalling pathway

1 Introduction

According to Reuter et al. [1], in conditions characterised by oxidative stress (OS), free radical production goes out of balance with the production of reactive metabolites. Antioxidants remove reactive oxygen species, also known as oxidants. The term "*Reactive Oxygen Species (ROS)*" was initially used in the 1950's. It seems more

D. Anbumani · K. V. Dhandapani · B. Ranganathan (✉)
Nano and Energy Biosciences Laboratory, Department of Biotechnology, Thiruvalluvar University, Serkkadu, Vellore, Tamilnadu, India
e-mail: babukmg@gmail.com

K. Kasinathan and A. Muthukumaran (eds.), *Recent Advances in Oxidative Stress Associated Chronic Diseases Volume 2*, https://doi.org/10.1007/978-981-95-3758-7_14

likely that reactive oxygen species will be harmful. However, we now have a knowledge of ROS after more than 50 years. The various oxygen-containing species that are unavoidably produced during cellular metabolism are typically referred to as ROS. Because of their extreme reactivity, ROS are more reactive than molecular oxygen, which can damage lipids, proteins and DNA. To maintain a state of redox equilibrium, cells normally use defence or antioxidative systems to balance these harmful products [2]. It is possible to experience damage from superoxide anion radicals, peroxide radicals, as well as hydroxyl radical's species, as a result of its unpaired electrons, this molecular system is extremely reactive produced during regular physiological activities like inflammation or aerobic metabolism. Several antioxidant enzymes, such as glutathione peroxidase, play a critical role in anti-ageing, and thioredoxin reductase, as well as antioxidant scavengers such ascorbate, glutathione and thioredoxin, are cellular defences against ROS. Since oxidation–reduction (*redox*) events affect metabolic and other signal transduction pathways, they are essential to a cell. Damage to cells occurs when ROS production surpasses the antioxidant defences of the cell. A cell's capacity to protect itself from ROS is linked to chemotherapy resistance, and it is becoming more and more obvious that the production of ROS can be used therapeutically to treat cancer [3].

2 A Comparison of Free Radical Properties

2.1 Radical of Oxidative Stress

Electrons in antibond π^* orbitals is received to produce the reduced form of molecular oxygen known as the superoxide anion ($O_2^{\cdot -}$) [4]. Despite its high reactivity with biomolecules, superoxide does not maintain a high degree of stability, and it is less radical than O_2 due to its single unpaired electron. Another electron is added to superoxide anion produces O_2^-. There are no unpaired electrons in the peroxide ion, which has a weaker oxygen–oxygen bond than a radical. The bond is totally broken when two more electrons are added to O_2^{2-}, resulting in the production of two O_2^- (oxide ions). In biology, O_2 reduces to H_2O_2, which has two electrons, and water, which has four electrons. It is mostly produced during, as a result of oxidative phosphorylation in the mitochondria, adenosine triphosphate (ATP) is produced [5]. Biochemical systems generate superoxide anion primarily by enzymatic or non-enzymatic means. Oxygen molecules are directly electronic or photochemically reduced [6]. Superoxide is produced by enzymes such as cytochrome P450-dependent oxygenases, the oxidation of xanthine, cyclooxygenase and nicotinamide adenine dinucleotide phosphate (NADPH) are carried out by these enzymes. Superoxide radicals are moderately reactive free radicals, capable of forming hydrogen peroxide (H_2O_2). Having the unusual property of being able to be reduced into the extremely reactive hydroxyl radical either in full or in part. Superoxide radicals are capable of dismutate spontaneously or with the help of enzymes called superoxide dismutases. H_2O_2 can decompose into a hydroxyl radical through the Fenton reaction, which is catalysed, in the lower valence state of transition metals like Fe^{2+} and Cu^{+} or in the presence of superoxide radicals (*Haber–Weiss reaction*). By re-reducing oxidised transitional metals, the superoxide anion can repair the damage caused by Fenton reactions [7].

2.2 Hydroxyl Radical

Biologically active free radicals, chemically speaking, is the hydroxyl radical (HO·). It is created when superoxide combines with ferric iron (Fe^{3+}) to make Fe^{2+} and when free iron (Fe^{2+}) reacts with hydrogen peroxide in the Fenton reaction. Cu^{2+}, Fe^{3+}, Ti^{4+} and Co^{3+} are a few other ions that may be involved in the reaction in addition to iron. It is possible to recycle these ions into oxygen (O_2) through the interaction between superoxide anion and these ions. In an average cell, about 50 hydroxyl radicals are formed every second. Considering that radicals attack all molecules, they are considered the most dangerous free species. The distance between their source and their destination is less than a few nanometres (nm). As well as having the highest one-electron reduction potential (2310 mV), they are also among the most reactive. The reaction rate constant of this substance ranges from 109 to 1010/M/s for anything living [8]. More quickly than it is generated, the hydroxyl radical reacts powerfully with most organic and inorganic compounds, including deoxyribonucleic acid. In addition to vitamins, minerals, lipids, amino acids, carbohydrates and metals, there are also antioxidants [9]. In these reactions, electrons are added and transferred, and hydrogen is abstracted. A hydroxyl radical produces a free radical in saturated compounds by C-H bonds with weaker hydrogen atoms removed. Other free radicals may be produced when the resultant radicals interact with oxygen. Double bonds readily accept hydroxyl radicals. According to Davies et al. [10], hydroxyl radicals have the ability to oxidise amino acid residues of protein and inactivate all mitochondrial enzyme proteins. Additionally, hydroxyl radicals are implicated in conditions including cancer [11] and cardiovascular disease [12].

2.3 Hydroperoxyl Radical

A radical is protonated when it reacts with a superoxide anion hydroperoxide breakdown produces the simplest form of a peroxyl radical, known as a hydroperoxyl radical (HO_2·), also called a perhydroxyl radical or hydroperoxyl radical. As oxidants, protonated superoxide is present in the cytosol at a concentration of about 0.3% [13]. Fenton or Haber–Weiss reactions can be triggered by the hydroperoxyl radical's production of H_2O_2, among them are iron and copper, which both exhibit redox activity. Additionally, nicotinamide adenine dinucleotide hydrogen (NADH) or glyceraldehyde-3-phosphate dehydrogenase-NADH can hand are their hydrogen atoms extracted by the hydroperoxyl radical, producing H_2O_2. Its reactions are competitive with organic peroxyl radicals, but they happen more slowly than those of HO·. Lipid peroxidation chemistry plays an important role on the hydroperoxyl radical. Its capacity to take additionally, the fat contains lauric, homolenic and omega-3 fatty acids, far more potent oxidant than superoxide anion, indicating a potential function in the early stages of lipid oxidation [14].

2.4 Peroxyl Radical

In biological processes, oxygen produces the peroxyl radical (ROO·). The protonation of superoxide produces the perhydroxyl radical (HOO·), peroxyl radicals are the simplest type of radicals. Normal cells contain protonated forms of these molecules in their cytoplasms of superoxide anion ($O_2^{\cdot-}$) makes up around 0.3% of the total. It can also encourage the growth of tumours and initiates the peroxidation of fatty acids [15].

2.5 Singlet Oxygen

In addition to being highly reactive, it is toxic as well, singlet oxygen ($^1O_2$) is a meta-stable, electrically high excited state of molecular oxygen. The first state, during the process of oxygen molecule's excitement, it produces upon activation is 1Δg, followed by the excited singlet chain, 1Δg A pair of opposite-spin electrons occupy the same orbital in the first excited state. There are two electrons on the same electron that spin in opposite directions π^* orbital in the first excited state, 1Δg, while one electron with opposing spins is present in each degraded π^* orbital in the second excited state, 1Δg. According to Agnez-Lima et al. [16], comparing 1G to other electronically excited states, it is incredibly responsive. It is created in vivo by activating eosinophils and neutrophils ($HOCl + H_2O_2 \rightarrow 1O_2 + H_2O + Cl^-$) [17]. Additionally, it is created by some enzymatic reactions Lactoperoxidase, for example, facilitates the process, dioxygenases and lipoxygenases. DNA and tissue damage can result from this extremely strong oxidising agent [18].

2.6 Hydrogen Peroxide

The primary source of hydrogen peroxide (H_2O_2) is enzyme reactions, which can also be produced through the direct decrease of O_2 as well as the dismutation of superoxide anion. According to Winterbourn [19], oxidases, which are present in mitochondria, peroxisomes and microsomes, can directly synthesise hydrogen peroxide is formed when atomic oxygen accepts two electrons. As well as urea and glucose, oxidases are used to oxidise them, a type of enzyme, because it is liposoluble, hydrogen peroxide can permeate through cell membranes. The most of lipids, proteins and nucleic acids cannot be easily oxidised by this non-free radical due to its poor reactivity. H_2O_2 is dangerous because it can undergo homolytic fission, which is caused by UV light or as a result of a reaction with transition metal ions, and hydroxyl radical is produced. An oxidising reaction between hydrogen peroxide and a superoxide anion, HOCl, or chloramine in a biological system is called an oxidising reaction, singlet oxygen can be produced. When H_2O_2 directly attacks the structure of heme proteins, iron is released, enzymes are inactivated, and deoxyribonucleic acid, lipids, keto-acids and -SH groups are oxidised [20].

2.7 *Ozone*

Throughout Earth's history, high-energy electromagnetic rays and electrical impulses combined to create ozone (O_3) from O_2. According to Altenhofer et al. [21], it is a much stronger oxidising agent than oxygen and slightly less reactive than HO·. By oxidising biological molecules, it can produce free radicals and damage proteins, nucleic acids and lipids [22, 23]. Additionally, inflammatory processes are significantly influenced by ozone [24].

2.8 *Carbonate Radical Anion*

Aqueous solutions of bicarbonate/carbonate can be radiolysed to produce the carbonate radical anion ($CO_3^{\cdot-}$) or ·OH can react with carbonate or bicarbonate ions to form it. The process is made possible by high blood plasma bicarbonate levels (25 mM) [25]. Through electron transfer and hydrogen abstraction, radical anion of carbonates functions, in its capacity to oxidise with one electron, but not as potent as the hydroxyl radical. Hydroxyl has a longer half-life, so it can last longer oxidatively alter distant cell targets and propagate farther. $CO_3^{\cdot-}$ can oxidise a broad range of biomolecules. It is thought to be a significant in addition to oxidising DNA guanine bases, this substance also oxidises proteins, produces stable guanine oxidation compounds. A major Increases production of peroxynitrite, causing cells to oxidise oxidising them, the metabolism of xanthine oxidase, and the synthesis of superoxide dismutase has been suggested to be the carbonate radical anion. It is known to have a significant impact on how certain amino acids in proteins are altered in oxidative stress, ageing and inflammatory situations. A distinct function of $CO_3^{\cdot-}$ in MnSOD (Manganese superoxide dismutase) nitration by peroxynitrite is suggested by kinetic analysis of nitrogen nitration of tyrosine of CO_2. Diabetes, cardiovascular diseases and neurodegenerative diseases have all been linked to tyrosine nitration [26].

2.9 *Hypochlorous Acid*

One extremely reactive species that plays a role, as a result of oxidation processes, proteins and lipid molecules are chlorinated hypochlorous acid (HOCl). Myeloperoxidase in neutrophils and macrophages in inflammatory areas catalyses a reaction between hydrogen peroxide and the chloride anion that produces it. It has the ability to oxidise thiols as well as other biological substances like tryptophan, urate, ascorbate, along with pyridine nucleotides. Among the substances that HOCl chlorinates include, in the case of unreacted amino acids, chloramines are formed, tyrosyl residues are converted to chlorohydrins, and DNA may be changed to chlorohydrins as well [27].

2.10 *Nitric Oxide*

Reactive nitrogen species (RNS) include non-radicals like HNO_2 (nitrous acid) and N_2O_4 (dinitrogen tetroxide), additionally, peroxynitrite and nitric oxide (NO^+) are present. Nitrogen monoxide, often known as nitric oxide, is a free radical that has one unpaired electron. Nitric oxide's direct toxicity is lower than that of ROS because of its relatively low chemical reactivity. But when it reacts with superoxide anion, it produces peroxynitrite anion, which is damaging to DNA, lipids and proteins. Additionally, nitric oxide can, as it combines with oxygen and nitrogen, it creates produce the hazardous oxidising and nitrosating substances nitrogen dioxide or dinitrogen trioxide. Certain nitric oxide synthases [28], H_2O_2 reacts with arginine in order to form, breakdown of S-nitrosothiols ionised metals are the ways that nitric oxide is produced in cells and tissues. Since nitric oxide dissolves in both fat and water, it easily permeates the plasma membrane and cytoplasm. When NO· is introduced to human blood plasma, lipid peroxidation is initiated, and ascorbic acid and uric acid contents are reduced. In cell membranes and lipoproteins, species formed from nitric oxide react rapidly, by reacting with lipid peroxyl radicals, fatty acids transform into oxidised lipids, producing esterified cholesterol, lipid products that have been oxidised and nitrated [29]. In addition, nitric oxide plays a role in a number of physiological processes, including neurotransmission, smooth muscle relaxation, vasodilation. The regulation of blood pressure, the regulation of genes, the promotion of immunity, the maintenance of cellular functions and the regulation of inflammatory and immune responses are among the functions of this hormone. A variety of pathological processes are also influenced by it, including heart disease, neurodegenerative disorders [30].

2.11 Peroxynitrite

Under conditions of superoxide anion and nitric oxide, peroxynitrite ($ONOO^-$) is created. It can immediately combine with CO_2 to produce additional highly reactive nitrosoperoxo-carboxylates or peroxynitrous acid, which can then be further homolysis produces OH and N_2O while rearranging N_2O into OH NO_3 (nitrate ion). It is extremely poisonous. Diffusing easily through a cell membrane can be transformed into nitroguanine by peroxynitrite by converting proteins and DNA methionine residues and tyrosine to nitroguanine. It functions as an oxidant similarly to the radical hydroxyl. Peroxynitrite-induced cell damage is shown by nitrotyrosine residues, which have also been linked to tissue ageing. Peroxynitrite appears to be produced at sites of inflammation and oxidises low-density lipoprotein (LDL), causing tissue damage [31].

2.12 Nitrogen Dioxide

Since the electrons in nitrogen dioxide ($NO_2\cdot$) are not paired, it can be regarded as a free radical in contrast to nitrous oxide. It is formed when nitric oxide and the peroxyl radical react in smoke and contaminated air. With a reactivity between NO· and $ONOO^-$, nitrogen dioxide is a moderately strong oxidant. Depending on pH, the rate at which nitrogen dioxide reacts with organic molecules ranges from about 104–106 M/s. The highly reactive dinitrogen tetroxide can be produced by dimerising two nitrogen dioxide radicals. Ascorbic acid oxidation, which results in lipid peroxidation and the generation of free radicals, is one way that nitrogen dioxide might impact antioxidant systems [32].

2.13 Reactive Sulphur Species

Sulphur has been linked to the beginning of life and is found in large quantities in both nature and the human body. Thiols, disulphides, it includes sulphenic acids (sulfoxides), sulphinic acids and sulphonic acids examples of organic sulphur derivatives. These substances are classified as reactive sulphur species, in similarity to additionally, reactive nitrogen species exist as well [33]. As a consequence of thiols, in addition to reactive oxygen species, thiyl radicals are produced as well, produced when thiol compounds are exposed to traces of transition metal ions. Thiyl radicals and a superoxide radical anion are produced when transition metals like Fe^{3+} are reduced to Fe^{2+}, which is how pro-oxidative action occurs. Hydrogen sulphide, thiols, and their sulphane sulphur derivatives play important roles in metabolism, in protein structure and folding, signalling. A number of different factors such as metal ligation, cell defence, enzymes, metabolism and mitochondria can be found in

Table 1 List of common radical and non-radicals of biological system

Radicals	Superoxide anion Hydroxyl Hydroperoxyl Peroxyl Carbonate radical anion Nitric oxide Nitrogen dioxide Reactive sulphur species
Non-radicals	Singlet oxygen Hydrogen peroxide Ozone Hypochlorous acid Peroxynitrite

these cells made possible by their biochemistry. There are several exogenous thiols and associated disulphides that both humans and animals are constantly exposed to. According to Paul and Snyder [34], thiol compounds are present in food, pollutants and goods that include sulphur. The lowest oxidation state of hydrogenated sulphur is hydrogen sulphide (H_2S). It diffuses rapidly between compartments and is soluble and slightly hydrophobic in lipid membranes. There are numerous physiological effects of hydrogen sulphide that may be advantageous to health. It is produced via a variety of routes in the brain, kidneys and mitochondria, and is predominantly synthesised enzymatically and non-enzymatically in tissues of mammalian [35]. Because hydrogen sulphide inhibits cytochrome c oxidase and prevents oxidative phosphorylation, it has been considered toxic to mammals. H_2S has been identified as the third gasotransmitter since carbon monoxide and nitric oxide were identified as gasotransmitters [36]. We are summarise the radical and non-radicals of biological system in Table 1.

3 Sources of Reactive Oxygen Species

Both sources from within and outside the body can produce the reactive oxygen species. Several cellular organs with significant oxygen consumption, including the cytosol, mitochondria, peroxisomes and endoplasmic reticulum, are endogenous producers of ROS. In plants and animals, external sources such as tobacco, pollution, smoke, pharmaceuticals, xenobiotics, or ionising radiation can have irreversible impacts on tissue development. These emissions can be affected by abiotic factors like high temperatures or lack of water [37].

3.1 Cytosolic Sources

3.1.1 Source of NADPH Oxidases

The transmembrane enzymes known as NADPH oxidases (NOXs) are mostly found in the plasma membrane and are especially designed to generate cytosolic ROS (cytROS) [38]. By moving one electron from NADPH to oxygen, NOXs catalyse the generation of superoxide [39]. Either spontaneous dismutation or superoxide dismutase catalyse the conversion of superoxide to H_2O_2. Three of the family's members—as well as NOX1, NOX2 and NOX4, phagocytes also produce NOX, and seven members—NOX1-5 and Duox1-2 have been described thus far. Because of its significance in phagocytic activity, NOX2 is a characterised isoform. The plasma membrane is internalised and turns into the phagocytic vesicle's interior wall during phagocytosis. The internalised target is then killed by the release of the superoxide anion (also known as an "*oxidative burst*") into the vesicle by NOX2. Chronic granulomatous disease, a hereditary illness marked by gene mutations that encode NOX2 components, has shown the importance of cytROS in host immunity [40]. More recently, it has been demonstrated that the host's defence against *Toxoplasma gondii* infection also selectively depends on NOX4-mediated ROS generation [41].

3.1.2 Source of Xantine Oxidase

Under oxidative stress, the enzyme xanthine dehydrogenase can change into its oxygenase form, xantine oxidase (XO). Although XO also produces ROS, both forms use hypoxanthine or xanthine to produce uric acid [42]. Reversible sulfhydryl modification or irreversible proteolytic cleavage may be the cause of xanthine dehydrogenase's conversion to XO [43]. Although its function has been extensively researched for several years, a limited amount of information is available on innate immunity, however. Remarkably, strong evidence indicates the XO system results in ROS generation that activates the NLRP3 inflammasome, which is composed of nucleotide-binding domains, leucine-rich domains and pyrin domains. In macrophages and is a crucial factor in an inflammation-inducing agent in response to parasitic infection [44].

3.2 Mitochondrial Sources

Since mitochondria are the primary oxygen consumers in cells, they are significant sources of reactive oxygen species [45]. According to Wong et al. [46], these organelles produce ROS by a variety of processes, such as electron leakage, a succession of oxidases catalyses the transfer of electrons to oxygen or produces by-products because of electron transfer to oxygen. Remarkably, it has been proposed that mitochondrial ROS (mtROS) also play a significant role in antibacterial responses, exposing a new mechanism that links mitochondria with innate immune signalling. In particular, it was discovered that activation, involved in the activation of Toll-like receptors (TLR1, TLR2 and TLR4) increases mtROS and attracts mitochondria to macrophage phagosomes, although there is still much to learn about the mechanism [47]. The overproduction of pro-inflammatory cytokines cells from patients with TNF receptor-associated periodic syndrome (TRAPS) exhibit autoinflammatory responses disease brought on by missense mutations in the type-1 TNF (Tumour Necrosis Factor) receptor (TNFR1), is attributed to mtROS, according to Bulua et al. [48]. The authors postulated a number of ways that TNFR1 mutat ions improve mitochondrial respiration, but more precise investigation is needed to resolve this problem. Signals to initiate mitochondrial respiration can be sent from the endoplasmic reticulum, which is home to mutant TNFR1. According to Cardenas et al. [49], it has also been suggested mutant TNFR1 may activate riboflavin kinase, a protein that is linked to TNFR1 and improve the charging of enzymes in the mitochondria that are dependent on flavin adenine dinucleotide (FAD).

3.2.1 Effect of Transportation Chain for Electrons

An intramembrane protein group that transports electrons of the mitochondria make up the electron transport chain (ETC). It moves electrons to molecular oxygen from the catabolically produced reduced coenzymes $FADH_2$ (flavin adenine dinucleotide) and NADH. In order to sustain the proton motive force that drive ATP synthesis, three of the four protein complexes involved in this process pump protons throughout mitochondrial inner membrane. Mitochondrial ROS production is a physiological effect of the electron transport. In the mitochondria, complexes I, II and III hydrolyse oxygen to provide superoxide that can be dismutated by manganese superoxide dismutase can quickly convert to H_2O_2. Protein thiol oxidation can then be initiated by H_2O_2 freely diffusing throughout the cell. As a result of the reverse electron transfer (RET), can also result in the production of mtROS. Recent research has shown that signalling of mitochondrial redox in both conditions of physiological and pathological is mediated by RET at complex I [50]. Unlike conventional forward transport, it happens when electrons return through complex I due to a high mitochondrial membrane potential and a substantially reduced coenzyme Q. As a result, the electrons can induce the formation of superoxide by reducing NAD + to NADH. According to Mills et al. [51], RET at complex, it is believed that I contribute to the formation of mtROS in macrophages activated by lipopolysaccharide due to the concentration of succinate increases, which is oxidised by complex II (succinate dehydrogenase).

3.2.2 Effect of Monoamine Oxidases

Monoamine oxidase (MAO) is a significant mitochondrial-specific generator of ROS, although its effect on inflammation has been quite overlooked. The outer mitochondrial membrane contains MAO, which catalyses the oxidative deamination of dietary amines and neurotransmitters (catecholamines), synthesis H_2O_2, ammonia and aldehydes. The substrate specificity and inhibitor sensitivity of in addition to MAO-A, there are also MAO-B isoforms, are different. MAO-A is more attracted to hydroxylated amines, such as serotonin and noradrenaline, while MAO-B is more attracted to non-hydroxylated amines, such as β phenylethylamine. Despite this, their affinity for tyramine and dopamine is comparable. In the central nervous system, MAO has a well-established physiological role by blocking neurotransmitter signalling, which produces H_2O_2, which is continuously eliminated by endogenous scavengers [52]. Conversely, in pathological situations, such as muscular dystrophy and heart injury, as a result, antioxidative defences of the cell are compromised by the enzyme's increased activity, changing the redox homeostasis and producing harmful impacts [53]. The properties, there have been a few studies on MAO's role in innate immunity. Inflammatory mediators such as LPS or IL-4 and IL-13 have been demonstrated to upregulate MAO-A in phagocytic cells. According to Bhattacharjee et al. [54], these data also suggest that MAO-A mediated ROS production was generated by IL-13 and IL-4 via Jak signalling pathways. A study conducted a few years ago by Tschopp's lab demonstrated the critical function that mtROS plays in NLRP3 inflammasome activation, even though it was unclear where the ROS were produced. H_2O_2 generated by MAO-B has recently been demonstrated to have a regardless of whether it is in human or murine macrophages, this enzyme maintains NLRP3 inflammasome activity. By activating NF-κB and causing mitochondrial malfunction, MAO-B-dependent the mechanism by which ROS are produced leads to upregulation of pro-IL-1β and NLRP3. MAO-B defective mice exhibit a poor response to LPS-mediated endotoxemia, and the clinical-grade medication rasagiline inhibits MAO-B, preventing IL-1β release in vitro as well as in vivo [55].

3.2.3 Effect of p66Shc

A cytosolic adaptor protein called p66Shc controls apoptosis and the redox status of cells. p66Shc is phosphorylated by protein kinase C-β (PKCβ), which is activated by oxidative stress. Consequently, p66Shc translocate into mitochondria and produces H_2O_2. In fact, this redox enzyme catalyses the reduction in portion of molecular oxygen by oxidising cytochrome c, which uses reducing equivalents of electron transport chain in mitochondria [56]. Long-lived mice with p66Shc-deficient cells are less likely to create ROS and are more resilient to oxidative stress. The role of p66Shc several factors contribute to the development of atherosclerotic lesions, including a high-fat diet and diabetes was the primary focus of a few research that described the protein in the immune system. In wild-type mice, in animals with p66Shc knockouts, chronic high-fat diets accelerated atherosclerosis. The number of macrophage-derived foam cells in the early lesions of p66Shc knockout mice was lower than that of wild-type animals. In murine macrophages, it has been observed that p66Shc and NOX crosstalk because p66Shc-deficient mice have impaired NOX activation, which results in a reduction in superoxide generation [57].

3.3 Effect of Peroxisomes

Instead of being involved in ATP synthesis through phosphorylation through oxidation, the respiratory route in peroxisomes releases H_2O_2 is created when electrons are transferred between different metabolic products, creating free energy as heat. The peroxisome also generates other free radicals NO·, O_2·–OH· and H_2O_2. The primary metabolic mechanism that generates H_2O_2 in the peroxisome is a body responsible for converting fat into energy. Different reactive oxygen species must be produced. The process involves peroxisomal enzymes including acyl CoA oxidases, acyl CoA oxidases, L-α-hydroxy oxidases, urate oxidases and D-aspartate oxidases [58].

3.4 Effect of Endoplasmic Reticulum

Diamine oxidase and other endoplasmic reticulum enzymes, including oxidative stress is caused by cytochromes p-450 and b5 enzymes. H_2O_2 is created when it is the molecular oxygen that transfers electrons from dithiols enzyme Erop1p, another crucial thiol oxidase [59]. ROS can also be produced endogenously, phagocytic cells, decreased riboflavin, cytochrome P450, prostaglandin inflammation, mental stress, excessive exercise, infection, cancer, ageing and ischaemia are all caused by adrenaline metabolism, adrenaline auto-oxidation, immune cell activation, and adrenaline synthesis and more [60].

3.5 Effects of Enzyme System

Lipoxygenase produces free radicals and, when Fe^{2+} has been oxidised to Fe^{3+}, it can convert polyunsaturated fatty acids (PUFA) to hydroperoxides. The central nervous system contains arachidonic acid, which can be oxidised to hydroperoxyeicosatetraenoic acid by the three main mammalian lipoxygenases, 5-, 12- and 15-lipoxygenase. Furthermore, atherosclerotic lesions have been found to contain 15-lipoxygenase, suggesting that the enzyme could contribute to the in vivo formation of oxidised lipids [61]. Immune cells (phagocytes, granulocytes and lymphocytes) that protect the body from invasive microbes also produce high quantities of reactive oxygen species [62]. The NADPH oxidase complex, found in neutrophils and macrophages, produces hydrogen peroxide and superoxide radicals when it is activated. Hypochlorite is produced when the latter reacts with intracellular chloride ions, killing microorganisms. Individuals suffering from chronic granulomatous illness, where the NADPH oxidase complex significantly lowers ROS production, are extremely vulnerable to infections and typically pass away early [63]. Myeloperoxidase is the primary enzyme that neutrophils express. It converts hydrogen peroxide and chloride anion into hypochlorous acid with heme as a cofactor. Tyrosine is also converted to the tyrosine radical by oxidation. Neutrophils use the cytotoxic tyrosine radical and hypochlorous acid to destroy pathogenic organisms [64]. In their metabolic activities, cytochrome P450 enzymes use oxygen and produce small amounts of reactive oxygen species. Depending on the cytochrome P450 molecule and the substance being broken down, different amounts of ROS are generated. Cytochrome P450 2E1 is a molecule that is especially active in the generation of ROS [65].

3.6 Effects of Metals

It is commonly known that transition metal-mediated processes can produce free radicals [9]. A transition metal ion can be in any of several oxidation states, the autoxidation of numerous biomolecules may be facilitated by these ions when they are in the active redox state. According to Bokare and Choi [66], in Fenton reactions and reactions like Fenton, hydroxyl radicals are produced, typically starts the oxidation of biomolecules. When metal ions, particularly free iron, which is more significant due to in biological systems, hydroxyl radicals can be formed because of its abundant component, copper, which plays a role in two-step chemical reactions. Oxygen radicals can be found in created by hydrogen peroxide by taking metal ions are separated by electrons, that is involved. Following that, the initial metal ions are renewed by the superoxide radical, which makes them fresh for the hydrogen peroxide reaction. Both reactions take place chemically, provide evidence for the part that metals like copper and iron play in oxidative stress and ROS-induced cell damage. Once more, the concentration of the transition metal has less bearing on activation of pro-oxidants redox state. The ferrous ion has pro-oxidant activity only when a reducing agent, like ascorbic acid, is present, whereas the ferric ion is a weaker pro-oxidant. Lipid hydroperoxides are broken down by transition metals producing free radicals that can start or spread lipid peroxidation as part of their pro-oxidant action [67]. The metals can accelerate lipid oxidation by breaking down hydroperoxides into peroxyl and alkoxyl radicals. Although both ferric and ferrous ions can catalyse the breakdown of lipid hydroperoxides into free radicals formed from hydroperoxide, the ferrous ion has a higher catalytic activity than the ferric ion. Furthermore, when a labile hydrogen atom is abstracted, compared to peroxyl radicals, alkoxy radicals are highly reactive. Since iron cells, increased concentrations of ferrous ions contribute to the production of hydroxyl radicals and oxidative stress [68].

4 Molecular Effects of Antioxidants

Oxidative and nitrosative stress result from the generation of free radicals in excess when antioxidant defences and free radical production are an imbalance. These highly reactive an element of biological life can be damaged by free radicals like proteins, lipids and nucleic acids [69].

4.1 *Deoxyribonucleic Acid (DNA)*

Since mitochondrial DNA a short distance away from ROS's location are formed, it is more susceptible to ROS attack than nuclear DNA, it is possible for ROS to damage nucleic acids through oxidative reactions. The most significant of these ROS is the OH· (hydroxide) various alternations are caused by this radical, which attacks DNA's purine and pyrimidine bases, as well as its deoxyribose sugar backbone. By abstracting hydrogen atoms, the OH· radical creates DNA–protein cross linkages and different range of modified purine and pyrimidine base by-products. Several adducts of pyrimidines, including thymine glycol, uracil glycol, 5-hydroxydeoxyuridine, 5-hydroxydeoxycytidine and hydantoin, are produced when OH· attacks pyrimidines. Purine adducts such as 2,6-diamino-4-hydroxy-5-formamidopyrimidine, 8-hydroxydeoxy guanosine and 8-hydroxydeoxy adenosine hydroxyl radicals generate them. The free radicals cause a lot of DNA adducts, including alloxan and 5-hydroxy-6-hydrocytosine [70]. Eight-hydroxy deoxyguanosine has been associated with tumorigenesis, carcinogenesis and ageing. In fact, Barja [71] in mitochondrial DNA, 8-hydroxy-2′-deoxyguanosine (8-OHdG) is more abundant than in nuclear DNA.

4.2 *Ribonucleic Acid (RNA)*

The body produces a variety of RNAs, some of which are targets of ROS. Due to its single-stranded nature, the lack of an active mechanism for repairing oxidised RNA, and reduced protection by proteins, RNA is more vulnerable to oxidative damage than DNA. A large amount of reactive oxygen species is produced by mitochondria close to its location. According to Hofer et al. [72], human RNA is more vulnerable to oxidative damage than DNA. There is a wide range of pathological conditions that are associated with 7-dihydro-8-oxoguanosine (8-oxoG), including myopathies [73], Alzheimer's disease [74], Parkinson's disease [75], atherosclerosis [76] and hemochromastosis [77].

4.3 Lipid Oxidation

Free radicals are more likely to oxidise membrane lipids, and particularly polyunsaturated fatty acids found in phospholipids, according to Siems et al. [78]. The role that lipid peroxidation plays in clinical disorders makes it one of the most important in vivo factors. Because of lipid peroxidation, membrane fluidity is reduced and enzymes and receptors bound to membranes become inactive. Free radicals attack methylene groups on fatty acids, removing hydrogen from them, resulting in carbon-centred lipid radicals (L·). By interacting with molecular oxygen, lipid radicals produce lipid peroxyl radicals (LOO·). It is believed that the endoperoxides formed by cyclization of the resulting peroxyl radical cause protein and DNA damage [79]. This lipid peroxyl radical can extract hydrogen atoms from other lipid molecules, thereby accelerating peroxidation. By oxidising arachidonic acid, isoprostanes are produced by the body through esterification and are thought to cause oxidative lipid damage, according to Aruoma et al. [80].

4.4 Protein Oxidation

During protein oxidation, both radical and nonradical species, such as H_2O_2, O_3, HOCl, singlet oxygen and OONO-, may participate. When ROS oxidise, various amino acids found in proteins, they create the cross-linking of proteins impairs their function and denaturizes them, enzyme activity, receptor function and transport protein function. Amino acids that contain sulphur, like cysteine and methionine, are more vulnerable to ROS oxidation, which can result in the formation of sulphides and sulfoxides of methionine [81]. However, only these two oxidised protein forms can be restored to their original form in biological systems by two distinct enzymes called methionine sulfoxide reductases and disulphide reductases. When ROS attack amino acids, they produce oxidation products that include, among the synthesis products of tryptophan are nitrotryptophan, kynurenine and formylkynurinine. Among the products of phenylalanine are 2,3, 4-hydroxyphenylalanine, 2 and 3 hydroxyphenylalanine. Tyrosine produces tyrosine–tyrosine crosslinks, Tyr-O-Tyr and cross-linked nitrotyrosine. 2-Oxohistidine, asparagine and aspartic acid. Lysine produces a-aminoadipic semialdehyde, which is formed by glutamic semialdehyde,because of 2-amino-3-ketobutyric acid formation in threonine, hydroxyl residues are formed in leucine and valine residues, while 4-hydroxyproline pyroglutamic acid as well as glutamic semialdehyde are produced in proline residues [82]. An amino acid residue, such as arginine, proline, lysine, or threonine, becomes oxidatively damaged by ROS. These residues form carbonyl derivatives. A carbonyl group in proteins, which is a marker for protein oxidation by ROS, is found in proteins undergoing oxidative reactions [83], a distinction is also made between three-nitrotyrosine and one-nitrotyrosine. Dystrophies, rheumatoid arthritis, diabetes, dyslipidemia, respiratory dyspraxia and Werner's syndrome are just some of the conditions we can encounter, and ageing are among the clinical disorders that are associated with elevated levels of protein carbonyls [18].

5 Pathways of Reactive Oxygen Species

5.1 Pathway of Nuclear Factor-Kappa B (NF-κB)

Numerous physiological functions, including cell adhesion, inflammation, differentiation, apoptosis, immunology, autophagy, ageing and proliferation, depend on the transcription factor NF-κB [84]. Furthermore, NF-κB controls pro-inflammatory cytokines and can control inflammation associated with high H_2O_2 generation. Asthma, cancer, rheumatoid arthritis (RA), heart disease, neurological ill neurodegenerative ailments, and inflammation are among the diseases that have been linked to disruption of NF-κB [85]. A nuclear factor-Kappa B family member is RelB, which belongs to NF-kappa B2, RelA (p65), Rel (c-Rel) and p50/p105 (NF-kappa B1). There are different ways that nuclear factor-Kappa B is activated: canonical way and non-canonical way. Viral infections can cause infections, phosphorylation-dependent cytokines and stress, that cause inflammation to trigger the canonical route. After proteasomal degradation of IκBa, NF-κB penetrates the nucleus and stimulates specific genetic transcription. On the other hand, Viral diseases associated with Epstein-Barr, including human T-cell leukaemia, B-cell activating factor (BAF), lymphotoxin β (LTβ), cluster of differentiation 40 (CD40) ligand and 27 (CD27) ligands all activate the non-canonical Nuclear factor-Kappa B pathway [86]. Additionally, by triggering a proteolytic process is involved in the processing of the NF-κB2/p100 precursor, it activates NF-κB2/RelB complexes and is dependent on IKKα. According to Pajares et al. [87], coordinating the activity of NF-KB and NRF2 signalling pathways is crucial for balancing inflammation and antioxidation in cells. Inflammatory mediators and the NF-B pathway have a complicated and wide-ranging interaction, according to Hayes et al. [88]. It is evident that ROS can both stimulate and inhibit NF-B-mediated transcription depends upon the cell type, where they are located inside the cell. When ROS are triggered, the NF-κB pathway is initiated. This involves, because of ROS—dependent phosphorylation, p65 interacts with CBP/300 and subsequently functions as a transcription factor activity. Furthermore, by causing a specific phosphorylation of IkBα, exogenous ROS, especially H_2O_2, can stimulate NF-κB activation. IkBα is then degraded within the proteasome because of this modification. Dimerised NF-κB proteins can then go to the nucleus because of the ensuing degradation, which makes it easier for their target genes to start transcription. There is evidence that H_2O_2 can contribute to inflammatory response to phosphorylation of the IKKβ subunit and improve the dimerisation of IKKγ/NEMO. This suggests that either IKKβ is directly oxidised, which deactivates it and activates the NF-κB pathway, or an inhibitor of nuclear factor-kB (IkB) kinase (IKK) phosphatase is inhibited.

5.2 A Pathway of Mitogen-Activated Protein Kinases

MAK family proteins play a crucial role in supporting cell survival and accelerating tumour development. Protein members of this class include c-Jun N-terminal kinase (JNK), p38 MAPK and extracellular signal-regulated kinases (ERK1/2). Redox signalling activates MAPK family proteins via the action of upstream kinases, Protein kinases, such as mitogen-activated protein kinase kinases (MAPKKs) and mitogen-activated protein kinases (MAPKs). Certain phosphatases deactivate MAPK family proteins. Depending on the cellular location and stimuli, MAPK family proteins have two functions. When ROS are present, ERK1/2 helps cancer cells survive, while JNK activation promotes autophagy and death. Activation of ERK1/2 in the cytoplasm is caused by oncogenic mutations in the RAS gene and overexpression of receptor tyrosine kinase under chronic ROS stress [89]. As a result, cytoprotective mechanisms are activated, which promotes the spread of cancer. Additionally, ROS can activate the epidermal growth factor receptor (EGFR), which causes ERK1/2 to translocate to the nucleus and improves cells that grow cancerous. Apoptosis signal-regulating kinase 1 (ASK1) a high-oxygen environment causes p38 MAPK to be activated and attenuates ERK1/2's growth stimulating effects. Numerous studies have verified that p38 MAPK is downregulated in cancer pathogenesis [90].

5.3 Signalling Pathway of PI3K-Akt

The progression of cell cycle, induced apoptosis, autophagy, protein synthesis and drug resistance are influenced by cytokines (IL2, IL6 and IL17). In addition to growth factors (Vascular Endothelial Growth Factors (VEGF), nerve growth factors, platelet-derived growth factors, epidermal growth factors and growth factors derived from platelets), PI3K pathways also influence hormones (prostaglandin: PGE2) [91]. By directly stimulating class 1A PI3K in response to its binding factor found in growth factor receptors, the protein is stimulated by its regulatory unit or insulin receptor-like adapters. PI3K is activated by proteins called insulin receptor substrates (IRS). In the process, active PI3K updates phosphatidylinositol 2,4,5-bisphosphate (PIP2) to phosphatidylinositol 3,4,5-triphosphate (PIP3). Phosphorylinositide-dependent protein kinase (PDK), protein kinase B (Act) and glycogen synthase kinase-3 (GSK3) are activated in this process to stimulate mTOR1 and improve the activity of PDK and FOXO. Accordingly, ROS suppress Akt activation by oxidising cysteine residues inside the active centre and directly activating PI3K and downregulating PTEN (phosphatase and tensin homolog), deficiency of this enzyme inhibits the production of PIP3 [92].

5.4 Signalling Pathway of Keap1-Nrf2-ARE

Keap1-Nrf2-ARE is essential for maintaining metabolic and redox cellular balance. It also promotes oxidative stress can lead to several problems dangerous diseases, like cancer and diabetes. According to Xu et al. [93], Nrf2 is an essential an antioxidative protein transcription factor that prevents the damage caused by oxidative stress. Three cell components make up this signalling pathway: Kelch like ECH-associated protein 1 (KEAP1), antioxidant reactions (ARE) and nuclear factor erythroid 2-related factor 2 (Nrf2). In addition to being a primary sensor of oxidative and electrophilic stress, KEAP1 also functions as a marker of inhibits NRF2 and contains various cysteine residues that cause oxidation. Accordingly, Cys151 disulphide creation changes in conformation of KEAP1 dimer, preventing NRF2 from being ubiquitinated and acting as a transcription factor for the expression of antioxidative defence proteins. KEAP1's many cysteine thiol groups react differently to various oxidants. Thioredoxin reductase 1 and the sirtuin family of deacetylases control the NRF2-KEAP1 system. Additionally, the NRF2 system is modulated by redox-sensitive microRNAs [94]. Numerous human cancers, consisting of the skin, breast, lung, prostate and ovaries, have been shown to have persistent NRF2 activation. It has been observed that NRF2 activation is triggered by KEAP1 mutations, Myc and KRAS and other dominant oncogenes. In different cancer cells, abnormalities there have been studies linking NRF2-KEAP1 pathway to metastasis, genomic instability, drug resistance and apoptosis resistance [95]. Through an increase in oxidative stress-dependent cellular death, depleted NRF2 demonstrated decreased tumour growth. Because NRF2 can interact with KEAP1 (Kelch like ECH-associated protein 1) and the Cullin 3 (Cul3) E3 ubiquitin ligase, it undergoes proteasomal degradation under normal physiological conditions. In contrast, caused oxidative stress increases Reactive Oxygen Species, oxidises KEAP1 and prevents NRF2 from interacting with the compound responsible for KEAP1 degradation [96].

5.5 Pathway of Factor Hypoxia-Inducible

An HIF is a transcription factor that is activated by hypoxia, which are made up of the constitutively stable subunit (HIF-β) and the oxygen-labile subunit (HIF-α), are essential for controlling immune cell effector activities and triggering biological reactions to hypoxia. By binding to hypoxia-responsive elements (HREs), they generate a heterodimeric complex that promotes metabolic and functional cell reprogramming by stimulating the transcription of their genes of interest. Under homeostatic conditions, prolyl hydroxylase (PHD), an enzyme that is a member of the families of dioxygenases that consume 2-oxoglutarate (2-OG), hydroxylates HIF by using molecular oxygen to catalyse the transformation of 2-OG to succinate [97]. Upon ubiquitylation, the von Hippel-Lindau tumour suppressor protein (VHL) inhibits the activation of HIF due to its hydroxylation, and the proteasome subsequently degrades it. When macrophages are stimulated by LPS, they reprogram their metabolism to favour glycolysis, which causes ROS to be produced. This can sustain hypoxic adaption through HIF stabilisation [98] and simultaneously promote the expression of cytokines, including IL-1 β, through HIF-1 α. ROS plays an important role in the development of cells, reducing ROS generation by expressing the alternative oxidase (AOX) or uncoupling mitochondria prevents macrophages from exhibiting an inflammatory phenotype. According to Kaelin and Ratcliffe [99], ROS also help stabilise HIF-1a by (I) oxidising the PHD cofactor Fe^{2+} to Fe^{3+} and (II) 2-OG is redirected towards non-enzymatic decarboxylation in PHD. Lastly, reactive oxygen species perform the role of FIH (Factor Inhibiting HIF), an enzyme that hydroxylates asparaginyl residues, which is a member of the 2-OGDD family. By decreasing the activity of its C-terminal transactivation domain, FIH hydroxylation, which takes place at unlike PHD hydroxylation, this occurs at a different place, compromises HIF-1 α role. It is important to note In contrast with prolyl hydroxylation, HIF asparaginyl hydroxylation (OH) exhibits a higher degree of susceptibility H_2O_2 levels, while prolyl OH is more efficiently inhibited in moderate hypoxia. This suggests that ROS and hypoxia may regulate HIF transcriptional output to varying degrees [100]. In support of this observation, that mtDNA deletion is adequate to prevent HIF-1 α stabilisation under hypoxia in cancer cell lines. Spermidine-generated mtROS in bone-derived murine macrophages improve mitochondrial activity by activating AMPK (AMP-activated protein kinase) HIF-1 α. Additionally, HIF-1 α limits oxygen consumption by reducing mitochondrial mass by mitophagy. Targeting SDH subunit A (SDHA), mtROS inhibits its enzymatic activity, stabilising HIF-1 α and a TCA intermediate is formed as a result and subsequent, prolonged production of IL-1 β [101].

5.6 Pathway of Transcriptional Activator that Transduces and Activates Signals

A transcription factor called STAT (signal stimulator and activator of transcription) plays a crucial role in transcription are crucial for the way cells react to growth stimuli and cytokines. When the cytoplasm is at rest, STATs are dormant. Cytokines and other extracellular stimuli a cytokine is a protein that binds to cell surface receptors and activates a tyrosine kinase called Jak (also known as Janus kinase), allowing STAT proteins to be phosphorylated and translocated to the nucleus, where they trigger transcription of several cytokines. An essential component of the immune system is the JAK-STAT signalling pathway [102]. It has been possible to identify the STAT1-4, STAT5A, STAT5B and STAT6 family members so far. In the same way that interferon plays a role in antiviral responses, STATs are essential for wound healing and the inflammatory/anti-inflammatory response. It is essential to have a functional JAK-STAT pathway, macrophage polarisation because STAT1 causes macrophages to adopt an M1 profile that is pro-inflammatory and triggered. It is activated by interferon gamma (IFN-g) while STAT6 is activated by interleukin-4 (IL-4) to control inflammation in M2 cells. It is yet unclear how ROS contribute to macrophage STAT activation. Since STAT1 mice exhibited ROS-dependent impairment in p38MAPK activation, researchers have observed a positive correlation between ROS, p38MAPK and STAT1. Reduced ROS levels in NOX-deficient diabetic mice were discovered aiming at inhibiting STAT1 activation while promoting STAT6 activation in diabetes progression (M2). This finding further suggests ROS and STAT1 activate each other cooperatively [103].

5.7 Pathway of Stimulator of Interferon Genes

Detecting double-stranded DNA in cytosol is brought about by the Stimulator of Interferon Genes (STING) pathway and can indicate damage to nuclear DNA, infection with microbes, or cell injury. As a result of cytosolic dsDNA activating cyclic-GMP-AMP synthase (cGAS), cyclic GMP-AMP (cGAMP) is produced endogenously. A STING transmembrane receptor is activated by cGAMP when it binds and binds to STING, causing them to be released as strong antiviral and anticancer cytokines [104]. However, the STING pathway's hyperactivity has been linked to both acute and chronic inflammation as well as several debilitating autoimmune disorders, such as systemic lupus erythematosus. In the absence of ligands, STING forms a domain-swapped homodimer; nevertheless, when cGAMP binds, it experiences significant conformational changes that result in oligomerisation [105].

5.8 Role of Thioredoxins and ASK1

The disulphide reductase activity of thioredoxins (TRXs), which are tiny proteins, makes them an essential defence mechanism against oxidative stress. Reversible oxidation of cysteines within TRXs keeps redox equilibrium within cells by maintaining redox equilibrium. In addition to their reducing action, TRXs play a significant role in redox signalling pathways. In fact, through direct physical interaction, TRXs regulate the function of many proteins. For example, TRX forms an inactive TRX-ASK1 complex by binding the apoptotic signal-regulating kinase 1. Disulphide bond formation in TRX is induced by an increase in intracellular ROS. ASK1 release is made possible by this conformational change, which initiates a program of cell death by activating the p38MAPK and NF-kB pathways [106]. Because LPS stimulation causes the generation of H_2O_2 to activate NF-kB in a Myd88-dependent way, macrophages promote this axis. Remarkably, during TB infection, TLR2's detection of the tuberculin protein led to rise in reactive oxygen species generation, which enhanced boosted a positive remark burst of reactive oxygen species by activating TRX-ASK1-p38. According to a different mechanism that has been reported in microglia, ASK1 activation happens when extracellular ATP is sensed by the

P2X7 receptor. A product of this process is ROS, which creates TRX-ASK1, which releases reactive oxygen species into the environment. Additionally, TRX attaches itself to the thioredoxin interacting protein (TXNIP). Thioredoxin interacting protein activation seems to be necessary for NLRP3 assembly by many inflammasome agonists, including ATP, high glucose concentrations and monosodium urate crystals (MSU). The production of ROS is increased by these stimuli, which leads to TRX cysteine oxidation, the disruption of the TRX-TXNIP association, and the binding of NLRP3 by TXNIP, which stabilises the inflammasome assembly. Furthermore, ROS-mediated IRE1α induces by endoplasmic reticulum stress raises the stability of TXNIP mRNA and improving NLRP3 activation in sterilised inflammation [107].

5.9 p53 and FOXO Are Transcription Factors Involved in This Pathway

Transcriptional factors such as FOXOs and p53 contribute to the suppression of tumours are essential components of the downstream MAPK signalling cascade because they largely control the development of proapoptotic and cell cycle pathways. By triggering both TNFα-FasL-mediated extrinsic apoptotic pathways and caspase-dependent intrinsic apoptotic pathways, these molecules can cause apoptosis [108]. In addition, FOXO and p53 molecules could respond to OS stimuli by upregulating several antioxidant genes, like glutathione peroxidase, catalase and superoxide dismutase. Cell proliferation is aided by the formation of intracellular reactive oxygen species and PI3K-mediated signalling pathways activity, which downregulate (phosphorylate) the two transcription factors. The FOXO proteins are phosphorylated by the IKK/NFκB molecules, which facilitates their nuclear exclusion and proteasomal degradation. By facilitating the genes that regulate invasion and migration, FOXO family proteins can promote the survival of cancer cells. In addition to causing the wild-type p53 molecule decreases, oncogenic mutations in p53 protein also disrupt the balance between the Bcl2, Bim, Bax and p21 are apoptosis-regulating proteins. Both cell survival and cell death depend on the interaction between the members of the FOXO family and p53. Proteins from the FOXO family can dramatically control how p53 activates IGF-1-mediated signalling pathways that determine a cell's destiny. Several FOXO family members are activated post translationally in mammals, can also be controlled by p53 [109].

5.10 Advanced Glycation End Products Pathway

A diverse collection proteins, lipids and nucleic acids that make up DNA produced non-enzymatically are referred to as Advanced Glycation End products (AGEs) [110]. In both oxidative stress and hyperglycemia, AGE is produced in increased amounts. It is in this binding process that AGE meet their specific receptor on cell surface, known as RAGE, that they are released, post-receptor signalling is triggered, intracellular oxygen free radicals are produced and gene expression is activated [111]. The AGE-RAGE interaction increases the expression of VEGF in the retina, which is a mediator of the delay issues of diabetes. Accordingly, AGE act as both a marker and a mediator of chronic vascular illnesses and late diabetes complications [112].

5.11 Protein Kinase C (PKC) Pathway

Diacylglycerol, an allosteric PKC activator, is more abundant in tissues where diabetic complications arise [113]. The rise in diacylglycerol causes the activation of several PKC isoforms. In the kidney, retina and vasculature, Protein Kinase C-β is the main isoform that is stimulated. Chronic hyperglycemia causes, there is a correlation between diabetes complications and an increase in PKC activity, which regulates blood flow and vascular permeability. A particular synthetic inhibitor of PKC-β has been shown to improve aberrant retinal and renal haemodynamics in diabetic rats, which clearly supports the importance of an activation of the PKC pathway is one of the leading causes of diabetes problems. Moreover, hyperglycemia-induced PKC pathway activation works in concert with another kinase pathways. For instance, hyperglycemia strengthened the vasoactive peptide endothelial-1's activation of MAPK in A PKC-dependent mechanism mediates the interaction between mesangial cells and macrophages. Interestingly, [114] suggest that the long-term consequences of hyperglycemia may be significantly these pathways may interact with other stress-activated pathways.

5.12 Role of Polyol Pathway

Increases in intracellular glucose promote aldose reductase activity, which catalyses the synthesis of sorbitol. Sorbitol dehydrogenase can then oxidise this sorbitol to fructose [115]. Cell damage results from intracellular sorbitol accumulation. Furthermore, sorbitol strongly activates signalling pathways that are susceptible to stress, such as p38 MAPK and JNK. A transgenic mouse model overexpressing aldose reductase gene showed that the polyol pathway was important in the development of diabetic issues [116]. Further research shows these mice do not develop neuropathy, nephropathy, retinopathy, or cataracts when this enzyme is inhibited.

5.13 Pathway of Hexosamine

There are numerous lines of evidence that demonstrate the hexosamine biosynthesis pathway is activated when there is an excessive flow of glucose into different cell types [117]. This pathway has been associated with insulin resistance and the emergence of diabetes-related complications in the late stages. Hexosamine biosynthesis's rate-limiting enzyme, GFAT (glutamine-fructose-6-phosphate amidotransferase) is overexpressed in transgenic mice, insulin resistance is the result of this. When transgenic mice's livers overexpress GFAT, their phenotype changes towards energy storage, leading to obesity and hyperlipidaemia. GFAT overexpression increased NFκB-dependent promoter activation in mesangial cells [118]. Leptin and other gene products are produced by the hexosamine system, a cellular "*sensor*" of energy availability. According to recent findings, the hexosamine pathway is activated when hyperglycemia ROS are produced in greater quantities. It depends on the presence of electron transport inhibitors, mitochondrial uncoupling agents (CCCP), or UCP1 or SOD2 expression, were able to prevent the hexosamine pathway from being significantly increased in bovine endothelial cells by hyperglycemia [119].

5.14 Role of Calcium

Numerous processes related to cellular life and death include divalent calcium ions (Ca^{2+}). In secretion, metabolism, cell survival and death, Ca^{2+} signalling plays a crucial role [120]. The calcium ion gradient across the plasma membrane is quite considerable (about 10,000). Through the production and removal of ROS, intracellular Ca^{2+} regulate cellular redox balance, enabling cells to change the environment of redox to a high reducing or oxidising state. Through oxidative phosphorylation and enzyme activation, Ca^{2+} assists in the synthesis of ATP and produces ROS when activated mitochondria [121]. Ca^{2+} activates citric acid cycle dehydrogenases, causing reactive oxygen species to be generated. ROS synthesis is stimulated by Ca^{2+}-regulated extra mitochondrial enzymes, including nitric oxide synthase and NADPH oxidases, under both conditions of physiological and pathological [122]. ROS clearance is achieved by the direct activation enzymes that act as antioxidants GSH reductase, catalase and SOD by Ca^{2+}. Furthermore, a class of Ca^{2+} transporters' cysteine thiols are oxidised by ROS. In general, ROS-induced thiol oxidation increases channel function and improves the efflux of Ca^{2+}. There is an association between L-type channels and thiol group oxidation, which alters calcium entrance into the cytoplasm [123].

5.15 *Role of Protein Kinases Family*

A wide family of kinases known as the protein kinase family catalyses the phosphorylation of proteins. As part of their role in regulating metabolic processes, apoptosis and proliferation, they play a crucial role in cell proliferation [124]. Many malignancies, including lung, pancreatic and breast cancers, as well as autoimmune conditions like psoriasis and rheumatoid arthritis, are associated with disturbed regulation of protein kinases. These proteins include the receptor tyrosine kinase (RTK) and calcium/calmodulin-independent protein kinase II, all have –SH groups on cysteine residues that ROS oxidise. These proteins then phosphorylate target proteins involved in different signalling pathways. The phosphorylation of protein kinase A can be altered by ROS, which affects ROS equilibrium. Protein kinase C activity can be stimulated by low ROS levels, while the activity can be reduced when high ROS levels oxidise important cysteine residues. Hydrogen peroxide, for instance, was able to control the phosphorylation of protein kinase in the residues 512 and 523, there are tyrosine residues. Just like protein kinase C, ROS activates kinase D by phosphorylating the tyrosine 463 residue. It has been demonstrated that low to intermediate ROS levels activate both Ca/calmodulin and receptor tyrosine kinase, the latter of which is activated by the oxidation of methionine residues 281 and 282 [125].

6 Conclusion

The principal molecular redox targets, including as proteins, lipids and nucleic acids, as well as the many forms of free radicals and the locations where they are generated, are compiled in this chapter along with the corresponding physiological reaction. Both chronic or excessive inflammation-related disorders and the inflammatory response can be triggered by ROS's signalling and harmful characteristics. ROS and important signalling pathways like PI3K/Akt, NF-κB, MAPK/ERK and Nrf2/Keap1 have complex connections that highlight how important redox management is to cellular function. More studies focussing on certain ROS sources and well-defined processes are eagerly awaited.

References

1. Reuter, S., Gupta, S. C., Chaturvedi, M. M., & Aggarwal, B. B. (2010). Oxidative stress, inflammation, and cancer: How are they linked? *Free Radical Biology and Medicine, 49*(11), 1603–1616.
2. Lau, A. T., Wang, Y., & Chiu, J. F. (2008). Reactive oxygen species: Current knowledge and applications in cancer research and therapeutic. *Journal of Cellular Biochemistry, 104*(2), 657–667.

3. Renschler, M. F. (2004). The emerging role of reactive oxygen species in cancer therapy. *European Journal of Cancer, 40*(13), 1934–1940.
4. Halliwell, B., & Gutteridge, J. M. (1985). Free radicals in biology and medicine, 331–332.
5. Brand, M. D. (2016). Mitochondrial generation of superoxide and hydrogen peroxide as the source of mitochondrial redox signaling. *Free Radical Biology and Medicine, 100*, 14–31.
6. Min, D. B., & Boff, J. M. (2002). Chemistry and reaction of singlet oxygen in foods. *Comprehensive Reviews in Food Science and Food Safety, 1*(2), 58–72.
7. Sayre, L. M., Moreira, P. I., Smith, M. A., & Perry, G. (2005). Metal ions and oxidative protein modification in neurological disease. *Annali dell'Istituto superiore di sanita, 41*(2), 143–164.
8. Korycka-Dahl, M. B., Richardson, T., & Foote, C. S. (1978). Activated oxygen species and oxidation of food constituents. *Critical Reviews in Food Science & Nutrition, 10*(3), 209–241.
9. Min, B., & Ahn, D. U. (2005). Mechanism of lipid peroxidation in meat and meat products—A review. *Food Science and Biotechnology, 14*(1), 152–163.
10. Davies, M. J. (2005). The oxidativde environment and protein damage. *Biochimica et Biophysica Acta, 1703*, 249–260.
11. Dizdaroglu, M., & Jaruga, P. (2012). Mechanisms of free radical-induced damage to DNA. *Free Radical Research, 46*(4), 382–419.
12. Lipinski, B., & Pretorius, E. (2012). Hydroxyl radical-modified fibrinogen as a marker of thrombosis: The role of iron. *Hematology, 17*(4), 241–247.
13. De Grey, A. D. (2002). $HO_2\cdot$: The forgotten radical. *DNA and Cell Biology, 21*(4), 251–257.
14. Bielski, B. H., Cabelli, D. E., Arudi, R. L., & Ross, A. B. (1985). Reactivity of HO_2/O^- 2 radicals in aqueous solution. *Journal of Physical and Chemical Reference Data, 14*(4), 1041–1100.
15. Cerruti, P. A. (1985). Prooxidant states and tumor promotion. *Science, 227*(4685), 375–381.
16. Agnez-Lima, L. F., Melo, J. T., Silva, A. E., Oliveira, A. H. S., Timoteo, A. R. S., Lima-Bessa, K. M., et al. (2012). DNA damage by singlet oxygen and cellular protective mechanisms. *Mutation Research/Reviews in Mutation Research, 751*(1), 15–28.
17. Hampton, M. B., Kettle, A. J., & Winterbourn, C. C. (1998). Inside the neutrophil phagosome: Oxidants, myeloperoxidase, and bacterial killing. *Blood, The Journal of the American Society of Hematology, 92*(9), 3007–3017.
18. Phaniendra, A., Jestadi, D. B., & Periyasamy, L. (2015). Free radicals: Properties, sources, targets, and their implication in various diseases. *Indian journal of clinical biochemistry, 30*, 11–26.
19. Winterbourn, C. C. (2013). The biological chemistry of hydrogen peroxide. *Methods in Enzymology, 528*, 3–25.
20. Choe, E., & Min, D. B. (2006). Mechanisms and factors for edible oil oxidation. *Comprehensive Reviews in Food Science and Food Safety, 5*(4), 169–186.
21. Altenhofer, S., Radermacher, K. A., Kleikers, P. W., Wingler, K., & Schmidt, H. H. (2015). Evolution of NADPH oxidase inhibitors: Selectivity and mechanisms for target engagement. *Antioxidants & Redox Signaling, 23*(5), 406–427.
22. Goldstein, B. D., Lodi, C., Collinson, C., & Balchum, O. J. (1969). Ozone and lipid peroxidation. *Archives of Environmental Health: An International Journal, 18*(4), 631–635.
23. Sharma, V. K., & Graham, N. J. (2010). Oxidation of amino acids, peptides and proteins by ozone: A review. *Ozone: Science & Engineering, 32*(2), 81–90.
24. Lerner, R. A., & Eschenmoser, A. (2003). Ozone in biology. *Proceedings of the National Academy of Sciences, 100*(6), 3013–3015.
25. Meli, R., Nauser, T., Latal, P., & Koppenol, W. H. (2002). Reaction of peroxynitrite with carbon dioxide: Intermediates and determination of the yield of $CO_3^{\cdot-}$ and $NO_2^{\cdot}$. *JBIC Journal of Biological Inorganic Chemistry, 7*, 31–36.
26. Surmeli, N. B., Litterman, N. K., Miller, A. F., & Groves, J. T. (2010). Peroxynitrite mediates active site tyrosine nitration in manganese superoxide dismutase. Evidence of a role for the carbonate radical anion. *Journal of the American Chemical Society, 132*(48), 17174–17185.
27. Winterbourn, C. C., & Kettle, A. J. (2000). Biomarkers of myeloperoxidase-derived hypochlorous acid. *Free Radical Biology and Medicine, 29*(5), 403–409.

28. Ghafourifar, P., & Cadenas, E. (2005). Mitochondrial nitric oxide synthase. *Trends in Pharmacological Sciences, 26*(4), 190–195.
29. Repetto, M., Semprine, J., & Boveris, A. (2012). Lipid peroxidation: Chemical mechanism, biological implications and analytical. *Lipid Peroxidation, 1.*
30. Moncada, S. R. M. J., Palmer, R. M. L., & Higgs, E. (1991). Nitric oxide: Physiology, pathophysiology, and pharmacology. *Pharmacological Reviews, 43*(2), 109–142.
31. Beckman, J. S., & Koppenol, W. H. (1996). Nitric oxide, superoxide, and peroxynitrite: The good, the bad, and ugly. *American Journal of Physiology-Cell Physiology, 271*(5), C1424–C1437.
32. Papas, A. M. (2019). Diet and antioxidant status. *Antioxidant Status, Diet, Nutrition, and Health, 89–106.*
33. Giles, G. I., & Jacob, C. (2002). Reactive sulfur species: an emerging concept in oxidative stress.
34. Paul, B. D., & Snyder, S. H. (2015). H_2S: A novel gasotransmitter that signals by sulfhydration. *Trends in Biochemical Sciences, 40*(11), 687–700.
35. Kabil, O., & Banerjee, R. (2014). Enzymology of H_2S biogenesis, decay and signaling. *Antioxidants & Redox Signaling, 20*(5), 770–782.
36. Shibuya, N., Koike, S., Tanaka, M., Ishigami-Yuasa, M., Kimura, Y., Ogasawara, Y., & Kimura, H. (2013). A novel pathway for the production of hydrogen sulfide from D-cysteine in mammalian cells. *Nature Communications, 4*(1), 1366.
37. Andres Juan, C., Perez de Lastra, J. M., Plou Gasca, F. J., & Perez-Lebeia, E. (2021). The chemistry of reactive oxygen species (ROS) revisited: Outlining their role in biological macromolecules (DNA, lipids and proteins) and induced pathologies. *International Journal of Molecular Science, 22*, 4642.
38. Brandes, R. P., Weissmann, N., & Schroder, K. (2014). Nox family NADPH oxidases: Molecular mechanisms of activation. *Free Radical Biology and Medicine, 76*, 208–226.
39. Panday, A., Sahoo, M. K., Osorio, D., & Batra, S. (2015). NADPH oxidases: An overview from structure to innate immunity-associated pathologies. *Cellular & Molecular Immunology, 12*(1), 5–23.
40. Lekstrom-Himes, J. A., & Gallin, J. I. (2000). Immunodeficiency diseases caused by defects in phagocytes. *New England Journal of Medicine, 343*(23), 1703–1714.
41. Kim, J. H., Lee, J., Bae, S. J., Kim, Y., Park, B. J., Choi, J. W., & Yuk, J. M. (2017). NADPH oxidase 4 is required for the generation of macrophage migration inhibitory factor and host defense against *Toxoplasma gondii* infection. *Scientific Reports, 7*(1), 6361.
42. Bortolotti, M., Polito, L., Battelli, M. G., & Bolognesi, A. (2021). Xanthine oxidoreductase: One enzyme for multiple physiological tasks. *Redox Biology, 41*, 101882.
43. Nishino, T., Okamoto, K., Kawaguchi, Y., Hori, H., Matsumura, T., Eger, B. T., & Nishino, T. (2005). Mechanism of the conversion of xanthine dehydrogenase to xanthine oxidase: Identification of the two cysteine disulfide bonds and crystal structure of a non-convertible rat liver xanthine dehydrogenase mutant. *Journal of Biological Chemistry, 280*(26), 24888–24894.
44. Ty, M. C., Zuniga, M., Gotz, A., Kayal, S., Sahu, P. K., Mohanty, A., & Rodriguez, A. (2019). Malaria inflammation by xanthine oxidase-produced reactive oxygen species. *EMBO Molecular Medicine, 11*(8), e9903.
45. Turrens, J. F. (2003). Mitochondrial formation of reactive oxygen species. *The Journal of Physiology, 552*(2), 335–344.
46. Wong, H. S., Dighe, P. A., Mezera, V., Monternier, P. A., & Brand, M. D. (2017). Production of superoxide and hydrogen peroxide from specific mitochondrial sites under different bioenergetic conditions. *Journal of Biological Chemistry, 292*(41), 16804–16809.
47. West, A. P., Brodsky, I. E., Rahner, C., Woo, D. K., Erdjument-Bromage, H., Tempst, P., & Ghosh, S. (2011). TLR signalling augments macrophage bactericidal activity through mitochondrial ROS. *Nature, 472*(7344), 476–480.
48. Bulua, A. C., Simon, A., Maddipati, R., Pelletier, M., Park, H., Kim, K. Y., & Siegel, R. M. (2011). Mitochondrial reactive oxygen species promote production of proinflammatory

cytokines and are elevated in TNFR1-associated periodic syndrome (TRAPS). *Journal of Experimental Medicine, 208*(3), 519–533.
49. Cardenas, C., Miller, R. A., Smith, I., Bui, T., Molgo, J., Müller, M., & Foskett, J. K. (2010). Essential regulation of cell bioenergetics by constitutive InsP3 receptor Ca^{2+} transfer to mitochondria. *Cell, 142*(2), 270–283.
50. Chouchani, E. T., Pell, V. R., Gaude, E., Aksentijevic, D., Sundier, S. Y., Robb, E. L., & Murphy, M. P. (2014). Ischaemic accumulation of succinate controls reperfusion injury through mitochondrial ROS. *Nature, 515*(7527), 431–435.
51. Mills, E. L., Kelly, B., Logan, A., Costa, A. S., Varma, M., Bryant, C. E., & O'Neill, L. A. (2016). Succinate dehydrogenase supports metabolic repurposing of mitochondria to drive inflammatory macrophages. *Cell, 167*(2), 457–470.
52. Youdim, M. B., Edmondson, D., & Tipton, K. F. (2006). The therapeutic potential of monoamine oxidase inhibitors. *Nature Reviews Neuroscience, 7*(4), 295–309.
53. Kaludercic, N., Takimoto, E., Nagayama, T., Feng, N., Lai, E. W., Bedja, D., & Paolocci, N. (2010). Monoamine oxidase A-mediated enhanced catabolism of norepinephrine contributes to adverse remodeling and pump failure in hearts with pressure overload. *Circulation Research, 106*(1), 193–202.
54. Bhattacharjee, A., Shukla, M., Yakubenko, V. P., Mulya, A., Kundu, S., & Cathcart, M. K. (2013). IL-4 and IL-13 employ discrete signaling pathways for target gene expression in alternatively activated monocytes/macrophages. *Free Radical Biology and Medicine, 54*, 1–16.
55. Sanchez-Rodriguez, R., Munari, F., Angioni, R., Venegas, F., Agnellini, A., Castro-Gil, M. P., & Canton, M. (2021). Targeting monoamine oxidase to dampen NLRP3 inflammasome activation in inflammation. *Cellular & Molecular Immunology, 18*(5), 1311–1313.
56. Giorgio, M., Migliaccio, E., Orsini, F., Paolucci, D., Moroni, M., Contursi, C., & Pelicci, P. G. (2005). Electron transfer between cytochrome c and p66Shc generates reactive oxygen species that trigger mitochondrial apoptosis. *Cell, 122*(2), 221–233.
57. Tomilov, A. A., Bicocca, V., Schoenfeld, R. A., Giorgio, M., Migliaccio, E., Ramsey, J. J., & Cortopassi, G. A. (2010). Decreased superoxide production in macrophages of long-lived p66Shc knock-out mice. *Journal of Biological Chemistry, 285*(2), 1153–1165.
58. Schrader, M., & Fahimi, H. D. (2006). Peroxisomes and oxidative stress. *Biochimica et Biophysica Acta (BBA)-Molecular Cell Research, 1763*(12), 1755–1766.
59. Gross, E., Sevier, C. S., Heldman, N., Vitu, E., Bentzur, M., Kaiser, C. A., & Fass, D. (2006). Generating disulfides enzymatically: Reaction products and electron acceptors of the endoplasmic reticulum thiol oxidase Ero1p. *Proceedings of the National Academy of Sciences, 103*(2), 299–304.
60. Cheeseman, K. H., & Slater, T. F. (1993). An introduction to free radical biochemistry. *British Medical Bulletin, 49*(3), 481–493.
61. Tian, S., Qin, G., & Li, B. (2013). Reactive oxygen species involved in regulating fruit senescence and fungal pathogenicity. *Plant Molecular Biology, 82*, 593–602.
62. Rosen, G. M., Pou, S., Ramos, C. L., Cohen, M. S., & Britigan, B. E. (1995). Free radicals and phagocytic cells. *The FASEB Journal, 9*(2), 200–209.
63. Kohchi, C., Inagawa, H., Nishizawa, T., & Soma, G. I. (2009). ROS and innate immunity. *Anticancer Research, 29*(3), 817–821.
64. Heinecke, J. W., Li, W., Francis, G. A., & Goldstein, J. A. (1993). Tyrosyl radical generated by myeloperoxidase catalyzes the oxidative cross-linking of proteins. *The Journal of Clinical Investigation, 91*(6), 2866–2872.
65. Lieber, C. S. (1997). Cytochrome P-4502E1: Its physiological and pathological role. *Physiological Reviews, 77*(2), 517–544.
66. Bokare, A. D., & Choi, W. (2014). Review of iron-free Fenton-like systems for activating H2O2 in advanced oxidation processes. *Journal of hazardous materials, 275*, 121–135.
67. Faustman, C., Sun, Q., Mancini, R., & Suman, S. P. (2010). Myoglobin and lipid oxidation interactions: Mechanistic bases and control. *Meat science, 86*(1), 86–94.

68. Tsukamoto, H., & Lu, S. C. (2001). Current concepts in the pathogenesis of alcoholic liver injury. *The FASEB Journal, 15*(8), 1335–1349.
69. Droge, W. (2002). Free radicals in the physiological control of cell function. *Physiological Reviews.*
70. Dizdaroglu, M., Jaruga, P., Birincioglu, M., & Rodriguez, H. (2002). Free radical-induced damage to DNA: Mechanisms and measurement. *Free Radical Biology and Medicine, 32*(11), 1102–1115.
71. Barja, G. (2000). The flux of free radical attack through mitochondrial DNA is related to aging rate. *Aging Clinical and Experimental Research, 12*, 342–355.
72. Hofer, T., Badouard, C., Bajak, E., Ravanat, J. L., Mattsson, A., & Cotgreave, I. A. (2005). Hydrogen peroxide causes greater oxidation in cellular RNA than in DNA. *Biological Chemistry, 386*(4), 333–337.
73. Tateyama, M., Takeda, A., Onodera, Y., Matsuzaki, M., Hasegawa, T., Nunomura, A., & Itoyama, Y. (2003). Oxidative stress and predominant Aβ42 (43) deposition in myopathies with rimmed vacuoles. *Acta Neuropathologica, 105*, 581–585.
74. Abe, T., Tohgi, H., Isobe, C., Murata, T., & Sato, C. (2002). Remarkable increase in the concentration of 8-hydroxyguanosine in cerebrospinal fluid from patients with Alzheimer's disease. *Journal of Neuroscience Research, 70*(3), 447–450.
75. Kikuchi, A., Takeda, A., Onodera, H., Kimpara, T., Hisanaga, K., Sato, N., & Itoyama, Y. (2002). Systemic increase of oxidative nucleic acid damage in Parkinson's disease and multiple system atrophy. *Neurobiology of Disease, 9*(2), 244–248.
76. Martinet, W., De Meyer, G. R. Y., Herman, A. G., & Kockx, M. M. (2004). Reactive oxygen species induce RNA damage in human atherosclerosis. *European Journal of Clinical Investigation, 34*(5), 323–327.
77. Broedbaek, K., Poulsen, H. E., Weimann, A., Kom, G. D., Schwedhelm, E., Nielsen, P., & Boger, R. H. (2009). Urinary excretion of biomarkers of oxidatively damaged DNA and RNA in hereditary hemochromatosis. *Free Radical Biology and Medicine, 47*(8), 1230–1233.
78. Siems, W. G., Grune, T., & Esterbauer, H. (1995). 4-Hydroxynonenal formation during ischemia and reperfusion of rat small intestine. *Life Sciences, 57*(8), 785–789.
79. Marnett, L. J. (1999). Lipid peroxidation—DNA damage by malondialdehyde. *Mutation Research/Fundamental and Molecular Mechanisms of Mutagenesis, 424*(1–2), 83–95.
80. Aruoma, O. I. (1998). Free radicals, oxidative stress, and antioxidants in human health and disease. *Journal of the American Oil Chemists' Society, 75*(2), 199–212.
81. Pryor, W. A., Jin, X., & Squadrito, G. L. (1994). One-and two-electron oxidations of methionine by peroxynitrite. *Proceedings of the National Academy of Sciences, 91*(23), 11173–11177.
82. Berlett, B. S., & Stadtman, E. R. (1997). Protein oxidation in aging, disease, and oxidative stress. *Journal of Biological Chemistry, 272*(33), 20313–20316.
83. Chevion, M., Berenshtein, E., & Stadtman, E. R. (2000). Human studies related to protein oxidation: Protein carbonyl content as a marker of damage. *Free Radical Research, 33*, S99–S108.
84. Bonizzi, G., & Karin, M. (2004). The two NF-κB activation pathways and their role in innate and adaptive immunity. *Trends in Immunology, 25*(6), 280–288.
85. Baldwin, A. S. (2012). Regulation of cell death and autophagy by IKK and NF-κB: Critical mechanisms in immune function and cancer. *Immunological Reviews, 246*(1), 327–345.
86. Elmetwali, T., Young, L. S., & Palmer, D. H. (2014). Fas-associated factor (Faf1) is a novel CD40 interactor that regulates CD40-induced NF-κB activation via a negative feedback loop. *Cell Death & Disease, 5*(5), e1213–e1213.
87. Pajares, M., Cuadrado, A., Engedal, N., Jirsova, Z., & Cahova, M. (2018). The role of free radicals in autophagy regulation: Implications for ageing. *Oxidative medicine and cellular longevity, 2018*(1), 2450748.
88. Hayes, J. D., Dinkova-Kostova, A. T., & Tew, K. D. (2020). Oxidative stress in cancer. *Cancer Cell, 38*(2), 167–197.

89. Chio, I. I. C., & Tuveson, D. A. (2017). ROS in cancer: The burning question. *Trends in Molecular Medicine, 23*(5), 411–429.
90. Liu, H., Nishitoh, H., Ichijo, H., & Kyriakis, J. M. (2000). Activation of apoptosis signal-regulating kinase 1 (ASK1) by tumor necrosis factor receptor-associated factor 2 requires prior dissociation of the ASK1 inhibitor thioredoxin. *Molecular and Cellular Biology, 20*(6), 2198–2208.
91. Qiu, X., Cheng, J. C., Chang, H. M., & Leung, P. C. (2014). COX2 and PGE2 mediate EGF-induced E-cadherin-independent human ovarian cancer cell invasion. *Endocrine-Related Cancer, 21*(4), 533–543.
92. Innocenti, M., Frittoli, E., Ponzanelli, I., Falck, J. R., Brachmann, S. M., Di Fiore, P. P., & Scita, G. (2003). Phosphoinositide 3-kinase activates Rac by entering in a complex with Eps8, Abi1, and Sos-1. *The Journal of Cell Biology, 160*(1), 17–23.
93. Xu, M., Che, L., Gao, K., Wang, L., Yang, X., Wen, X., & Jiang, Z. (2023). Taurine alleviates oxidative stress in porcine mammary epithelial cells by stimulating the Nrf2-MAPK signaling pathway. *Food Science & Nutrition, 11*(4), 1736–1746.
94. Sies, H., & Jones, D. P. (2020). Reactive oxygen species (ROS) as pleiotropic physiological signalling agents. *Nature Reviews Molecular Cell Biology, 21*(7), 363–383.
95. Kerins, M. J., & Ooi, A. (2018). A catalogue of somatic NRF2 gain-of-function mutations in cancer. *Scientific Reports, 8*(1), 12846.
96. Kirtonia, A., Sethi, G., & Garg, M. (2020). The multifaceted role of reactive oxygen species in tumorigenesis. *Cellular and Molecular Life Sciences, 77*, 4459–4483.
97. Santos, S. A. D., & Andrade Júnior, D. R. D. (2017). HIF-1alpha and infectious diseases: A new frontier for the development of new therapies. *Revista do Instituto de Medicina Tropical de Sao Paulo, 59*, e92.
98. Freemerman, A. J., Johnson, A. R., Sacks, G. N., Milner, J. J., Kirk, E. L., Troester, M. A., & Makowski, L. (2014). Metabolic reprogramming of macrophages: Glucose transporter 1 (GLUT1)-mediated glucose metabolism drives a proinflammatory phenotype. *Journal of Biological Chemistry, 289*(11), 7884–7896.
99. Kaelin, W. G., & Ratcliffe, P. J. (2008). Oxygen sensing by metazoans: The central role of the HIF hydroxylase pathway. *Molecular Cell, 30*(4), 393–402.
100. Masson, N., Singleton, R. S., Sekirnik, R., Trudgian, D. C., Ambrose, L. J., Miranda, M. X., & Ratcliffe, P. J. (2012). The FIH hydroxylase is a cellular peroxide sensor that modulates HIF transcriptional activity. *EMBO Reports, 13*(3), 251–257.
101. Fuhrmann, D. C., Wittig, I., & Brune, B. (2019). TMEM126B deficiency reduces mitochondrial SDH oxidation by LPS, attenuating HIF-1α stabilization and IL-1β expression. *Redox Biology, 20*, 204–216.
102. Salas, A., Hernandez-Rocha, C., Duijvestein, M., Faubion, W., McGovern, D., Vermeire, S., & Vande Casteele, N. (2020). JAK–STAT pathway targeting for the treatment of inflammatory bowel disease. *Nature Reviews Gastroenterology & Hepatology, 17*(6), 323–337.
103. Padgett, L. E., Burg, A. R., Lei, W., & Tse, H. M. (2015). Loss of NADPH oxidase–derived superoxide skews macrophage phenotypes to delay type 1 diabetes. *Diabetes, 64*(3), 937–946.
104. Burdette, D. L., & Vance, R. E. (2013). STING and the innate immune response to nucleic acids in the cytosol. *Nature Immunology, 14*(1), 19–26.
105. Shang, G., Zhang, C., Chen, Z. J., Bai, X. C., & Zhang, X. (2019). Cryo-EM structures of STING reveal its mechanism of activation by cyclic GMP–AMP. *Nature, 567*(7748), 389–393.
106. Shu, N., Hägglund, P., Cai, H., Hawkins, C. L., & Davies, M. J. (2020). Modification of Cys residues in human thioredoxin-1 by p-benzoquinone causes inhibition of its catalytic activity and activation of the ASK1/p38-MAPK signalling pathway. *Redox Biology, 29*, 101400.
107. Lerner, A. G., Upton, J. P., Praveen, P. V. K., Ghosh, R., Nakagawa, Y., Igbaria, A., & Papa, F. R. (2012). IRE1α induces thioredoxin-interacting protein to activate the NLRP3 inflammasome and promote programmed cell death under irremediable ER stress. *Cell Metabolism, 16*(2), 250–264.
108. Shrestha, A., Kim, M., Kim, S. H., Jeong, C. H., Jung, S., Cho, E., & Park, P. H. (2016). Critical role of AMPK/Fox03A axis in globular adiponectin-induced cell cycle arrest and apoptosis in cancer cells. *Journal of Cellular Physiology, 26*, 45–45.

109. Coomans de Brachene, A., & Demoulin, J. B. (2016). FOXO transcription factors in cancer development and therapy. *Cellular and Molecular Life Sciences, 73*, 1159–1172.
110. Bierhaus, A., Hofmann, M. A., Ziegler, R., & Nawroth, P. P. (1998). AGEs and their interaction with AGE-receptors in vascular disease and diabetes mellitus. I. The AGE concept. *Cardiovascular Research, 37*(3), 586–600.
111. Schmidt, A. M., & Stern, D. M. (2000). RAGE: A new target for the prevention and treatment of the vascular and inflammatory complications of diabetes. *Trends in Endocrinology & Metabolism, 11*(9), 368–375.
112. Lu, M., Kuroki, M., Amano, S., Tolentino, M., Keough, K., Kim, I., & Adamis, A. P. (1998). Advanced glycation end products increase retinal vascular endothelial growth factor expression. *The Journal of Clinical Investigation, 101*(6), 1219–1224.
113. Koya, D., & King, G. L. (1998). Protein kinase C activation and the development of diabetic complications. *Diabetes, 47*(6), 859–866.
114. Glogowski, E. A., Tsiani, E., Zhou, X., Fantus, I. G., & Whiteside, C. (1999). High glucose alters the response of mesangial cell protein kinase C isoforms to endothelin-1. *Kidney International, 55*(2), 486–499.
115. Stevens, M. J., Obrosova, I., Feldman, E. L., & Greene, D. A. (2000). The sorbitol-osmotic and sorbitol-redox hypothesis. In *Diabetes mellitus: a fundamental and clinical text* (pp. 972–983).
116. Singh, S. B., Malamas, M. S., Hohman, T. C., Nilakantan, R., Carper, D. A., & Kitchen, D. (2000). Molecular modeling of the aldose reductase-inhibitor complex based on the X-ray crystal structure and studies with single-site-directed mutants. *Journal of Medicinal Chemistry, 43*(6), 1062–1070.
117. Hawkins, M., Barzilai, N., Liu, R., Hu, M., Chen, W., & Rossetti, L. (1997). Role of the glucosamine pathway in fat-induced insulin resistance. *The Journal of Clinical Investigation, 99*(9), 2173–2182.
118. James, L. R., Tang, D., Ingram, A., Ly, H., Thai, K., Cai, L., & Scholey, J. W. (2002). Flux through the hexosamine pathway is a determinant of nuclear factor κB–dependent promoter activation. *Diabetes, 51*(4), 1146–1156.
119. Evans, J. L., Goldfine, I. D., Maddux, B. A., & Grodsky, G. M. (2002). Oxidative stress and stress-activated signaling pathways: A unifying hypothesis of type 2 diabetes. *Endocrine Reviews, 23*(5), 599–622.
120. Tarasov, A. I., Griffiths, E. J., & Rutter, G. A. (2012). Regulation of ATP production by mitochondrial Ca2+. *Cell Calcium, 52*(1), 28–35.
121. Lewis, A., Hayashi, T., Su, T. P., & Betenbaugh, M. J. (2014). Bcl-2 family in inter-organelle modulation of calcium signaling; roles in bioenergetics and cell survival. *Journal of Bioenergetics and Biomembranes, 46*, 1–15.
122. Roe, N. D., He, E. Y., Wu, Z., & Ren, J. (2013). Folic acid reverses nitric oxide synthase uncoupling and prevents cardiac dysfunction in insulin resistance: Role of Ca^{2+}/calmodulin-activated protein kinase II. *Free Radical Biology and Medicine, 65*, 234–243.
123. Nishida, M., Ishikawa, T., Saiki, S., Sunggip, C., Aritomi, S., Harada, E., & Kim-Mitsuyama, S. (2013). Voltage-dependent N-type Ca^{2+} channels in endothelial cells contribute to oxidative stress-related endothelial dysfunction induced by angiotensin II in mice. *Biochemical and Biophysical Research Communications, 434*(2), 210–216.
124. Elkoshi, Z. (2021). The binary classification of protein kinases. *Journal of Inflammation Research*, 929–947.
125. Aicart-Ramos, C., Sanchez-Ruiloba, L., Gomez-Parrizas, M., Zaragoza, C., Iglesias, T., & Rodriguez-Crespo, I. (2014). Protein kinase D activity controls endothelial nitric oxide synthesis. *Journal of Cell Science, 127*(15), 3360–3372.

MIX
Papier aus verantwortungsvollen Quellen
Paper from responsible sources
FSC® C105338

If you have any concerns about our products,
you can contact us on
ProductSafety@springernature.com

In case Publisher is established outside the EU,
the EU authorized representative is:
Springer Nature Customer Service Center GmbH
Europaplatz 3, 69115 Heidelberg, Germany

Printed by Libri Plureos GmbH
in Hamburg, Germany